FORTSCHRITTE DER CHEMIE ORGANISCHER NATURSTOFFE

PROGRESS IN THE CHEMISTRY OF ORGANIC NATURAL PRODUCTS

PROGRÈS DANS LA CHIMIE DES SUBSTANCES ORGANIQUES NATURELLES

HERAUSGEGEBEN VON   EDITED BY   RÉDIGÉ PAR

L. ZECHMEISTER

CALIFORNIA INSTITUTE OF TECHNOLOGY, PASADENA

FÜNFTER BAND FIFTH VOLUME CINQUIÈME VOLUME

VERFASSER   AUTHORS   AUTEURS

G. W. BEADLE · F. E. BRAUNS · V. DEULOFEU · M. DOUDOROFF
D. L. FOX · E. GEIGER · A. J. HAAGEN=SMIT · W. Z. HASSID
T. P. HILDITCH · P. KARRER · E. PACSU · R. S. RASMUSSEN

MIT 34 ABBILDUNGEN IM TEXT WITH 34 ILLUSTRATIONS AVEC 34 ILLUSTRATIONS

Springer-Verlag Wien GmbH 1948

ISBN 978-3-7091-7181-3   ISBN 978-3-7091-7179-0 (eBook)
DOI 10.1007/978-3-7091-7179-0

TITEL NR. 8211

Inhaltsverzeichnis.
Contents. — Table des matières.

Infrared Spectroscopy in Structure Determination and its Application to Penicillin. By R. S. RASMUSSEN, Shell Development Company, Emeryville, California

Carotinoid-epoxyde und furanoide Oxyde von Carotinoidfarbstoffen.

Von **P. Karrer**, Zürich.

I. Vorkommen, Konstitution und Partialsynthesen der Carotinoid-epoxyde.

Die Entdeckung der Carotinoid-epoxyde schließt sich an eine Untersuchung des Violaxanthins an. Bei der Einwirkung von äußerst verdünnter Mineralsäure auf dieses Pigment wurde die Bildung dreier Farbstoffe beobachtet: des Auroxanthins, das kurz vorher in gelben Blüten von *Viola tricolor* entdeckt worden war (*30*), eines Pigments, das den Namen Mutatoxanthin erhielt und von Zeaxanthin. Diese Feststellung zeigte, daß sehr säureempfindliche Carotinoide existieren, die sich in gut charakterisierte neue Produkte überführen lassen und die daher zu einer eingehenden Bearbeitung einluden.

In jener Zeit der Carotinoidforschung waren eine ganze Reihe natürlicher Vertreter dieser Gruppe mit drei und mehr Sauerstoffatomen bekannt, in deren Konstitution man noch keinerlei Einblick gewonnen hatte. Die wichtigsten dieser Pigmente waren folgende:

Flavoxanthin	$C_{40}H_{56}O_3$,
Chrysanthemaxanthin	$C_{40}H_{56}O_3$,
Antheraxanthin	$C_{40}H_{56}O_3$,
Violaxanthin	$C_{40}H_{56}O_4$,
Taraxanthin	$C_{40}H_{56}O_4$,
Auroxanthin	$C_{40}H_{56}O_4$,
Fucoxanthin	$C_{40}H_{56}O_6$.

In der Absicht, die Konstitutionsaufklärung dieser Verbindungen zu fördern, haben wir den Weg der Synthese beschritten und einfacher gebaute Carotinoide der Oxydation mit Persäuren (Phthalmonopersäure, seltener Benzopersäure) unterworfen.

Hierbei ließ sich feststellen, daß alle Carotinoide, welche einen β-Iononring enthalten, in übersichtlicher Weise zu gut definierten *1,2-Epoxyden* oxydiert werden können, in denen der Oxido-sauerstoff an jenen C-Atomen fixiert ist, welche vordem die Kohlenstoff-doppelbindung des β-Iononringes gebildet hatten. So entsteht beispielsweise aus Xanthophyll (I) Xanthophyll-epoxyd (II) [KARRER und JUCKER (*12*)].

$$\text{CH}_3 \quad \text{CH}_3$$
$$\backslash \quad /$$
$$\text{C} \qquad \text{CH}_3 \qquad \text{CH}_3 \qquad \text{CH}_3 \qquad \text{CH}_3 \qquad \text{H}_3\text{C} \quad \text{CH}_3$$
$$/ \ \backslash \qquad | \qquad | \qquad | \qquad | \qquad \backslash \ /$$
$$\text{CH}_2 \quad \text{CCH}=\text{CHC}=\text{CHCH}=\text{CHC}=\text{CHCH}=\text{CHCH}=\text{CCH}=\text{CHCH}=\text{CCH}=\text{CH}-\text{CH} \quad \text{CH}_2$$
$$| \qquad || \qquad\qquad\qquad\qquad | \qquad |$$
$$\text{HOCH} \quad \text{C} \qquad \text{Xanthophyll (I.)} \qquad\qquad \text{C} \quad \text{CHOH}$$
$$\backslash \ /\ \backslash \qquad\qquad\qquad\qquad\qquad / \ \backslash /$$
$$\text{CH}_2 \quad \text{CH}_3 \qquad\qquad\qquad\qquad\qquad \text{H}_3\text{C} \quad \text{CH}$$

$$\downarrow$$

$$\text{CH}_3 \quad \text{CH}_3 \qquad\qquad\qquad\qquad\qquad\qquad \text{H}_3\text{C} \quad \text{CH}_3$$
$$\backslash \ / \qquad\qquad\qquad\qquad\qquad\qquad\qquad \backslash \ /$$
$$\text{C} \qquad \text{CH}_3 \qquad \text{CH}_3 \qquad \text{CH}_3 \qquad \text{CH}_3 \qquad \text{C}$$
$$/ \ \backslash \qquad | \qquad | \qquad | \qquad | \qquad \backslash$$
$$\text{CH}_2 \quad \text{C}-\text{CH}-\text{CHC}=\text{CHCH}=\text{CHC}=\text{CHCH}=\text{CHCH}=\text{CCH}=\text{CHCH}=\text{CCH}=\text{CH}-\text{CH} \quad \text{CH}_2$$
$$| \qquad | \ \backslash / \qquad\qquad\qquad\qquad \text{Xanthophyll-epoxyd (II.)} \qquad | \qquad |$$
$$\text{HOCH} \quad \text{C} \quad \text{O} \qquad\qquad\qquad\qquad\qquad\qquad \text{C} \quad \text{CH}_2$$
$$\backslash \ /\ \backslash \qquad\qquad\qquad\qquad\qquad\qquad\qquad / \ \backslash /$$
$$\text{CH}_2 \quad \text{CH}_3 \qquad\qquad\qquad\qquad\qquad\qquad \text{H}_3\text{C} \quad \text{CH}$$

Für die genannte Stellung des Oxido-sauerstoffs in diesen Epoxyden spricht zunächst der Umstand, daß sie sich nur aus solchen Carotinoiden gewinnen ließen, welche einen β-Iononring enthalten; sind zwei β-Iononringe im Pigment vorhanden, so können Di-epoxyde erhalten werden. Eine weitere Grundlage für die Formulierung ist die Tatsache, daß die Einführung der Epoxygruppe eine Verlagerung des längstwelligen Absorptionsmaximums um nur zirka 8 mμ (in Schwefelkohlenstofflösung) gegenüber dem Ausgangspigment in Richtung kürzerer Wellenlängen bewirkt; zwei Epoxygruppen haben ungefähr den doppelten Effekt. Wäre der Oxido-sauerstoff an irgendeiner anderen, innerhalb der Konjugation liegenden Doppelbindung fixiert worden, so müßte der Unterbruch der Konjugation einen stärkeren optischen Effekt, eine bedeutend größere Violett-Verschiebung der Absorptionsmaxima, zur Folge haben. Die Absorptionsmaxima des Xanthophyll-epoxyds (II) fallen fast genau mit jenen eines Dioxy-α-carotins (III) zusammen, dessen chromophores System jenem des Xanthophyll-epoxyds entspricht; darin kann eine weitere Stütze unserer Formulierung der Carotinoidepoxyde erblickt werden [KARRER, SOLMSSEN und WALKER (*33*); KARRER, VON EULER und SOLMSSEN (*8*)].

$$\text{CH}_3 \quad \text{CH}_3 \qquad\qquad\qquad\qquad\qquad\qquad\qquad \text{H}_3\text{C} \quad \text{CH}_3$$
$$\backslash \ / \qquad\qquad\qquad\qquad\qquad\qquad\qquad\qquad\qquad \backslash \ /$$
$$\text{C} \quad \text{OH} \quad \text{CH}_3 \qquad \text{CH}_3 \qquad \text{CH}_3 \qquad \text{CH}_3 \qquad \text{C}$$
$$/ \ \backslash \ / \qquad | \qquad | \qquad | \qquad | \qquad / \ \backslash$$
$$\text{CH}_2 \quad \text{C}-\text{CH}=\text{CHC}=\text{CHCH}=\text{CHC}=\text{CHCH}=\text{CHCH}=\text{CCH}=\text{CHCH}=\text{C}-\text{CH}=\text{CH}-\text{CH} \quad \text{CH}_2$$
$$| \qquad |$$
$$\text{CH}_2 \quad \text{C}-\text{OH} \qquad\qquad \text{(III.) Dioxy-}\alpha\text{-carotin.} \qquad\qquad\qquad \text{C} \quad \text{CH}_2$$
$$\backslash \ /\ \backslash \qquad\qquad\qquad\qquad\qquad\qquad\qquad\qquad / \ \backslash /$$
$$\text{CH}_2 \quad \text{CH}_3 \qquad\qquad\qquad\qquad\qquad\qquad\qquad \text{H}_3\text{C} \quad \text{CH}$$

Die bisher partialsynthetisch, d. h. durch Oxydation der entsprechenden Carotinoide mit. Phthalpersäure gewonnenen Carotinoidepoxyde
sind in Tabelle 1 enthalten. Die Formeln (II) sowie (IV) bis (XII) geben
ihre Konstitution wieder.

Tabelle 1. Partialsynthetisch bereitete Epoxyde.

Formel	Epoxyd (und Literaturhinweis)	Abs. Maxima in CS_2		Schmelz-punkt	Blaufärbung mit konz. Salzsäure
(IV)	α-Carotin-mono-epoxyd (14) ..	503	471 mμ	175°	schwach, unbeständig
(II)	Xanthophyll-mono-epoxyd (12)	501,5	472 mμ	192°.	ziemlich beständig
(V)	β-Carotin-mono-epoxyd (13) ..	511	479 mμ	160°	schwach, unbeständig
(VI)	Kryptoxanthin-mono-epoxyd (16, 6)	512	479 mμ	154°	blau, unbeständig
(VII)	Zeaxanthin-mono-epoxyd (12).	510	478 mμ	205°	blau, unbeständig
(VIII)	Rubixanthin-mono-epoxyd (21)	526	491 mμ	171°	blau, unbeständig
(IX)	Capsanthin-mono-epoxyd (15).	534	499 mμ	189°	blau, unbeständig
(X)	β-Carotin-di-epoxyd (13)	502	470 mμ	184°	tiefblau, beständig
(XI)	Kryptoxanthin-di-epoxyd (16, 6)	503	473 mμ	194°	tiefblau, beständig
(XII)	Zeaxanthin-di-epoxyd (12) ...	500	469 mμ	200°	tiefblau, beständig

(IV.) α-Carotin-mono-epoxyd.

(V.) β-Carotin-mono-epoxyd.

(VI.) Kryptoxanthin-mono-epoxyd.

1*

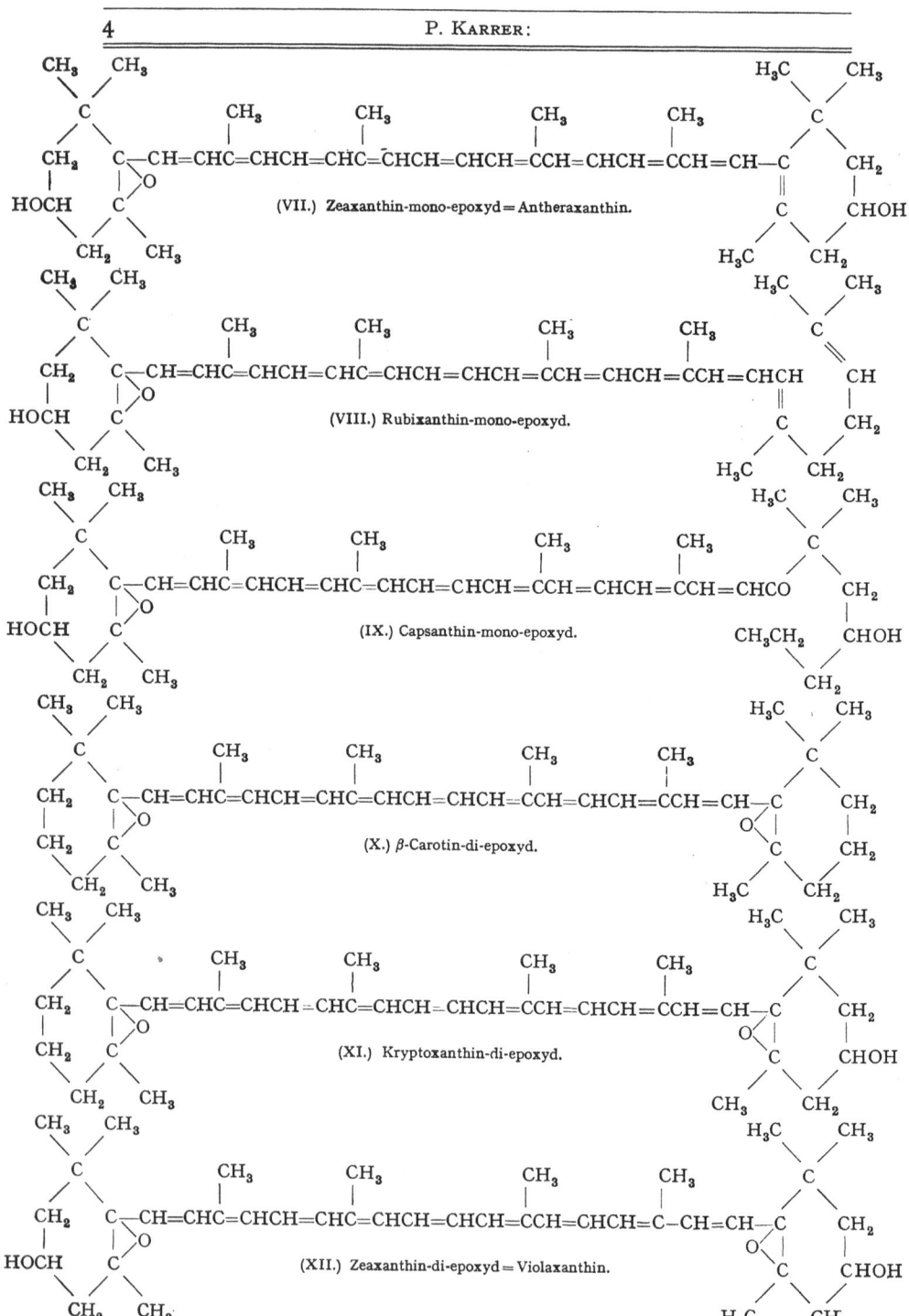

(VII.) Zeaxanthin-mono-epoxyd = Antheraxanthin.

(VIII.) Rubixanthin-mono-epoxyd.

(IX.) Capsanthin-mono-epoxyd.

(X.) β-Carotin-di-epoxyd.

(XI.) Kryptoxanthin-di-epoxyd.

(XII.) Zeaxanthin-di-epoxyd = Violaxanthin.

Alle bekannten Carotinoid-epoxyde geben mit konzentrierter wäßriger Salzsäure eine Farbreaktion: wenn man ihre ätherischen Lösungen mit konzentrierter Salzsäure durchschüttelt, färbt sich diese mehr oder weniger intensiv blau. Di-epoxyde veranlassen tiefere Blaufärbung als Mono-epoxyde; auch Hydroxylgruppen verstärken die Beständigkeit und Intensität deutlich (vgl. Tabelle 1).

Die Adsorbierbarkeit der Carotinoidepoxyde ist natürlich verschieden stark und wird insbesondere durch die Anzahl der in ihnen auftretenden Hydroxylgruppen wesentlich beeinflußt. Im Chromatogrammrohr ordnen sich diese Oxyde in nachstehender Reihenfolge:

(oben) Violaxanthin
Antheraxanthin
Xanthophyllepoxyd
Rubixanthin-mono-epoxyd
Kryptoxanthin-mono-epoxyd
β-Carotin-di-epoxyd
β-Carotin-mono-epoxyd
(unten) α-Carotinepoxyd

In der Natur wurden bisher die folgenden Epoxyde von Carotinoiden aufgefunden:

α-Carotin-epoxyd in den Blüten von *Tragopogon pratensis*, *Ranunculus acer* (*20*) usw.

Xanthophyll-epoxyd, sehr verbreitet im Pflanzenreich. Das Pigment ist ein Hauptfarbstoff der grünen und etiolierten Blätter (*23*); ferner fand man es z. B. in den Blüten von Winterastern (*9*), von *Sarothamnus scoparius* (*10*), von *Tragopogon pratensis* (*20*), von *Ranunculus acer* (*20*), von *Trollius europaeus* (*17*), von *Laburnum anagyroides* (*17*), von *Kerria japonica* (*17*), von *Elodea canadensis* (*31*, *7*) und anderen Orts.

Zeaxanthin-mono-epoxyd = Antheraxanthin kommt in den Staubbeuteln von *Lilium tigrinum* (*26*, *12*) und in Früchten von *Evonymus europaeus* (*22*) vor.

Zeaxanthin-di-epoxyd = Violaxanthin gehört zu den sehr verbreiteten Carotinoid-pigmenten. Blüten, in denen es festgestellt wurde, sind u. a. folgende: *Viola tricolor* (*36*, *29*), *Tulipa* (*42*), *Ranunculus acer* (*34*), *Sinapis officinalis* (*25*), *Laburnum anagyroides* (*25*), *Tragopogon pratensis* (*25*), *Ulex europaeus* (*42*), *Calendula officinalis* (*46*), *Tagetes grandiflora* (*31*), *Taraxacum officinale* (*35*), *Tussilago farfara* (*24*), *Cytisus laburnum* (*25*). Auch in den folgenden Früchten ist es nachgewiesen: *Citrus aurantium* (*47*, *43*), *Carica Papaya* (*44*), *Citrus pooensis* Hort. (*45*), *Cucurbita maxima* (*48*), *Arbutus Unedo* L. (*40*), *Diospyros costata* (*40*), *Iris pseudacorus* (*3*).

Trollixanthin, in Blüten von *Trollius europaeus* (*17*).

Carotinoid-epoxyde gehören demnach zu den häufigen Pigmenten im Pflanzenreich. Dagegen ist man ihnen bisher noch nie im tierischen

Organismus begegnet. Ob darin eine tiefere biologische Ursache verborgen liegt, werden vielleicht spätere Forschungen entscheiden können.

II. Die furanoid gebauten Carotinoid-epoxyde.

Die Epoxyde der Carotin-Pigmente zeichnen sich durch große chemische Reaktionsfähigkeit aus, was ihrem Studium einen besonderen Reiz verleiht. Vor allem fällt ihre Empfindlichkeit gegen Säuren auf. Schon Spuren von Mineralsäure, z. B. Chlorwasserstoff, wie er sich im Chloroform findet, das einige Zeit aufbewahrt worden war, genügen, um die Carotinoid-epoxyde innerhalb weniger Sekunden zu verändern und in neue Stoffe umzuwandeln. Diese Reaktionen spielen sich bei allen Gliedern dieser Klasse prinzipiell in derselben Art ab und verlaufen so, daß ein Teil des Epoxyd-farbstoffs zum ursprünglichen Carotinoid, von dem er sich ableitete, reduziert und ein anderer (größerer) Teil in ein isomeres, *furanoides* Oxyd umgelagert wird. Die folgenden Teilformeln veranschaulichen diese Vorgänge:

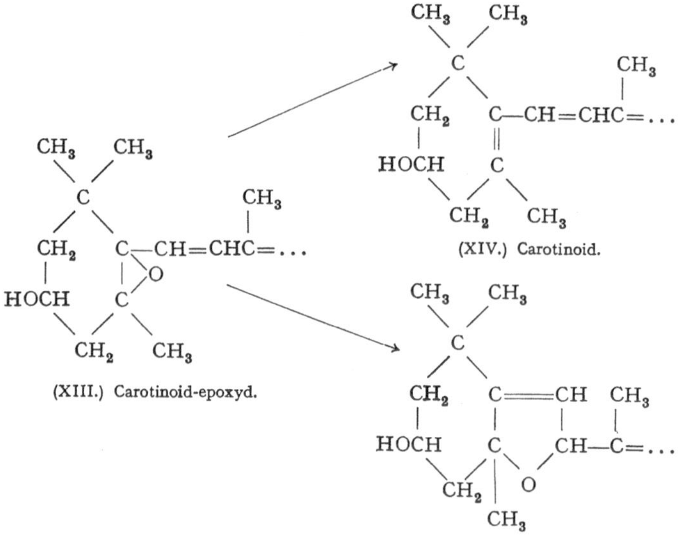

$$\begin{array}{c} CH_3 \quad CH_3 \\ \backslash \quad / \\ C \qquad\qquad CH_3 \\ / \ \backslash \qquad\qquad | \\ CH_2 \quad C{-}CH{=}CHC{=}\ldots \\ | \qquad \| \\ HOCH \quad C \\ \backslash \ / \ \backslash \\ CH_2 \quad CH_3 \\ \text{(XIV.) Carotinoid.} \end{array}$$

$$\begin{array}{c} CH_3 \quad CH_3 \\ \backslash \quad / \\ C \qquad\qquad CH_3 \\ / \ \backslash \qquad\qquad | \\ CH_2 \quad C{-}CH{=}CHC{=}\ldots \\ | \qquad | \ \rangle O \\ HOCH \quad C \\ \backslash \ / \ \backslash \\ CH_2 \quad CH_3 \\ \text{(XIII.) Carotinoid-epoxyd.} \end{array}$$

$$\begin{array}{c} CH_3 \quad CH_3 \\ \backslash \quad / \\ C \\ / \ \backslash \\ CH_2 \quad C{=\!=\!=}CH \quad CH_3 \\ | \qquad | \qquad | \qquad | \\ HOCH \quad C \quad CH{-}C{=}\ldots \\ \backslash \ / \backslash \ / \\ CH_2 \ | \ O \\ CH_3 \\ \text{(XV.) Furanoides Oxyd des Carotinoids.} \end{array}$$

Es war infolgedessen möglich, auf diesem Weg aus jedem Carotinoid-mono-epoxyd ein neues furanoides Oxyd zu gewinnen und aus den Di-epoxyden zwei, bisweilen sogar drei Verbindungen mit Furanringen: ein Pigment, in welchem beide Epoxydgruppen in furanoide Struktur umgelagert sind; ein zweites Pigment, in dem nur die eine Epoxygruppe diese Umlagerung erlitten hat, und schließlich ein dritter Farbstoff, in dem eine furanoide Gruppe vorkommt, während das zweite Sauer-

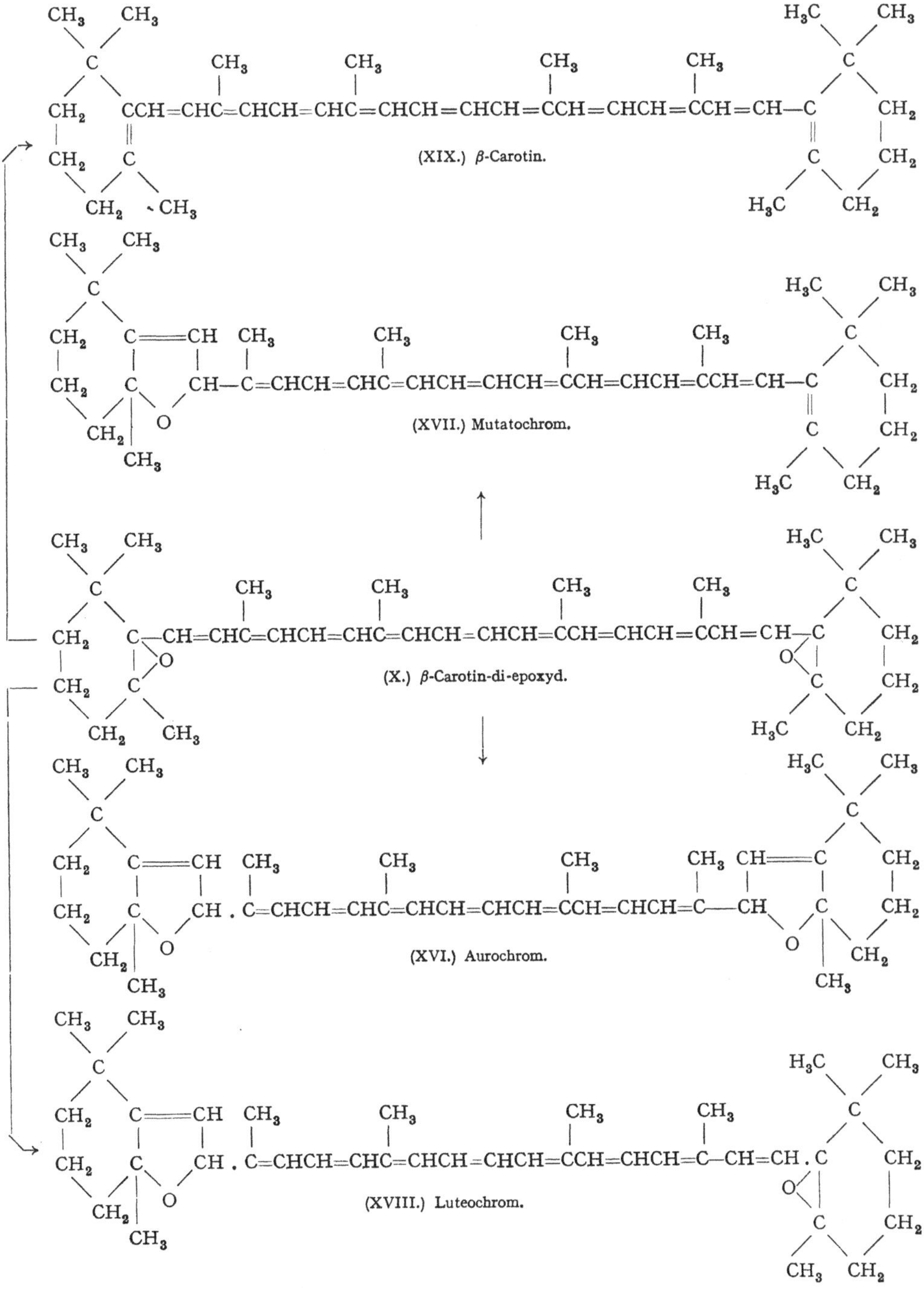

CH₃　CH₃

CH₃　　　CH₃　　　　CH₃　　　CH₃

H₃C　CH₃

$$CH_3 \quad CH_3$$

$$CCH=CHC=CHCH=CHC=CHCH=CHCH=CCH=CHCH=CCH=CH-C$$

(XIX.) β-Carotin.

(XVII.) Mutatochrom.

(X.) β-Carotin-di-epoxyd.

(XVI.) Aurochrom.

(XVIII.) Luteochrom.

stoffatom reduktiv entfernt worden war. Als Beispiel für das Gesagte kann die Umwandlung des β-Carotin-di-epoxyds (X) in Aurochrom (XVI), Mutatochrom (XVII), Luteochrom (XVIII) und β-Carotin (XIX) aufgeführt werden, wie sie sich abspielt, wenn man chlorwasserstoff-haltiges Chloroform auf das Di-epoxyd einwirken läßt.

Durch solche Umlagerungsreaktionen sind bisher die in der Tabelle 2 enthaltenen furanoiden Carotinoid-oxyde aus den entsprechenden Epoxyden partialsynthetisch erhalten worden.

Tabelle 2. Partialsynthetisch bereitete furanoide Oxyde.

Formel	Furanoides Oxyd (und Literaturhinweis)	Abs. Maxima in CS_2	Schmelz-punkt	Blaufärbung mit konz. Salzsäure
(XX)	Flavochrom (*14*) (aus α-Carotin-epoxyd)	482 451 mμ	189°	schwach, unbeständig
(XXI)	Flavoxanthin (*12*) (aus Xanthophyll-epoxyd)..	479 449 mμ	180°	blau, zieml. beständig
(XXI)	Chrysanthemaxanthin (*12*) (aus Xanthophyll-epoxyd).............	479 449 mμ	185°	blau, zieml. beständig
(XVII)	Mutatochrom (*13*) (aus β-Carotin-mono-epoxyd)	489 459 mμ	164°	schwach, unbeständig
(XXII)	Kryptoflavin (*16*) (aus Kryptoxanthin-mono-epoxyd).............	490 459 mμ	171°	blau, unbeständig
(XXIII)	Mutatoxanthin (*12*) aus Zeaxanthin-mono-epoxyd).............	488 459 mμ	177°	blau, unbeständig
(XXIV)	Rubichrom (*21*) (aus Rubi-xanthin-epoxyd)	506 476 mμ	154°	blau, beständig
(XXV)	Capsochrom (*15*) (aus Capsanthin-epoxyd) ...	515 482 mμ	195°	blau, unbeständig
(XVI)	Aurochrom (*13*) (aus β-Carotin-di-epoxyd) ..	457 426 mμ	185°	tiefblau, beständig
(XXVI)	Kryptochrom (*16*) (aus Kryptoxanthin-di-epoxyd	456 424 mμ	?	tiefblau, beständig
(XXVII)	Auroxanthin (*12*) (aus Violaxanthin)	454 423 mμ	203°	tiefblau, beständig
(XVIII)	Luteochrom (*13*) (aus β-Carotin-di-epoxyd) ..	482 451 mμ	176°	tiefblau, beständig

Die Konstitution der meisten furanoiden Carotinoid-oxyde, welche in der Tabelle 2 aufgeführt sind, geht aus ihrer Herstellung bzw. aus der Struktur ihrer Ausgangsstoffe hervor. Die folgende Zusammenstellung enthält ihre Konstitutionsformeln (XX—XXVII), soweit diese nicht schon weiter oben wiedergegeben wurden.

```
CH₃   CH₃
   \  /
    C
   /                                                      H₃C   CH₃
 CH₂   C═══CH  CH₃        CH₃            CH₃      CH₃        \  /
  |     |    |  |          |              |        |          C
 CH₂    C    CH─C=CHCH=CHC=CHCH=CHCH═CCH=CHCH=C─CH=CH─CH    CH₂
   \   / \  /O                                              |    |
   CH₂|  O                                                  C    CH₂
      CH₃                      (XX.) Flavochrom.           //  \\
                                                         H₃C    CH
```

```
CH₃   CH₃
   \  /
    C
   /                                                      H₃C   CH₃
 CH₂   C═══CH  CH₃        CH₃            CH₃      CH₃        \  /
  |     |    |  |          |              |        |          C
HOCH    C    CH─C=CHCH═CHC=CHCH=CHCH═CCH=CHCH═CCH═CH─CH    CH₂
   \   / \  /O                                              |    |
   CH₂|  O                                                  C    CHO]
      CH₃              (XXI.) Flavoxanthin. Chrysanthemaxanthin.  //  \\
                                                               H₃C    CH
```

```
CH₃   CH₃
   \  /
    C
   /                                                      H₃C   CH₃
 CH₂   C═══CH  CH₃        CH₃            CH₃      CH₃        \  /
  |     |    |  |          |              |        |          C
 CH₂    C    CH . C=CHCH=CHC=CHCH=CHCH═CCH=CHCH═CCH═CH─C    CH₂
   \   / \  /O                                              ||    |
   CH₂|  O                                                  C    CHO:
      CH₃                    (XXII.) Kryptoflavin.         //  \\
                                                         H₃C    CH₂
```

```
CH₃   CH₃
   \  /
    C
   /                                                      H₃C   CH₃
 CH₂   C═══CH  CH₃        CH₃            CH₃      CH₃        \  /
  |     |    |  |          |              |        |          C
HOCH    C    CH . C=CHCH=CHC=CHCH=CHCH═CCH=CHCH═CCH═CH─C    CH₂
   \   / \  /O                                              ||    |
   CH₂|  O                                                  C    CHO
      CH₃                   (XXIII.) Mutatoxanthin.        //  \\
                                                         H₃C    CH₂
```

```
CH₃   CH₃
   \  /
    C
   /                                                      H₃C   CH₃
 CH₂   C═══CH  CH₃        CH₃            CH₃      CH₃        \  /
  |     |    |  |          |              |        |          C
HOCH    C    CH . C=CHCH=CHC=CHCH═CHCH═CCH═CHCH═CCH═CHCH    CH
   \   / \  /O                                              ||    |
   CH₂|  O                                                  C    CH₂
      CH₃                    (XXIV.) Rubichrom.            //  \\
                                                         H₃C    CH₂
```

CH₃ CH₃
\ /
C
/ \
CH₂ C══CH CH₃   CH₃   CH₃   CH₃   H₃C   CH₃
| | | | | | | \   /
HOCH C   CH . C=CHCH=CHC=CHCH=CHCH=CCH=CHCH=CCH=CHCO   C
 \ / \ / O   / \
 CH₂ | CH₂ CH₂
      CH₃   (XXV.) Capsochrom. | |
                                                    H₃C CHOH
                                                          CH₂

CH₃ CH₃
\ /
C
/ \
CH₂ C══CH CH₃   CH₃   CH₃   CH₃ CH══C   CH₂
| | | | | | | | | |
CH₂ C   CH . C=CHCH=CHC=CHCH=CHCH=CCH=CHCH=C──CH C CHOH
 \ / \ / O   O | \ /
 CH₂ | | CH₂
      CH₃   (XXVI.) Kryptochrom.   CH₃

CH₃ CH₃
\ /
C
/ \
CH₂ C══CH CH₃   CH₃   CH₃   CH₃ CH══C   CH₂
| | | | | | | | | |
HOCH C   CH . C=CHCH=CHC=CHCH=CHCH=CCH=CHCH=C──CH C CHOH
 \ / \ / O   O / \
 CH₂ | | CH₂
      CH₃   (XXVII.) Aurochrom.   CH₃

Die furanoiden Carotinoid-oxyde sind durchwegs gut kristallisierte, relativ beständige Verbindungen, die mit Salzsäure Blaufärbungen geben, welche von derselben Intensität und Beständigkeit sind wie diejenigen, welche die zugehörigen Epoxyde unter denselben Bedingungen zeigen. Dies ist darauf zurückzuführen, daß die Epoxyde durch die konzentrierte Salzsäure zunächst in die furanoiden Oxyde umgelagert werden, so daß die blaue Farbreaktion in beiden Fällen von demselben Stoff hervorgerufen wird.

Beim Übergang eines Carotinoid-epoxydes in das isomere, furanoide Oxyd tritt eine Aufhellung der Farbe ein; die Absorptionsmaxima erfahren eine Blauverschiebung. Diese Verlagerung nach kürzeren Wellenlängen beträgt für die längstwellige Bande in Schwefelkohlenstofflösung 19—22 mµ. Beim Übergang eines Di-epoxydes in das Isomere mit zwei Furanringen ist die Verschiebung etwa doppelt so groß, d. h. etwa 40—45 mµ:

β-Carotin-mono-epoxyd (V),   längstwelliges Maximum in CS_2 511 mμ,
Mutatochrom   (XVII),   „   „   „ CS_2 489 „ ,

β-Carotin-di-epoxyd   (X),   „   „   „ CS_2 502 „ ,
Aurochrom   (XVI),   . „   „   „ CS_2 457 „ .

Die furanoiden Carotinoidoxyde werden von Adsorbentien stets etwas stärker adsorbiert als die entsprechenden isomeren Epoxyde. So haftet Auroxanthin etwas fester als Violaxanthin, Flavoxanthin fester als Xanthophyllepoxyd usw. Liegt ein Gemisch eines Epoxyds und des zugehörigen furanoiden Oxyds vor, wie man es häufig in einer Pflanze trifft, so wird man also letzteres im Chromatogramm stets oberhalb des zugehörigen Epoxydes finden.

Die Reihenfolge einiger furanoider Oxyde in der Chromatogrammröhre ist folgende:

(oben) Auroxanthin
       Mutatoxanthin
       Flavoxanthin
       Chrysanthemaxanthin
       Rubichrom
       Aurochrom
       Mutatochrom (Citroxanthin)
(unten) Flavochrom

Mehrere furanoide Carotinoid-oxyde hat man in Pflanzen angetroffen. Sie begleiten gewöhnlich die zugehörigen Epoxyde und es ist sehr wahrscheinlich, daß sie aus letzteren in der pflanzlichen Zelle auf demselben Weg entstehen wie in vitro, d. h. durch Einwirkung von Pflanzensäuren auf die Epoxyde.

Da schon bei der Aufarbeitung pflanzlichen Materials Carotinoidepoxyde leicht in furanoide Oxyde umgelagert werden können, wenn man nicht Säuren aller Art peinlichst ausschließt, so sind manche frühere Angaben über das Vorkommen von Pigmenten, die später als furanoide Carotinoidoxyde erkannt worden sind, mit Vorsicht aufzunehmen. Es besteht die Möglichkeit, daß in der Pflanze in manchen solchen Fällen genuin nur das Epoxyd vorkommt und sich das entsprechende furanoide Oxyd erst während der Aufarbeitung des Pflanzenmaterials gebildet hatte.

Folgende Carotinoid-oxyde furanoïder Struktur sind bisher als natürlich vorkommende Pigmente beschrieben worden:

Flavochrom in Blüten von *Ranunculus acer* und *Tragopogon pratensis* (*20*).

Mutatochrom = Citroxanthin in Orangenschalen (*18, 11*).

Flavoxanthin findet sich in vielen Blüten, z. B. solchen von *Ranunculus acer* (*34, 20*), Löwenzahn (*28*), *Viola tricolor* (*30*), *Sarothamnus scoparius* (*10*), *Ulex europaeus* (*41*), *Senecio vernalis* (?) (*34*).

Chrysanthemaxanthin in den Blüten von Winterastern (*9*), denjenigen des Besenginsters (*10*) und von *Ranunculus acer* (*20*).

Auroxanthin in gelben Blüten des Stiefmütterchens, *Viola tricolor* (*29*).

Rubichrom in Blüten von *Tagetes patula (nana flora plenum)* (*21*).

III. Die Molekularstruktur der Carotinoid-epoxyde.

Das eigenartige Verhalten der Carotinoid-epoxyde zu Säuren, das sich von demjenigen einfacher 1,2-Epoxyde wesentlich unterscheidet, ist durch die übliche Formulierung der 1,2-Epoxyde nicht verständlich und bedarf einer näheren Betrachtung. Der leichte Übergang in die furanoiden Oxyde einerseits, die Abgabe des Oxido-sauerstoffs anderseits, die sich unter der Einwirkung von Säuren, z. B. Salzsäure, vollziehen, wird leichter verständlich, wenn man annimmt, daß *diese* Epoxyde in gewissen Umsetzungen in einer polaren Form reagieren, im Sinne des Gleichgewichtes: (XXVIII) $\rightleftharpoons$ (XXIX). Der Übergang in die furanoide Form könnte sich im Sinn der Reaktion *a* über ein chlorhaltiges Zwischenprodukt (XXX) vollziehen; man kann aber auch annehmen, daß unter der Wirkung des positiven C-Atoms 1 ein Elektronenpaar, das zwischen den C-Atomen 2 und 3 liegt, zwischen die C-Atome 1 und 2 verschoben wird (XXXI). Dadurch erhält C-Atom 3 positiven Charakter und nun schließt sich zwischen diesem positiven Kohlenstoffatom und dem negativen Sauerstoffatom der Ring zum furanoiden Oxyd (Reaktion *b*).

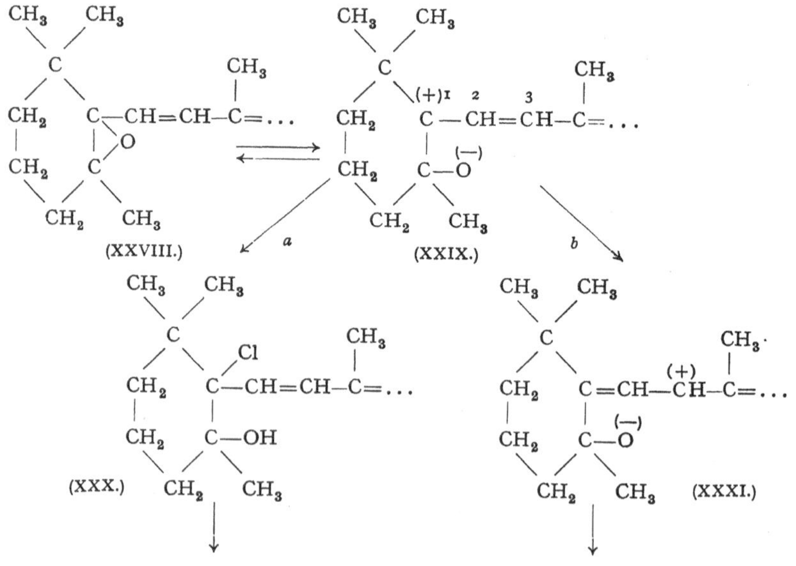

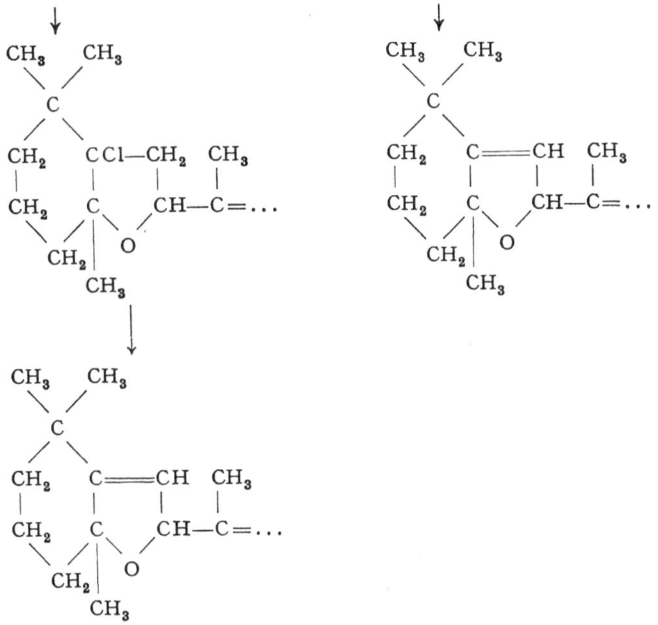

$$
\begin{array}{c}
\overset{\displaystyle\downarrow}{} \\
CH_3 \quad CH_3 \\
\diagdown \; C \; \diagup \\
CH_2 \quad CCl{-}CH_2 \quad CH_3 \\
| \qquad | \qquad | \quad | \\
CH_2 \quad C \quad\;\; CH{-}C{=}\ldots \\
\diagdown\;CH_2\;\diagup\; O \\
\qquad CH_3
\end{array}
\qquad
\begin{array}{c}
\overset{\displaystyle\downarrow}{} \\
CH_3 \quad CH_3 \\
\diagdown \; C \; \diagup \\
CH_2 \quad C{=\!=}CH \quad CH_3 \\
| \qquad | \qquad | \quad | \\
CH_2 \quad C \quad\;\; CH{-}C{=}\ldots \\
\diagdown\;CH_2\;\diagup\; O \\
\qquad CH_3
\end{array}
$$

$$
\begin{array}{c}
\downarrow \\
CH_3 \quad CH_3 \\
\diagdown \; C \; \diagup \\
CH_2 \quad C{=\!=}CH \quad CH_3 \\
| \qquad | \qquad | \quad | \\
CH_2 \quad C \quad\;\; CH{-}C{=}\ldots \\
\diagdown\;CH_2\;\diagup\; O \\
\qquad CH_3
\end{array}
$$

Die „polare" Formel der Carotinoid-epoxyde, in welcher der Sauerstoff durch semipolare Bindung am Kohlenstoffatom gebunden ist, macht auch die leichte Abgabe dieses Sauerstoffatoms verständlich. Denn die „polaren" Epoxyde werden damit zu Analoga der Aminoxyde, Jodosoverbindungen und Azoxyde, die alle semipolar gebundenen Sauerstoff enthalten und als Oxydationsmittel bekannt sind, d. h. ihr Sauerstoffatom leicht abspalten.

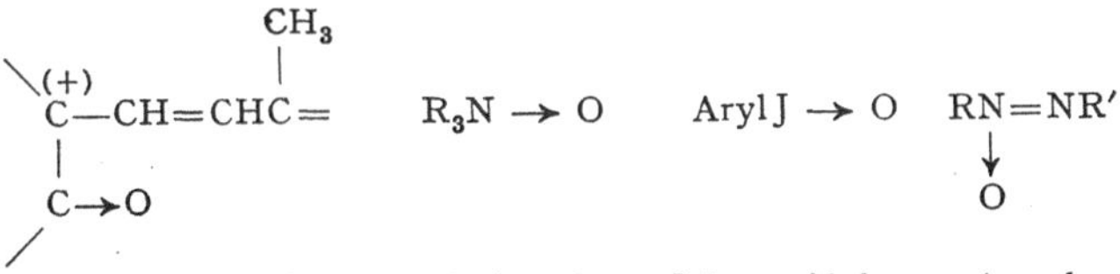

$$
\begin{array}{c}
\qquad\qquad CH_3 \\
\qquad\qquad | \\
\diagdown\;\overset{(+)}{C}{-}CH{=}CHC{=} \qquad R_3N \to O \qquad Aryl\,J \to O \qquad RN{=}NR' \\
| \qquad\qquad\qquad\qquad\qquad\qquad\qquad\qquad\qquad\qquad\qquad \downarrow \\
C \to O \qquad\qquad\qquad\qquad\qquad\qquad\qquad\qquad\qquad\qquad\; O \\
\diagup
\end{array}
$$

„Polares" Epoxyd.   Aminoxyd.   Jodoso-verbindung.   Azoxyd.

Die klassische Formel für 1,2-Epoxyde [(XXXII) für Äthylenoxyd] stellt für zahlreiche Substanzen dieser Verbindungsgruppe zweifellos den treffendsten Ausdruck in unserer Formelsprache dar. Sie kann das allgemeine Verhalten und die chemischen Umsetzungen zahlreicher 1,2-Epoxyde zwanglos deuten und steht auch mit der Tatsache in Übereinstimmung, daß die meisten Vertreter dieser Verbindungsklasse keine oxydierenden Wirkungen besitzen. Es gibt indessen eine Reihe von 1,2-Epoxyden, die von diesem allgemeinen Verhalten abweichen, z. B. die Glycidsäureester (XXXIII) sowie alle Oxido-verbindungen, die sich von α,β-ungesättigten Ketonen ableiten (XXXV). Diese letzteren

Epoxyde sind Oxydationsmittel und oxydieren z. B. Jodwasserstoff-säure quantitativ zu Jod. Sie entsprechen somit in ihrem Verhalten den Aminoxyden; polare Strukturformeln (XXXIV, XXXVI) mit semipolar gebundenem Sauerstoff können diese Analoga und das genannte Verhalten am besten zum Ausdruck bringen.

$$CH_2\text{---}CH_2 \qquad CH_2\text{---}CH \cdot COOR \qquad \overset{(+)}{CH_2}\text{---}\overset{|}{CH} \cdot COOR \qquad R \cdot CH\text{---}CH \cdot CO \cdot R' \qquad R \cdot \overset{(+)}{CH}\text{---}\overset{|}{CH} \cdot CO \cdot R'$$

(XXXII.)　　　(XXXIII.)　　(XXXIV.　　　(XXXV.)　　　(XXXVI.)

Die Epoxyde aus α,β-ungesättigten Ketonen, die Glycidsäureester und die Carotinoid-epoxyde haben gemeinsam, daß eines der beiden C-Atome, das den Oxido-sauerstoff trägt, mit einer stark ungesättigten (negativen) Gruppe verbunden ist (—COOR, —C=O oder ein System konjugierter Doppelbindungen). Diese negativen Substituenten polari-sieren die Äthylenoxydgruppe, und den betreffenden Epoxyden werden damit Eigenschaften verliehen, wie sie den Aminoxyden, Jodosover-bindungen und Azoxykörpern zukommen.

Sofern diese Auffassung richtig ist, kann man vermuten, daß die Polarisation der Epoxydgruppe geringer oder vollkommen aufgehoben sein wird, wenn von beiden Seiten derselbe polarisierende Einfluß auf die Epoxydgruppe einwirkt. Ein geeignetes Objekt zur Prüfung dieser Frage schien die Äthylenoxyd-dicarbonsäure zu sein (XXXVII). Tat-sächlich hat dieser Stoff keine oxydierenden Eigenschaften [Karrer und Rodmann (27)]; er macht aus HJ kein Jod frei. Sein Verhalten wird somit am besten durch die Formel (XXXVII) zum Ausdruck ge-bracht.

$$HOOC \cdot CH\text{---}CH \cdot COOH \qquad\qquad CH_3 \cdot CO \cdot CH_2 \cdot CH_2 \cdot \overset{(5)}{C} \cdot CO \cdot CH\text{---}CH \cdot \overset{(2)}{CO} \cdot CH_3$$

(XXXVII.)　　　　　　　　　H_3C　　CH_3　O
　　　　　　　　　　　　　　　　(XXXVIII.)

Anderseits wirkt das Triketon-epoxyd (XXXVIII) oxydierend (27), vermutlich als Folge seines in bezug auf die Epoxydgruppe unsymmetri-schen Baues, der einen verschieden großen Einfluß der beiden Carbonyl-gruppen 2 und 5 auf die Epoxydgruppe bedingt.

IV. Biologische Bedeutung der Carotinoid-epoxyde.

1,2-Epoxyde hatte man früher in Naturprodukten nur vereinzelt angetroffen. Natürlich vorkommende Vertreter dieser Gruppe sind das Aurapten (1, 2) (XXXIX), das in Orangen vorkommt, Linalool-epoxyd (XL) (38) im Linaloeöl, Elaidinsäure-epoxyd (XLI) in roher Elaidinsäure (4), Äthylenoxyd-α,β-dicarbonsäure (XXXVII) in *Monilia formosa* und *Aspergillus fumigatus* (39) u. a. m. Über deren Bedeutung für die betreffenden Organismen ist nichts bekannt.

CH
CH
CO
CH_3O O
CH_2
CH
O
C
CH_3 CH_3
(XXXIX.) Aurapten.

$(CH_3)_2C$————$CH \cdot CH_2 \cdot CH_2 \cdot C—CH=CH_2$
O OH
CH_3

(XL) Linalool-epoxyd.

$CH_3 \cdot (CH_2)_7 \cdot CH$————$CH \cdot (CH_2)_7 \cdot COOH$
O

(XLI.) Elaidinsäure-epoxyd.

Die Auffindung der Carotinoid-epoxyde in den Pflanzen hat die Zahl der natürlichen Vertreter stark vermehrt und die Frage nach ihrer biologischen Bedeutung aufgeworfen. Diesbezügliche Untersuchungen stehen noch in ihren Anfängen. Wir können heute über die Rolle, welche diese Pigmente im Pflanzenreich spielen, noch nichts Bestimmtes aussagen, wenn auch die Leichtigkeit der Sauerstoffabgabe, die diesen Verbindungen eigen ist, zu der Vermutung Anlaß gibt, daß sie bei Sauerstoffübertragungen in der Zelle eine Rolle spielen könnten.

In der Vermutung einer biologischen Bedeutung der Carotinoidepoxyde für die Pflanze wird man bestärkt durch die Tatsache ihrer weiten Verbreitung im Pflanzenreich und ihres Auftretens in besonders wichtigen pflanzlichen Organen. So wurde Antheraxanthin (VII) in den Antheridien von *Lilium tigrinum* gefunden, während in den Staubgefäßen von *Lilium candidum* und solchen von *Clivia miniata* Xanthophyll-epoxyd (II) vorhanden war.

Besonders Xanthophyll-epoxyd ist ein überaus verbreitetes Pigment dieser Gruppe. Neben Carotin und Xanthophyll findet es sich in grünen Blättern, und zwar in Mengen, die von derselben Größenordnung wie diejenigen des Carotins und Xanthophylls sind [KARRER, KRAUSE-VOITH und STEINLIN (23)].[1] Es ist demnach ein Hauptfarbstoff des Blattes.

---

[1] In einer Erwiderung auf die Mitteilung von KARRER, KRAUSE-VOITH und STEINLIN (23), in welcher auf die Verbreitung des Xanthophyllepoxydes in allen untersuchten grünen Blättern hingewiesen worden ist, hat STRAIN (42a) die Ansicht vertreten, dieses Pigment sei nicht Xanthophyllepoxyd, sondern identisch mit dem u. a. in gelben Stiefmütterchen vorkommenden Violaxanthin. Weiterhin schreibt der genannte Autor, daß sowohl Violaxanthin aus Stiefmütterchen wie „Blattviolaxanthin" aus grünen Blättern unter der Einwirkung von Säuren zuerst in Flavoxanthin und hierauf in Auroxanthin übergehen.

In Wirklichkeit führen, wie wir gezeigt haben (30, 12,), Säuren das richtige Violaxanthin aus *Viola tricolor* in Auroxanthin, aber niemals in Flavoxanthin über Letzteres wäre schon aus dem Grunde unmöglich, weil Violaxanthin und Auro-

In diesem erscheint es sehr frühzeitig. *Avena*-Primärblätter, die im Dunkeln gewachsen und geerntet worden waren und sich als chlorophyll-frei erwiesen hatten, enthielten neben Carotin und Xanthophyll bereits große Mengen Xanthophyll-epoxyd (*23*) (Menge des Xanthophylls zu Xanthophyll-epoxyd etwa wie 2 : 3). Die Entstehung des Xanthophyll-epoxydes im Primärblatt von *Avena* ist somit nicht an die Einwirkung des Lichtes gebunden. In *Avena*-Koleoptilen ließ sich dagegen kein Xanthophyll-epoxyd, sondern nur Xanthophyll nachweisen (*23*).

Etiolierte Kressekeimlinge *(Lepidium sativum)*, die 8 Tage im Dunkeln gewachsen waren, enthielten ebenfalls neben Carotin und Xanthophyll beträchtliche Mengen Xanthophyll-epoxyd (*23*).

Die Frage der Teilnahme von Carotin und Xanthophyll beim *Assimilationsprozeß* im grünen Blatt ist in den letzten Jahrzehnten immer wieder aufgeworfen worden, ohne daß sie sich einer Klärung entgegen-führen ließ. Inskünftig muß man sie erweitern und auf das Xanthophyll-epoxyd ausdehnen, das dank seiner Eigenschaft, Sauerstoff leicht abzu-spalten, vielleicht eine besondere Funktion zu erfüllen hat.

Im *Tierreich* konnten Carotinoid-epoxyde bisher nicht gefunden werden und es scheint daher, daß sie für die tierische Zelle entbehrlich sind. Von Interesse war aber die Frage, was mit Carotinoid-epoxyden geschieht, wenn sie in den tierischen Organismus gelangen.

Einigen Aufschluß darüber gaben Versuche mit α-Carotin-epoxyd (IV), β-Carotin-di-epoxyd (X) und Luteochrom (XVIII), die keinen unsubsti-tuierten β-Jononring enthalten, dagegen eine bzw. zwei aus dem un-substituierten β-Jononring entstandene Epoxyd-gruppierungen. Da Vitamin-A-Wirkung an das Vorhandensein eines unsubstituierten β-Jononringes geknüpft ist, so mußte die Prüfung der genannten drei Verbindungen auf Vitamin-A-Wirkung bei der Ratte entscheiden, ob diese Verbindungen im Organismus dieses Tieres unter Abspaltung des Oxido-sauerstoffs zum ursprünglichen Carotinoid reduziert werden.

Der biologische Versuch zeigte, daß den genannten drei Epoxyden starke *Vitamin-A-Wirkung* zukommt. α-Carotin-epoxyd war in 10-γ-Dosen voll wirksam, β-Carotin-di-epoxyd in 17-γ-Dosen und Luteochrom in Tagesdosen von 18 γ [VON EULER (*5*)]. Wahrscheinlich liegen die

---

xanthin die Formeln $C_{40}H_{56}O_4$, d. h. vier Sauerstoffatome besitzen, während Flavo-xanthin $C_{40}H_{56}O_3$, nur deren drei hat. Anderseits wird Xanthophyllepoxyd $C_{40}H_{56}O_3$ durch Säuren in das furanoide Isomere, Flavoxanthin $C_{40}H_{56}O_3$ umge-lagert; Auroxanthin kann hier nicht entstehen, besitzt es doch im Gegensatz zu Xanthophyllepoxyd und Flavoxanthin vier Sauerstoffatome. Alle diese Zusammen-hänge sind auf den vorhergehenden Seiten nochmals ausführlich geschildert worden.

Der Haupt-carotinoidfarbstoff der grünen Blätter, welcher die Blaufärbung mit konz. wäßriger Salzsäure liefert, ist, wie wir früher zeigten (*23*), Xanthopyllep-oxyd und nicht — wie STRAIN glaubt — Violaxanthin. Ob letzteres in untergeordneter Menge in einzelnen Blättern vorkommt, ist nicht ausgeschlossen, aber vorläufig unbewiesen.

minimalen, voll wirksamen Mengen noch etwas niedriger. Man kann daraus mit Sicherheit schließen, daß diese Carotinoid-epoxyde im Körper der Ratte mindestens teilweise zu den Oxidosauerstoff-freien Carotinoiden reduziert wurden, daß sich somit derselbe Vorgang abspielt, der auch in vitro zu beobachten ist. Vielleicht liegt in dieser Fähigkeit des tierischen Körpers, aus den Carotinoid-epoxyden den Sauerstoff reduktiv abzuspalten, der Grund, daß man diese Verbindungen bisher im tierischen Organismus nicht nachweisen konnte.

Die furanoiden Oxyde der Carotinoide erwiesen sich der Erwartung gemäß nicht vitamin-A-wirksam, wenn sie nicht einen unsubstituierten β-Jononring in der anderen Molekülhälfte besitzen (geprüft wurde z. B. Aurochrom). Die tierische Zelle ist nicht imstande, die beständige furanoide Atomgruppierung reduktiv aufzuspalten.

Das Studium des biologischen Verhaltens der Carotinoid-epoxyde steht erst in seinen Anfängen. Es kann erwartet werden, daß seine Weiterführung neue Erkenntnisse zeitigen wird.

Literaturverzeichnis.

1. Böhme, H. u. G. Pietsch: Zur Kenntnis des Auraptens. Ber. dtsch. chem. Ges. **72**, 773 (1939).

2. Böhme, H. u. E. Schneider: Oxydativer Abbau und Konstitution des Auraptens. Ber. dtsch. chem. Ges. **72**, 780 (1939).

3. Drumm, P. J. and W. F. O'Connor: The Pigments of the Yellow Iris *(Iris pseudocorus)*. Biochemic. J. **39**, 211 (1945).

4. Ellis, G. W.: Autoxydation of the Fatty Acids I. The Oxygen Uptake of Elaidic, Oleic and Stearic Acids. Biochemic. J. **26**, 791 (1932).

5. Euler, H. von: vgl. *(20)*.

6. Euler, H. von, P. Karrer u. E. Jucker: Zur Konstitution des Kryptoxanthin-monoepoxyds und Kryptoflavins. Helv. chim. Acta **30**, 1159 (1947).

7. Hey, D.: Eloxanthin—a new Carotenoid Pigment from the Pondweed, *Elodea canadensis*. Biochemic. J. **31**, 532 (1937).

8. Karrer, P., H. von Euler u. U. Solmssen: Oxydationsprodukte des α-Carotins; über Zusammenhänge zwischen Konstitution und Vitamin-A-Wirkung. Helv. chim. Acta **17**, 1169 (1934).

9. Karrer, P. u. E. Jucker: Carotinoide aus den Blüten von Winterastern. Chrysanthemaxanthin. Helv. chim. Acta **26**, 626 (1943).

10. — — Über Carotinoide aus Blüten des Besenginsters *(Sarothamnus scoparius)*. Chrysanthemaxanthin. Helv. chim. Acta **27**, 1585 (1944).

11. — — Vorläufige Mitteilung über ein neues Carotinoid aus Orangenschalen: Citroxanthin. Helv. chim. Acta **27**, 1695 (1944).

12. — — Partialsynthesen des Flavoxanthins, Chrysanthemaxanthins, Antheraxanthins, Violaxanthins, Mutatoxanthins und Auroxanthins. Helv. chim. Acta **28**, 300 (1945).

13. — — Oxyde des β-Carotins: β-Carotin-mono-epoxyd, β-Carotin-di-epoxyd, Mutatochrom, Aurochrom, Luteochrom. Helv. chim. Acta **28**, 427 (1945).

14. — — α-Carotin-mono-epoxyd und Flavochrom. Helv. chim. Acta **28**, 471 (1945).

15. — — Capsanthin-epoxyd und Capsochrom. Helv. chim. Acta **28**, 1143 (1945).

16. Karrer, P. u. E. Jucker: Vom Kryptoxanthin sich ableitende Epoxyde und furanoide Oxyde. Helv. chim. Acta **29**, 229 (1946).

17. — — Über weitere Vorkommen von Carotinoid-epoxyden. Trollixanthin und Trollichrom. Helv. chim. Acta **29**, 1539 (1946).

18. — — Die Konstitution des Citroxanthins. Helv. chim. Acta **30**, 536 (1947).

19. Karrer, P., E. Jucker u. E. Krause-Voith: Zur Verbreitung der Carotinoide, insbesondere Carotinoid-epoxyde in Blüten. Helv. chim. Acta **30**, 537 (1947).

20. Karrer, P., E. Jucker, J. Rutschmann u. K. Steinlin: Zur Kenntnis der Carotinoid-epoxyde. Natürliches Vorkommen von Xanthophyll-epoxyd und α-Carotin-epoxyd. Helv. chim. Acta **28**, 1146 (1945).

21. Karrer, P., E. Jucker u. K. Steinlin: Rubichrom, ein neues, natürlich vorkommendes Carotinoid mit furanoidem Ringsystem. Helv. chim. Acta **30**, 531 (1947).

22. Karrer, P. u. E. Krause-Voith: Einige weitere Beobachtungen bezüglich der Verbreitung der Carotinoide, insbesondere Carotinoidepoxyde. Helv. chim. Acta **31**, 802 (1948).

23. Karrer, P., E. Krause-Voith u. K. Steinlin: Ein neuer Blattfarbstoff: Xanthophyllepoxyd. Helv. chim. Acta **31**, 113 (1948).

24. Karrer, P. u. R. Morf: Taraxanthin aus *Tussilago farfara* (Huflattich). Helv. chim. Acta **15**, 863 (1932).

25. Karrer, P. u. A. Notthafft: Zur Kenntnis der Carotinoide aus Blüten. Helv. chim. Acta **15**, 1195 (1932).

26. Karrer, P. u. A. Oswald: Carotinoide aus den Staubbeuteln von *Lilium tigrinum*. Ein neues Carotinoid: Antheraxanthin. Helv. chim. Acta **18**, 1303 (1935).

27. Karrer, P. u. E. Rodmann: Zur Kenntnis der 1,2-Epoxyde und verwandter Verbindungen. Helv. chim. Acta **31**, 1074 (1948).

28. Karrer, P. u. J. Rutschmann: Über die Phytoxanthine der Löwenzahnblüten. Flavoxanthin. Helv. chim. Acta **25**, 1144 (1942).

29. — — Auroxanthin, ein kurzwellig absorbierender Carotinfarbstoff. Helv. chim. Acta **25**, 1624 (1942).

30. — — Über Violaxanthin, Auroxanthin und andere Pigmente der Blüten von *Viola tricolor*. Helv. chim. Acta **27**, 1684 (1944).

31. — — Über die Carotinoide aus *Elodea canadensis*. Helv. chim. Acta **28**, 1526 (1945).

32. Karrer, P. u. K. Steinlein: Zur Kenntnis der Carotinoid-epoxyde. Natürliches Vorkommen von Xanthophyll-epoxyd und α-Carotin-epoxyd. Helv. chim. Acta **28**, 1146 (1945).

33. Karrer, P., U. Solmssen u. O. Walker: Vorläufige Mitteilung über neue Oxydationsprodukte aus α-Carotin und Physalien. α-Carotindijodid. Helv. chim. Acta **17**, 417 (1934).

34. Kuhn, R. u. H. Brockmann: Flavoxanthin. Hoppe-Seyler's Z. physiol. Chem. **213**, 192 (1932).

35. Kuhn, R. u. E. Lederer: Taraxanthin, ein neues Xanthophyll mit vier Sauerstoffatomen. Hoppe-Seyler's Z. physiol. Chem. **200**, 108 (1931).

36. Kuhn, R. u. A. Winterstein: Violaxanthin, das Xanthophyll des gelben Stiefmütterchens *(Viola tricolor)*. Ber. dtsch. chem. Ges. **64**, 326 (1913).

37. Kuhn, R., A. Winterstein u. E. Lederer: Zur Kenntnis der Xanthophylle. Hoppe-Seyler's Z. physiol. Chem. **197**, 141 (1933).

38. Naves, Y. R. et P. Bachmann: Études sur les matières végétales volatiles. XXXV. Sur la constitution du soi-disant «linaloloxyde» (époxylinalol). Helv. chim. Acta **28**, 1227 (1944).

39. SAKAGUCHI, K., T. INOUE u. S. TADA: Über die Bildung von Äthylenoxyd-α,β-dicarbonsäure durch Pilze. Zbl. Bakteriol., Parasitenkunde, Infektionskrankh., Abt. II **100**, 302 (1939); Chem. Zbl. **1940** I, 68.

40. SCHÖN, K.: Studies on Carotenoids. I. The Carotenoids of *Diospyros* Fruits. II. The Carotenoids of *Arbutus* Fruits *(Arbutus unedo)*. Biochemic. J. **29**, 1779 (1935).

41. — Studies on Carotenoids. III. An Isomeride of Lutein isolated from the Furze *(Ulex europaeus)*. Biochemic. J. **30**, 1960 (1936).

42. SCHUNCK, C. A.: The Xantophyll Group in Yellow Pigments. Proc. Roy. Soc. [London] **72**, 165 (1903).

42a. STRAIN, H. H.: Leaf Xanthophylls, J. Amer. chem. Soc. **70**, 1672 (1948).

43. VERMAST, P. G. F.: Die Carotinoide von *Citrus aurantium*. Naturwiss. **19**, 442 (1931).

44. YAMAMOTO, R. u. S. TIN: Über die Farbstoffe von *Carica papaya* L. Chem. Zbl. **1933** I, 3090.

45. — — Über die Carotinoide in der Frucht von *Citrus pooenensis* HORT. Chem. Zbl. **1934** I, 1660.

46. ZECHMEISTER, L. u. L. v. CHOLNOKY: Über den Farbstoff der Ringelblume *(Calendula officinalis)*. Ein Beitrag zur Kenntnis des Blüten-Lycopins. Hoppe-Seyler's Z. physiol. Chem. **208**, 26 (1932).

47. ZECHMEISTER L. u. P. TUZSON: Über das Pigment der Orangenschale. Naturwiss. **19**, 307 (1931).

48. — — Das Pigment der *Cucurbita maxima* DUCH. (Riesenkürbis). Ber. dtsch. chem. Ges. **67**, 842 (1934).

(Eingelaufen am 25. März 1948.)

Some Biochemical Aspects of Marine Carotenoids.

By **D. L. Fox**, La Jolla, California.

Introduction.

The chemical unsaturation of carotenoids renders these biochromes particularly susceptible to degradation with bleaching by atmospheric oxygen and other oxidizing agents, augmented by light of certain wave lengths and by elevated temperatures. In the terrestrial world direct sunlight, wide changes of temperature, relatively high oxygen tension and desiccation are therefore effective factors in the rapid destruction of carotenoids, e. g., in fallen leaves, dying organisms, soils and the feces of herbivorous animals.

In contrast to conditions on land, many deep-lying regions of the ocean floor are characterized by perpetual darkness, low temperatures and little or no free oxygen. All of these features favor the preservation of carotenoids in bottom sediments for centuries of time [Fox (23, 24); Fox, UPDEGRAFF and NOVELLI (34)]. But there are nevertheless chemical and biochemical factors which effect the gradual destruction of carotenoids in the ocean and in the muds of the sea floor. Certain biological agencies therein also bring about a striking reversal in the ratio of carotenes (hydro-carbon) to xanthophylls (oxygen containing carotenoids). The present survey is devoted to a consideration of the marine animals which metabolize carotenoids in such a way as to contribute to the latter processes.

In the green parts of land plants xanthophylls predominate in quantity considerably over carotenes.

Some herbivorous or omnivorous land animals selectively assimilate the xanthophylls (hens and other birds), others the carotenes (horses and cattle); still others absorb neither kind (swine, sheep, goats, rodents, etc.), while certain species are able to absorb and store any or all types of carotenoids [man, frog; ZECHMEISTER (66); Fox and CRANE (27); CRANE, unpublished]. Whether or not one or the other type of carotenoid is absorbed, studies on several land animals have indicated that substantial fractions of both carotenes and xanthophylls are destroyed or rejected in the gut. This is true of sheep [ROGOZINSKI (51), horses [ZECHMEISTER· and TUZSON (67); ZECHMEISTER, TUZSON and ERNST (68)], and rats [HOVE (37); SHERMAN (55)].

At least one species of marine worm is able to assimilate β-carotene and destroy dietary xanthophylls in its alimentary tract [FOX, CRANE and McCONNAUGHEY (28)].

Marine muds.

In the marine world, as on land, we find a great preponderance of xanthophylls over carotenes, e. g., in marine algae, in colloidal and other finely particulate organic detritus suspended in the water, and in beach sand or freshly deposited sediment at greater depths. Ratios may be as high as 90 : 10 in marine plants, or even 95 : 5 in suspended or freshly deposited detritus. But when mud cores are collected from the unlighted depths of the ocean floor, contrasting circumstances are encountered. The carotenoids at the mud surface may be some two-thirds xanthophylls, but beneath the top few inches the polyene biochromes are often from two-thirds to more than four-fifths carotenes. While the actual concentrations of total carotenoids tend to decrease with depth (amounting at most to some tenths of a milligram per 100 grams of mud on the dry weight basis), it is manifest that biochemical processes contribute to this gradual change, favoring the relative survival of the hydrocarbon type as the organic matter is utilized [FOX, UPDEGRAFF and NOVELLI (34)].

The absence of both free oxygen and light in the sediments themselves, the prevailing chemical reducing conditions there, and the occurrence therein of unusual carotenes which typify certain bacteria, all lend support to the writer's view that the degradation of carotenoids in the muds of the ocean floor must occur largely at the agency of saprophytic plants and detritus-feeding animals.

Among carotenoids of the hydrocarbon class, β-carotene, $C_{40}H_{56}$, has been found as a prominent constituent of the lipochromes from deep marine sediments. α-Carotene, $C_{40}H_{56}$, occurs in secondary concentrations, frequently accompanied by various unusual carotenes. Some of the latter may be isomers, formed slowly in the long-standing muds or produced as artifacts during chemical separation in the laboratory. The absorption spectra of other members resemble those of some rarer carotenes. A recurring fraction spectroscopically close to leprotene, $C_{40}H_{56}$, characteristic of certain acid-fast pathogens, has been encountered, while other fractions similarly resembling rhodopurpurin and flavorhodin, previously isolated from *Rhodovibrio* bacteria, were frequently conspicuous. Torulene, $C_{40}H_{54}(OCH_3)_2$, a red carotene from yeast-like fungi of the *Torula* group, has often similarly been recovered upon laboratory chromatograms.

Of the alcoholic carotenoids, zeaxanthin, $C_{40}H_{54}(OH)_2$, or more probably diatoxanthin from marine diatoms, seems to be the chief representative, but is accompanied by other hydroxypolyenes, resembling antheraxanthin $C_{40}H_{56}O_3$, petaloxanthin $C_{40}H_{56}O_3$, sulcatoxanthin (peridinin) $C_{40}H_{52}O_8$, lutein $C_{40}H_{54}(OH)_2$, fucoxanthin $C_{40}H_{56}O_6$ (or $C_{40}H_{60}O_6$) and other carotenoids described by STRAIN, MANNING and HARDIN (61) as characteristic of marine phytoplankton.

Relatively heavily oxygenated acidogenic carotenoids (e. g., astaxanthin and similar compounds which occur in crustaceans, echinoderms and other invertebrates) were very rare or completely absent in the numerous sediments examined. Since these animal carotenoids occur in the vast populations of oceanic microcrustaceans, their singular absence from buried marine sediments would appear to emphasize the highly reducing character of the mud.

In general, the quantities of recoverable carotenoids decrease with the age (depth of strata) of mud in any given area.

Examples. The carotenoids decreased from 0,24 mg. per 100 g. (dry weight) in the first foot (30,5 cm.) to about an eighth of this value at the 3 to 4 foot levels, under 106 fathoms of water (1 fathom = 183 cm.). But concentrations may differ by several fold in different areas, although retaining the same low order of magnitude. Thus, one area off the Southern California coast, 650 fathoms deep, yielded 0,29 mg. of carotenoid pigments per 100 g. of mud at the 8-foot level, representing an estimated age of some *2500 years*.

Still another area, in the Gulf of California whose waters are particularly rich in organic content, yielded a mud core beneath a water depth of 364 fathoms, containing as much as 0,61 mg. of carotenoids per 100 g. of mud at a sediment depth of 15 to 16 feet, estimated at an age of some *6000 to 7000 years*. The polyene pigment of this sample contained 69% carotenes, while proportions in various other samples reached 80% or higher.

Values of approximately 2,5% organic matter in such muds would mean that a sediment containing say 0,25 mg. of carotenoids per 100 g. of the dry solid would reach concentration of 10 mg. of the pigments per 100 g. organic matter. Average figures of about 0,05% of lipid material in dry sediments, or about 2% of the total organic matter therein, would give an equivalent of 500 mg. of carotenoids (approximately 400 mg. of carotenes and 100 mg. of xanthophylls) in 100 g. of sedimentary lipid material. The latter seems to contain no true fats or typical sterols; there may be traces of higher fatty acids, some heavy aliphatic alcohols and some hydrocarbon material present.

Analyses of intertidal beach sand at La Jolla (California) gave approximately 1% organic matter, and about 0,07 mg. of carotenoids per 100 g. of the dry sand, corresponding to about 7 mg. of polyenes per 100 g. of organic matter. Chlorophylls and their breakdown products are present in somewhat higher concentrations. Of the sand carotenoids, some 95% are xanthophylls, and the remaining 5% carotenes [FOX, CRANE and McCONNAUGHEY (28)]. Mud cores from the depths of the ocean have not been observed to yield unaltered chlorophyll, but only varying concentrations of its numerous degradation products. The latter have not been given thorough qualitative or quantitative study, but such investigations will be of considerable importance and are anticipated.

Further consideration will be given to the biochemical position of the marine lipochromes after an examination of some of the marine animal phyla,[1] their food and the carotenoids of their tissues.

Marine animals in general.

In the fleshy parts of animals, carotenoids are usually dissolved in microscopic or colloidally dispersed oil droplets which, accordingly, assume yellow, orange or red colors. The highest concentrations of these biochromes are generally encountered in gonads, digestive or glandular tissues, and notably in integumentary structures. In many invertebrate animals, the integumentary carotenoids may occur as blue, green, purple, pink, brown, grey or other chromoproteins. Upon heating or treating with alcohol or other protein coagulants, the protein conjugant is split off and coagulated, thus unmasking the red or orange color of the free carotenoid. Examples of this effect may be seen in boiled lobsters and in alcohol-preserved crustaceans, echinoderms and some similarly preserved mollusks and coelenterates.

In spite of numerous investigations which have been devoted to the subject, no proof yet exists for the biosynthesis of carotenoids from colorless precursors in animals. Carotenoids almost certainly originate exclusively in plants, both green and saprophytic. In passing, it may be remarked that the inability of fungi and nearly all bacteria to synthesize chlorophyll, as do all green plants, is still somewhat arresting. For the saprophytes, like all protoplasm indeed, are capable of elaborating other such tetrapyrrole compounds as cytochrome, catalase, etc., which posses the same fundamental molecular pattern as chlorophyll. Similarly, it is of interest to recall that there is considerable inconsistency among animals regarding the capacity for organizing tetrapyrroles into hemo-globins, synthesized by all vertebrate and numerous invertebrate species. The comparative biochemistry of tetrapyrrole biochromes calls urgently for further studies.

In surveying the marine organisms with respect to their carotenoids, it seems logical to consider the animals in the order of their phylogenic position. It should also be pointed out that preliminary investigations in the writer's and other divisions of the Scripps Institution have shown that numerous chromogenic marine bacteria owe their yellow, red and orange colors to the presence of carotenoids. The same is doubtless true of some of the highly colored marine fungi.

---

[1] *Phylum*, meaning race or tribe, refers to the large primary groups of organisms in zoological and botanical classification.

Protista.

The unicellular, chlorophyll-bearing flagellates, while naturally classed as plants, are included with the true Protozoa or single-celled animals for the purpose of this discussion for several reasons. In the first place, certain euglenoid species are known to share some biochemical characteristics of both plants and animals, photosynthesizing their own food in the light, but readily ingesting by mouth finely divided organic matter when cultured in the dark. Furthermore, numerous species of flagellated phytozoans are encountered within the tissues of some Protozoa and multicellular invertebrates, whence such commensal hosts may derive much or all of their natural color [FRITSCH (35)].

According to YONGE (63, 64), numerous Protozoa may carry non-motile algae of the green zoochlorellae or yellow or brown zooxanthellae classes in their tissues. This is notably true of some radiolarians. The same applies to such multicellular invertebrates as sponges, anemones, corals, turbellarian and annelid worms, univalve and bivalve mollusks, colonial ascidians, rotifers and some bryozoans and echinoderms.

Algal cells occurring in the tissues of the sea anemone, Cribrina xanthogrammica confer upon the host the brown and greenish colors which characterize the tentacles and other parts in lighted waters. STRAIN, MANNING and HARDIN (61) found several chlorophylls, phaeophytin a, carotenes and numerous xanthophylls in extracts of this animal with its commensal algae. They also obtained a similar series of pigments from a free-living species of fresh water dinoflagellate, Peridinium cinctum. The mixture of green and yellow pigments within the algae are responsible for the complementary brown colors observed. STRAIN and his colleagues established the identity of peridinin (a prominent xanthophyll in Peridinium cinctum) with sulcatoxanthin, $C_{40}H_{52}O_8$, isolated by HEILBRON, JACKSON and JONES (36) from the sea anemone, Anemonia sulcata, which harbors commensal zooxanthellae.

Extensive colored patches in the ocean (so-called "red", "yellow" or "brown water") owe their appearance to intense blooms of dinoflagellate algae such as Gonyaulax polyedra, Prorocentrum micans, Ceratium furca or C. tripos, Noctiluca scintillans and others. Noctiluca is named for the remarkable luminescence which it imparts at night to breaking waves of other mechanical disturbances of the water. Numerous protistan species share this property of imparting conspicuous colors by day, followed by brilliant luminescence in the night hours. The carotenoids which contribute to the color are not known to be related in any direct way with the nocturnal manifestation of luciferin.

At certain seasons of the year, countless millions of zoospores from the giant kelp, Macrocystis pyrifera and related seaweeds contribute yellow-green hues to areas near the kelp beds and inshore from them, off the coast of California. CARTER, HEILBRON and LYTHGOE (8) have examined the carotenoids of many algal species. They report the close relationship between the green and red algae and higher plants in the kinds of lipochromes synthesized, while the brown and some of the blue-green forms indicate more primitive relationships in elaborating fucoxanthin. Carotenes occur in all the species, accompanied by any of several classes of xanthophylls.

Ellipsoid biflagellated cells of the green algae, *Dunaliella viridis* and *D. salina* occur in natural or man-made, land-locked salterns the world over. As the summer season advances, bringing, through evaporation a steady increase in the salt concentration of these natural basins or the salt vats of industry, the color of the water changes from the green of spring, through murky intermediate aspects, to the brick-red hues which persist until the autumn rains. During these progressive color changes, decreasing numbers of the smaller, green form, *D. viridis*, can be recognized, while the larger variety, *D. salina*, undergoes an interesting change in its successive generations. The originally substantial quantities of chlorophyll disappear, the small amounts of xanthophyll becomes greatly reduced, while carotene, especially the β-isomer, undergoes a conspicuous increase and is responsible for the red hue of the cells and macroscopically observed color of the ponds [Fox and SARGENT (32)].

These extensive biochemical changes suggest a possible tendency for the cells to substitute a saprophytic for the photosynthetic mode of life pursued in the less concentrated medium. It should be added, however, that even these red organisms cannot live in the dark. Perhaps, therefore, minute amounts of chlorophyll in the cells act synergistically with large quantities of carotenoids in carrying out photosynthesis [DUTTON and MANNING (14)]. BLUM and FOX (6) have shown that the ready positive phototaxis of *Dunaliella salina*, in either the green or the red phase, is limited to regions of the visible spectrum between about 330 and 550 mμ., with a decided maximum response in the blue to blue-green between 480 and 530 mμ., wherein are located the principal absorption maxima of β-carotene.

Some true protozoans exhibit rich red, orange or yellow colors which are probably due to carotenoids. Examples are to be found in floating scarlet colored masses of *Globigerina*, long ago discussed by AGASSIZ, and in red foraminifera such as *Myxotheca arenilega*.

Sponges.

A number of species from the relatively simple poriferan phylum exhibit conspicuous orange, red, yellow or purple carotenoid coloration, reflecting the animals' consumption of finely divided detritus and minute organisms. The red sponge, *Axinella crista-galli* yields astacin, $C_{40}H_{48}O_4$ (or more probably its precursor, astaxanthin, $C_{40}H_{52}O_4$, the red acidogenic carotenoid of many crustaceans), according to KARRER and SOLMSSEN (39). The same investigators described a similar red carotenoid with a single absorption maximum at 500 mμ. in CS_2, isolated from another red sponge, *Suberites domuncula*. LEDERER (46), on the other hand, failed to detect such a compound in a sponge which he believed to belong to the above species. No xanthophylls or acidogenic carotenoids were encountered, but instead two suspected hydrocarbons believed to be torulene and lycopene. *Ficulina ficus* also yielded a lycopene-like carotenoid. Both species contained β- and α-carotene.

According to DRUMM and O'CONNOR (11) as well as DRUMM, O'CONNOR and RENOUF (12), another red sponge, *Hymeniacidon sanguinea*, contains no astaxanthin but crystallizable echinenone (a ketonic provitamin A from echinoids, probably $C_{40}H_{58}O$) and γ-carotene, $C_{40}H_{56}$.

The seeming preponderance of carotenes over xanthophylls in members of the sponge phylum is strikingly in contrast to relative carotenoid ratios in most other marine animals.

Coelenterates.

Members of this carnivorous phylum, e. g., jellyfishes and especially sea anemones and corals, exhibit their assimilation of carotenoids by their brilliant pigmentation and by the colors imparted to organic extracts of their fleshy parts [FOX and PANTIN (31)].

Red, brown and green color-variants of the anemone, *Actinia equina* derive their carotenoids from the diet, and exhibit a remarkable specificity either in the initial selective absorption of given carotenoid molecules or in the subsequent metabolism of a common carotenoid, assimilated by all variants. The red variety possesses an unstable acidogenic carotenoid, actinioerythrin, while the green form yields also a red-orange xanthophyll whose protein-conjugant is green. Both β- and α-carotene are present in each color-variety of the species [ABELOOS-PARIZE and ABELOOS-PARIZE (1); LEDERER (43); FABRE and LEDERER (20); HEILBRON, JACKSON and JONES (36)].

Some unique xanthophylls, occasionally accompanied by small amounts of familiar carotenes, occur in *Anemonia sulcata*, *Metridium senile (= Actinoloba dianthus)* and *Tealia felina* [HEILBRON et al. (36)].

Acidogenic carotenoids are probably nearly as common in colored anemones as in crustaceans. FOX and MOE (29) encountered in a small Pacific Coast anemone, *Epiactis prolifera*, one such compound which closely resembled a similar biochrome isolated by HEILBRON et al. (36) from *Tealia felina*. Both the natural ester and the free acidic product of hydrolysis and oxidation showed a single broad maximum at 500 mμ.

in CS_2. The plumose anemone, *Metridium senile* occurs in numerous color variants, reflecting wide differences in kinds and quantities of carotenoids as well as in their melanin content [Fox and PANTIN (*30*)]. While *Metridium senile* harbors no commensal algae, but carries the pigments in its own tissues, we have noted that the common Pacific anemone, *Cribrina xanthogrammica* owes its kinds and quantities of carotenoids to the commensal zooxanthellae within its body [STRAIN *et al.* (*61*)]. This is true of numerous other coelenterate and other invertebrate species, and must be taken into account in any investigations relating to the comparative metabolism of carotenoids in marine animals. Species harboring commensal algae invariably yield chlorophyll, while those carrying no such plant organisms do not.

The small blue siphonophore, *Velella lata*, a floating colonial coelenterate known as "by-the-wind-sailer", is an oceanic species which is occasionally carried by winds to shore in vast numbers. The association harbors brownish zooxanthellae ventrally, but the pneumatophore or disk-shaped air bladder carries a bright blue pigment, especially in the integument near the boundaries. Sail, tentacles and other parts are paler blue. The finely ground, acetone-extracted colonies yield chlorophyllic and carotenoid pigments from the commensal algae, while the blue pigment is a protein-conjugated carotenoid which turns red-orange upon denaturation.

Specimens of *Velella lata*, carried to shore at La Jolla, California, in April, 1948, yielded a number of carotenoids and non-carotenoid chromolipoids. The three principal carotenoids were β-carotene [maxima at *507,5* and *485* mμ. in CS_2, and at *476* and *449* mμ. in petroleum ether "Skellysolve"]; a rather unstable xanthophyll with maxima at 505 and 474 mμ. in CS_2, and at 470 and 442 mμ. in Skellysolve; and an acidogenic carotenoid with a single broad maximum at 470 mμ. in Skellysolve. (All readings were made with a BECKMAN photoelectric spectrophotometer.) The acidogenic carotenoid of *Velella* is spectroscopically identical with crustacean astaxanthin, and is undoubtedly the component involved in the blue chromoprotein of this coelenterate [CRANE (*9*)].

Worms.

The carotenoids of platyhelminths (flat worms), nemerteans (ribbon worms) annelids (segmented worms) and members of the gephyrean or sipunculid group have not received much study, although the nemerteans often exhibit the conspicuous colors of these pigments in their integument [Fox (*25*)]. Investigations by the writer and his colleagues [SUMNER and Fox (*62*); Fox (*22*); Fox, CRANE and McCONNAUGHEY (*28*)] have shown that the beach-worm, *Thoracophelia mucronata*, which occurs in vast numbers in the beach sand of the California coast, stores appreciable quantities of β-carotene but no other carotenoid. This selectivity is the more remarkable since the worm, consuming sand and the organic detritus adhering thereto, ingests nearly ten times as much xanthophyllic material as it does carotenes. The xanthophylls seem to be largely destroyed, as in some terrestrial herbivores, while the α-carotene is apparently rejected in the gut. β-Carotene, on the other hand, seems to be concentrated by some five-fold in the worm, which stores roughly 1,4 mg. of this biochrome per 100 g. of its own organic matter, originating from a diet containing only some 0,28 mg. of carotene per 100 g. of organic matter, associated with the sand. In its consumption of large quantities of sand, this worm is probably typical of other invertebrate detritus-eaters. Its relationships to the substrate in which it lives will therefore receive further consideration below.

Echinoderms.

In this phylum, which is exclusively marine, the asteroid (sea-star) and ophiuroid (serpent-star or brittle star) classes contain the richest and most conspicuous supplies of carotenoids. The pigments are especially concentrated in the integument and digestive gland, while the gonads of these classes and of the echinoids (sea-urchins and sand-dollars) likewise contain substantial quantities. The thin skin of echinoids, however, usually carries little carotenoid material, but instead, large amounts of naphthoquinone pigments of the echinochrome type.

Thus, it is in the carnivorous classes of echinoderms that the highest concentrations of carotenoids are encountered, while the herbivorous or omnivorous echinoids and holothurians (sea-cucumbers) generally yield far smaller quantities, save perhaps seasonally, when the sex-products are ripe. Moreover, the oxygenated carotene derivatives, comprising acidogenic pigments and neutral xanthophylls or their esters, are greatly preponderant over the hydrocarbon type in asteroids, while some of the ophiuroids assimilate no carotenes at all. Conversely, several echinoid species appear to store no acidogenic carotenoids, and contain as much or more carotenes than xanthophylls. A mud-and-sand eating holothurian, *Stichopus californicus*, contains very small and approximately equal amounts of carotenes and xanthophylls, in spite of the fact that it subsists upon the same ultimate organic detritus as is consumed by carotenoid-rich species in other phyla (e. g., some bivalve mollusks, certain sponges, worms, and tunicates). Echinenone has been found in several echinoid species, and was suspected in small quantities in the holothurian, *Stichopus*. It was absent in the asteroids examined, except, possibly in the sand-star, *Astropecten californicus* [Fox and Scheer (33)]. The crinoids or sea-lilies, which belong to a sessile, particulate-feeding class inhabiting greater depths, display colors of the carotenoid type, but have not yet become available to the writer for investigation.

Like numerous Crustacea, the asteroid echinoderms carry much of their skin-carotenoids as blue, purple, orange, red, pink, grey or brown chromoproteins [EULER and HELLSTRÖM (16); EULER, HELLSTRÖM and KLUSSMANN (17); KARRER and BENZ (38); FOX and SCHEER (33)].

Some species of ophiuroids yield substantial quantities of free and esterified xanthophylls and acidogenic carotenoids. Many of the fractions seem to be unique, but very unstable in air [Fox and SCHEER (33)].

Among the sea-urchins and sand-dollars, relatively large quantities of β- and α-carotenes and some unusual xanthophylls have been found, notably in the gonads of both sexes, and secondarily in the digestive tract [FOX and SCHEER (33)].

Mollusks.

Various investigators have encountered carotenoids in the ripe eggs, developing sperm, digestive gland, mantle and other tissues of mollusks. Xanthophylls seem to preponderate greatly over carotenes, which are entirely absent in some species. Among the xanthophylls are some unusual ones encountered rarely if at all elsewhere, e. g., glycymerin from the ripe gonads of the scallop, *Pectunculus glycymeris*; pectenoxanthin, $C_{40}H_{54}O_3$, from the red gonads and mantle of another scallop, *Pecten maximus*; and mytiloxanthin from the California mussel, *Mytilus californianus*. Astacin has been obtained from a few bivalve and gastropod species [LEDERER (47); FABRE and LEDERER (20); SCHEER (52); FOX (25)].

The carotenoid content of mollusks appears to vary somewhat with season, sex and reproductive activity. The bivalves, feeding characteristically by means of mucous films and beating cilia, upon minute particles of organic detritus suspended in the water, are rather outstanding in their carotenoid assimilation. This is likewise true of some of the plant-eating gastropods, while exclusively carnivorous species, notably cephalopods (octopus, squid) are far less conspicuous in this respect. A chief site of carotenoid storage in mollusks is the digestive gland or so-called "liver". Among the cephalopods, carotenoids are not present in such relative abundance, nor are they permanent constituents. In some species, for example, *Octopus bimaculatus*, carotenoids are found only in the hepatopancreas, where they may be stored in considerable but highly variable quantities. The kinds and amounts of carotenoids there deposited reflect the current diet of the animal, which is not selective in its assimilation of these biochromes. Small amounts of xanthophylls, acidogenic carotenoids and esters of both, but none of the carotenes, are secreted from the hepatopancreas into the ink-sac, which lies in attached juxtaposition to the organ. Starvation or subsistence for a few weeks upon carotenoid-free flesh (e. g., hog liver) result in the disappearance of all carotenoids from the liver [Fox and Crane (27); Fox (25); Crane (10)].

Crustaceans.

The crustacean members of the vast arthropod phylum have long attracted attention because of their brilliant pigments and wide distribution. Zoologists and biochemists have devoted special attention to the pigments in the colored cells or chromatophores of the true skin (hypoderm) underlying the hard shell or carapace, and in the shell itself. Much investigation has also been given the carotenoids of the blood, hepatopancreas, eyes and eggs of numerous species of crabs, lobsters, shrimps and allied decapod crustaceans. Blue, green, purple and other carotenoid protein complexes are encountered widely in this group.

Although much research has been devoted to the question of the possible synthesis of carotenoids by animals, primarily crustacean species, no positive evidence has yet been found [Fox (25)]. The crab, *Carcinus maenas* absorbs its carotenoids directly from its food, concentrating the pigments in the hepatopancreas which serves as a temporary storehouse for materials destined for the hypoderm, eggs and other tissues via the blood [Abeloos and Fischer (2)].

Astaxanthin (3,3'-dihydroxy-4-4'-diketo-β-carotene), the red acidogenic carotenoid which occurs in many crustaceans, was first isolated as its oxidation product, astacin (3,3',4,4'-tetraketo-β-carotene) from the deep brown chromoprotein of the shell, the red complexes of the underlying skin, and the blue to green carotenoid proteins of the eggs of the crayfish, *Astacus gammarus*; likewise from the eggs of the spider-crab, *Maja squinado* and other tissues of several other Crustacea [Kuhn and Lederer (41); Kuhn, Lederer and Deutsch (42); Fabre and Lederer (19, 20)].

"Ovoverdin", the green astaxanthin-protein complex from the eggs of the lobster, *Homarus americanus*, retains its natural color between p_H 4 and p_H 8, exhibits an isoelectric point close to p_H 7, possesses a molecular weight of approximately 300,000, and shows absorption maxima in the red and violet regions at 640 and 470 mμ. respectively. This interesting chromoprotein is reversibly dissociable into its carotenoid and protein constituents by warming for short periods in the presence of neutral salts [Stern and Salomon (59, 60)]. Ball (5) investigated a similar blue chromoprotein from the eggs of two species of goose-barnacle, *Lepas fascicularis* and *L. anatifera* (?). He found the same reversible dissociability, induced by heat and by careful acidification of chilled systems of the chromoprotein in the presence of ammonium sulfate.

Like the echinoderms, mollusks and others of the larger invertebrate phyla, the crustaceans assimilate and store principally the oxygenated type of carotenoids rather than the hydrocarbon class. Carotenes are, however, frequently encountered in lower concentrations.

Tunicates.

The solitary and the colonial forms of ascidians, or sea-squirts, which belong to the primitive chordate group, are mucous-and-ciliary feeders, ingesting finely particulate suspended matter, as do sponges, bivalve mollusks and numerous other invertebrates. Many members of the ascidian group have rich stores of carotenoids in the outer cloak, or tunic, as well as in their eggs and body tissues.

Major investigations of conspicuous pigments in ascidians have been carried out by LEDERER (44, 47), who gave special attention to four species, viz. two solitary types, *Halocynthia (= Cynthia) papillosa* and *Microcosmus sulcatus*, the "social" form *Dendrodoa grossularia* and the compound species, *Botryllus Schlosseri*. The dark red tunic and orange internal organs of *Halocynthia* yielded rich quantities of astacin, accompanied by moderate amounts of a new xanthophyll, cynthiaxanthin, and traces of α- and β-carotene. Much free and esterified xanthophyllic material and smaller amounts of α-carotene and echinenone were reported in the violet-cloaked *Microcosmus* which yielded, among its alcoholic carotenoids, fractions resembling cynthiaxanthin (or zeaxanthin or diatoxanthin), lutein, and fucoxanthin (prominent in brown algae).

The rose-colored tunicate, *Dendrodoa*, like its solitary relative *Halocynthia*, was found to contain much astaxanthin with lesser quantities of neutral xanthophylls and traces of α- and β-carotenes.

Considerable variation was reported to occur in the types and relative quantities of carotenoids in the brown-red colonial ascidian, *Botryllus*. Lycopene and β-carotene were found, while capsanthin, $C_{40}H_{58}O_3$, capsorubin, $C_{40}H_{60}O_4$, and pectenoxanthin were present in some but not all specimens. This variability probably reflects differences in the diet; capsanthin, capsorubin and allied carotenoids were believed to be acquired by the animals' ingestion of industrial pimento pepper wastes discharged into the harbor.

The preponderance of astaxanthin and its heavily oxygenated homologues in numerous invertebrate animals suggests special oxidative metabolic features, relating to carotenoids, which the vertebrates, save perhaps for some fish species, seem to have largely lost in their evolutionary development. The carotenoid precursor of astaxanthin has not yet been established.

Fishes.

Members of the bony fish class possess and display the richest stores of carotenoids met with in the vertebrate subphylum. Many bright red colored species deposit in skin and muscles astaxanthin or related pigments which are in all probability derived from a diet of crustaceans and other invertebrates. Examples are to be found in the salmon, *Salmo salar* [EMMERIE, VAN EEKELEN, JOSEPHY and WOLFF (*15*); EULER, HELL-STRÖM and MALMBERG (*18*)], and *Onchorhynchus nerka* [BAILEY (*4*)], the red liver-oils of the oar-fish, *Regalecus glesne* and of *Cyclopterus lumpus*. These fishes apparently transfer astaxanthin from liver to skin and flesh during the spawning period of the summer [SCHMIDT-NIELSEN, SÖRENSEN and TRUMPY (*53*). The same pigment occurs in the red skin, oral and gill mucus and eyes of the marine dorado, *Beryx dedactylus* and in the skin of the red variety of the common goldfish, *Carassius auratus*; in some but not all varieties of the freshwater perch, *Perca fluviatilis*, and in the rock cod *Sebastes marinus* [LEDERER (*47*)]. The sea-devil or angler-fish, *Lophius piscatorius* yields astaxanthin and neutral tara-xanthin-like xanthophylls from its liver oil [BURKHARDT, HEILBRON, JACKSON and PARRY (*7*); SÖRENSEN (*57*)]; while the same acidogenic biochrome has been reported also in the flesh of trout [SÖRENSEN and STENE (*58*)]. It is accompanied by α-carotene in the oily orange liver of the sunfish, *Orthagoriscus mola* [SÖRENSEN (*56*)].

Numerous other species contain no astaxanthin or other acidogenic carotenoids in any of their tissues, but instead copious amounts of other xanthophylls and sometimes traces of carotenes. Examples are the pike, *Esox lucius*, the eel-pout, *Eleginus* and others. Among the xanthophylls occurring in fishes, taraxanthin-like compounds and their esters are very common. To name a few species which are so characterized, and normally yield no carotenes, reference is made to the Pacific killifish, *Fundulus parvipinnis*; the opal-eye, *Girella nigricans*; the long-jawed goby, *Gillichthys mirabilis*; the Garibaldi, *Hypsypops rubicunda*; and the surf-perch, *Cymatogaster aggregatus* [FOX (*25*)].

The female *Fundulus* remains drab in color, transferring free xantho-phylls to her ripening eggs as do many other species, while the male exhibits increasing quantities of yellow xanthophyll esters in skin and fins, during the reproductive season. Visible chromatic adaptation by the species to variously colored backgrounds involves gradual changes in the quantities of black or brown melanin, and alterations in shape or size of the yellow pigment cells, but no change in xanthophyll content. *Fundulus* retains its total xanthophyll complement, which is deposited almost entirely in the skin, even after long periods of subsistence upon carotenoid-free food. The xanthophyll supply is somewhat augmented by a diet containing the same pigment, and even when carotene is substituted for dietary xanthophyll. How this species may be able to oxidize carotenes to xanthophylls remains obscure.

Gillichthys mirabilis manifests a retention of xanthophyll similar to that demonstrated by *Fundulus*, when maintained in tanks of various colors. *Girella*, however, undergoes a steady and extensive loss in xanthophylls when kept in laboratory aquaria. This loss seems to be implemented in some way by optic stimuli, since it occurs more rapidly in specimens living in white tanks, wherein reflected light is maximal, than in those kept in black or even in yellow containers, while food and incident light from above are identical.

The esters of both neutral and acidogenic carotenoids occurring in the shrimp, *Hippolyte californiensis* are hydrolyzed in the gut of this animal's predator, the surf perch, *Cymatogaster aggregatus*. But the fish rejects quantitatively the acidogenic compounds and the carotenes of the diet, assimilating only part of the xanthophylls, which are re-esterified and stored in the skin. Excess quantities of the xanthophyll are deposited unesterified in the short terminal rectal segment of the intestine which consequently assumes a bright orange color. The pigment in the rectal segment disappears soon after the perches are captured and maintained in the laboratory, but the stores of esterified xanthophylls in the skin are retained for weeks or months [YOUNG and FOX (65)].

The Garibaldi, *Hypsypops rubicunda*, retains the brilliant orange or yellow color of its skin after months of confinement in laboratory aquaria. Like the opal-eye and killifish, it is omnivorous but partakes of much plant material in nature, assimilating no carotenes from the diet. The copious quantities of xanthophylls are nearly entirely esterified and are stored in the integument. Amounts of xanthophylls in the skin are somewhat variable, and increase with age. Fresh specimens of large size yield an average quantity of about 95 mg. per 100 g. of skin. When maintained for some months on a diet free from xanthophylls but containing large amounts of carotene (9 parts of β- to 1 part of α-carotene), this species deposits very small amounts of β-carotene in its skin. It thus seems possible to induce at least one habitual xanthophyll-selecting, carotene-rejecting species, when denied xanthophyll and provided with carotene, to assimilate hydrocarbon type polyenes [KRITZLER, FOX, HUBBS and CRANE (40)].

Much evidence has been accumulated for the occurrence of vitamins A in the liver oils of fishes, and for the derivation of these vitamins from carotenes, as in higher vertebrates. The richness in vitamin A of the liver of the soup-fin shark, *Galeorhynchus zyopterus*, is well known, and has been summarized by RIPLEY, BOLOMEY and others (50). The manner of accumulation of vitamin A in the liver of this carnivorous animal, and its possible functions therein, are problems yet awaiting solution. LOVERN, MORTON and IRELAND (48) encountered substantial quantities of vitamin A in the sturgeon, *Acipenser sturio*. The oil from liver and from intestine yielded respectively 1% and 10% of the vitamin.

Marine fishes are doubtless capable of converting carotene into vitamin A, as do freshwater species such as perch and dace [MORTON and CREED (49)].

Mammals.

Marine mammals have received relatively little study as regards the content and metabolism of carotenoids in their tissues. It has been reported, however, that astacin may be isolated in large amounts from the liver oil and feces of certain whales which in turn derive this carotenoid from the consumption of countless numbers of euphausids and kindred crustaceans or so-called "krill" [SCHMIDT-NIELSEN *et al.* (53); BURKHARDT *et al.* (7); DRUMMOND and MACWALTER (13)]. SCHMIDT-NIELSEN and his colleagues (54) report the appearance of an occasional "red" blue whale, wherein the lipochrome has become distributed in the tissues generally rather than remaining concentrated solely in the liver.

Conclusions.

Copious amounts of carotenoids are present in the sea, whence they may be recovered by (a) gathering large marine plants and animals; (b) straining small planktonic organisms with the use of special nets; (c) collecting the feces of herbivorous or filter-feeding animals; (d) passing large volumes of sea water through fine laboratory filters designed to retain particulate detritus; and (e) collecting sediments from the moist beach of from deeper areas or the ocean floor.

In marine as in terrestrial green plants, xanthophylls are present in greater quantities than are carotenes. This condition is reflected in mixed plankton-catches, in filter-residues of sea water and in extracts of damp beach sand. Xanthophylls may preponderate over carotenes likewise in the sapropel, or very top layer of fresh, loose organic detrital mud on the ocean floor. In the cold, unlighted, oxygen-poor depths of the sea, carotenoids remain undestroyed for long periods and within the chemically reducing mud of the ocean floor, carotenoids may persist for thousands of years. In this environment, carotenes outlast xanthophylls, and hence are found in predominant quantities.

The great majority of marine animals examined prove to absorb and store in their tissues more xanthophylls than carotenes. Such animals, voiding the unchanged carotenes in their feces, but assimilating some of the xanthophylls from their diet of plants, plankton or detritus, could readily contribute to the observed condition. Microorganisms inhabiting the anerobic environment of bottom mud may also exercise some differential utilization of the oxygenated over the hydrocarbon type of lipochrome. Circumstantial evidence further indicates the contribution of unusual carotenes to the environment by mud-dwelling bacteria and fungi.

To provide a summarizing example, reference is made to the beach-worm, *Thoracophelia mucronata*, studied by Fox, Crane and Mc Connaughey (28). Great colonies of this polychaete burrow to depths of a foot or more in the damp, intertidal sand of local beaches. Studies made upon their populations, area inhabited, diet, rate of filling and emptying the gut, organic content and carotene concentrations in sand and in the animals' tissues give some interesting, roughly approximate figures.

A typical wave-washed worm-bed one mile long, ten feet wide and one foot thick contains some $2,32 \times 10^3$ tons of sand, 28 tons of organic matter, some 225 kg. of mixed xanthophylls (largely fucoxanthin), and about 10,8 kg. of β- and α-carotene in a ratio of approximately 2 : 1. Worms living in this bed number about 3000 per cubic foot, weight some 40 mg. each; and each animal cycles at least 84 g. of sand per year. Thus, there are about 7 tons of these worms, cycling some $1,46 \times 10^3$ tons of sand, or the quantity of sand in their own bed 5,2 times per year, i. e., once every 10 weeks. Since the organic content of the sand is about 1%, and that of the worm tissues approximates 20%, the relative efficiency of the animals in maintaining a standing population, assuming an annual life cycle, is $7 \times 20/146 = 0,96\%$.

In cycling the sand and organic matter, the animals reject and largely destroy the xanthophyllic material, reject also and perhaps partially destroy the α-carotene, and store β-carotene in amounts of about 0,25 mg. per 100 g. of their own weight, or some 16 g., representing a relative efficiency of 0,042% in β-carotene assimilation. These values are of the general order of magnitude which might be expected in a sand-eating invertebrate. The low values pertaining to the assimilation of organic matter doubtless reflect the fact that considerable amounts of such material ingested is inert or metabolically refractory.

The feeding and carotenoid metabolism of the mussel [Scheer (52); Fox and Coe (26)] which captures the finely particulate matter in its mucus sheet, and the beach-worm, which obtains similar materials adsorbed to the ingested sand grains, are doubtless typical of vast numbers of marine detritus feeders which resolve a fraction of the organic matter ingested, and assimilate or destroy portions of one or another kind of available carotenoid.

From the data on hand at present, we may conclude that the majority of marine animals tend to select dietary xanthophylls in the gut, rejecting the carotenes; certain others assimilate both kinds, and a few species may, like the sponges and the beach-worm, store in part only carotene materials, rejecting and even destroying ingested xanthophylls.

References.

1. ABELOOS-PARIZE, M. et R. ABELOOS-PARIZE: Sur l'origine alimentaire du pigment carotinoïde d'*Actinia equina*. C. R. Séances Soc. Biol. Filiales Associées **94**, 560 (1926).

2. ABELOOS, M. et E. FISCHER: Sur l'origine et les migrations des pigments carotinoïdes chez les Crustacés. C. R. Séances Soc. Biol. Filiales Associées **95**, 383 (1926).

3. — — Les pigments Carotinoïdes chez les Crustacés: Sur l'origine des pigments de la carapace. C. R. Séances Soc. Biol. Filiales Associées **96**, 374 (1927).

4. BAILEY, B. E.: The Pigment of Salmon. J. biol. Board Canada **3**, 469 (1937).

5. BALL, E. G.: A Blue Chromoprotein Found in the Eggs of the Goose-barnacle. J. biol. Chemistry **152**, 627 (1944).

6. BLUM, H. F. and D. L. FOX: Light Responses in the Brine Flagellate *Dunaliella salina* with Respect to Wave Length. Univ. of Calif. Publ. in Physiol. **8**, 21 (1933).

7. BURKHARDT, G. N., I. M. HEILBRON, H. JACKSON and E. G. PARRY: Pigmented Animal Oils. 1. Pigments from the Angler Fish *(Lophius piscatorius)*, the Prawn *(Nephrops norvegicus)*, and the Whale. Biochemic. J. **28**, 1698 (1934).

8. CARTER, P. W., I. M. HEILBRON and B. LYTHGOE: The Lipochromes and Sterols of the Algal Classes. Proc. Roy. Soc. [London], Ser. B **128**, 82 (1939).

9. CRANE, S. C.: Astaxanthin in the Colonial Jellyfish, *Velella lata* (in preparation).

10. — Metabolic Studies of Carotenoids in the Octopus, *Octopus bimaculatus*. Thesis, Univ. of Calif. (1949).

11. DRUMM, P. J. and W. F. O'CONNOR: Pigments of Sponges. Nature [London] **145**, 425 (1940).

12. DRUMM, P. J., W. F. O'CONNOR and L. P. RENOUF: The pigments of sponges. 1. The Lipid Pigments of the Sponge *Hymeniacidon sanguineum* (GRANT). Biochemic. J. **39**, 208 (1945).

13. DRUMMOND, J. C. and R. J. MACWALTER: Provitamin A in the Food of Whales. J. exp. Biology **12**, 105 (1935).

14. DUTTON, H. J. and W. M. MANNING: Carotenoid-sensitized Photosynthesis in the Diatom *Nitzschia*. Amer. J. Bot. **28**, 516 (1941).

15. EMMERIE, A., M. VAN EEKELEN, B. JOSEPHY and L. WOLFF: Salmon Acid, a Carotenoid from Salmon. Acta brevia néerl. Physiol., Pharmacol., Microbiol. E. A. **4**, 139 (1934); Chem. Abstr. **29**, 2190 (1935).

16. EULER, H. v. u. H. HELLSTRÖM: Über Asterinsäure, eine Carotinoidsäure aus Seesternen. Hoppe-Seyler's Z. physiol. Chem. **223**, 89 (1934).

17. EULER, H. v., H. HELLSTRÖM u. E. KLUSSMANN: Über den Carotinoidgehalt einiger Evertebraten. Hoppe-Seyler's Z. physiol. Chem. **228**, 77 (1934); Chem. Abstr. **29**, 184 (1935).

18. EULER, H. v., H. HELLSTRÖM and M. MALMBERG: Salmon Acid, a Carotenoid from Salmon. Svensk kem. Tidskr. **45**, 151 (1933); Chem. Abstr. **27**, 5563 (1933).

19. FABRE, R. et E. LEDERER: Note sur la présence de l'astacin chez les Crustacés. C. R. Séances Soc. Biol. Filiales Associées **113**, 344 (1933).

20. — — Contributions à l'études des lipochromes des animaux. Bull. Soc. Chim. biol. **16**, 105 (1934).

21. FISCHER, E.: Sur l'absorption digestive chez les Crustacés Decapodes: les pigments carotinoïdes. C. R. Séances Soc. Biol. Filiales Associées **95**, 438 (1926).

22. Fox, D. L.: Further Studies of the Carotenoids of two Pacific Marine Fishes, *Fundulus parvipinnis* and *Hypsypops rubicunda*, and of a Marine Annelid, *Thoracophelia sp.* Proc. nat. Acad. Sci. U.S.A. **22**, 50 (1936).

23. — Biochemical Fossils. Science [New York] **100**, 111 (1944).

24. — Fossil Pigments. Sci. Monthly **59**, 394 (1944).

25. — Carotenoid and Indolic Biochromes of Animals. Ann. Rev. Biochem. **16**, 443 (1947).

26. Fox, D. L. and W. R. Coe: Biology of the California Sea-mussel *(Mytilus californianus)*. II. Nutrition, Metabolism, Growth and Calcium Deposition. J. exp. Zoology **93**, 205 (1943).

27. Fox, D. L. and S. C. Crane: Concerning the Pigments of the Two-spotted Octopus and the Opalescent Squid. Biologic. Bull. **82**, 284 (1942).

28. Fox, D. L., S. C. Crane and B. H. McConnaughey: A Biochemical Study of the Marine Annelid worm *Thoracophelia mucronata*. Its Food, Biochromes, and Carotenoid Metabolism. J. Marine Res. (in press).

29. Fox, D. L. and C. R. Moe: An Astacene-like Pigment from a Pacific Coast Anemone, *Epiactus prolifera*. Proc. nat. Acad. Sci. U.S.A. **24**, 230 (1938).

30. Fox, D. L. and C. F. A. Pantin: The Colors of the Plumose Anemone, *Metridium senile*. Philos. Trans. Roy. Soc. London, Ser. B **230**, 415 (1941).

31. — — Pigments in the Coelenterata. Biol. Rev. Cambridge philos. Soc. **19**, 121 (1944).

32. Fox, D. L. and M. C. Sargent: Variations in Chlorophyll and Carotenoid Pigments in the Brine Flagellate, *Dunaliella salina*, Induced by Environmental Concentrations of Sodium Chloride. Chem. and Ind. **57**, 1111 (1938).

33. Fox, D. L. and B. T. Scheer: Comparative Studies of the Pigments of Some Pacific Coast Echinoderms. Biologic. Bull. **80**, 441 (1941).

34. Fox, D. L., D. M. Updegraff and G. D. Novelli: Carotenoid Pigments in the Ocean Floor. Arch. Biochem. **5**, 1 (1944).

35. Fritsch, F. E.: The Structure and Reproduction of the Algae. Vol 1. Cambridge University Press, 1935.

36. Heilbron, I. M., H. Jackson and R. N. Jones: The Lipochromes of Sea Anemones. 1. Carotenoid Pigments of *Actinia equina*, *Anemonia sulcata*, *Actinoloba dianthus* and *Tealia felina*. Biochemic. J. **29**, 1384 (1935).

37. Hove, E. L.: The *in-vitro* Destruction of Carotene by Water Extracts of Minced Rat Stomachs in the Presence of Methyl Linolate. Science [New York] **98**, 433 (1943).

38. Karrer, P. u. F. Benz: Über ein neues Vorkommen des Astacins. Ein Beitrag zu dessen Konstitution. Helv. chim. Acta **17**, 412 (1934).

39. Karrer, P. u. U. Solmssen: Über das Vorkommen von Carotinoiden bei einigen Meerestieren. Helv. chim. Acta **18**, 915 (1935).

40. Kritzler, H., D. L. Fox, C. L. Hubbs and S. C. Crane: A Study of Carotenoid Pigmentation in the Pomacentrid Fish *Hypsypops rubicunda*. Copeia (in press).

41. Kuhn, R. u. E. Lederer: Über die Farbstoffe des Hummers *(Astacus gammarus)* und ihre Stammsubstanz, das Astacin. Ber. dtsch. chem. Ges. **66**, 488 (1933).

42. Kuhn, R., E. Lederer u. A. Deutsch: Astacin aus den Eiern der Seespinne *(Maja squinado)*. I. Hoppe-Seyler's Z. physiol. Chem. **220**, 229 (1933).

43. Lederer, E.: Note sur les Lipochromes d'*Actinia equina* (L.). C. R. Séances Soc. Biol. Filiales Associées **113**, 1391 (1933).

44. — Sur les caroténoïdes de trois Ascidies *(Halocynthia papillosa, Dendrodoa grassularia, Botryllus Schlosseri)*. C. R. Séances Soc. Biol. Filiales Associées **117**, 1086 (1934).

45. LEDERER, E.: Les Caroténoïdes des Animaux. Paris: Hermann et Cie. 1935.

46. — Recherches sur les Caroténoïdes des Invertebrés. Bull. Soc. Chim. biol. **20**, 567 (1938).

47. — Recherches sur les Caroténoïdes des Animaux Inférieurs et des Cryptogames. Paris: Lons-le-Saunier. 1938.

48. LOVERN, J. A., R. A. MORTON and J. IRELAND: The Distribution of Vitamin A and A_2. II. Biochemic. J. **33**, 325 (1939).

49. MORTON, R. A. and R. H. CREED: The Conversion of Carotene to Vitamin A_2 by Seven Freshwater Fishes. Biochemic. J. **33**, 318 (1939).

50. RIPLEY, W. E., R. A. BOLOMEY, *et al.*: The Biology of the Soupfin, *Galeorhynchus zyopterus*, and Biochemical Studies of the Liver. Fish Bull. No. 64. State of California. Deptm. of Natural Resources, Div. of Fish a. Game, Bureau of Marine Fishes (1946).

51. ROGOZINSKI, F.: Carotenoids and Chlorophyll in the Digestion of the Ruminant. Bull. int. Acad. polon. Sci. Lettres, Cl. Sci. math. natur., Ser. B **2**, 183 (1937); Chem. Abstr. **32**, 1760 (1938).

52. SCHEER, B. T.: Some Features of the Metabolism of the Carotenoid Pigments of the California Sea Mussel *(Mytilus californianus)*. J. biol. Chemistry **136**, 275 (1940).

53. SCHMIDT-NIELSEN, S., N. A. SÖRENSEN and B. TRUMPY: The Pigments of the Oil of *Regalecus glesne*. Lipochromes in the Fat of Marine Animals. I. Kong. norske Vidensk. Selsk. Skr. **4**, 114 (1932).

54. — — — A Red Colored Whale Oil. Lipochromes in the Fats of Marine Animals. II. Kong. norske Vidensk. Selsk. Skr. **5**, 118 (1932).

55. SHERMAN, W. C.: Relative Gastro-intestinal Stability of Carotene and Vitamin A and Protective Action of Xanthophylls. Proc. Soc. exp. Biol. Med. **65**, 207 (1947).

56. SÖRENSEN, N. A.: The Pigments of the Oil of *Orthagoriscus mola*. Lipochromes in the Fats of Marine Animals. III. Kong. norske Vidensk. Selsk. Skr. **6**, 154 (1933).

57. — The Liver Pigments of the Sea Devil *(Lophius piscatorius)*. Kong. norske Vidensk. Selsk. Skr. **6**, 1 (1934).

58. SÖRENSEN, N. A. and J. STENE: Lipochromes of Marine Animals. VIII. The Carotenoids of Trout. Kong. norske Vidensk. Selsk. Skr. **9**, 1 (1939).

59. STERN, K. G. and K. SALOMON: Ovoverdin, a Pigment Chemically Related to Visual Purple. Science [New York] **86**, 310 (1937).

60. — — On Ovoverdin, the Carotenoid-protein Pigment of the Egg of the Lobster. J. biol. Chemistry **122**, 461 (1938).

61. STRAIN, H. H., W. M. MANNING and G. HARDIN: Xanthophylls and Carotenes of Diatoms, Brown Algae, Dinoflagellates, and Sea-anemones. Biologic. Bull. **86**, 169 (1944).

62. SUMNER, F. B. and D. L. FOX: A Study of Variations in the Amount of Yellow Pigment (Xanthophyll) in Certain Fishes, and of the Possible Effects upon this of Colored Backgrounds. J. exp. Zoology **66**, 263 (1933).

63. YONGE, C. M.: "Symbiosis" between Invertebrates and Unicellular Algae. Proc. Linn. Soc. London, Session **147**, Pt. IV, 90 (1935).

64. — Experimental Analysis of the Association between Invertebrates and Unicellular Algae. Biol. Rev. Cambridge philos. Soc. **19**, 68 (1944).

65. YOUNG, R. T. and D. L. FOX: The Structure and Function of the Gut in Surf Perches (Embiotocidae) with Reference to their Carotenoid Metabolism. Biologic. Bull. **71**, 217 (1936).

66. ZECHMEISTER, L.: Die Carotinoide im tierischen Stoffwechsel. Ergebn. Physiol. biol. Chem. exp. Pharmakol. **39**, 117 (1937).

67. ZECHMEISTER, L. u. P. TUZSON: Beitrag zum Lipochrom-Stoffwechsel des Pferdes. Hoppe-Seyler's Z. physiol. Chem. **226**, 255 (1934).

68. ZECHMEISTER, L., P. TUZSON u. E. ERNST: Selektive Speicherung von Lipochrom. Nature [London] **135**, 1039 (1935).

(Received, June 7, 1948.)

Azulenes.

By A. J. HAAGEN-SMIT, Pasadena, California.

With 1 Figure.

Introduction.

We repeatedly find in the literature observations on blue and green colors in distillation products of plant materials (*37, 69, 73*). Well known examples are found in the oils of camomile, spike, blue camphor and cubeb. PIESSE (*36, 75, 76*) in 1864 studying a blue fraction from oil of wormwood introduced the name *Azulene* for the compound responsible for this color. Other cases are known where blue and green colorations appear on heating of the colorless or yellow volatile oils with acids in the presence of air. Probably of similar origin are some blue products, $C_{15}H_{18}$, isolated from "Braunkohlengenerator-teer" (*18*). We owe to WILLSTAEDT (*86*) the discovery that the same type of substances is involved in the blue discoloration which appears on the cut surface of certain mushrooms.

SABETAY et. al. (*71a*) used the dehydrogenating action of bromine in demonstrating the presence of azulenogenic compounds in essential oils, while MÜLLER (*29a, 29b, 29c*) applied dimethylamino-benzaldehyde in glacial acetic acid for the same purpose. In all these cases oxidation processes convert a naturally occurring precursor to the intensely blue pigment. Even a formation of such compounds in vitro had been recorded; SCHLÄPFER and STADLER (*72*) and others (*80a*) showed the presence of blue hydrocarbons among the polymerization products of acetylene.

HENTZSCHEL and WISLICENUS (*17*) and others (*52, 71*) describe the accumulation of blue and green substances in the higher boiling fraction of distillates obtained from heating calcium adipate.

The remarkable but scattered observations mentioned passed largely unnoticed and it was the interest in the structure of the precursors of some of these blue substances which stimulated an intense study of this field and furnished the explanation of many of the observed phenomena. SHERNDAL (*75, 76*) recognized that the structure of the blue hydrocarbons could be related to those of the *sesquiterpenes*; and the dehydrogenation of a pure crystalline sesquiterpene alcohol, *guaiol*, by RUZICKA and RUDOLPH (69) to a blue hydrocarbon proved this hypothesis. The azulenes formed a welcome addition to the two other well defined dehydrogenation products: *eudalene* and *cadalene*. The formation of these naphthalene hydrocarbons had proved to be of decisive help in unraveling the structural chemistry of two large groups of sesquiterpenes, and it was hoped that a knowledge of the azulene structure would clarify the carbon skeleton of a third group of volatile oil components.

Isolation and Purification.

SHERNDAL (75, 76) in a study of the blue colors of camphor oil, oil of cubeb, and acid-treated gurjun oil succeeded in purifying the pigment by using its property of forming addition products with concentrated acids such as phosphoric acid and sulfuric acid. While this behaviour is common to a number of unsaturated compounds, it is so pronounced in this instance that even a 50% aqueous sulfuric acid is able to bind the azulene (25, 28). These compounds can then be washed free from impurities, mostly more saturated derivatives, with inert solvents. The extraction with concentrated acids and also the subsequent decomposition with water are preferably carried out at 0° to prevent polymerization of the highly unsaturated azulene. Further purification is obtained by preparing the nearly black addition compounds with picric acid. This method has been extended by using other aromatic nitro compounds, especially sym. trinitrobenzene. Important improvements have also been made in the regeneration of the azulenes from their addition products. The picrate in ether solution is usually split by. washing with alkali. For the recovery of the azulene from the trinitrobenzene addition product, sublimation and also reduction with ammonium sulfide were tried, but these methods were abandoned when PLATTNER and PFAU (52) found in chromatography a most effective way of purifying the azulenes and also of separating and splitting the nitro addition compounds. The trinitrobenzene or other nitro compounds are strongly absorbed at the top of an alumina column while the freed azulene is adsorbed in lower layers or is washed through completely when the chromatogram is developed with solvents such as hexane.

Physical Properties.

The azulenes were first obtained in crystalline form by BIRREL (6), who noticed that the oily azulene obtained by dehydrogenation of guaiol with sulfur ("S-guaiazulene") was actually in the supercooled state. After cooling in liquid air the substance remained solid indefinitely at room temperature; recrystallized from alcohol, it had a melting point

Table 1. Melting points of azulenes, their picrates and trinitrobenzene addition products and literature references covering syntheses and spectral data.

Compound	m.p.	m.p. Picrate	m.p. Trinitrobenzene add. product	References
Azulene	98.5–99.5°	—	166.5–167.5°	(37, 38, 43, 48a, 52, 79, 80a)
1-Methylazulene	—	134–135°	154°	(37, 38, 43, 47, 48a, 57)
2-Methylazulene	47–48°	130–131°	140–141°	(37, 43, 47, 48, 48a)
4-Methylazulene	—	144°	177.5–178°	(1, 32, 37, 47, 48a)
5-Methylazulene	27°	—	142–143°	(2, 37, 45, 47, 48, 48a, 53)
6-Methylazulene	—	125°	140–141°	(1, 47, 48, 48a, 56)
1-Isopropylazulene	—	—	114–115°	(3)
2-Ethylazulene	44–45°	110–111°	107°	(39, 42, 82)
4-Ethylazulene	—	128,5°	147,5°	(32)
2-n-Propylazulene	—	—	118–119°	(42)
2-Isopropylazulene	31°	—	113–114°	(37, 48a)
2-Phenylazulene	230°	—	—	(48a, 55)
4-Phenylazulene	—	80–81°	86–87°	(32)
1,2-Dimethylazulene	58–59°	129–130°	166–167°	(37, 43, 48a, 58)
1,3-Dimethylazulene	54°	134–136°	164–166°	(43)
1,4-Dimethylazulene	69–70°	142–143°	177–178°	(37, 48a, 49, 58)
4,8-Dimethylazulene	69–70°	157–158°	179–180°	(35, 48a, 53a, 58)
4-Methyl-2-ethylazulene	—	—	112°	(82)
1,2-Benzazulene	176°	—	155°	(40)
Indeno-[2',1':1,2] azulene	—	—	—	(48a)
1-Isopropyl-6-methylazulene	—	—	98–98,5°	(3)
1-Isopropyl-5- (or 7-) methylazulene	—	—	149–150°	(3)

1,2,3-Trimethylazulene	—	160°	181–182°	(43)
1,3,7-Trimethylazulene	—	—	166°	(84)
1,4,7-Trimethylazulene	—	—	177–178°	(37, 51)
1,6,8-Trimethylazulene	—	—	163–164°	(83)
2,4,8-Trimethylazulene (?) (pyrethazulene)	—	—	167–168°	(71a)
4,8-Dimethyl-2-ethylazulene	—	118–119°	—	(82)
4,7- (4,6?-) Dimethyl-1-isopropylazulene	30,5–31°	117–118°	—	(54, 81)
1,4-Dimethyl-7-isopropylazulene (S-guaiazulene)	—	122–123°	151,5°	(32, 37, 65, 79, 86)
3,7-Dimethyl-1-isopropylazulene	32–33°	—	147,5°	(84)
4,8-Dimethyl-2-isopropylazulene (vetivazulene)	39°	122–122,5°	151,5–152°	(32, 33, 35, 79)
4,8-Dimethyl-6-isopropylazulene	70–71°	145°	173–173,5°	(54)
4,8-Dimethyl-6-isopropenylazulene	—	—	132°	(54)
Se-Guaiazulene (2,4-dimethyl-7-isopropylazulene?)	—	115°	123–124°	(32, 37, 65)
Chamazulene	—	110°	132°	(37, 65, 69, 81, 86)
Elemazulene	—	—	—	(65, 70)
Lactarazulene	88–89°	—	122–123°	(86)
1,3,4,8-Tetramethylazulene	—	—	194°	(44, 48a)
4,8-Dimethylazulene-6-carboxylic acid	230–265° (dec.)	—	215°	(53a)
4,8-Dimethylazulene-6-carboxylic ethyl ester	60–60,5°	82–83°	92°	(53a)
4,8-Dimethylazulene methyl ester	66,5°	103°	127–128°	(53a)
1,2,3-Trimethylazulene-carboxylic acid	245° (dec.)	—	—	(43)
2-Ethylazulene-6-carboxylic ethyl ester	—	—	71°	(82)
4,8-Dimethyl-2-isopropylazulene-6-carboxylic ethyl ester	54°	—	—	(82)
4,8-Dimethyl-6-(hydroxy-isopropyl-) azulene	—	—	170°	(54)
2-Ethyl-4-methyl-8-methoxy-azulene	53°	—	—	(82)
Lactaroviolin ($C_{15}H_{24}O$; aldehyde)	—	—	—	(48a, 86)

of 31,5°. A number of synthetic azulenes have since been obtained in crystalline form. By characterizing azulenes more sharply Ruzicka et al. (65) and Plattner (37) considerably reduced the number of the natural azulenes known at that time. The azulenes formed in dehydrogenating processes from kessylalcohol, ledol, gurjunene, and aromadendrene proved to be identical with S-*guaiazulene* from guaiacum wood. *Chamazulene* prepared from the oil of camomile is identical with that from milfoil oil; however, the picrates and styphnates showed distinct melting point depressions with those of S-guaiazulene. Melting point depressions were also observed with a violet azulene, "*Se-guaiazulene*", obtained by dehydrogenation of guaiol with selenium (65) as well as with *vetivazulene* originating from vetiver oil (32). Different from the compounds mentioned are, *lactarazulene* from the fungus *Lactarius deliciosus* (86) and *elemazulene* from elemi oil (65, 70).

Some of the physical constants of natural as well as synthetic azulenes and their derivatives are listed in Table 1.

Chemical Properties.

S-guaiazulene when shaken with powdered sodium, gives rise to a grayish brown powder. Air causes the immediate reappearance of the blue color but hydroxylic solvents do not regenerate the azulene (25, 28, 7). Attempts to use the reaction products for obtaining an insight in the chemical structure of the azulenes have failed.

Hydrogenation experiments demonstrated the strongly *unsaturated character* of the azulenes; Sherndal (75, 76), Augsburger (4) and Kremers (25) prepared the octahydro and decahydro products by catalytic hydrogenation in the presence of colloidal palladium. The formation of decahydro-azulenes was confirmed by Ruzicka et al. (65) in hydrogenation experiments with platinum in glacial acetic acid. It was also observed that partly hydrogenated, colorless azulenes turned blue again upon shaking with platinum black in the presence of air.

The unsaturated character of the azulenes is clearly shown in the vigorous action of bromine, nitrogen trioxide or nitrosylchloride, whereby no well defined products can be obtained. The action of oxidizing agents, such as nitric acid, chromic acid or permanganate causes far reaching degradation, and only small fragments of the original molecule could be isolated. These fragments consisted of acetic acid, oxalic acid, carbon dioxide, acetone and isobutyric acid (65).

Interesting is the reaction with sulfuric acid in acetic anhydride whereby sulfonic acids of the azulenes are formed (76). These acids give violet, water soluble salts which change to light green on acidification. The free acids taken up in ether show violet color and can be returned to the water phase by extraction with alkali. The relative thermostability of the azulene structure can be demonstrated by heating the sodium salt of a sulfonic acid. Blue fumes are produced which condense to a blue oil, while sulfur dioxide is liberated. This stability, which is noticeable in dehydrogenation experiments, is also shown in the behavior of the azulenes towards formic acid with which the free azulenes can be heated at 90° for four to five hours before the blue color disappears. Boiling with hydriodic acid and phosphorus for several hours is necessary before a conversion to colorless naphthalene derivatives occurs.

Structure Determination.

The physical constants of the octa- and decahydro-derivatives are very similar for the different azulenes. Furthermore, it can be seen from the molecular refraction values that in all cases we are dealing with *bicyclic structures*. There was the possibility that the difference between the individual azulenes merely consisted in the distribution of the double bonds. However, octahydro- and decahydro-S-guaiazulene as well as decahydro-chamazulene regenerated the original guaiazulene and chamazulene upon dehydrogenation. The five double bonds present must be confined to twelve of the fifteen carbon atoms since RUZICKA et al. (65) found isobutyric acid among the oxidation products. In an effort to distribute the five double bonds over the molecule correctly several formulae were recommended. The most probable seemed those which contain a fulvene ring system, accounting for the light extinction in the orange region of the visible spectrum. The investigations on a natural precursor of the azulenes, the sesquiterpene alcohol, guaiol, $C_{15}H_{26}O$, by SEMMLER and MAYER (74) showed that it contains a bicyclic structure with one double bond and a tertiary hydroxyl group. Upon oxidation with permanganate or ozone a crystalline compound was obtained which was thought to be a trihydroxy derivative. After dehydrogenation of this substance, a compound, $C_{15}H_{22}O$, appeared which contained two double bonds and, because of its passivity towards hydroxyl and carbonyl reagents, was held for an oxide.

Since further oxidation experiments, and those on guaiol itself, did not seem too promising, PFAU and PLATTNER (32) turned their attention to the conversion of guaiol to naphthalene derivatives. While dehydrogenation with sulfur or heating with formic acid had proven unsuccessful, a treatment with phosphorus and hydriodic acid resulted in the production of 1,4-dimethyl-6-isopropylnaphthalene (II). In a similar way vetivone gave 1,5-dimethyl-7-isopropylnaphthalene (V). The oxidation of both guaiol and vetivone, followed by ring closure and catalytic dehydrogenation, yielded phenols.

Since during the opening and subsequent closing of the ring one ring carbon atom was lost, it was concluded that the starting materials contained a seven-membered ring system. The phenol obtained from vetivone was shown by synthesis (*32*) to be identical with 2-isopropyl-4,7-dimethyl-indanol (III). The isomerisation of the azulene to a naphthalene derivative could then be formulated as a retro-pinacolin rearrangement as indicated in the skeleton formulas (I) and (II) for guaiol and in (IV) and (V) for vetivone.

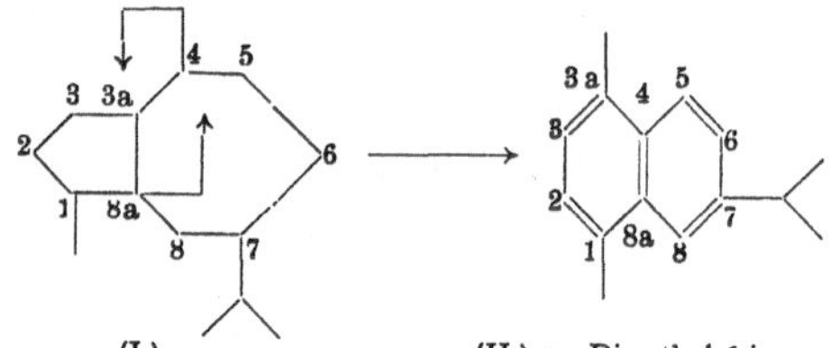

(I.)   (II.) 1,4-Dimethyl-6-isopropylnaphthalene.

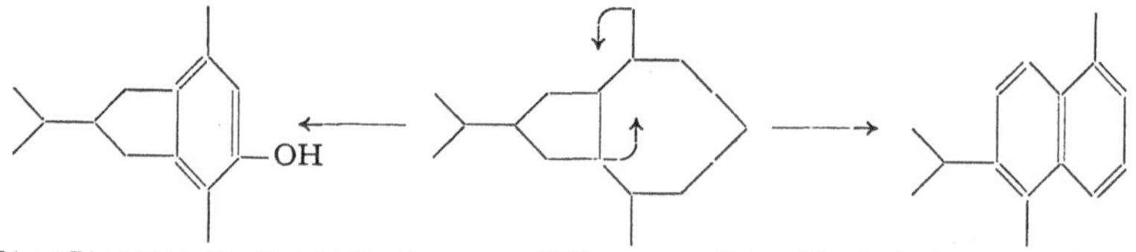

(III.) 4,7-Dimethyl-2-isopropyl-6-indanol.   (IV.)   (V.) 1,5-Dimethyl-7-isopropylnaphthalene.

On the basis of the results obtained in the oxidation of guaiol, the double bonds formed in the dehydrogenation of this alcohol would probably be confined to the rings. Comparison of the spectral characteristics of the natural azulenes with those of synthetic unsubstituted azulene, $C_{10}H_8$, proved this point beyond doubt (*52*), and we can assign the resonating structures (VI) to S-guaiazulene and (VII) to vetivazulene.

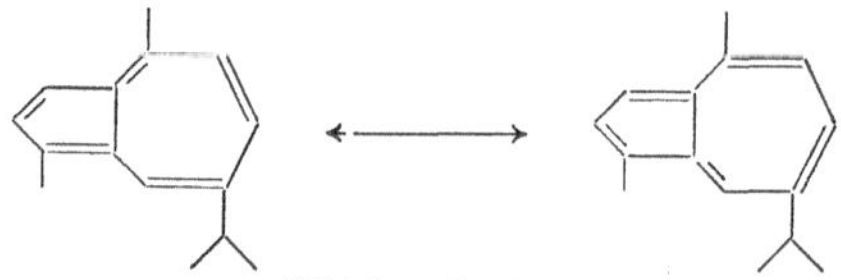

(VI.) S-guaiazulene.

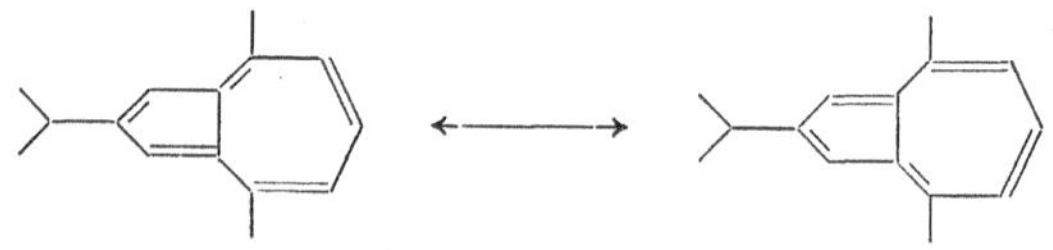

(VII.) Vetivazulene.

Nomenclature.

The parent hydrocarbon, $C_{10}H_8$, has been named azulene (*32*) and the numbering of the carbon atoms has been adopted as listed in (VIII) (*52, 32, 79*). The Ring Index (*30*) and the Chemical Abstracts prefer the name *cyclopentacycloheptene*.

For the hydrogenated derivatives it is convenient to use the name decahydro-azulene (or decahydro-cyclopentacycloheptene) with the same numbering of the C-atoms as in the azulenes. This has the obvious advantage that a conversion of hydrogenated derivatives into azulenes leaves the numbering of substituents unchanged.

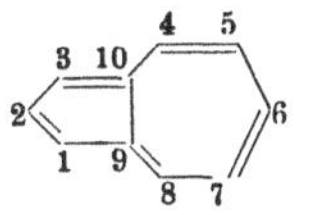

(VIII)
Azulene (52).
Cyclopentacycloheptene (30).

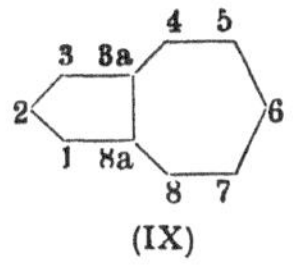

(IX)
Decahydro-azulene (52).
Cyclopentano-cycloheptane (24).
Decahydro-cyclopentacycloheptene (30).

Hückel et al. (24) used the name cyclopentano-cycloheptane, in order to indicate complete saturation. Octahydro-azulene with a double bond between the carbon atoms 3a and 8a was named cyclopenteno-cycloheptane. When the double bond was located between the 7 and 8 positions, the name cyclopentano-cycloheptene was chosen.

If we would adopt Baeyer's nomenclature, the name bicyclo-[0,3,5-] decane should be used with the numbering as in (X). The bracketted figures indicate the numbers of members in each one of the three bridges situated between the bridge heads formed by the carbon atoms 1 and 7. The Ring Index prefers to arrange these numbers in descending order, viz. bicyclo-[5,3,0-]decane.

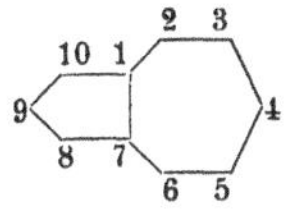

(X.) Bicyclo-[0,3,5-]decane.

In the present review the shortest names, viz. azulene and decahydro-azulene, and the numbering as given in symbol (IX) will be used. The structures are represented by the bonds between the carbon atoms and only in exceptional cases are hydrogen and carbon atoms indicated by C and H.

Structure of Naturally Occuring Precursors of Azulenes.

In 1926 Ruzicka and Rudolph (69) expressed the opinion that the knowledge of the azulene structure would lead to a better insight into the structure of the sesquiterpenes from which they originated. Their prediction was fully substantiated in the work of Pfau and Plattner (32). A reconsideration of the data on the sesquiterpene alcohol, guaiol, led these authors to a number of interesting conclusions regarding the course of the oxidation and dehydration of this sesquiterpene alcohol. The crystalline oxidation product of guaiol is the dihydroxyketone (XIII) formed by oxidation and opening of the C—C bridge, followed by ring closure. The position of this new C—C bridge is clear since one of the carbonyl groups reacts with the methylene hydrogens adjacent to the second carbonyl. The compound, $C_{15}H_{22}O$, which previously had been assumed to be an oxide (65) is in reality a doubly unsaturated ketone (XIV). The proof of its structure was given by the dehydrogenation to 1-hydroxy-cadalene as illustrated in the symbols (XI) to (XVI).

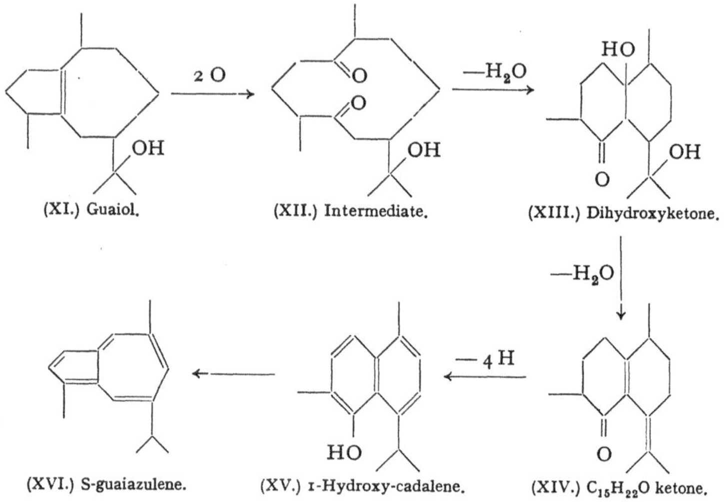

(XI.) Guaiol.   (XII.) Intermediate.   (XIII.) Dihydroxyketone.

(XVI.) S-guaiazulene.   (XV.) 1-Hydroxy-cadalene.   (XIV.) $C_{15}H_{22}O$ ketone.

These formulations were further confirmed by degradations of dihydro-guaiol. After removal of the hydroxy-isopropyl group by oxidation, the remaining 1,4-dimethyl-cyclopentanoheptanone (XIX) gave 1,4-dimethyl-azulene, $C_{12}H_{12}$, (XX) upon reduction followed by dehydrogenation.

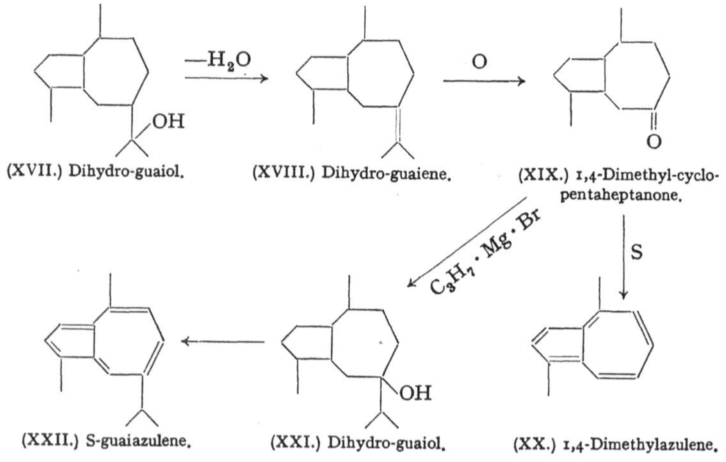

(XVII.) Dihydro-guaiol.   (XVIII.) Dihydro-guaiene.   (XIX.) 1,4-Dimethyl-cyclopentaheptanone.

(XXII.) S-guaiazulene.   (XXI.) Dihydro-guaiol.   (XX.) 1,4-Dimethylazulene.

The isopropyl group lost in the oxidation could be re-introduced into the cyclopentano-heptenone with isopropyl-magnesium-bromide. Upon dehydrogenation and dehydratation of the reaction product (XXI) an azulene (XXII) appeared which proved to be identical with that obtained from dihydroguaiene (XXIII) and from guaiol itself. The results of these investigations are in harmony with the guaiol formula (XI). For β-vetivone PFAU and PLATTNER (34, 35) were able to establish the structure (XXIII). Both structures have since been supported by synthesis of the corresponding azulenes (9).

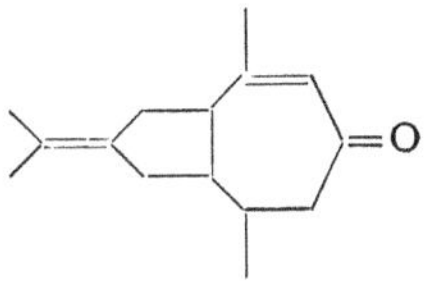

(XXIII.) β-Vetivone.

Synthesis of Azulenes.

The period of structure determination is now being followed by one of synthesis of numerous azulene derivatives. The main object at the present time is the study of the relationship between the structure and color in this field; and it is also to be expected that it will clarify the structures of Se-guaiazulene, chamazulene, elemazulene, and lactarazulene.

HÜCKEL and SCHNITZSPAHN (22, 23, 24) performed the oxidative fission of the carbon bridge in partly hydrogenated naphthalene (XXIV). This gave rise to a ten-membered ring diketone (XXV) which, upon treatment with alkali or acid, formed a bicyclic ketone (XXVI) containing a seven- and a five-membered ring system.

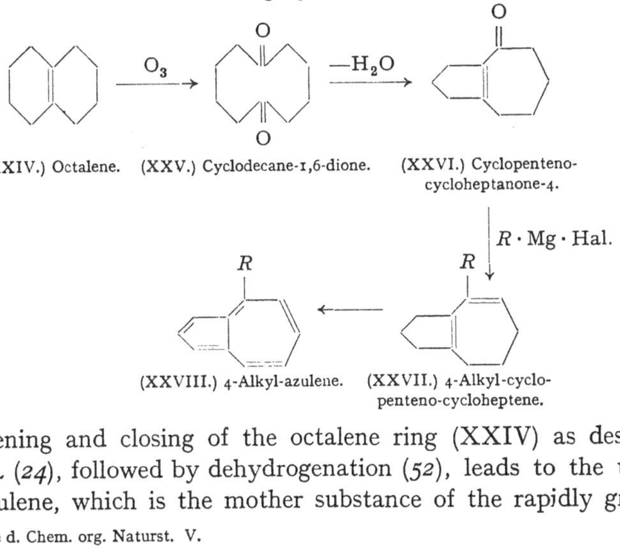

(XXIV.) Octalene.   (XXV.) Cyclodecane-1,6-dione.   (XXVI.) Cyclopenteno-cycloheptanone-4.

(XXVIII.) 4-Alkyl-azulene.   (XXVII.) 4-Alkyl-cyclopenteno-cycloheptene.

The opening and closing of the octalene ring (XXIV) as described by HÜCKEL (24), followed by dehydrogenation (52), leads to the unsubstituted azulene, which is the mother substance of the rapidly growing

series of blue and violet compounds. This parent hydrocarbon forms
azure blue crystals, m. p. 98,5–99°, having an odor like that of naphthalene
of which it is an isomer (*29*, *52*).

In the course of this type of synthesis, substitutions in the 4 position
can easily be accomplished with the help of Grignard reagents (XXVII).
On dehydrogenation with palladium or selenium, substituted azulenes
(XXVIII) are obtained which closely resemble the corresponding natural
products (*52*).

The most frequently used method in the past was the widening of
the ring with diazoacetic ester (*8*, *33*) which adds to a double bond in
conjugated or aromatic systems and forms a cyclopropane ring. Such
a three-membered ring is easily disrupted and can open in different
ways. When the double bonds to which the ester was added were situated
in a ring system, one of the three possible openings would lead
to a ring with one more member than that of the starting material.

If the ester group is hydrolysed, the carboxyl group can be removed
upon dehydrogenation, and then the number of the alkyl substituents
remains the same as in the starting material. An example of such a
synthesis, starting from a substituted indan (XXIX), was given in the
preparation of 4,8-dimethylazulene by PLATTNER and WYSS (*58*) (XXI
to XXXII).

The ester group which marks the position of the added ring carbon
atom, may serve for the introduction of alkyl groups. GRIGNARD's reagent,
e. g. CH_3MgI, converts this ester to a hydroxy-isopropylsubstituted
derivative from which isopropyl-azulene can be prepared. Reduction
of the ester group with sodium and alcohol followed by dehydrogenation
produces methyl substituted azulenes (*2*).

(XXIX.) 4,7-Dimethylindan.   (XXX.)   (XXXI.)

(XXXII.) 4,8-Dimethylazulene.

An uncertainty in the interpretation of conversions of this kind is
the place where the diazo-acetic ester adds to the benzene ring. In general,

the addition takes place by preference on unsubstituted double bonds (*8*); however, when substituents are already present, there is considerable doubt about the position of the introduced group (*2*). PLATTNER and RONIGER (*53*) illustrate these possibilities by the addition reactions on 5-methyl-indan. Assuming that the addition takes place on the double bonds of the two KEKULÉ structures, 5-methylazulene as well as 6-methylazulene would be formed as shown in (XXXIII) to (XLI).

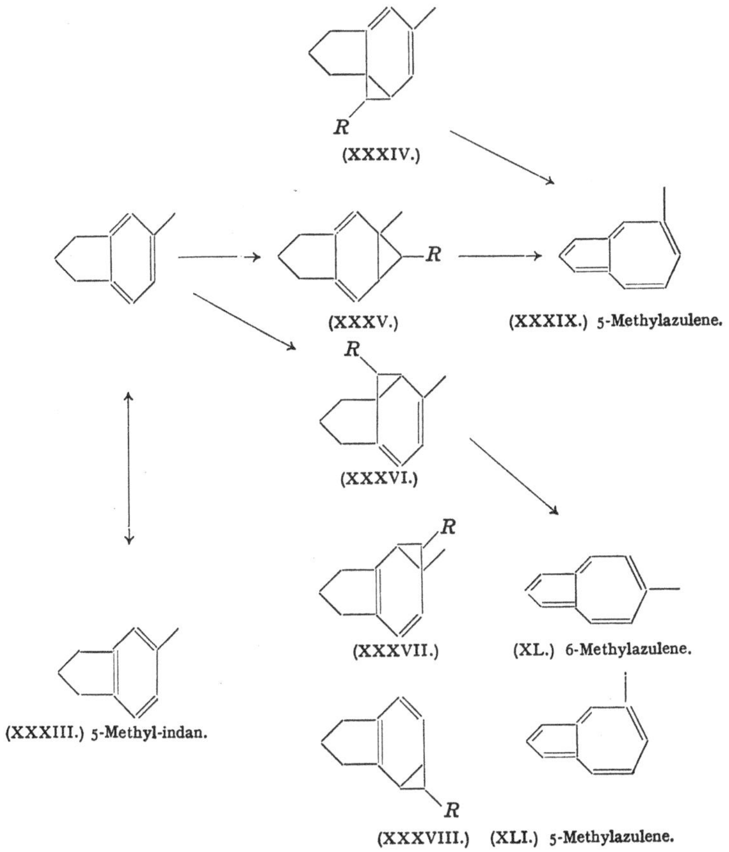

(XXXIV.)

(XXXV.)

(XXXIX.) 5-Methylazulene.

(XXXVI.)

(XXXVII.)

(XL.) 6-Methylazulene.

(XXXIII.) 5-Methyl-indan.

(XXXVIII.) (XLI.) 5-Methylazulene.

Similar doubts arose in the case of the substitution in 1,5,8-trimethyl-indan (*83*). The azulene actually isolated has been assigned the 1,6,8-trimethyl- rather than the 1,5,8-trimethyl structure on the basis of its spectrum.

An interesting attempt of HIPPCHEN (*20*) to prepare S-guaiazulene starting from *m*-cymene, via 1,4-dimethyl-6-isopropyl-indan (XLII), led also to a mixture of azulenes in which S-guaiazulene probably pre-

4*

dominates as shown by the similarity of the absorption spectra of the synthetic and natural products.

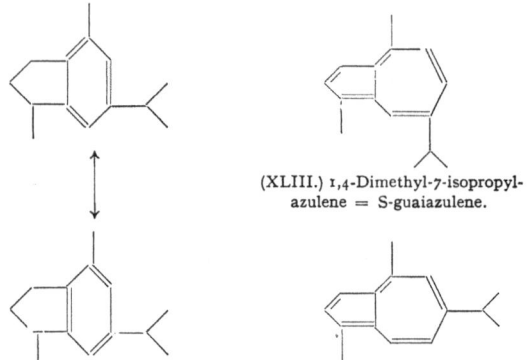

(XLIII.) 1,4-Dimethyl-7-isopropyl-
azulene = S-guaiazulene.

(XLII.) 1,4-Dimethyl-6-isopropylindan.   (XLIV.) 1,4-Dimethyl-6-isopropylazulene.

In the majority of cases it will be hard to avoid the formation of isomers since the quantity of each one will depend on the rates at which the different bonds react with diazoacetic acid. To overcome this difficulty ARNOLD (1) introduced in this field of synthesis the ring enlargement method of DEMJANOV and LUSCHNIKOW (11, 12, 13, 14).

In a study of the nitroso compounds of terpene amines the russian workers noticed in 1901 a peculiar effect of silver nitrite on cyclobutyl-methyl-amine (XLVII). Besides the expected reaction products (primary alcohol and unsaturated hydrocarbon) a secondary alcohol, viz. cyclopenta-nol (XLVIII) was formed. In a later study (13) this discovery was extended to the formation of cycloheptanol (LII) from cyclohexyl-methylamine (LI), as well as to that of cyclobutanol (XLVI) from cyclo-propyl-methylamine (XLV). WALLACH repeated these experiments and obtained from cycloheptyl-methylamine (LIII) a still higher membered ring system, viz. cyclo-octanol (LIV) and its ketone. He recognized the importance of this type of ring widening (85) as can be seen from the following quotation (1907):

„Wenn man bei niederen Temperaturen einen Kohlenstoffsechsring zu einem Siebenring und einen Siebenring zu einem Achtring erweitern kann, so wird man mit der Ansicht brechen müssen, daß die höheren Ringsysteme unter allen Umständen die unbeständigeren sind."

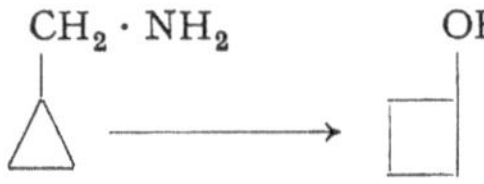

$CH_2 \cdot NH_2$ OH $CH_2 \cdot NH_2$

                                                                                OH

(XLV.) Cyclopropyl-   (XLVI.) Cyclobutanol. 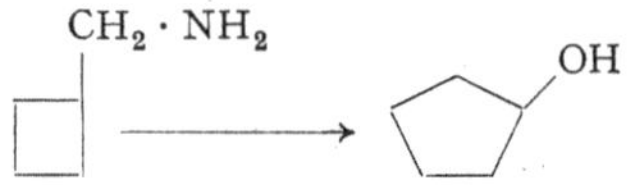   (XLVII.) Cyclobutyl-   (XLVIII.) Cyclopentanol.
methylamine.   methylamine.

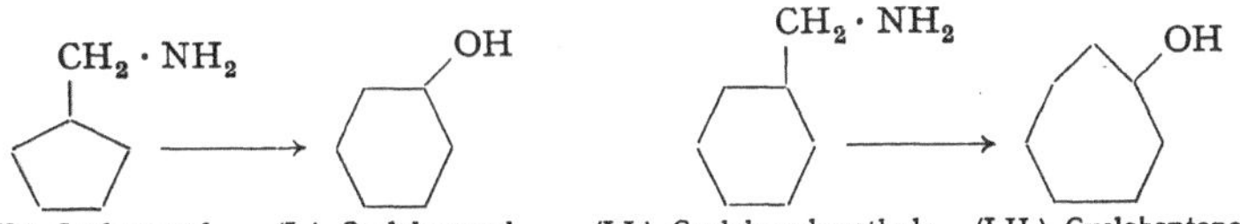

(XLIX.) Cyclopentyl-amine.   (L.) Cyclohexanol.   (LI.) Cyclohexyl-methyl-amine.   (LII.) Cycloheptanol.

(LIII.) Cycloheptyl-methylamine.   (LIV.) Cyclooctanol.

The hold of the BAEYER strain theory was, however, still so strong at that time, that no serious attempt was made to test such ideas experimentally. Twenty years elapsed before RUZICKA's pioneer experiments on macro ring systems (*61, 63, 64*) defined clearly the limits of the BAEYER theory.

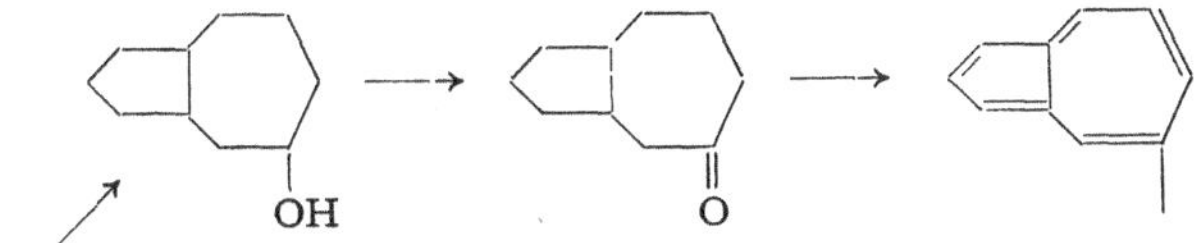

(LVI.) 7-Hydroxy-decahydro-azulene.   (LVII.) 7-Oxo-decahydro-azulene.   (LVIII.) 5-Methylazulene.

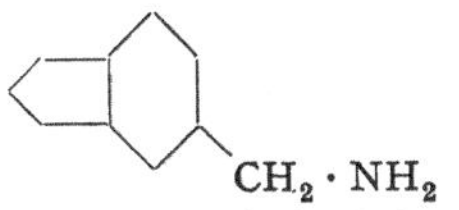

(LV.) 6-Methylamino-indan.

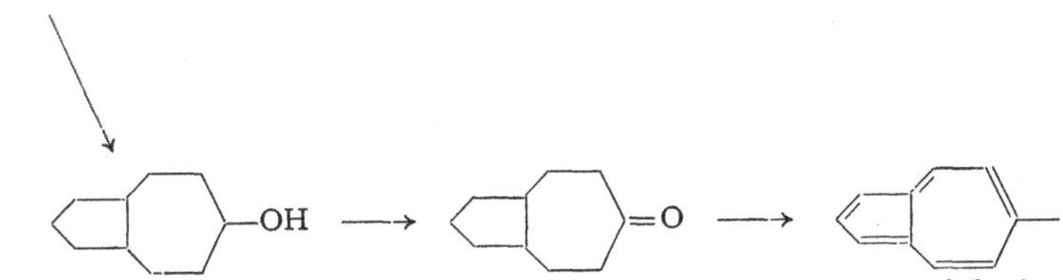

(LIX) 6-Hydroxy-decahydro-azulene.   (LX.) 7-Oxo-decahydro-azulene.   (LXI.) 6-Methylazulene.

The method of DEMJANOW has been applied with success by RUZICKA (*63*) and also by BARBIER (*5*). The latter author showed that substitution in the ring as in 1,1,3-trimethylcyclohexyl-methylamine does not prevent the application of this method. As an example the preparation of 5-methylazulene can be cited (*1, 45*). The starting material is indan which, after chloromethylation, is converted into the corresponding methylamine (LV). One ring is then enlarged with nitrous acid to give a mixture of the *cis* and *trans* forms of the isomeric alcohols (LVI) and (LIX). Through oxidation a mixture of isomeric ketones (LVII) and (LX) is obtained differing in position of the keto group and in the *cis-trans* position of the bridge hydrogen atoms. From these ketones the respective methyl derivatives can be prepared with Grignard reagent which gives (upon dehydrogenation) a product composed of 88% 5-methyl-azulene (LVIII) and 12% 6-methyl-azulene (LXI) (45, 48).

The doubt about the position of the methyl group was caused by the uncertain position of the entering hydroxyl group. If, however, the position of the methyl has been determined beforehand by its introduction with the starting material (indan), the nitrous acid method should lead to a single substituted azulene only. Arnold (1) carried out such a synthesis, starting with 5-methylindan from which 5-methyl-6-amino-methylhexahydro-indan (LXII) was prepared. The nitrous acid treatment resulted in the formation of a primary alcohol (LXV) and two secondary alcohols (LXIII–LXIV) which, upon removal of the hydroxyl group followed by dehydration, yielded 5-methylazulene (LXVI).

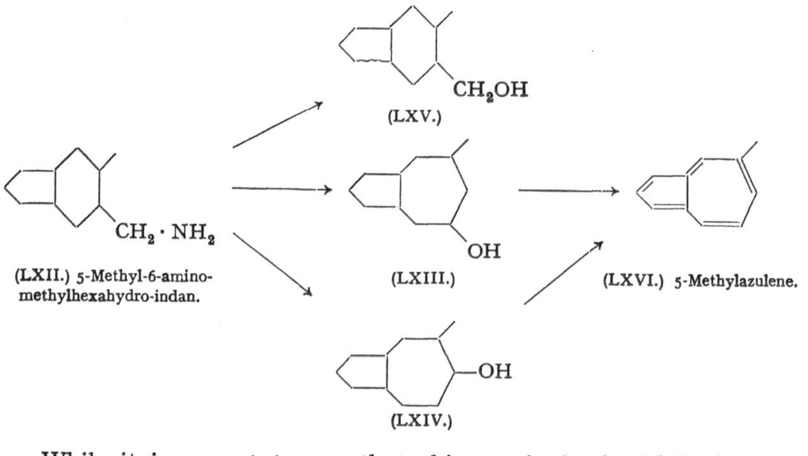

(LXV.)

(LXII.) 5-Methyl-6-amino-
methylhexahydro-indan.

(LXIII.)

(LXVI.) 5-Methylazulene.

(LXIV.)

While it is correct to say that this synthesis should lead to pure 5-methyl-azulene, any impurity present in the methyl-indan will appear in the resulting azulene. That this was actually the case in the synthesis of the 5- and 6-methyl-azulenes was shown in accurate purity tests by Plattner and his collaborators (48).

A different method of ring enlargement was followed by Plattner et al. (45) incorporating an extra carbon atom in the ring by means of diazomethane (68). This procedure had been originally proposed by Meerwein (27), and in a German patent the formation of cyclo-heptanone and cyclo-octanone from cyclo-hexanone is given as an example (15). The application of this method for the synthesis of 5-methylazulene started with the preparation of 5-indanol (LXVII). The corresponding ketone (LXIX) reacting with diazomethane gave rise to two structures,

(LXX) and (LXXI) which, after introduction of methyl groups, lead to a mixture of 5- and 6-methylazulenes.

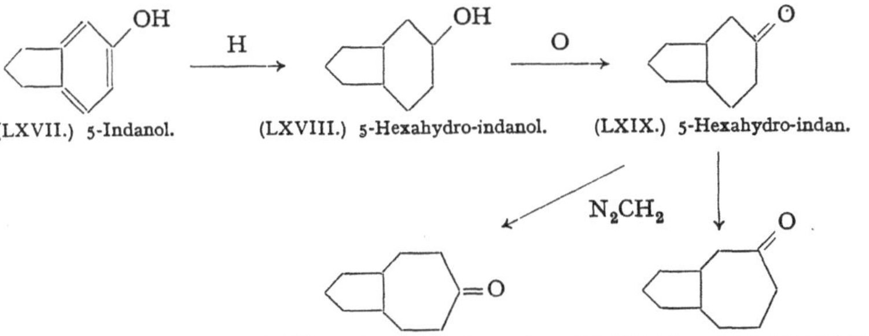

(LXVII.) 5-Indanol.　(LXVIII.) 5-Hexahydro-indanol.　(LXIX.) 5-Hexahydro-indan.

(LXX.) 5-Oxo-decahydro-azulene. (LXXI.) 6-Oxo-decahydro-azulene.

If the methyl group is introduced before the ring enlargement, a single end product might be expected. Coats and Cook (9) used this method in synthesizing vetivazulene.

The position of the hydroxyl group in the reaction product of diazomethane is not controlled, and mixtures of substituted azulenes may be formed. Sorm (78, 45), avoiding this uncertainty by ring closure of the proper octane-dicarboxylic acid (LXXII) and subsequent introduction of the methyl group with Grignard reagent, prepared pure 5-methylazulene via decahydro-azulene-5-ketone (LXXIII).

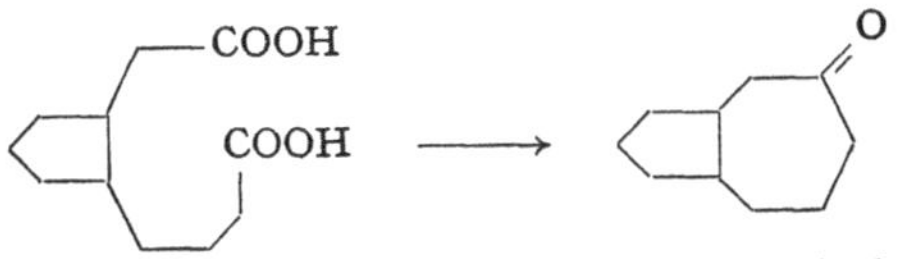

(LXXII.) Cyclopentano-octane-　(LXXIII.) 5-Oxo-decahydro-
dicarboxylic acid.　azulene.

Still another approach to the synthesis of the azulenes was followed recently by Plattner, Fürst and Jirasek (41) as well as by Plattner and Büchi (38). While in the previous syntheses the seven-membered ring was built onto the five-membered one, in these new syntheses a cyclo-heptanone is the starting material. Cyclo-heptanone-2-carboxylic acid (LXXIV) was condensed with bromoacetic acid to the next homologue, viz. cyclo-heptanone-2-acetic acid (LXXV). A Reformatski reaction led to the lactone (LXXVI) and to the unsaturated dicarboxylic acid (LXXVII). After ring closure, the ketone (LXXVIII) was converted in the usual way into a 2-alkylazulene (LXXIX). This synthesis permits the preparation of a number of substituted azulenes which would be otherwise difficult to obtain. An added advantage is the certainty of the position of the entering groups.

Plattner and Büchi (38) condensed cyclo-heptanone with succinic ester in the presence of potassium tert. butylate and, by dehydration with ZnCl$_2$, they obtained readily a cyclohepteno-cyclopentanone which could be converted into 1-methylazulene. A similar road was followed by Cook et al. (10a) in the preparation of unsubstituted azulene and hydroazulene ketones.

(LXXIV.) Cyclohepta-
none-2-carboxylic acid-
alkylester.

(LXXV.) Cycloheptanone-2-acetic
acid-alkylester.

(LXXVI.) Lactone.

hydrolysis

(LXXIX.) 2-Methyl-
azulene.

(LXXVIII.) 2-Oxo-decahydro-
azulene.

(LXXVII.) Unsaturated
dicarboxylic acid.

After these recent, successful syntheses it is interesting to look back to the formation of the blue hydrocarbons obtained by heating calcium adipate by HENTZSCHEL and WISLICENUS (17) in 1893, and by SABATIER and MAILHE (71) in 1914, during the distillation of adipic acid with thorium oxide. PLATTNER and PFAU (52) carried out similar experiments

(LXXX.) Adipic acid.

(LXXXI.) 6-Oxo-undecane-dicarboxylic
acid.

(LXXXII.) Cyclopenteno-
octane-dicarboxylic acid.

(LXXXIV.) Azulene.

(LXXXIII.) Cyclopenteno-
heptanone(4).

using calcium adipate with a small amount of nickel catalyst. The blue distillate obtained in this case showed an absorption spectrum identical with that of azulene; consequently, these authors were justified in proposing the reaction sequence (LXXX–LXXXIV).

Absorption Spectra.

Because of the importance of a comparative study of the spectra PLATTNER et al. (48) examined critically the purity of a number of methylazulenes. The purity of the samples was estimated by measuring the extinction coefficient and by determining the melting point diagrams in combinations of various azulenes. On this basis estimates can be made of the amount and nature of some impurities of azulenes prepared by different procedures. It is of practical significance and should also contain a warning, that the trinitrobenzene addition products of 4-methylazulene and 5-methylazulene do not give melting point depressions with each other.

With the help of spectroscopic measurements the interesting observation was made that the dehydrogenation of substituted hydroazulenes with palladium gives less pure products. Since this dehydrogenation is carried out at a higher temperature, it is suspected that under these conditions the substituents may change their position (26). Similar changes were observed by PLATTNER et al. (55) in the case of the preparation of phenylazulene whereby the phenyl group migrated from the 1-position; thus, 2-phenylazulenes were formed.

The spectrum of azulene and its derivatives in solution is characterized by several strong bands below 360 mμ. and, at a higher concentration, by a series of bands between 480 mμ. and 700 mμ. The spectrum of azulene vapor shows similar though less sharp bands than that of solutions. A structure analysis of the individual absorption bands is only possible with spectrographs of very large dispersion. Such investigations were, however, not attempted by the workers in this field. The primary stimulus for a better knowledge of the spectra came from the necessity of characterizing the azulenes. In a systematic survey of this kind carried out by PLATTNER et al. (37, 43, 47, 48), it was possible to detect certain regularities, due to the position of a substituent in the molecule.

A beginning has also been made with a quantitative photoelectric estimation of the azulene spectra which is far superior to the visual method. PLATTNER was able to classify the alkyl substituted azulenes in two main groups (37, 43). Within each group the relative positions and intensities of corresponding bands were very similar (Figure 1, p. 58).

The spectra of unsubstituted azulenes and of the azulenes which carry a methyl group in the 1-, 4-, 5- or 6-position show from eight to

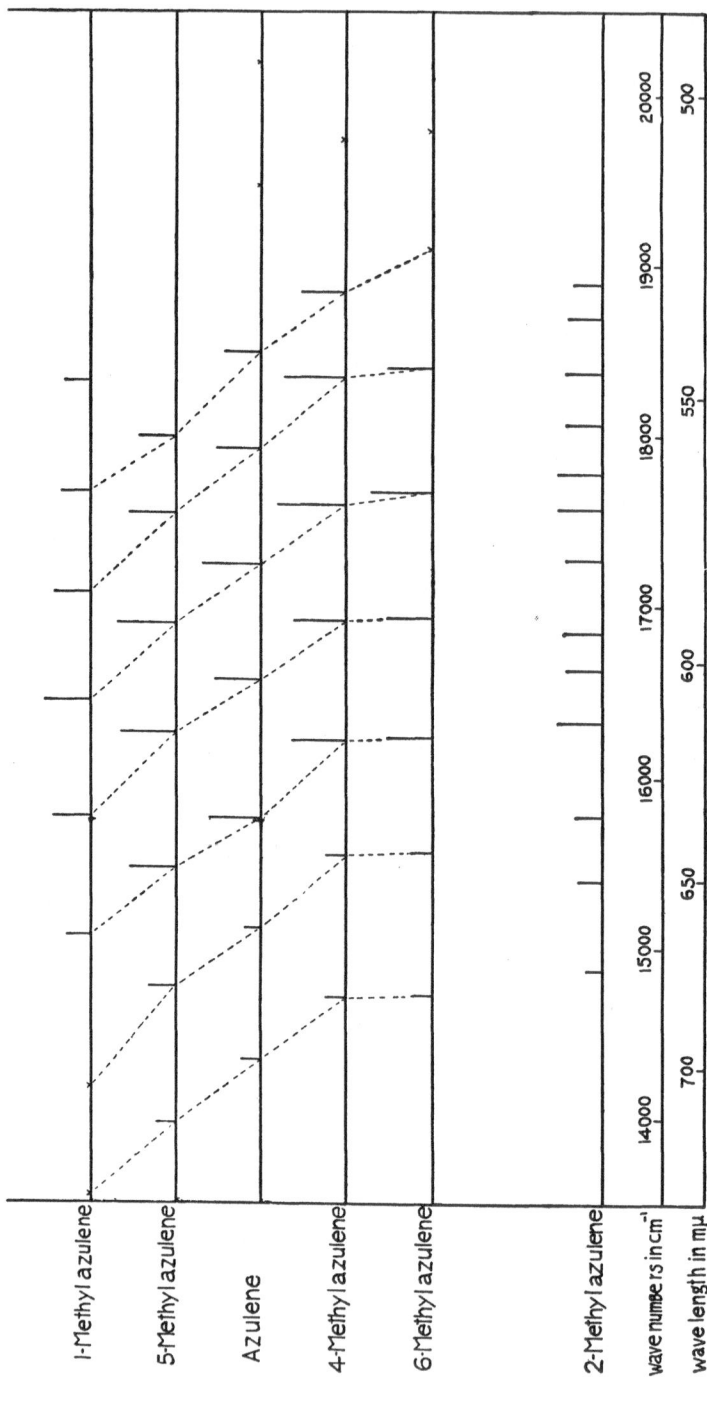

Fig. 1. Absorption spectra of methyl substituted azulenes. The strength of the bands is indicated with small vertical lines, the length of which is proportional to the photometrically determined extinction coefficient. The dotted lines connect corresponding spectral bands.

nine bands with a wave length difference from 16 to 36 mμ. in the region of 500–700 mμ. (solvent, hexane). The spectrum of 2-methylazulene, however, is more compressed and thus, fifteen bands appear within the region mentioned; they are about 11 mμ. apart. Lengthening of the side chain does not influence the type to which the spectrum belongs. For example, the spectra of 2-ethyl-, 2-propyl- and 2-isopropyl-azulene are practically identical with that of 2-methyl-azulene. Even 2-phenyl-azulene shows the large number of bands which characterizes the 2-alkyl-azulenes. All bands, however, have shifted by 20 mμ. to shorter wave lengths in this case. The shifts in the spectral bands which are caused by the substitution in the azulene ring system show clearly an alternating effect. Substitution in the 1-, 3-, 5- or 7-position causes a shift to longer wave lengths, whereas substitution in the 2-, 4- or 8-position has the opposite effect and, consequently, produces a more reddish color. This alternating effect is demonstrated in formula (LXXXIV) for the extinction band with the longest wave length, and is expressed in wave numbers (cm.$^{-1}$).

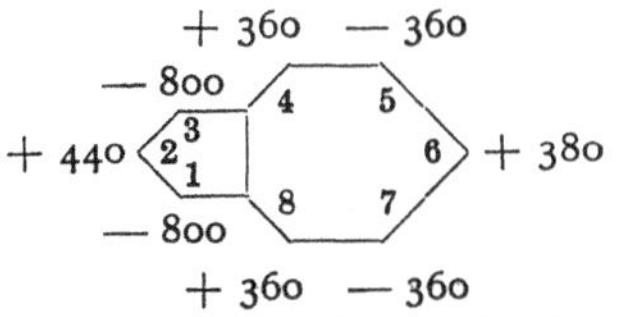

$$+ 360 \quad - 360$$
$$- 800$$
$$+ 440 \; \left\langle {}^{3}_{2}{}_{1} \quad 4 \quad 5 \quad 6 \right\rangle + 380$$
$$- 800$$
$$+ 360 \quad - 360$$

(LXXXIV.) Wave number shift in alkylazulenes (cm.$^{-1}$) for each position of the alkyl group (see Table 3).

Of interest is the numerically nearly equal shift of 360 to 440 cm.$^{-1}$ as a result of attaching one or more carbon atoms to the resonating system. The only exception to this rule is the substitution at the carbon atom 1 or 3 which causes a shift about twice that shown in other positions (54).

Table 2. Calculation of the wave number of the longest wave length spectral band of 1,3,4,8-tetramethylazulene.

	Wave number (cm.$^{-1}$)	Shift (difference from 14 350 cm.$^{-1}$)
Azulene.....................	14350	—
1- (or 3-) Methylazulene	13550	— 800
3- (or 1-) Methylazulene	13550	— 800
4- (or 8-) Methylazulene	14710	+ 360
8- (or 4-) Methylazulene	14710	+ 360
		Total shift: — 880 (calculated)
1,3,4,8-Tetramethylazulene	13410	Total shift: — 940 (found)

These spectral shifts are typical for each ring carbon atom and are to a large extent independent of substitutions at other carbon atoms and of the size of the entering group. Consequently, we are able to calculate with considerable accuracy the wave length position of the spectral bands of poly-substituted azulenes from the values obtained with monomethyl derivatives.

For example, in order to calculate the position of the longest wave length band of 1,3,4,8-tetramethylazulene, we add the effect of the four single shifts as caused by mono-substitution of azulene in the respective positions (Table 2).

If we add the shifts caused by 1- and 3-substitution to the + 660 shift caused by 4,8-dimethylazulene (Table 3), the agreement is complete, since a shift of — 940 is found as well as calculated. The calculated and observed values of a number of poly-substituted azulenes are listed in Table 3. In general the agreement is quite satisfactory, especially, since in a number of cases the shifts in absorption bands of the iso-propylazulenes had to be compared with those of methylazulenes.

Table 3. Observed and calculated shifts for the longest wave length spectral bands in the visible region when azulene is alkylated (solvent, hexane).

	Wave number (cm.$^{-1}$)	Shift (cm.$^{-1}$)		Color
		Found	Calculated	
Azulene..........................	14350	—	—	blue violet
1- (or 3-) Methylazulene	13550	— 800	—	blue
2-Methylazulene	14790	+ 440	—	violet
4- (or 8-) Methylazulene	14710	+ 360	—	violet
5- (or 7-) Methylazulene	13990	— 360	—	blue
6-Methylazulene	14730	+ 380	—	violet
1,2-Dimethylazulene...............	13950	— 400	— 360	blue
1,3-Dimethylazulene...............	12990	—1360	—1600	blue green
1,4-Dimethylazulene...............	13870	— 480	— 480	blue
4,8-Dimethylazulene...............	15010	+ 660	+ 720	violet
1,2,3-Trimethylazulene.............	13230	—1120	—1160	blue
1,4,7-Trimethylazulene.............	13530	— 820	— 800	blue
1,4-Dimethyl-7-isopropylazulene (S-Guaiazulene).................	13660	— 790		blue
Chamazulene	13660	— 750		blue
2,4,7-Trimethylazulene.............			+ 440	
2,4-Dimethyl-7-isopropylazulene (Se-Guaiazulene?)..............	14990	+ 640		violet
4,8-Dimethyl-2-isopropylazulene (Vetivazulene)	15460	+1110	+1140	red violet
4,6,8-Trimethylazulene.............			+1040	
4,8-Dimethyl-6-isopropylazulene	15360	+1010		red violet
1,3,4,8-Tetramethylazulene	13410	— 940	— 880	blue

PLATTNER et al. (*37, 43*) have based similar calculations on shifts relative to a strong band in the azulene spectrum at 17270 cm.$^{-1}$. Since the number of bands in the spectra of 2-substituted azulenes is often nearly double of that of azulenes which are substituted in other positions, it is difficult to determine which bands are corresponding in the different spectra. The assumption therefore, is made that some of the bands are suppressed or weakly developed in differently substituted azulenes. We may also maintain that whenever we find a larger number of bands, they were formed by splitting of equivalent bands in azulene spectra with less bands. Such "twin" bands are probably present in the spectrum of 2-methylazulene, since 1,2,3-trimethylazulene and 1,2-dimethylazulene show only a single band in the corresponding position. PLATTNER et al. (*43*) have, therefore, chosen the average position of two neighboring strong bands in 2-methylazulene for calculations in the class of 2-substituted azulenes. In comparing this average value with the position of the respective single bands in 1,2,3-trimethyl- and 1,2-dimethylazulene, good agreement of calculated and found values is obtained.

The choice of the strongest bands as the reference band is of advantage when the spectrum has only few and indistinct bands. However, it is difficult in such cases to point with certainty to comparable bands in the respective spectra. In Table 3 the bands of longest wave length were arbitrarily selected for comparison. The establishment of the exact equivalence of the absorption bands in the different types of spectra has to await a mathematical treatment.

In recent investigations PLATTNER and HEILBRONNER (*48a*) have extended their optical measurements of the alkyl-substituted azulenes to the ultraviolet region. In this part of the spectrum absorption curves are obtained resembling the curves of aromatic hydrocarbons, especially those of the naphthalene series. In the ultraviolet as well as in the visible region we find that the spectrum is determined by the position of the substituent rather than by the size of the latter. The introduction of an alkyl group causes a shift of the bands to longer wave lengths amounting to 250–700 cm.$^{-1}$ or 2–7 mμ. Substitution in the five-membered ring causes a greater spectral shift than similar substitution in the seven-membered ring. In the ultraviolet spectra of the alkylazulenes the individual bands are not shifted by equal amounts when compared to the corresponding bands of unsubstituted azulene. For example, in the case of 5-methylazulene these shifts range from zero for band 1 at 42017 cm.$^{-1}$ (238 mμ.) to 788 cm.$^{-1}$ (6 mμ.) for band 2 at 35842 cm.$^{-1}$ (279 mμ.).

These irregularities would not make it reasonable to calculate an "average" shift of the whole spectrum. Therefore, it is, also impossible to calculate the approximate position of bands of poly-substituted azulenes by adding up the respective shifts caused by mono-substitution as was done for the visible spectrum. This makes the ultraviolet spectra less useful for the identification of azulenes than the bands in the visible region.

The peculiar and surprising deep blue color of the azulenes naturally attracted the attention of physicists, and attempts have been made to explain and calculate the position of the spectral bands. SKLAR (77) following the HEITLER, LONDON, SLATER and PAULING method calculated the position of the longest wave length band of azulene and found good agreement with the experimentally determined value (found, 700 mμ.; calculated, 690 mμ.). HEILBRONNER and WIELAND (16) using SKLAR's approximation method calculated 43 kcal. per mol for the resonance energy of azulene. The same authors using the heat of combustion values as determined by PERROTET et al. (31) found 46 kcal. per mol. For comparison, the thermochemically determined resonance energy values of a few other compounds are shown in Table 4 (16).

Table 4. Some resonance energy values.

	Resonance energy (kcal. per mol.)
Benzene.............	41
Naphthalene.........	77
Azulene	46
Anthracene..........	116
Cyclo-heptatriene.....	7
Cyclo-pentadiene	3
Indene..............	39
Indan...............	37

Some Biochemical Aspects.

It is likely that the azulenes arise in the processing of the essential oils through oxidation of natural oil components. These precursors belong to the group of sesquiterpenes, $C_{15}H_{24}$, and their oxygenated derivatives. The natural azulenes are built along the same pattern as the sesquiterpenes from which they are derived. It is well known that the formulas of the terpenes can be represented as built of isopentane chains. In most cases we can detect a regular arrangement of these chains as in the sesquiterpene alcohol, farnesol (LXXXV).

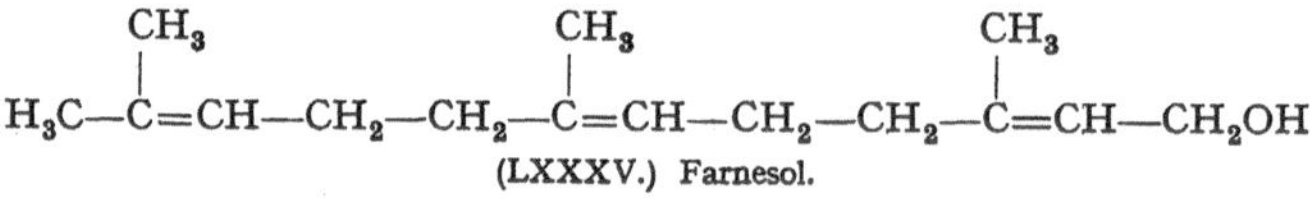

$$CH_3 \qquad\qquad CH_3 \qquad\qquad CH_3$$
$$H_3C-C=CH-CH_2-CH_2-C=CH-CH_2-CH_2-C=CH-CH_2OH$$

(LXXXV.) Farnesol.

(LXXXVI.)

This structure has often been described as a "head to tail" arrangement of the C_5 building units. It is found in a group very closely related to the azulenes, viz. the bicyclic sesquiterpenes of the cadalene, eudalene, vetivazulene and S-guaiazulene type. We can visualize these sesquiterpenes as formed by rolling up of a regularly built C_{15}-chain (LXXXVI), followed by secondary ring closures. In (LXXVII) to (XC) the solid lines represent the C_{15}-chain, while the dotted lines indicate the secondarily formed linkages between carbon atoms of this chain.

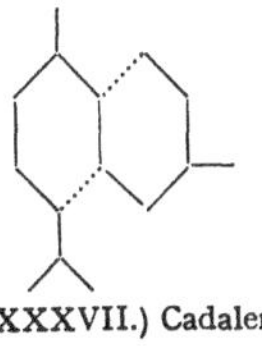

(LXXXVII.) Cadalene   (LXXXVIII.) Eudalene   (LXXXIX.) Vetivazulene   (XC.) S-guaiazulene
type.   type.   type.   type.

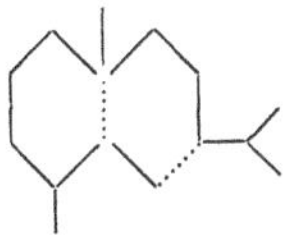

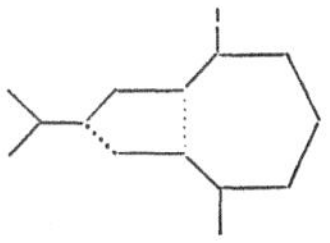

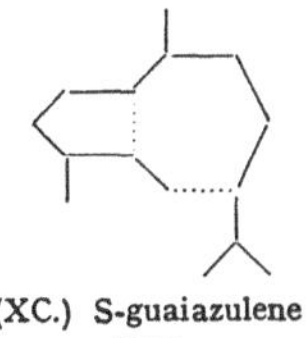

These different types yield, upon dehydrogenation, the naphthalene hydrocarbons: cadalene and eudalene and, furthermore, the azulenes: vetivazulene and S-guaiazulene, all of which can be characterized readily through their picrates and other crystalline nitro addition products. The reaction has been of great help in the structure determination of the sesquiterpenes. The lack of well defined dehydrogenation products from the sesquiterpenes, cedrenes, caryophyllene and longifolene is largely responsible for the unsatisfactory state of our knowledge regarding their structure. The positions of the substituents in some azulenes such as elemazulene, chamazulene and lactarazulene still have to be clarified before these azulenes too can serve for the identification of groups of parent sesquiterpenes.

In an attempt to predict the structure of azulenes which might be found in the natural oils, PFAU and PLATTNER (32) list the following variations in folding up a regular "head to tail" isoprene chain (XCI–IC). Sesquiterpenes built according to this scheme may give azulenes upon dehydrogenation.

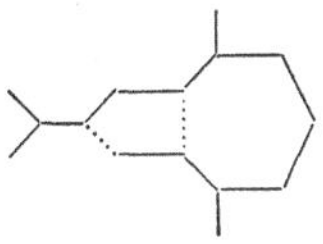

(XCI.) Vetivazulene type. 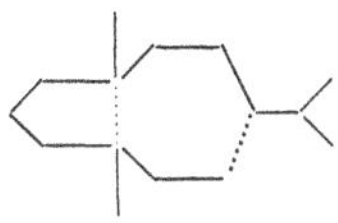   (XCII.) 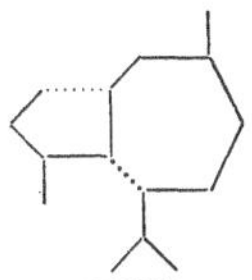   (XCIII.)

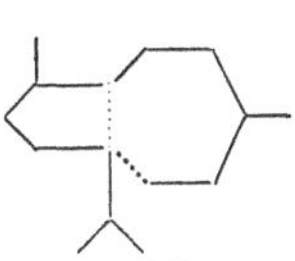

(XCIV.) 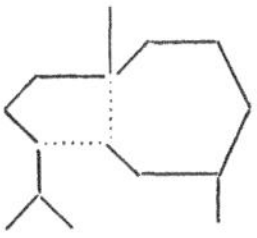   (XCV.) 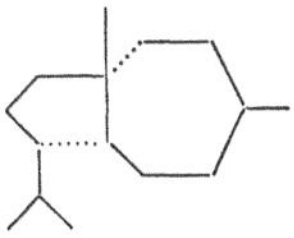   (XCVI.)

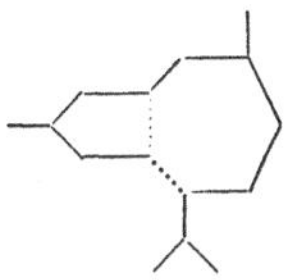

(XCVII.) 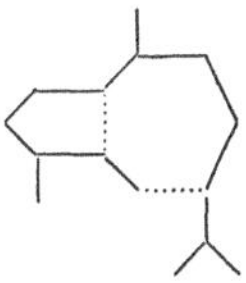   (XCVIII.) S-guaiazulene type. 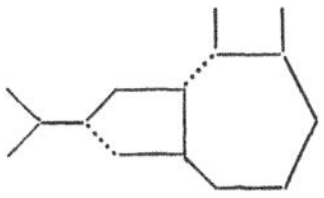   (IC.)

In some of the cases listed alkyl groups have to be split off before azulene formation can take place. We are well acquainted with the removal of methyl groups under similar circumstances in the aromatization of eudesmol to eudalene (26). If we admit this possibility, the occurrence of $C_{14}H_{16}$ and even lower homologues might be expected among the dehydrogenation products of some of the sesquiterpenes.

Of the different possibilities listed, only (XCIII), (XCVIII), (XCI), (XCVII) and (IC) will give azulenes containing fifteen carbon atoms. From a calculation of the absorption bands as shown earlier the first two of these structures should cause a blue color and the others a violetblue one. The violet-blue *elemazulene* the picrate and styphnate of which give melting point depressions with those of the other natural azulenes, could have the structure (XCVII) or (IC), since (XCI) must be reserved for vetivazulene. Elemazulene is formed on dehydrogenation of elemene with selenium at high temperature; and here as in the case of Se-guaiazulene we have to count with the possibility of a migration of the alkyl groups. For the assignment of a structure to the blue chamazulene and lactarazulene only 1,5-dimethyl-8-isopropylazulene (XCIII) is left, since (XCVIII) represents S-guaiazulene. It is therefore necessary to find at least one other arrangement of the isoprene chain units than those already listed.

The tabulation of the different possibilities postulates a regular "head to tail" union of the isoprenes. Although this type of linkage seems predominant in the lower terpenes, it is by no means the exclusive pattern on which the terpene molecules are built. Several examples of deviation from this common pattern are known, such as *artemisia ketone* and *lavendulol* in the monoterpenes and *abietic acid* and triterpene saponines in the higher terpenes. Also in the group of sesquiterpenes we can draw a number of isoprenic structures, which are not necessarily built in "head to tail" fashion. In formulas (C) to (CIII) a few of such structures are presented. These examples have been chosen in such a way that they may represent one of the blue azulenes: chamazulene or lactarazulene. The similarity of the absorption spectra of chamazulene, lactarazulene and S-guaiazulene requires that the alkyl groups be in 1- (or 3-), 4- (or 8-) and 5- (or 7-)position.

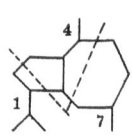 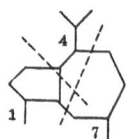 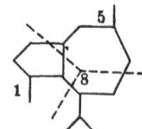 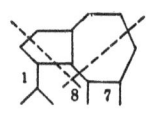

(C.) 4,7-Dimethyl-1-isopropylazulene.   (CI.) 1,7-Dimethyl-4-isopropylazulene.   (CII.) 1,5-Dimethyl-8-isopropylazulene.   (CIII.) 7,8-Dimethyl-1-isopropylazulene.

According to a suggestion of PFAU and PLATTNER (*34*), based on spectrum regularities (Table 3), Se-guaiazulene could have the structure of 2,4-dimethyl-7-isopropylazulene (CV), which would be an example of an irregularly built azulene. However, in this case the deviation from the normal pattern might well have been caused by methyl migration, since the parent substance gives, when dehydrogenated with sulfur at a lower temperature, the "regularly" built S-guaiazulene (CIV).

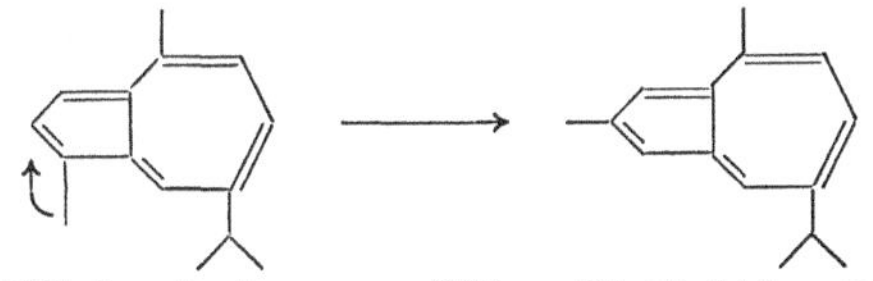

(CIV.) S-guaiazulene. (CV.) 2,4-Dimethyl-7-isopropylazulene.

The formulas (XCI) to (IC) are based on the assumption that the regular isoprene chain is folded in such a manner that the resulting cyclic structure never contains two alkyl groups on the same carbon atom of the ring. Two methyl groups on one cyclic carbon atom occur frequently in some isoprene type compounds such as ionone, resin acids and carotinoids. Formula (CVI) illustrates one of these many possibilities. Dehydrogenation would remove one of the methyl groups with the formation of a lower homologue, viz. a C_{14} azulene (CVII).

Degradation and rearrangements of this type might be involved when crystalline pyrethrosin, $C_{17}H_{22}O_5$, isolated from *Pyrethrum* flowers, is subjected to zinc-dust distillation. The pyrethazulene, $C_{13}H_{14}$, thus formed has been assigned the structure of a 2,4,8-trimethylazulene in view of the resemblance of its spectrum with that of vetivazulene.

(CVI.) (CVII.) Pyrethazulene.

Such considerations lead us to expect a much greater variety of sesquiterpene and azulene structures than had been postulated earlier. At the rate at which the azulenes are synthesized at present, a large number of them will soon be available for comparison with azulenes which originate from natural sources.

Further possible types of azulenes.

The natural azulenes built of fused five and seven membered rings suggest the synthesis of analogues of equally intense coloration. The five and seven membered rings, for example, may be expanded to *larger cyclic systems* containing an uneven number of carbon atoms. Other possibilities are the inclusion of hetero atoms in the ring. Attempts in this general direction have been made by HORN et al. (*21*) who tried to prepare heptalene (CIX), and by BARRETT et al. (*5a, 4a; cf.* also *10b*) who were unsuccessful in synthesizing pentalene (CIX).

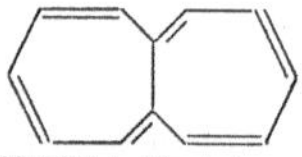

(CVIII.) Heptalene. 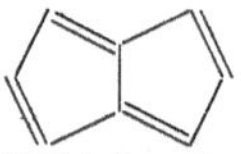　　　　　(CIX.) Pentalene.

The syntheses of azulenes with benzene rings fused onto an azulene structure in 5,6 position were not successful (*10*). However, PLATTNER et al. (*40*), TREIBS (*80*) and HORN et al. (*21*), using diazoacetic acid, enlarged one of the six-membered rings in fluorene (CX) to 1,2-benzazulene (CXI).

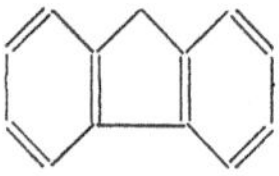

(CX.) Fluorene. 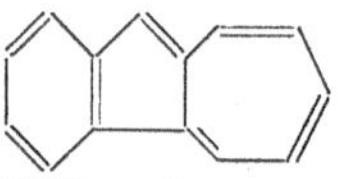　　　　　(CXI.) 1,2-Benzazulene.

The spectrum of 1,2-benzazulene shows stronger and broader bands than those present in the azulene spectrum. These bands are shifted considerably more to the longer wave length region than would be expected from 1,2-alkyl- or phenyl-substituted azulene (longest wave length band at 790 mμ. or 12660 cm.$^{-1}$). This behavior is undoubtedly connected with the presence of three resonating structures in 1,2-benzazulene (CXII–CXIV) against two in alkylazulenes (VI). Recently, the synthesis and the ultraviolet spectrum of an indeno- [2′,1′ : 1,2-]azulene (CXV) containing four rings has been published by PLATTNER et al. (*48a, 59a*).

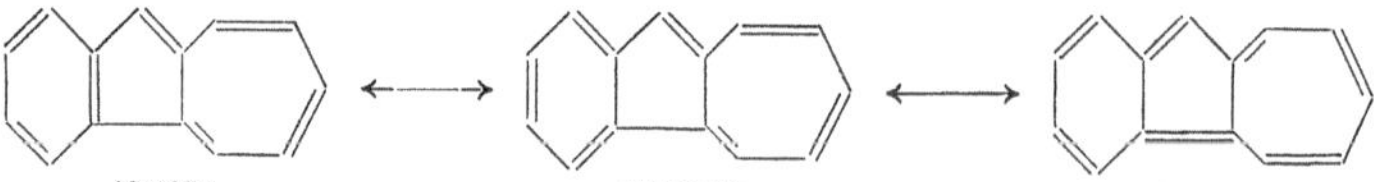

(CXII.)　　　　　(CXIII.)　　　　　(CXIV.)

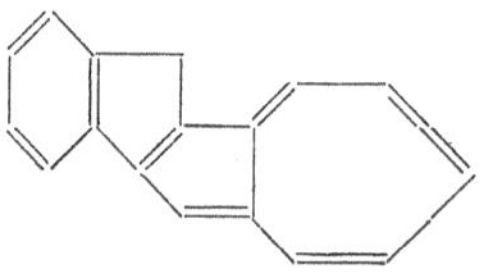

(CXV.) Indeno- [2′,1′ : 1,2-]azulene.

Pharmacological Properties of Azulenes.

The azulenes offer still another interesting angle. A beginning has been made with a study of the physiological effects of the azulenes and their derivatives. Reports have been published on the anti-inflammatory action of pure chamazulene by HEUBNER and ALBATH (*19*) and by POMMER (*59*). WAGNER-JAUREGG et al. (*82*) report negative findings in the chemotherapeutic treatment of lepra in rats with ethyl-2-isopropyl-4,8-dimethylazulene-6-carboxylate. The presence of a conjugated system of double bonds in the azulenes invited tests on vitamin A activity. However, no effects on xerophthalmia in mice and on kolpokeratosis in castrated female vitamin A-depleted cats were observed (65). With normal diets no influence of added azulene was noticed.

References.

1. ARNOLD, H.: Über die Ringerweiterung bei der Umsetzung von Aminomethyl-hexahydroindanen mit salpetriger Säure und ihre Bedeutung für die Azulen-synthese. Ber. dtsch. chem. Ges. **76**, 777 (1943).

2. — Untersuchungen zur Darstellung 6-alkylierter Azulene. Dehydrierung des Cyclopentano-cycloheptadienyl-methanols. Ber. dtsch. chem. Ges. **80**, 123 (1947).

3. — Untersuchungen zur Darstellung 6-alkylierter Azulene. Die Darstellung des 6-Methyl-1-isopropyl-azulens und des 5- (bzw. 7-) Methyl-1-isopropyl-azulens. Ber. dtsch. chem. Ges. **80**, 172 (1947).

4. AUGSBURGER, L.: The Isolation of a blue Hydrocarbon from Milfoil Oil. Science [N. S.] **42**, 100 (1915).

4a. BAKER, W.: Nonbenzenzoïd Aromatic Hydrocarbons. Tilden Lecture. J. chem. Soc. [London] **1945**, 258.

5. BARBIER, H.: Extension des cycles dans la série hydroaromatique. Essais avec la 1,1,3-triméthyl-cyclohexyl-méthylamine-5-(dihydro-isophoryl-méthyl-amine). Helv. chim. Acta **23**, 518 (1940).

5a. BARRET, J. W. and R. P. LINSTEAD: Fused Carbon Rings. Some Fundamental Properties of the o : 3 : 3-bicyclo-Octane Ring. J. chem. Soc. [London] **1936**, 611.

6. BIRREL, K. S.: Crystalline Guaiazulene. J. Amer. chem. Soc. **56**, 1248 (1934).

7. — Studies on the Chemistry of Azulenes. J. Amer. chem. Soc. **57**, 893 (1935).

8. BRAREN, W. u. E. BUCHNER: Über Pseudophenylessig- oder Norcaradiëncarbon-säure. Ber. dtsch. chem. Ges. **34**, 982 (1901).

9. COATS, R. R. and J. W. COOK: Synthesis of Vetivazulene. J. chem. Soc. [London] **1943**, 559.

10. COOK, J. W., N. A. McGINNIS and S. MITCHELL: Hydro-anthracenes and Hydro-phenantrenes. J. chem. Soc. [London] **1944**, 286.

10a. COOK, J. W., R. PHILIP and A. R. SOMERVILLE: Azulenes. Exploration of New Synthetic Routes. J. chem. Soc. [London] **1948**, 164.

10b. CRAIG, D. P. and A. MACCOLL: The Non-Benzenoid Aromatic Hydrocarbon, Pentalene. Nature **161**, 481 (1948).

11. DEMJANOW, N.: Die Ringerweiterung bei den cyclischen Aminen mit der Seitenkette CH_2NH_2. Über den Alkohol aus dem Amin $C_3H_5 \cdot CH_2 \cdot CH_2NH_2$. Ber. dtsch. chem. Ges. **40**, 4393 (1907).

12. DEMJANOW, N. u. M. LUSCHNIKOW: Über die Einwirkungsprodukte der sal-petrigen Säure auf Tetramethylenmethylamin. J. russ. physik.-chem. Ges. **35**, 26 (1903); Chem. Zbl. **1903** I, 828.

13. — — Über das Nitril der Hexamethylencarbonsäure, das Amin $C_6H_{11}CH_2NH_2$ und die Umwandlung des letzteren in Suberyl-alkohol. J. russ. physik.-chem. Ges. **36**, 166 (1904); Chem. Zbl. **1904** I, 1214.

14. — — Über die Einwirkung von salpetriger Säure auf Tetramethylenmethyl-amin und über ein Dibromid des Tetramethylenmethans. J. russ. physik.-chem. Ges. **33**, 279 (1901); Chem. Zbl. **1901** II, 335.

15. EISTERT, B.: in Neuere Methoden der präparativen organischen Chemie, S. 398. Berlin. 1943; siehe (*38*), und Angew. Chem. **54**, 99, 124 (1941); Chem. Abstr. **35**, 4731; Die Chemie **55**, 118 (1942) (Nachtrag); Chem. Abstr. **38**, 2003 (1944).

16. Heilbronner, E. u. K. Wieland: Spektrographische und thermochemische Untersuchungen an dampfförmigem Azulen. Helv. chim. Acta **30**, 947 (1947).

17. Hentschel, W. u. J. Wislicenus: Adipinketon. Liebigs Ann. Chem. **275**, 312 (1983).

18. Herzenberg, J. u. S. Ruhemann: Über das blaue Öl des Braunkohlenteers, Ber. dtsch. chm. Gees. **58**, 2249 (1925).

19. Huebner, W. and W. Albath: The Anti-inflammatory Action of Pure Azulene from *Matricaria chamomilla* L. Naunyn-Schmiedebergs Arch. exp. Pathol. Pharmakol. **192**, 383 (1939); Chem. Abstr. **34**, 2932 (1940).

20. Hippchen, H.: Darstellung eines Azulens vom Guajazulen-Typus aus Cymol. Z. Naturforsch. **1**, 325 (1946).

21. Horn, D. H. S., J. R. Nunn and W. S. Rapson: Synthesis of Cyclic Conjugated Polyolefins. Nature [London] **160**, 829 (1947).

22. Hückel, W., R. Danneel, A. Schwartz u. A. Gercke: Zur Stereochemie bicyclischer Ringsysteme. Octalin. Liebigs Ann. Chem. **474**, 121 (1929).

23. Hückel, W., A. Gercke u. A. Gross: Cyclodecan. Ber. dtsch. chem. Ges. **66**, 563 (1933).

24. Hückel, W. u. L. Schnitzspahn: Zur Stereochemie bicyclischer Ringsysteme, Derivate des Cyclopentano-cycloheptans. Liebigs Ann. Chem. **505**, 274 (1933).

25. Kremers, R. E.: Experiments on Azulene. J. Amer. chem. Soc. **45**, 717 (1923).

26. Linstead, R. P., A. F. Millidge, S. L. S. Thomas and A. L. Walpole: Dehydrogenation. The Catalytic Dehydrogenation of Hydronaphthalenes with and without an Angular Methyl group. J. chem. Soc. [London] **1937**, 1146.

27. Meerwein, H.: Verfahren zur Umsetzung organischer Verbindungen mit Diazomethan. D. R. P. 579309; Friedländer, **20**, 386; Chem. Zbl. **1933** II, 1758.

28. Melville, J.: Azulene. J. Amer. chem. Soc. **55**, 3288 (1933).

29. Misch, L. and A. van der Wyk: Structure of Crystallised Azulene. C. R. Séances Soc. Physiques Hist. natur. Genève **54**, 106 (1937); Chem. Abstr. **32**, 6523 (1938).

29a. Müller, A.: Beitrag zum Nachweis terpen-chromogener, insbesondere azulenogener Verbindungen und eine neue Farbreaktion der ätherischen Öle. J. prakt. Chem. **151**, 233 (1938).

29b. — Über terpen-chromogene bzw. terpenochrome Verbindungen. Eigenschaften der E. M. Farbsalze aus Absinth und Kamillenöl. J. prakt. Chem. **156**, 179 (1940).

29c. — Über terpen-chromogene bzw. terpenochrome Verbindungen. Spektroskopische Untersuchungen über die bei der E. M. Reaktion mit ätherischen Ölen auftretenden Farbstoffe. J. prakt. Chem. **153**, 77 (1939).

30. Patterson, A. M. and L. T. Capell: The Ring System. Amer. chem. Soc. Monograph; Reinhold Publishing Corp. New York. 1940.

31. Perrotet, E., W. Taub et E. Briner: Sur les états énergétiques comparatifs des noyaux azulénique et naphthalénique. Helv. chim. Acta **23**, 1260 (1940).

32. Pfau, A. St. u. Pl. A. Plattner: Zur Kenntnis der flüchtigen Pflanzenstoffe. Über die Konstitution der Azulene. Helv. chim. Acta **19**, 858 (1936).

33. — — Zur Kenntnis der flüchtigen Pflanzenstoffe. Synthese des Vetivazulens. Helv. chim. Acta **22**, 202 (1938).

34. — — Études sur les matières végétales volatiles. Sur les vétivones, constituants odorants des essences de vétiver. Helv. chim. Acta **22**, 640 (1939).

35. — — Études sur les matières végétales volatiles. Sur la constitution de la β-vétivone. Helv. chim. Acta **23**, 768 (1940).

36. Piesse, D.: Art of Perfumery, 1879, p. 57; C. R. hebd. Séances Acad. Sci. **57**, 1016 (1864), cf. (75).

37. PLATTNER, PL. A.: Zur Kenntnis der Sesquiterpene. Konstitution und Farbe der Azulene. Helv. chim. Acta **24**, 283 E (1941).

38. PLATTNER, PL. A. u. G. BÜCHI: Zur Kenntnis der Sesquiterpene. Über eine einfache, von Cycloheptanon ausgehende Azulen-Synthese. Helv. chim. Acta **29**, 1608 (1946).

39. PLATTNER, PL. A. u. A. FÜRST: Zur Kenntnis der Sesquiterpene. Über den Einfluß der Substitution auf die Farbe der Azulene: 2-Äthyl-azulen. Helv. chim. Acta **28**, 1636 (1946).

40. PLATTNER, PL. A., A. FÜRST, J. CHOPIN u. G. WINTELER: Zur Kenntnis der Sesquiterpene. 1,2-Benz-azulen. Helv. chim. Acta **31**, 501 (1948).

41. PLATTNER, PL. A., A. FÜRST u. K. JIRASEK: Zur Kenntnis der Sesquiterpene. Über das Bicyclo-(o,3,5-)decan. Helv. chim. Acta **29**, 730 (1946).

42. — — — Zur Kenntnis der Sesquiterpene. Über eine neue Azulen-Synthese. Helv. chim. Acta **29**, 740 (1946).

43. — — — Zur Kenntnis der Sesquiterpene. Im Fünfring mehrfach substituierte Azulene. Helv. chim. Acta **30**, 1320 (1947).

44. PLATTNER, PL. A., A. FÜRST u. H. SCHMID: Zur Kenntnis der Sesquiterpene. Über den Einfluß der Substitution auf die Farbe der Azulene: 1,3,4,8-Tetramethyl-azulen. Helv. chim. Acta **28**, 1647 (1945).

45. PLATTNER, PL. A., A. FÜRST u. A. STUDER: Zur Kenntnis der Sesquiterpene. Über neuere Synthesen des 5-Methyl-azulens. Helv. chim. Acta **30**, 1091 (1947).

46. PLATTNER, PL. A., A. FÜRST, J. WIJSS u. R. SANDRIN: Zur Kenntnis der Sesquiterpene. 2-Isopropyl-azulen. Helv. chim. Acta **30**, 689 (1947).

47. PLATTNER, PL. A. u. E. HEILBRONNER: Zur Kenntnis der Sesquiterpene. Die Adsorptionskurven des Azulens und der fünf Monomethyl-azulene im sichtbaren Bereich. Helv. chim. Acta **30**, 910 (1947).

48. PLATTNER, PL. A., E. HEILBRONNER u. A. FÜRST: Zur Kenntnis der Azulene. Die spektroskopische Prüfung verschiedener Präparate von 5-Methylazulen. Helv. chim. Acta **30**, 1100 (1947).

48a. PLATTNER, PL. A. u. E. HEILBRONNER: Zur Kenntnis der Sesquiterpene und Azulene. Die Ultraviolett-Adsorptionsspektren der fünf Monomethyl- und einiger mehrfach substituierter Azulene. Helv. chim. Acta **31**, 804 (1948).

49. PLATTNER, PL. A. u. L. LEMAY: Zur Kenntnis der Sesquiterpene. Über das Kohlenstoff-gerüst des Guajols und des Guajazulens. Helv. chim. Acta **23**, 897 (1904).

50. PLATTNER, PL. A. u. G. MAGYAR: Zur Kenntnis der Sesquiterpene. Synthese des 5-Oxy-1,6-dimethyl-4-isopropyl-naphthalins, ein Beitrag zur Konstitutionsaufklärung des Guajols. Helv. chim. Acta **24**, 1163 (1941).

51. — — Zur Kenntnis der Sesquiterpene. Abbau des Dihydroguajols mit Chromsäure. Bereitung des 1,4,7-Trimethyl-azulens. Helv. chim. Acta **25**, 581 (1942).

52. PLATTNER, PL. A. u. A. ST. PFAU: Zur Kenntnis der flüchtigen Pflanzenstoffe. Über die Darstellung des Grundkörpers der Azulenreihe. Helv. chim. Acta **20**, 224 (1937).

53. PLATTNER, PL. A. u. H. RONIGER: Zur Kenntnis der Sesquiterpene. Synthese des 5-Methyl-azulens. Helv. chim. Acta **25**, 590 (1942).

53a. — — Zur Kenntnis der Sesquiterpene. Herstellung der 4,8-Dimethyl-azulene-6-carbonsäure. Helv. chim. Acta **25**, 1077 (1942).

54. — — Zur Kenntnis der Sesquiterpene. 4,8-Dimethyl-6-isopropyl-azulen. Helv. chim. Acta **26**, 905 (1943).

55. PLATTNER, PL. A., R. SANDRIN u. J. WYSS: Zur Kenntnis der Sesquiterpene. 2-Phenyl-azulen, Beobachtungen über die Wanderung von Substituenten am Azulenkern. Helv. chim. Acta **29**, 1604 (1946).

56. Plattner, Pl. A. u. A. Studer: Zur Kenntnis der Sesquiterpene. Über 6-Methylazulen. Helv. chim. Acta 29, 1432 (1946).
57. Plattner, Pl. A. u. J. Wyss: Zur Kenntnis der Sesquiterpene. Synthese des 1,4-Dimethyl-azulens. Helv. chim. Acta 23, 907 (1940).
58. — — Zur Kenntnis der Sesquiterpene. Synthese einiger Mono- und Dimethyl-azulene. Helv. chim. Acta 24, 483 (1941).
59. Pommer, C.: Action of Azulenes on Inflammation. Naunyn-Schmiedebergs Arch. exp. Pathol. Pharmakol. 199, 74 (1942); Chem. Abstr. 37, 3828 (1943).
59a. Roniger, H.: Thesis. Eidg. Techn. Hochschule, Zürich, Switzerland (1943); cf. (48a).
60. Ruzicka, L.: The many membered Carbon rings. Chem. and Ind. 54, 2 (1935).
61. — Zur Kenntnis des Kohlenstoffringes. Über die Konstitution des Zibetons. Helv. chim. Acta 9, 230 (1926).
62. Ruzicka, L. u. W. Brugger: Zur Kenntnis des Kohlenstoffringes. Über die Gewinnung des Cyclononanons aus Sebacinsäure. Helv. chim. Acta 9, 389 (1926).
63. — — Zur Kenntnis des Kohlenstoffringes. Über die Ringerweiterung vom acht- zum neungliedrigen Kohlenstoffring. Helv. chim. Acta 9, 399 (1926)-
64. Ruzicka, L., W. Brugger, M. Pfeiffer, H. Schinz u. M. Stoll: Zur Kennt. nis des Kohlenstoffringes. Über die relative Bildungsleichtigkeit, die relative Beständigkeit und den räumlichen Bau des gesättigten Kohlenstoffringes. Helv. chim. Acta 9, 499 (1926).
65. Ruzicka, L. u. A. J. Haagen-Smit: Polyterpene und Polyterpenoide. Zur Kenntnis der Azulene. Helv. chim. Acta 14, 1104 (1931).
66. — — Polyterpene und Polyterpenoide. Zur Kenntnis des Guajols. Helv. chim. Acta 14, 1122 (1931).
67. Ruzicka. L. u. E. Peyer: Zur Kenntnis von Dehydrierungsvorgängen II. Einwirkung von Selen oder Palladiumkohle auf Cyclopentanderivate bei erhöhter Temperatur. Helv. chim. Acta 18, 676 (1935).
68. Ruzicka, L., Pl. A. Plattner u. H. Wild: Zur Kenntnis der Sesquiterpene. Über ein ergiebiges Verfahren zur Herstellung von Cyclohexanon. Helv. chim. Acta 26, 1631 (1943).
69. Ruzicka, L. u. E. A. Rudolph: Höhere Terpenverbindungen. Zur Kenntnis der Azulene. Helv. chim. Acta 9, 118 (1926).
70. Ruzicka, L. u. A. G. van Veen: Höhere Terpenverbindungen. Zur Kenntnis der Konstitution des Elemols. Liebigs Ann. Chem. 476, 70 (1929).
71. Sabatier, P. et A. Mailhe: Sur l'emploi de l'oxyde manganeux pour la catalyse des acides, préparation d'aldéhydes et d'acétones pentaméthyléniques. Formation des cyclopentylamines. C. R. hebd. Séances Acad. Sci. 158, 985 (1914).
71a. Sabetay, S. et H. Sabetay: Sur une réaction colorée des sesquiterpènes azulénogènes. C. R. hebd. Séances Acad. Sci. 199, 313 (1934).
71b. Schechter, M. S. and H. L. Haller: The Formation of an Azulene on Zinc-dust Distillation of Pyrethrosin. J. Amer. chem. Soc. 63, 3507 (1941).
72. Schläpfer, P. u. O. Stadler: Untersuchungen über den Cuprenteer. Beitrag zur Kenntnis pyrogener Acetylen-Kondensationen. Helv. chim. Acta 9, 185 (1926).
73. Semmler, F. W.: Ätherische Öle, Bd. III, S. 260. Leipzig. 1906.
74. Semmler, F. W. u. E. W. Mayer: Zur Kenntnis der Bestandteile ätherischer Öle, Pseudocedrol, ein physikalisches Isomeres des Cedrols. Notizen über einige Sesquiterpenalkohole. Tetrahydro-caryophyllen. Ber. dtsch. chem. Ges. 45, 1384 (1912).
75. Sherndal, A. E.: On the Blue Hydrocarbon Occurring in some Essential Oils. J. Amer. chem. Soc. 37, 167 (1915).

76. SHERNDAL, A. E.: Azulene, a blue Hydrocarbon. J. Amer. chem. Soc. **37**, 717 (1915).

77. SKLAR, A. C.: Theory of Color of Organic Compounds. J. chem. Physics **5**, 669 (1937).

78. SORM, F. u. J. FAJKOS: Chem. Obzor **19**, 181 (1944); **21**, 23 (1946); cf. (*39, 48*).

79. SUSZ, B., A. ST. PFAU et PL. A. PLATTNER: Études sur les matières végétales volatiles. Sur les spectres d'absorption de l'azulène, du gaiazulène et du vétivazulène. Helv. chim. Acta **20**, 469 (1937).

80. TREIBS, W.: Naturwiss. **33**, 371 (1946); cf. (*40*).

80a. U. S. Dept. of Commerce. Publications Board Rep. PB 520; PB 18852.

81. WAGNER-JAUREGG, TH., H. ARNOLD u. F. HÜTER: Synthese des 1-Isopropyl-4,7-(4,6-)dimethylazulens. Ber. dtsch. chem. Ges. **75**, 1293 (1942).

82. WAGNER-JAUREGG, TH., H. ARNOLD, F. HÜTER u. J. SCHMIDT: Über synthetische Azulene. Ber. dtsch. chem. Ges. **74**, 1522 (1941).

83. WAGNER-JAUREGG, TH., E. FRIESS, H. HIPPCHEN u. F. PRIER: Synthese des 1,6,8-Trimethylazulens. Über die Konstitution der natürlichen blauen Azulene. Ber. dtsch. chem. Ges. **76**, 1157 (1943).

84. WAGNER-JAUREGG, TH. u. H. HIPPCHEN: Synthese von 1,3,7-Trialkyl-azulenen. Über die Spektren mehrfach substituierter Azulene. Ber. dtsch. chem. Ges. **76**, 694 (1943).

85. WALLACH, O.: Zur Kenntnis der Terpene und ätherischen Öle. Überführung von Cyclohexanon in Cycloheptanon (Suberon). Liebigs Ann. Chem. **353**, 326 (1907).

86. WILLSTAEDT, H.: Über die Farbstoffe des echten Reizkers (*Lactaris deliciosus* L.). Ber. dtsch. chem. Ges. **68**, 333 (1935); **69**, 997 (1936); Atti X Congr. int. Chim., Roma **3**, 390 (1938).

(Received, April 21, 1948.)

Recent Advances in the Study of Component Acids and Component Glycerides of Natural Fats.

By **T. P. Hilditch**, Liverpool.

Introduction.
Discussion of the "Rule of even distribution".

The study of the chemical composition of natural fats resolves itself into the determination as far as may be experimentally possible of the proportions and compositions of the individual mixed glycerides present in any given fat, the latter being in general a somewhat formidable mixture of mixed triglycerides. This in turn involves the determination of the different fatty acids present in such individual mixed glycerides or rather, in practice, of those present in the relatively simpler mixtures of mixed glycerides into which it may have been possible to separate the fat by physical means. Further, the composition of the total fatty acids of a fat is in itself of considerable interest, since it has been shown that to a very large extent the fatty acids in different plants or animals, and in different tissues or organs of the same plant or animal, are often specific in character and depend, to a considerable degree, on the biological nature and the evolutionary development of the plant or animal. On the other hand, the general manner in which the fatty acid groups are woven into triglyceride molecules is very similar throughout nature, and is largely independent of the biological environment and also of the particular mixture of fatty acids which is characteristic for any given fat or group of fats.

The basis of mixed triglyceride formation in plants or animals appears to be the widest distribution of fatty acids amongst the glycerol molecules (rather than, for example, random distribution according to the laws of mathematical probability). This was pointed out in many earlier papers from the University of Liverpool, and by Hilditch and Lovern (52); and the so-called "rule of even (or widest) distribution" has more recently been expressed by Hilditch (41) as follows: If an acid A forms about

35 per cent. (mol.) or more of the total fatty acids $(A + X)$ in a fat, it will occur at least once, $G(AX_2)$, in practically all the triglyceride molecules in the fat. If it forms from about 35 to about 65 per cent (mol.) of the total fatty acids, it will occur twice, $G(A_2X)$, in some of the triglyceride molecules, and of course the more frequently the higher the proportion of the acid in the total fatty acids. If an acid A forms 70 per cent of the total fatty acids, the remaining fatty acids (X) can at most only form mixed triglycerides, $G(A_2X)$; and the excess of A in such a case, and broadly speaking then only, appears as a simple triglyceride, $G(A_3)$. This general tendency to concentrate upon the elaboration of mixed triglycerides of this form and to produce minimal amounts of simple triglycerides is extremely marked throughout the whole range of natural fats.

As would be expected in a group of natural products, the even distribution "rule" is not adhered to with mathematical rigidity: it is a generalisation which fits the facts observed in the great majority of the natural fats, not a mathematical law. In the vast majority of natural fats it expresses the facts quite closely; but in minor respects slightly greater amounts of one or other triglyceride type [e.g., $G(A_2X)$ or even the simple $G(A_3)$] than would be expected from complete adherence to the "rule" are often observed. Thus, in an earlier communication [HILDITCH (46)] it was recorded that seed fats in which saturated acids formed less than two-thirds of the total acids, had been observed not infrequently to contain up to 2 or 3 per cent. of fully-saturated triglycerides; and it has also been shown [HILDITCH and MADDISON (54); DUNN, HILDITCH and RILEY (35)] that cotton seed and certain *Citrus* seed fats which contain only about 25 per cent. of saturated acids, nevertheless contain small proportions of disaturated-mono-unsaturated triglycerides.

The principle of "even distribution" operates (apart from the minor reservations to which reference has just been made) in almost all land vegetable fats, in the fats of aquatic flora and fauna (including the larger fishes and marine mammals), and in many fats of land animals.

It is at first sight not followed in two important groups of land animal fats, namely, their milk fats and their body (depot) fats when the latter are rich in stearic acid. In these two groups the proportion of fully-saturated glycerides suggests, as first pointed out by BANKS and HILDITCH (6) and by BHATTACHARYA and HILDITCH (9), that the saturated acids are distributed in the triglyceride molecules at random, i. e., according to considerations of mathematical probability; so that it might seem at first sight that biosynthesis of the triglycerides in these fats must follow a radically different course from that in all other natural fats. Further study of this problem by the Liverpool school has led us to conclude, however, that the glyceride structure of both these classes of fats is consistent with the superposition of certain other chemical changes on an originally preformed mixture of (mainly) palmitic and oleic acids assembled as mixed glycerides on the usual lines of "even distribution". In the stearic-rich animal depot fats, there is evidence which suggests that preformed oleo-glycerides are in part converted into stearo-glycerides [HILDITCH and PAUL (69); HILDITCH and MURTI (65); HILDITCH and PEDELTY (71); HILDITCH and ZAKY (85); HILDITCH (42); ACHAYA and BANERJEE (1)], and that in milk-fats transformation of oleo-glycerides into the short-chain saturated and unsaturated glycerides characteristic of these fats may similarly take place [HILDITCH and SLEIGHTHOLME (81, 82); HILDITCH and THOMPSON (84); HILDITCH and LONGENECKER (51); SMITH and DASTUR (100); SHAW (97); SHAW, POWELL and KNODT (98); HILDITCH and MEARA (61); HILDITCH (43)].

In 1941, however, LONGENECKER (89) again stressed the apparent elaboration of the saturated glycerides of animal fats according to the application of simple considerations of the theory of probability, and reverted to our first impression that vegetable and animal fats might be built up by processes involving different mechanisms. More recently NORRIS and MATTIL (94) have supported this view, and have pointed out its corollary that this explanation involves the assumption that the enzymes responsible for the esterification of fatty acids with glycerol differ, or operate entirely differently, in animal and in plant systems. A further contribution from this laboratory will be published elsewhere in which we shall supply additional evidence, obtained with the use of some of the more recently developed techniques discussed in this paper, which confirms theconclusions at which we arrived earlier. As stated in the concluding part of this paper, we maintain that these classes of animal fatty glycerides are merely special cases of the general principle of "even distribution".

The foregoing survey will, it is hoped, serve to summarise the general position which has been reached as regards present knowledge of the glycerides and fatty acids contained in natural fats. In the past decade considerable improvements have been made in the methods of experimental attack in this field and it has therefore seemed appropriate in this article to give considerable prominence to these advances (which will be illustrated by reference to various specific problems), rather than to discuss in detail the composition of any one of the different groups of natural fats which have received investigation during this period.

Spectrophotometric Determination of Certain Unsaturated Higher Fatty Acids.

Higher fatty acids which contain conjugated systems of two or more carbon-carbon double bonds exhibit selective light absorption at specific wave lengths in the ultra-violet spectrum. Measurement of the extinction coefficients of these absorption bands therefore gives a direct measure (the extinction coefficient for the pure acid being known) of the amount of a conjugated acid present in a mixture of fatty acids. The proportion, in a mixture of fatty acids, of conjugated acids[1] such as elaeostearic (octadeca-9,11,13-trienoic) or parinaric (octadeca-9,11,13,15-tetraenoic) can thus be determined from the observed extinction coefficients of the maxima at, respectively, 268 mμ and 305 mμ, (the values of $E_{1\,cm.}^{1\%}$ for the pure acids being, respectively, 1790 and 2624).

More important is the observation [MOORE (92); KASS and BURR (87)] that acids containing the grouping —CH : CH · CH$_2$ · CH : CH— (or a multiple thereof) are partly transformed into *conjugated* isomers whe heated with excess of alkali hydroxide at about 180°, since this leads to the similar spectrophotometric determination of, amongst others, the frequently ocurring linoleic and linolenic acids of natural fats.* When linoleic acid is isomerised with alkali under carefully standardised conditions, a constant proportion of it is converted into the corresponding conjugated diene acid which has an absorption maximum at 234 mμ. Similarly, linolenic acid gives a constant proportion both of conjugated triene and diene acids. The procedure was recommended for the determination of linoleic and linolenic acids by MITCHELL, KRAYBILL and ZSCHEILE (91), who effected isomerisation with potassium hydroxide in glycol solution at 180° for 25 minutes. It has been thus applied by RUSOFF, HOLMAN and BURR (96), by BEADLE (8), by the Spectroscopy Committee of the American Oil Chemists' Society (3) and others to the rapid analytical determination of linoleic, linolenic and arachidonic acids. For investigational purposes, HILDITCH, MORTON and RILEY (64) preferred to base the linolenic acid determination on a separate alkali-isomerisation at 170° for 15 minutes, using a separate alkali-isomerisation at 180° for 60 minutes for linoleic acid. At Liverpool we have not yet employed the method to determine arachidonic acid or other poly-ethenoid acids of the C$_{20}$ and C$_{22}$ (fish oil) group, on account of uncertainty of the

---

[1] *Acid:*   *Constitution:*

Elaeostearic.. $CH_3 \cdot (CH_2) \cdot (CH:CH)_3 \cdot (CH_2)_7 \cdot COOH$

Parinaric .... $CH_3 \cdot CH_2 \cdot (CH:CH)_4 \cdot (CH_2)_7 \cdot COOH$

Linoleic ..... $CH_3 \cdot (CH_2)_4 \cdot CH:CH \cdot CH_2 \cdot CH:CH \cdot (CH_2)_7 \cdot COOH$

Linolenic ....$CH_3 \cdot CH_2 \cdot CH:CH \cdot CH_2 \cdot CH:CH \cdot CH_2 \cdot CH:CH \cdot (CH_2)_7 \cdot COOH$

Oleic........ $CH_3 \cdot (CH_2)_7 \cdot CH:CH \cdot (CH_2)_7 \cdot COOH$

values to be attributed to the individual acids, and of the number and complexity of conjugated di-, tri-, and perhaps tetra-ene acids resulting from the non-conjugated, multi-ethenoid acids.

By determining the extinction coefficients at 268 mμ for elaeostearic acid, and for this acid and linolenic acid after treatment with alkali, and also determining this coefficient at 234 mμ for linoleic, linolenic and elaeostearic acids after treatment with alkali, the composition of mixtures of any of these with oleic and saturated acids can be obtained as follows. Any *elaeostearic acid* present is determined from the extinction coefficient of the total fatty acids as such. *Linolenic acid* is determined after the acids have been isomerised with alkali at 170° for 15 minutes, the observed extinction coefficient at 268 mμ being corrected for elaeostearic acid from the preceding estimations and the known value of the extinction coefficient observed when elaeostearic acid has been heated with alkali, under the same conditions. *Linoleic acid* is similarly determined, after alkali isomerisation at 180° for 60 minutes, from the extinction coefficient then observed at 234 mμ, deductions being made for contributions to the extinction coefficient at 234 mμ due to linolenic and/or elaeostearic acid in consequence of the alkali treatment. *Oleic acid* is then estimated from the difference between the iodine value of the total acids and the sum of the increments of iodine value due to the observed proportions of the three poly-ethenoid acids mentioned. *Saturated acids* are finally obtained by difference.

The cumulative effect of experimental error renders this method, which is otherwise excellent, somewhat uncertain when applied to a complex mixture of acids, but this can be largely overcome by preliminary resolution of the total fatty acids of a fat into groups in each of which one, or at most two, of the saturated, mono-, di-, or tri-ethenoid acids largely predominate. For this purpose crystallisation from appropriate solvents at low temperatures, which will next be discussed, has proved extremely serviceable.

Resolution of Mixed Fatty Acids by Crystallisation from Solvents at Low Temperatures.

The final detailed analytical determination of the component acids in a natural fat depends upon the vacuum distillation of their methyl esters through a suitable fractionation column, in such a manner that a progressive series of ester-fractions is obtained, each of which consists of not more than two saturated and two unsaturated members of the homologous series of fatty acids [HILDITCH (45)]. In general the total fatty acids of a fat present too complex a mixture for satisfactory resolution in this way, and it is desirable to effect their preliminary resolution into less complicated mixtures by other means. Until comparatively recently this involved their separation in the form of certain metallic salts. The sparing solubility of the lead salts of the higher saturated fatty acids in alcohol has been the most useful means hitherto available for largely segregating these from the corresponding unsaturated acids and from saturated acids of lower molecular weight [TWITCHELL (*103*); HILDITCH (*44*)], although mono-ethenoid acids of the C_{20}, C_{22} and C_{24} series, a number of the solid isomers of oleic acid, and to some extent oleic acid itself, also form sparingly soluble lead salts. Highly unsaturated acids, present in quantity in fats from aquatic sources, give lithium salts which are readily soluble in acetone, and this can be utilised for their rough separation from accompanying mono-ethenoid and saturated acids [TSUJIMOTO (*102*); HILDITCH and MADDISON (*56*)].

J. B. BROWN and co-workers (*15, 16, 17, 18*) first drew attention to the use of solvents at low temperatures in connection with the isolation of pure oleic and linoleic acids; and later BROWN (*14*) stated that the method could well be applied to analytical separations of natural fatty acid mixtures, the physical method of crystallisation and the low working temperatures possessing advantages over chemical procedures (lead salt, etc.) owing to their simplicity and directness.

The method was probably first employed in a component fatty acid analysis by CRAMER and BROWN (*30*) who separated the *methyl ester fractions* of human body fat acids by crystallisation from solvents at — 20° to — 70° C., after they had been fractionally distilled in a vacuum. DE LA MARE and SHORLAND (*31*) crystallised the mixed *methyl esters* of a pig fat from acetone at — 35° C. prior to fractional distillation of the insoluble and soluble esters so obtained. FOREMAN and BROWN (*36*) made a thorough study of the solubilities of many individual fatty acids in acetone, methyl alcohol, or petrol at different temperatures down to — 50° C., in the course of which they made two important general observations: (*a*), it is necessary to crystallise for as long a period as possible because the solutions come to equilibrium exceedingly slowly at low temperatures; and (*b*) that, although the individual higher saturated acids are only very slightly soluble in most of the pure solvents at — 30° or — 40° C., the presence of other (more soluble) acids introduces mutual solubility effects which often cause small but appreciable amounts of the higher saturated acids to remain in solution even at — 60° or — 70° C.

In our laboratory at Liverpool we have been glad to avail ourselves of Professor BROWN's techniques, especially in the investigation of the more unsaturated kinds of vegetable and aquatic animal fatty oils, the mixed acids of which are least amenable to preliminary separation in the form of lead salts. It may be of interest to indicate the procedures which have been found most generally applicable, and to illustrate them by examples selected from some of the work which has recently been carried out, especially on vegetable drying oils and some marine animal oils.

At the outset it may be pointed out that, for fats rich in palmitic and stearic acid, it is still considered that a lead salt separation is desirable in the first instance. This causes more complete removal of the two acids than is obtainable (by reason of the "mutual solubility effects" already mentioned) if low temperature crystallisation from solvents is exclusively used. When, however, the saturated acids form less than about 20 per cent. of the total fatty acids, all the preliminary separations are conveniently made by the crystallisation process. After a lead salt separation has been applied (in the cases indicated), the acids recovered from the soluble lead salts may be usefully submitted to low temperature crystallisation from solvents.

Amongst the advantages of the latter procedure is its great flexibility: concentration, temperature, and time of crystallisation can all be varied if necessary, although circumstances may limit the maximum period of crystallisation to about five hours. Whilst the most suitable conditions can, therefore, be chosen for any particular case, it is proposed here to outline conditions which we have found appropriate for the majority of mixed fatty acids encountered in the more unsaturated natural fats, but which are open to modification to suit special circumstances. As stated, the maximum time of crystallisation may be limited by working conditions to five hours, and the operation is usually conducted with 10 per cent. solutions of the fatty acids in the solvent; higher concentrations in general seem to result in less efficient separations, but on the other hand in isolated instances a greater dilution (e. g. 5 per cent. solutions) may be found advantageous. Crystallisation is always commenced at the *lowest* temperature to be employed, the material left in solution being the most unsaturated portion of the whole. The deposited acids (perhaps after recrystallisation under the conditions initially employed) are next crystallised at a higher temperature, and so on, until, finally, from the last crystallisation both the deposited and soluble fractions are submitted to esterification and fractional distillation.

For a majority of *fatty oils* it has become possible to apply the following general procedure. The total fatty acids obtained by hydrolysis of the fatty oil are first crystallised from 10 per cent. solution in acetone at — 60° C. (or, in some instances, at — 70° C.); the acids left in solution are recovered, whilst those deposited may be recrystallised from acetone at — 60° C., or, in other cases, may then be crystallised from 10 per cent. solution in ether at — 40° C. The acids then deposited will consist of saturated acids accompanied by small proportions (up to 10 or sometimes 20 per cent.) of oleic or other mono-ethenoid acid. The acids left in solution in ether at — 40° C. are mainly those of the mono-ethenoid series, accompanied by some saturated acids and also by small proportions of linoleic or other poly-ethenoid acids. The greater part of any linoleic acid present, and almost all of any linolenic acid or other poly-ethenoid acids (if present in the original fat) are left in solution in the first crystallisations from acetone at — 60° C. Owing to the mutual solubility effect already mentioned, however, small proportions of oleic and saturated acids invariably appear in this group, in which the most highly unsaturated acids are nevertheless concentrated.

Application of the spectrophotometric method of analysis to each group of acids furnishes more accurate data as to their contents of linoleic and linolenic acids than that obtainable by similar analysis of the total fatty acids. The proportions of the saturated and other acids present are obtained in further detail by fractional distillation of the methyl esters of each group of acids.

Some further general indications of the techniques employed for different categories of fats may be added. For the more unsaturated (liquid) types of *vegetable oils*, such as linseed, rubber seed, and many others, and for *marine animal oils*, separation into three groups as described (first from acetone at — 60° C. and then from ether at — 40° or — 30° C.) is in general most suitable (*39, 67*). For less unsaturated ("semi-drying") *vegetable oils*, such as sunflower seed, groundnut and other oils, separation into two fractions (using acetone at — 50° or — 60° C.) may be sufficient (*74*). For *vegetable oils and fats* containing high proportions of palmitic and or stearic acids (such as palm oil, cottonseed oil, cacao butter) and for *animal body fats* with high contents of these acids (tallows, lards, etc.) it is best first to employ a lead salt separation; the "liquid" acids from the soluble lead salts may then be further separated from acetone at — 50° to — 60° C. (*63, 80*). In other cases, other modifications are desirable. Thus, after removal of steam-volatile lower fatty acids from the acids of *animal milk fats*, the residual acids are conveniently treated by a single crystallisation from ether at — 40° C. (*2*). For *oils of the Cruciferae* (such as rape oils) or other oils rich in erucic as well as in unsaturated acids of the C_{18} series, crystallisation from ether at — 40° C. has also been found to give the most useful resolution (*5*). Finally, *vegetable oils which contain conjugated acids* (e. g., tung oil with a high content of elaeostearic acid) require different handling: the conjugated acid is best deposited (with any saturated acids) by preliminary crystallisation from petrol at — 60° or — 70° C., the acids left in solution (still including appreciable amounts of conjugated acid owing to strong "mutual solubility" effects) being further crystallised from acetone at similar low temperatures. It may be remarked in passing that application of these techniques [HILDITCH

and RILEY (75)] to various seed oils which contain elaeostearic acid
has led to the observation that linolenic acid and elaeostearic acid do
not seem to occur together in one and the same seed fat.

Some illustrations of the separations effected by use of these methods
are given in Table 1, in which the percentage proportions and iodine
values of the groups of acids separated from the total fatty acids of
different fats are shown.

Table 1. Separation of mixed fatty acids by low-temperature
crystallisation.

Fat (reference)	Total acids		Least soluble		Intermediate		Most soluble	
	Iodine value	Proce-dure	%	Iodine value	%	Iodine value	%	Iodine value
Palm oil (63)	55,3	(a)	44,7	nil.	41,9	83,3	13,4	151,0
Sheep body fat (80)	41,2	(b)	60,4	10,1	21,6	76,2	18,0	97,3
Neat's foot oil (79)	76,4	(c)	15,2	2,7	57,1	85,2	27,7	101,0
Cow milk fat (2)	34,3	(d)	41,7	3,1	—	—	51,1	65,0
Groundnut oil (78)	90,8	(e)	81,8	76,7	—	—	18,2	153,0
Lime seed oil (35)	116,4	(a)	29,1	6,4	10,8	120,2	60,1	170,4
Sunflower seed oil (7)	129,8	(c)	11,2	23,9	49,7	135,2	39,1	167,1
Niger seed oil (34)	134,9	(c)	20,4	33,1	43,3	169,5	36,3	165,8
Soya bean oil (47)	139,2	(c)	13,3	1,8	14,2	104,5	72,5	170,7
Rubber seed oil 77)	131,5	(c)	18,6	9,9	32,3	127,5	49,1	198,2
Linseed oil (77)	184,1	(c)	12,7	10,4	16,2	104,9	71,1	233,2
Conophor oil (77)	201,7	(c)	4,7	6,9	13,6	121,8	81,7	236,0
Rape oil (5)	107,0	(f)	28,1	67,9	8,8	84,9	63,1	126,8
Seal oil (67)	170,3	(c)	18,3	16,7	30,5	107,2	51,2	256,5
Whale oil (57)	121,0	(c), (g)	22,5	6,4	42,2	99,8	33,1	217,3

(a) Most soluble acids from acetone at — 40° C; least soluble from ether at
— 30° C.

(b) Lead salt separation of least soluble acids; remainder crystallised from
acetone at — 50° C.

(c) Most soluble acids from acetone at — 60° C; least soluble from ether at
— 40° C.

(d) Crystallisation from ether at — 40° C. only (after removal of 7,2% of lower
saturated acids by steam distillation).

(e) Crystallisation from acetone at — 60° C. only.

(f) Crystallisation from ether at — 40° C. only, followed by recrystallisation
from ether at — 40° C.

(g) 2,2% of unsaponifiable matter removed from mixed acids prior to crystallisation of the latter.

Although the purpose of Table 1 is to illustrate the operation of
the crystallisation techniques, and much subsequent detail (which cannot
be quoted here) is involved in the complete data for the ester-fractionation
of each group of acids, it may be of interest to give the final

component acid figures (Table 2, p. 82) thus obtained for each of the fats mentioned in Table 1.

In all other instances the lower saturated acid is myristic, and the higher saturated acid arachidic, except in soya bean and rape oils, which contain also small amounts of behenic and lignoceric acids.

Partial Resolution of Mixed Glycerides in Natural Fats by Crystallisation from Solvents at Low Temperatures.

The crystallisation procedure which has facilitated the study of the component acids present in fats can also be applied with advantage to the fats themselves. Since, however, the number of mixed glycerides present is usually very large, and especially since acids of very different character (e. g., saturated, mono-ethenoid and poly-ethenoid) may all be in combination in a single triglyceride molecule, it is clearly not possible—except in comparatively rare instances when the fatty acids present are restricted to two, three or four in number—to arrive, in most cases, at any profound degree of separation of the mixed glycerides.

It is rarely possible to isolate any individual mixed glyceride in quantity. This feature was remarked by BULL and WHEELER (*19*) who described the crystallisation of soya bean oil from acetone and other solvents at temperatures from — 76° to 15° C. It was also evident in the earlier classical work of BÖMER and his colleagues (*12*, *13*), in their attempts to separate the glycerides of the solid fats, coconut and palm kernel oils, by crystallisation from different solvents at and above 0° C. Nevertheless, it is practicable, by repeated crystallisation (usually from acetone) first at a low temperature (usually — 60° C.) and thereafter at successively higher temperatures, to effect considerable separation of the mixed glycerides in a fat or fatty oil into a number of fractions, each of which represents a much simpler mixture of mixed glycerides than that in the whole fat. Determination of the component acids present in each of these simpler mixtures permits an approximate estimate to be made of the component mixed glycerides therein, and thus to obtain similar data for the original fat. When the fatty acids present are relatively few in number, the individual component glycerides of a fat can be stated with some degree of precision. In the more frequent cases where the major component acids exceed three or four, the results are often less definite but it is nevertheless possible by this means to arrive at the approximate proportions of the main classes of glycerides present and, to some extent, of the chief individual mixed glycerides concerned.

To illustrate some recent applications of this method, the proportions (% wt.) and iodine values of the groups of glycerides thus separated from

Table 2. Component fatty acids (% wt.) of the fats listed in Table 1 (p. 80).

	Saturated				Hexadecenoic	Unsaturated			
	Lower	Palmitic	Stearic	Higher		Oleic	Linoleic	Linolenic	Higher
Vegetable fats:									
Palm oil (63)	3,1	42,4	3,6	0,2	0,9	39,5	10,0	0,3	—
Groundnut oil (78)	0,3	19,2 →		0,5	—	61,4	19,4	—	—
Lime seed oil (35)	—	26,1	9,6	2,0	—	11,1	39,3	13,1	—
Sunflower seed oil (7)	—	6,1	6,1	2,1	1,0	19,9	64,9	—	—
Niger seed oil (34)	—	9,1	10,5	2,1	—	2,7	74,4	1,2	—
Soya bean oil (47)	0,4	10,6	2,4	2,4	1,0	23,5	51,2	8,5	—
Rubber seed oil (77)	—	11,7	11,5	0,5	—	18,4	37,2	20,7	—
Linseed oil (77)	0,1	8,1	7,3	0,5	—	15,7	15,3	53,0	—
Conophor oil (77)									
Rape oil (4)	—	1,9	3,5	2,2	1,5	12,3	15,8	8,7	54,1 (Erucic, etc.)[1]
Land animal fats:									
Sheep body fat (80)	2,9	27,8	27,7	1,5	3,1	33,0	3,4	—	0,6 (Unsaturated C_{20} and C_{22})
Neat's foot oil (79)	0,7	16,9	2,7	0,1	9,4[2]	64,4	2,3	0,7	1,6 (″ C_{20} ″, C_{22})
Cow milk fat (2)	23,8[3]	31,3	8,3	0,9	1,6[3]	28,0	3,7	—	0,7 (″ C_{20} ″, C_{22})
Marine animal fats:									
Seal oil (67)	4,8	10,7	2,0	—	13,2[4]	{ 35,9[4]	}		30,7[4] (C_{20} ″, C_{22})
Whale oil (57)	9,2	15,6	1,9	0,6	13,9[5]	{ 37,2[5]	}		19,1[5] (C_{20} ″, C_{22})

[1] Higher unsaturated acids: eicosenoic 4,8%, erucic 47,8%, docosadienoic 1,5%.

[2] Also 1,2% tetradecenoic acid.

[3] Butyric 4,0%, hexoic 1,8%, octoic 1,0%, decoic 2,0%, decenoic 0,2%, lauric 2,2%, myristic 12,8%; also decenoic 0,2%, dodecenoic 0,3%, and tetradecenoic acid 1,2%.

[4] Unsaturated acids: C_{14} 2,7% (— 2,0 H); C_{16} 13,2% (— 2,2 H); C_{18} 35,9% (— 2,6 H); C_{20} 12,8% (— 6,4 H); C_{22} 17,9% (— 10,6 H); C_{24} 0,9 (— 11,0 H).

[5] Unsaturated acids: C_{14} 2,5% (— 2,0 H); C_{16} 13,9% (— 2,1 H); C_{18} 37,2% (— 2,4 H); C_{20} 12,0% (— 7,1 H); C_{22} 7,1% (— 10,5 H). (The terms (— 2,0 H), (— 7,1 H), etc., indicate the mean unsaturation of each group of acids expressed as the fractional number of atoms of hydrogen required for molecule to produce complete saturation.)

palm oil, soya bean oil, lime seed oil and seal blubber oil are summarised in Table 3. With the latter three oils, crystallisation from acetone commenced at — 60° C., the deposited glycerides (sometimes after recrystallisation at the same temperature) being crystallised at successively higher temperatures until, finally, a least soluble fraction (A) was obtained at — 30° C. (soya bean oil), at 0° C. (lime seed oil) or at — 10° C. (seal oil). With palm oil (which contains a much higher proportion of saturated acids, cf. Table 2) the oil was first crystallised from acetone at 0° C. several times, the solids finally deposited being then crystallised from acetone at room temperature. The acetone-soluble glycerides were again crystallised from 30 per cent. solution in acetone at 0° C. for a long period, the portion left in solution being then further resolved from acetone at — 50° C., and from ether at — 40° C. In Table 3, the list of fractions finally obtained in each case commences (A) with the least soluble and most saturated, and concludes with the most soluble and most unsaturated fraction, i. e., that left in solution at the lowest temperature employed (— 50° or — 60° C.).

Table 3. Resolution of mixed glycerides of natural fatty oils by crystallisation from solvents at low temperatures.

Iodine value	Palm oil 52,9		Soya bean oil 132,6		Lime seed oil 111,0		Seal oil 162,2	
	%	Iod. val.	%	Iod. val.	%	Iod. val.	%	Iod. val.
Fraction:								
A (least soluble) .	9,2	11,0	35,8	114,3	20,1	60,4	16,4	73,9
B	23,8	32,7	20,8	131,7	5,5	75,8	28,9	119,7
C	13,6	43,1	17,9	150,6	12,5	100,6	30,4	189,6
D	39,2	64,6	10,4	152,8	10,1	113,7	24,3	241,8
E	7,1	82,8	5,2	163,7	19,3	118,6	—	—
F	7,1	95,1	3,9	148,6	10,6	138,6	—	—
G	—	—	—	—	12,9	148,4	—	—
H (most soluble) .	—	—	—	—	9,0	156,5	—	—

The fractions obtained in this manner are in general much less complex mixtures of mixed glycerides than those in the original fats. The manner in which their composition progressively alters may be illustrated by comparing the relative amounts of saturated and unsaturated acids found to be present in each fraction (Table 4, p. 84).

In the least soluble fractions the proportion of any tri-saturated glycerides can be ascertained experimentally, and, consequently, the remainder of such fractions can be allocated as mono-unsaturated-di-saturated and di-unsaturated-mono-saturated glycerides. No independent determination of the latter groups, or of tri-unsaturated glycerides, is possible, and therefore it is necessary to consider how far tri-unsaturated

Table 4. Relative proportions (% mol.) of saturated and unsaturated acids in some glyceride fractions.

	A	B	C	D	E	F	G	H
Palm oil (63):								
Saturated................	88,2	66,1	62,1	41,2	27,6	23,9	—	—
Oleic	11,8	31,7	30,1	46,4	54,8	19,6	—	—
Linoleic.................	—	2,1	7,5	12,0	16,8	25,4	—	—
Linolenic................	—	0,1	0,3	0,4	0,8	1,1	—	—
Soya bean oil (47):								
Saturated................	26,1	21,8	12,4	10,3	7,8	7,8	—	—
Oleic	26,8	19,4	19,2	19,8	15,7	15,7	—	—
Linoleic.................	40,7	50,0	56,5	56,5	58,8	58,8	—	—
Linolenic................	6,4	8,8	11,9	13,4	17,7	17,7	—	—
Lime seed oil (35):								
Saturated................	68,2	57,4	40,1	36,4	32,8	21,3	13,7	8,5
Oleic	6,7	13,2	18,3	15,6	16,7	18,9	21,6	23,1
Linoleic.................	18,4	21,6	32,0	35,8	36,4	43,9	47,5	48,4
Linolenic................	6,7	7,8	9,6	12,2	14,1	15,9	17,2	20,0
Seal oil (67):								
Saturated................	44,8	21,0	12,8	5,4	—	—	—	—
Unsaturated C_{16} (+ C_{14})	15,2	25,8	22,4	24,0	—	—	—	—
,, C_{18}	26,6	37,0	32,6	28,1	—	—	—	—
,, C_{20} and C_{22}	13,4	16,2	32,2	42,5	—	—	—	—

glycerides are likely to be present in any of the more soluble fractions of glycerides. From the general trend of the relative proportions of saturated and unsaturated acids in the fractions of increasing solubility in each of the above examples, it is apparent that di-saturated glycerides are but slightly, if at all, soluble at the lower temperatures used and that they are largely concentrated in the least soluble fractions (cf., palm oil, A–D; lime seed oil, A–C; seal oil, A). Similarly, tri-unsaturated glycerides concentrate freely in the fractions most soluble at the lowest temperatures used (e. g., soya bean oil, D–F; lime seed oil, G–H; seal oil, C–D). Whilst it cannot be taken for granted that di-saturated glycerides are completely absent from even the most soluble fractions, it is certain that to a great extent saturated components therein will be mono-saturated-di-unsaturated glycerides. Much depends on the rigorous application of the crystallisation procedure at the lowest range of temperatures to be used, in order that the proportion of tri-unsaturated glycerides shall not be seriously underestimated. Fortunately, another factor operates here, namely, the circumstance that any one of the most soluble fractions is usually only a small part of the whole fatty oil; consequently, the differences in the tri-unsaturated glyceride content of such a fraction, when the latter is calculated alternatively to a mixture of (i) tri-unsaturated-(U_3)- and mono-saturated-(SU_2)-, or (ii) tri-unsaturated-(U_3)- and di-saturated-(S_2U)-glycerides, have less effect than may appear at first sight on the tri-unsaturated glyceride content of the whole fat.

This may be illustrated from Table 5 with reference to palm oil fraction F, soya bean oil fractions E and F, lime seed oil fractions G and H, and seal oil fractions C and D.

Table 5. Possible tri-unsaturated glyceride contents of certain fractions listed in Table 4.

Oil	Fraction No. % (mol)	% in fraction calculated from		Corresponding % in whole fat calculated from	
		(i) $U_4 + SU_2$	(ii) $U_3 + S_2U$	(i) $U_3 + SU_2$	(ii) $U_3 + S_2U$
Palm oil	F 6,9	28,3	64,1	2,0	4,5
Soya bean oil	E 5,1	76,6	88,3	3,9	4,5
,, ,, ,, ...	F 3,6	76,6	88,3	2,7	3,2
Lime seed oil.....	G 12,7	58,9	79,4	7,5	10,1
,, ,, ,, ...	H 8,8	74,5	87,2	6,6	7,7
Seal oil	C 30,1	61,6	80,8	18,6	24,3
,, ,, ...	D 23,4	83,8	91,9	19,6	21,5

Since the binary combination (ii) is certainly relatively small compared with (i), the true values for tri-unsaturated glycerides must lie closer to (i) than to (ii), whilst the differences (in terms of the whole fats) between the values of (i) and (ii) are in most instances less than 1 or 2 units per cent.

Table 6. Glyceride constituents (% mol.) of palm oil (63), soya bean oil (47), lime seed oil (35), and seal oil (67).

Glyceride	Palm oil % (mol.)	Soya bean oil % (mol.)	Lime seed oil % (mol.)	Seal oil % (mol.)
Tri-saturated (S_3)	6	—	2	—
Di-saturated mono-unsaturated (S_2U) ..	48	—	25	6
Mono-saturated-di-unsaturated (SU_2) ...	43	58	55	45
Tri-unsaturated (U_3)	3	42	18	49

By applying these principles to each of the fractions in the fat under investigation, the approximate proportions of numerous categories of mixed glycerides, and frequently of various individual mixed glycerides, present in the whole fat can be obtained. For the full results of studies of this kind which have so far been undertaken, the original publications should be consulted, since they are too lengthy to be reproduced here. However, for purposes of illustration the data for the four fatty oils which have been quoted above are reduced in Table 6 to an abbreviated form, showing the total proportions of tri-saturated, di-saturated-mono-unsaturated, mono-saturated-di-unsaturated, and tri-unsaturated glycerides estimated to be present in each of these fats.

The Present State of Knowledge of the Glyceride Structure of Natural Fats.

Although there remains much scope for further intensive study of the mixed glycerides which constitute the natural fats, sufficient data are now available to present a broad general picture of the nature of these compounds. Their composition can indeed best be understood in terms of the principle of maximum "even" (or "wide") distribution of the fatty acids amongst the triglyceride molecules as enunciated (*41*) at the commencement of this monograph (p. 72). The operation of this principle is most easily perceived when the number of major component acids in a fat is confined to three (or at most four); but studies with the aid of the improved methods described above have pointed to its similar application when the number of major component acids is considerably greater. The present position of our knowledge in this field may then be summarised briefly as follows.

1. Fats with three (or at most four) major component acids.

(a) Fats in which saturated acids predominate over unsaturated (oleic) acids.

In this large group no significant amount of tri-saturated glycerides appears unless the proportion of saturated acids is 60 per cent. or more of the total acids; with higher proportions of saturated acids the amount of tri-saturated glycerides is confined to the excess over that necessary for combination with unsaturated (oleic) acids as a mixture of about 80 per cent. of mono-oleo-di-saturated and 20 per cent. of di-oleo-mono-saturated glycerides (*46, 28*). This has been proved for very many fats, including those of coconut (*27*), palm kernel (*27*), dika (*21, 23*), nutmeg (*29*), Borneo tallow (*20, 72*), cacao butter (*83*), *Allanblackia* (*76, 90*), *Garcinia* (*33, 66*), shea (*37, 76*), mowrah (*48*), *Stillingia tallow* (*73*), palm oil (*49, 53, 63*), etc. Only three complete exceptions to this generalisation have yet been observed, namely, laurel kernel oil (*24, 29*), and the seed fats of *Myristica malabarica* (*26*) and of *Platonia insignis* (*22*). In this group, of course, the distribution of saturated and unsaturated acids in the glycerides is far removed from that which would result from "random distribution", i. e., that the amount of tri-saturated glycerides should be proportional to the cube of the proportion of saturated acids (Table 7, p. 87).

Table 7. Examples of the absence of "random" distribution in solid seed fats.

Fat	Saturated acids % (mol.)	Tri-saturated glycerides	
		Found	Calc. ("random")
Borneo tallow	62,8	4,5	24,8
Cacao butter	59,8	2,5	21,4
Allanblackia fat.....................	55,6	1,5	17,2
Piqui-a fat	53,1	2,5	15,0
Palm oil	52,4	6,2	14,4
Garcinia fat	50,3	2,7	12,7
Shea butter	45,1	2,5	9,2
Mowrah fat........................	43,4	1,2	8,2

It is self-evident, of course, that, whether "evenly distributed" or not, fats which contain very large proportions (80 per cent. or more) either of saturated or unsaturated acids are bound to possess respectively tri-saturated or tri-unsaturated glyceride contents which automatically approximate more and more to the figures calculated from probability considerations.

(b) *Fats in which oleic, linoleic and/or linolenic acid predominate.*

In this group again, recent studies of olive (55), cottonseed (54), groundnut (38), lime seed and sweet orange seed (35), soya bean (47), linseed (105, 106), and other fatty oils have uniformly indicated that tri-unsaturated glycerides are present in proportions tending towards the minimum, consistent with combination of all the (minor) saturated acids present as mono-saturated-di-unsaturated glycerides (occasionally with subsidiary proportions of di-saturated-mono-unsaturated glycerides, as in cottonseed and lime seed oils). Even when one unsaturated acid is present in large proportions, no great amount of the corresponding simple triglyceride has been observed; for instance, 30 per cent. and about 10 per cent. respectively of triolein in olive and neat's foot oils, the fatty acids of which contain respectively 77 per cent. and 64 per cent. of oleic acid. Similarly, trilinolein has not been detected in cottonseed or soya bean oils, the acids of which contain respectively 48 per cent. and 51 per cent. of linoleic acid; and trilinolenin is present, if at all, only in extremely small proportions in linseed oil or conophor oil, the acids of which contain respectively 55 per cent. and 65 per cent. of linolenic acid.

When more than one unsaturated acid is present, the proportion of tri-unsaturated glycerides may sometimes be fairly large owing to the operation of the "rule of even distribution" as stated. A major component acid (e. g., oleic) may be present in such proportions (60 per cent. or more) that most of the triglyceride molecules will contain two radicals of this acid, so that when one of the minor component unsaturated acids is also present, a mixed tri-unsaturated glyceride results. This has been observed in all the vegetable oils recently studied by the low temperature crystallisation technique in our laboratory. It is interesting to note that RIEMENSCHNEIDER, LUDDY, SWAIN and AULT (95), who applied the same technique to the resolution into seven fractions each of specimens of tallows and lards, reported a much higher content of tri-unsaturated glycerides in a lard than had hitherto been observed in pig fats; but their data show that the fat examined (containing 50 per cent. oleic acid) was unusually rich in linoleic acid (12,5 per cent.), so that the presence of the 18 per cent. of tri-unsaturated glycerides which they observed is readily accountable as linoleo-dioleins, in accordance with the above generalisation.

2. Computation of component glycerides from the proportions of the component fatty acids in a fat.

It was mentioned in the earlier paper (46) referred to (p. 73) that in simple cases the observed proportions of component glycerides could be reproduced by dividing the oleic acid content of a fat in arithmetical proportion to the palmitic and stearic acid contents, and then combining each portion with the latter so as to give arbitrarily calculated mixtures of monopalmito-dioleins and dipalmito-monooleins, monostearo-dioleins and distearo-monooleins. Later this method of computation was extended by HILDITCH and MEARA (58) to a number of other fats of both predominantly saturated and predominantly unsaturated types; and it was found that, providing the total fatty acids could be treated as comprising not more than three (or at most four) major components, good agreement with the observed glyceride composition followed so long as the acid arithmetically partitioned amongst the remaining acids was the *major component unsaturated acid* (oleic or linoleic, as the case might be) in the fat. This method of computation is in effect the arithmetical application of the principle of "even distribution" as defined earlier (p. 72), but is otherwise empirical in character. Nevertheless, it has consistently given results in general accordance with the figures deduced by experimental analysis, and appears to be capable of directly predicting the approximate proportions of at least the chief component glycerides of a fat from its fatty acid composition.

When more than four major component acids occur in a fat, or when, of four, two unsaturated acids predominante in approximately equal proportions, this arithmetical computation, however, fails to give figures in accordance with the experimental findings.

3. Fats with four or more component acids in similar proportions.

When the number of major component acids increases, it becomes increasingly difficult to follow the composition of the many mixed glycerides which will result from the principle of "even distribution" as defined in this paper. The components of a few fats and fatty oils of this kind have been studied experimentally. In certain seed fats, for example, instances are encountered in which four or five acids can be considered as major components (Table 8).

Table 8. Fats with four or five major component acids.

| Fat | Component fatty acids % | | | | | | | | |
| | Saturated | | | | | Unsaturated | | | |
	C_{14}	C_{16}	C_{18}	C_{20}	C_{22}	C_{16}	Oleic	Lin-oleic	C_{22}
Kepayang oil (62)	0,6	38,2	8,7	0,6	—	2,9	25,5	23,5	—
Mimusops elangi fat (86) .	—	11,0	10,1	—	0,4	—	64,0	14,5	—
Lophira alata fat (59)	1,9	27,1	—	—	16,5	1,5	14,5	33,3	5,2

In marine animal oils the number of major component acids is much greater, and probably amounts to at least seven or eight, each of which may contribute from about 12 per cent. upwards to the total fatty acids of the oils. Thus in herring (11), seal (67) and whale (57) oil, each of which has been studied by the low temperature crystallisation procedure, the component acids are as listed in Table 9 (the unsaturated C_{20} and C_{22} acids containing notable proportions of a mono-ethenoid and two poly-ethenoid acids).

Table 9. Component fatty acids of herring, seal, and whale oils.

| Fatty acid | Component fatty acids % | | |
	Herring oil	Seal oil	Whale oil
Saturated:			
Myristic.........................	7,1	4,8	9,2
Palmitic.........................	11,7	10,7	15,6
Stearic.........................	0,9	2,0	2,5
Unsaturated C_{14}.....................	1,2	2,7	2,5
,, C_{16}.....................	11,8	13,2	13,9
,, C_{18}.....................	19,6	35,9	37,2
,, C_{20}.....................	25,9	12,8	12,0
,, C_{22}.....................	21,8	17,9	7,1

The experimental results with all these six fatty oils indicate mixtures of mixed triglycerides of increasing complexity, as would be expected by the operation of the same principles which govern the assembly of smaller numbers of major component acids into mixed glycerides. Moreover, it is evident that the greater the number of component acids the more nearly will the final state of affairs come to resemble a merely random distribution of the acids amongst the glyceride molecules. That this is indeed what happens is indicated by Table 10, in which the observed proportions of the glyceride types "S_3", "S_2U", "SU_2" and "U_3" are compared with those calculated by probability considerations from the total proportions of saturated and unsaturated acids.

Table 10. Glyceride constituents (% mol.) of fatty oils containing more than four major component acids.

Glyceride	Kepayang		M. elangi		L. alata	
	Found	Calc.	Found	Calc.	Found	Calc.
Tri-saturated (S_3).....................	3	11	4	1	—	9
Di-saturated-mono-unsaturated (S_2U) ...	60	36	14	11	37	34
Mono-saturated-di-unsaturated (SU_2)....	24	39	?	48	62	41
Tri-unsaturated (U_3)..................	13	14	?	40	1	16

	Seal		Herring		Whale	
	Found	Calc.	Found	Calc.	Found	Calc.
Tri-saturated (S_3).....................	—	1	—	1	2	3
Di-saturated-mono-unsaturated (S_2U) ...	4	9	6	12	17	19
Mono-saturated-di-unsaturated (SU_2)....	61	37	45	41	50	44
Tri-unsaturated (U_3)..................	35	53	49	46	31	34

The three seed fats in Table 10, which contain four major component acids, are compared with the three marine animal fats in which the number of major component acids is much larger. The calculated figures, based on "random" distribution, for the three less complex fats show no marked resemblance to the values actually observed for the tri-saturated, etc., glycerides. However, with the more complex fatty acid mixtures present in the fatty oils of aquatic origin, the accordance between the observed and calculated figures becomes much closer, and in the cases of herring or whale oils is especially remarkable.

4. Animal body fats rich in stearic acid and animal milk fats containing notable proportions of butyric and/or other lower saturated acids.

(a) Animal body fats.

In animal depot fats in which large proportions (15–30 per cent.) of stearic acid are present, the proportion of tri-saturated glycerides is much greater than that which would be expected from the general principle of "even distribution". Depot fats of land animals, almost without exception, contain 30 ± 3 per cent. (mol.) of palmitic acid [BANKS and HILDITCH (6); HILDITCH and LONGENECKER (50)]. When the content of stearic acid is low, the proportion of tri-saturated glycerides is negligible and the usual principle of "even distribution" is closely followed. When the stearic acid content increases notably, the proportion of oleic acid falls to a corresponding extent, and the proportion of tri-saturated glycerides then also increases abnormally.

This feature, and consideration of the progressive increase, with augmenting stearic acid, not only in tri-saturated glycerides but also in oleo-disaturated glycerides (especially oleo-palmito-stearins) accompanied by corresponding diminution in palmito-dioleins (*42, 65, 71, 85*) has led us to conclude that a preformed, "evenly distributed" mixture of palmito-oleo-glycerides has become *hydrogenated* to varying extents in different animals. The effect of indiscriminate or "random" addition of hydrogen molecules to part of an "evenly distributed" palmito-oleo-glyceride mixture causes the proportion of tri-saturated glycerides to resemble that which might be expected by considerations of probability. When the amount of palmitic and stearic acids in the tri-saturated glycerides are allowed for, it is found that the same method of computation (*58*) which permits the prediction of the component glycerides of a fat from the proportions of its component acids, applied to the mixed saturated-unsaturated part of a stearic-rich animal depot fat, gives figures in close accordance with the experimental findings (*65, 71, 85*). It is interesting, on the other hand, to examine the accordance between the observed values for tri-saturated, tri-unsaturated glycerides, etc., and those calculated according to mathematical probability, in the instances of six body fats of this type which have been intensively studied at Liverpool (Table 11, p. 92).

As the stearic acid content increases, the observed proportions of tri-saturated glycerides approach more and more towards the mathematically calculated data, as pointed out by BANKS, BHATTACHARYA and HILDITCH (6, 9), and more recently by LONGENECKER (89) as well as by NORRIS and MATTIL (94). But the use of the more refined crystallisation procedures now permits at least approximate corresponding figures to be given for the tri-unsaturated and di-unsaturated-mono-saturated groups of glycerides: and here it will be seen that there is wide divergence from the "random" calculated values except in the last two instances, which are fats of extremely high stearic acid content. Any resemblances to a glyceride structure which follows the simple law of probability therefore appear to be only partial and very largely fortuitous. Indeed, inspection of Table 11 (p. 92) as a whole only reinforces

Table 11. Component glycerides and acids of pig, sheep, ox and cow body fats.

Body fat	Component acids % (mol.)			Component glycerides % (mol.)			
	Satd.	Unsatd.		Tri-saturated	Di-saturated-mono-unsaturated	Mono-saturated-di-unsaturated	Tri-unsaturated
Pig (back)	44,1	55,9	Found	5	32	60	3
			Calc.	8,5	33	41	17,5
Sheep (back)	50,2	49,8	Found	5	42	53	0
			Calc.	13	38	37	12
Pig (perinephric)	50,5	49,5	Found	9	43	45	3
			Calc.	13	38	37	12
Sheep (perinephric)	57,2	42,8	Found	14	48	38	0
			Calc.	19	42	31	8
Ox (English)	58,7	41,3	Found	17	49	34	0
			Calc.	20	43	30	7
Cow (Indian)	67,5	32,5	Found	28	51	20	0
			Calc.	31	44,5	21	3,5

the view which has been reached from other considerations, namely, the progressive changes in the chemical constitution of the glycerides as oleic acid is replaced by stearic acid in the component acids of the fats.

(b) Animal milk fats.

The milk fats of animals which elaborate important proportions of lower saturated fatty acids in their milk glycerides exhibit very similar aspects in their glyceride structure to the stearic-rich animal body fats just· discussed. Tri-saturated glycerides are present in much greater quantities than usual, and, whilst palmitic acid forms a more or less constant proportion (25–28 per cent.) of the total fatty acids, the oleic acid content varies inversely with the proportion of the lower saturated acids.

It is now generally accepted that milk fats are produced from the glycerides present in the blood passing through the mammary gland; and we have suggested (81, 82, 84) that the characteristic milk fat glycerides are produced from preformed "evenly distributed" palmito-oleo-glycerides by *oxidation-reduction changes* which involve the progressive shortening of oleic groups (in the form of glycerides) by two carbon atoms at a time.

This proceeds to different degrees in different animal species: in the ruminants it goes as far as the production of the C_4 (butyro-) group, in the human species [HILDITCH and MEARA (60, 61); BALDWIN and LONGENECKER (4)] it does not go beyond the C_{10} (deceno-) group; and in other animals (e. g., the pig or the whale) the milk fats differ little in composition from the body fats [DE LA MARE and SHORLAND (32); KLEM (88)]. Suggestive evidence which supports our explanation of the glyceride structure of milk fats lies in the concurrent production therein of very small proportions of mono-ethenoid C_{14}, C_{12} and C_{10} acids, in each of which the ethenoid bond is in the same (9 : 10) position, relative to the carboxyl group as in oleic acid [SMEDLEY (99); HILDITCH and PAUL (68); HILDITCH and LONGENECKER (51)]; and in the observation that the character of cow milk fat reverts towards that of a (comparatively unsaturated) body fat when the metabolism of the animal is disturbed by starvation [SMITH and DASTUR (100)] or by ketosis [Shaw et al. (97, 98)].

Detailed studies of cow milk fat glycerides by the use of crystallisation from solvents at low temperatures are not yet available, but preliminary studies have been reported by HENDERSON and JACK (40), whilst HILDITCH and PAUL (70) effected some partial separation by crystallisation from acetone at 0° and at − 20° C. The results of the latter workers indicate a complex mixture of mixed glycerides in which the acyl groups of the lower fatty acids were usually present in association with palmitic and/or oleic groups. Glycerides containing one palmito-group were present in much the same proportion (65 per cent.) as in the corresponding body fats. The more detailed data for human milk fat glycerides as observed by HILDITCH and MEARA (61) were of a similar nature, with monopalmito-glycerides forming about 60 per cent. of the whole. As in the case of the animal body fats, we may compare the observed values for tri-saturated, etc., glycerides with those calculated on the basis of "random distribution" (Table 12).

Table 12. Component glycerides and acids of milk fats.

	Cow milk fat		Human milk fat	
	Found	Calc.	Found	Calc.
Component acids:				
Saturated..............................	59,1	—	49,8	—
Unsaturated...........................	40,9	—	50,2	—
Component glycerides:				
Tri-saturated	19	21	9	12
Di-saturated-mono-unsaturated	57	43	43	37
Mono-saturated-di-unsaturated	20	29	40	38
Tri-unsaturated.......................	4	7	8	13

Resemblance between observed and calculated values is again moderately evident in the tri-saturated glycerides but much less so in the remaining groups. In earlier work (82, 84), the tri-saturated glyceride content of a number of cow milk fats had been determined; and a similar comparison of the observed values with those calculable from probability considerations is as follows (Table 13):

Table 13. Tri-saturated glycerides in cow milk fats.

Component acids % (mol.)		Tri-saturated glycerides	
Saturated	Unsaturated	Found	Calc.
61,9	38,1	27,2	23,7
66,1	33,9	31,5	28,9
67,3	32,7	33,8	30,5
70,2	29,8	39,6	34,6
72,4	27,6	41,3	37,9

Here there is again considerable superficial resemblance, but it is noticeable that the observed figures consistently exceed those calculated on the "random" basis, whereas the reverse was the case with the tri-saturated glycerides of animal body fat (Table 11). As already stated, such apparent resemblances appear to be fortuitous rather than connected with any general rule or mechanism governing the assemblage of fatty acids into mixed triglycerides.

5. Synthesis and transformation by chemical means of mixed glycerides in the laboratory.

Glycerol and fatty acids can be esterified to produce triglycerides by heating them together in a vacuum at 200° C. or above alone, or at somewhat lower temperatures in presence of traces of a catalyst such as an aromatic sulphonic acid. Mixed glycerides can also be rearranged or "inter-esterified" by heating them at above 200° C. with a small proportion of other catalysts such as sodium alkyloxide or a suitable tin salt.

The products obtained in these high-temperature processes are constituted in a manner which closely approaches "random" distribution as determined by probability considerations. Thus, BHATTACHARYA and HILDITCH showed in 1930 that esterification of various mixtures of saturated acids and oleic acids with glycerol (9) or with glycol (10) gave mixed esters in which the proportions of fully saturated and completely unsaturated esters approached closely to that calculated on a "random" distribution basis. Recently, NORRIS and MATTIL (93, 94) have demonstrated the corollary that an "evenly distributed" type of fat such as lard, soya bean oil or cottonseed oil undergoes rearrangement when heated at 225° C. in presence of stannous hydroxide; and that the resulting product has been altered in that its mixed glycerides approach much more closely to the composition indicated by "random" distribution.

Clearly, all of these investigations show that the most stable form of mixed glycerides produced at high temperatures tends towards "random" distribution; but it appears to the writer that this has no connection with or bearing upon the constitution of mixed glycerides as synthesised by enzyme action in the living cell at atmospheric temperature. Thermal syntheses or ester-interchange at high temperatures cannot usefully be employed to interpret the structure of the products of enzyme action at low temperatures.

To the writer it would appear that glyceride structure in the whole series of natural fats can at present best be interpreted on a consistent and logical basis by the considerations which have been discussed above, commencing from the standpoint that the fundamental tendency is the formation of mixed glycerides, in accordance with the principles enunciated in the introductory chapter of this monograph (p. 72). It is hoped that the description of the more recently available and increasingly useful techniques will have aided in illustrating the trends of recent work in this field.

References.

1. ACHAYA, K. T. and B. N. BANERJEE: Characteristics of Indian animal fats. Current Sci. **15**, 23 (1946).

2. ACHAYA, K. T. and T. P. HILDITCH: Indian cow and buffalo milk fats. (In preparation.) (1948.)

3. American Oil Chemists' Society (Spectroscopy Committee): Report, J. Amer. Oil Chem. Soc. **25**, 14 (1948).

4. BALDWIN, A. R. and H. E. LONGENECKER: Component fatty acids of early and mature human milk fat. J. biol. Chemistry **154**, 255 (1944).

5. BALIGA, M. N. and T. P. HILDITCH: The component acids of rape seed oils. J. Soc. chem. Ind. (in press) (1948).

6. BANKS, A. and T. P. HILDITCH: (*a*) The glyceride structure of beef tallows. Biochemic. J. **25**, 1168 (1931); (*b*) The body fats of the pig. II. Biochemic. J. **26**, 298 (1932).

7. BARKER, C. and T. P. HILDITCH: Component acids of sunflower seed oils. (In preparation.) (1948.)

8. BEADLE, B. W.: Applied ultra-violet spectrophotometry of fats and oils. Oil and Soap **23**, 140 (1946).

9. BHATTACHARYA, R. and T. P. HILDITCH: Structure of synthetic mixed glycerides. Proc. Roy. Soc. [London], Ser. A, **129**, 468 (1930).

10. — — Distribution of saturated and unsaturated higher fatty acids in mixed synthetic glycol esters. J. chem. Soc. [London] 901 (1931).

11. BJARNASON, O. B. and M. L. MEARA: Low temperature crystallisation of Icelandic herring oil. J. Soc. chem. Ind. **63**, 61 (1944).

12. BÖMER, A. u. J. BAUMANN: Beiträge zur Kenntnis der Glyceride der Fette und Öle. IX. Die Glyceride des Cocosfettes. Z. Unters. Nahrungs- u. Genußm. **40**, 97 (1920).

13. BÖMER, A. u. K. SCHNEIDER: Beiträge zur Kenntnis der Glyceride der Fette und Öle. XI. Die Glyceride des Palmkernfettes. Z. Unters. Nahrungs- u. Genußm. **47**, 61 (1924).

14. BROWN, J. B.: Low-temperature crystallisation of the fatty acids and glycerides. Chem. Reviews **29**, 333 (1941).

15. BROWN, J. B. and J. FRANKEL: Isolation of pure linoleic acid by crystallisation. J. Amer. chem. Soc. **63**, 1483 (1941).

16. BROWN, J. B. and G. Y. SHINOWARA: Preparation of pure oleic acid by a simplified method. J. Amer. chem. Soc. **59**, 6 (1937).

17. — — Purification of linolenic acid by fractional crystallisation of the fatty acids of linseed and perilla oils. J. Amer. chem. Soc. **60**, 2734 (1938).

18. BROWN, J. B. and G. G. STONER: Purification of linoleic acid by crystallisation methods. J. Amer. chem. Soc. **59**, 3 (1937).

19. BULL, W. C. and D. H. WHEELER: Low-temperature solvent crystallisation of soya bean oil and soya bean oil fatty acids. Oil and Soap **20**, 137 (1943).

20. BUSHELL, W. J. and T. P. HILDITCH: Fatty acids and glycerides of Borneo tallow. J. Soc. chem. Ind. **57**, 447 (1938).

21. — — Fatty acids and glycerides of Dika fat. J. Soc. chem. Ind. **58**, 24 (1939).

22. CHAVES, J. M. and E. PECHNIK: The glycerides of bacury fat (*Platonia insignis* MART.). Rev. quim. ind. **15**, No. 165, 5 (1946).

23. COLLIN, G.: Dika fat (*Irvingia* butter). J. Soc. chem. Ind. **49**, 138 T (1930).

24. — Some peculiarities in the glyceride structure of laurel fats. Biochemic. J. **25**, 95 (1931).

25. — The kernel fats of some members of the Palmae. Biochemic. J. **27**, 1366 (1933).

26. COLLIN, G.: The fatty acid and glyceride structure of the seed fat of *Myristica malabarica*. J. Soc. chem. Ind. **52**, 100 T (1933).

27. COLLIN, G. and T. P. HILDITCH: The component glycerides of coconut and palm kernel fats. J. Soc. chem. Ind. **47**, 261 T (1928).

28. — — Regularities in the glyceride structure of vegetable seed fats. Biochemic. J. **23**, 1273 (1929).

29. — — The fatty acids of nutmeg butter and of expressed oil of laurel. J. Soc. chem. Ind. **49**, 141 T (1930).

30. CRAMER, D. L. and J. B. BROWN: The component fatty acids of human depot fat. J. biol. Chemistry **151**, 427 (1943).

31. DE LA MARE, P. B. D. and F. B. SHORLAND: A detailed analysis of the back fat of the pig. Analyst **69**, 337 (1944).

32. — — The fat of sow's milk. Nature [London] **153**, 380 (1944).

33. DHINGRA, D. R., G. L. SETH and P. C. SPEERS: The study of some Indian seed fats. J. Soc. chem. Ind. **52**, 116 T (1933).

34. DUNN, H. C. and T. P. HILDITCH: Component acids of niger seed oil. (In preparation.) (1948).

35. DUNN, H. C., T. P. HILDITCH and J. P. RILEY: The composition of seed fats of West Indian *Citrus* fruits. J. Soc. chem. Ind. **67**, 199 (1948).

36. FOREMAN, H. D. and J. B. BROWN: Solubilities of the fatty acids in organic solvents at low temperatures. Oil and Soap **21**, 183 (1944).

37. GREEN, T. G. and T. P. HILDITCH: Fatty acids and glycerides of shea butter. J. Soc. chem. Ind. **57**, 49 (1938).

38. GUNDE, B. G. and T. P. HILDITCH: The mixed unsaturated glycerides of liquid seed fats. I. Some "non-drying" oils. J. Soc. chem. Ind. **59**, 47 (1940).

39. GUNSTONE, F. D. and T. P. HILDITCH: Use of low-temperature crystallisation in the determination of component acids of liquid fats. II. Fats which contain linolenic as well as linoleic and oleic acids. J. Soc. chem. Ind. **65**, 8 (1946).

40. HENDERSON, J. L. and E. L. JACK: The fatty acid components of glyceride fractions separated from milk fat. J. Dairy Sci. **28**, 65 (1945).

41. HILDITCH, T. P.: The Chemical Constitution of Natural Fats. 2nd ed., p. 16. London (Chapman & Hall) and New York (P. Wiley). 1947.

42. — Reference *41*, pp. 300–305 and 320–322.

43. — Reference *41*, pp. 300–304, 306–310 and 332–336.

44. — Reference *41*, pp. 468–470.

45. — Reference *41*, pp. 474–483.

46. — The component glycerides of vegetable fats. Fortschr. Chem. organ. Naturstoffe **1**, 24 (1938).

47. HILDITCH, T. P., J. HOLMBERG and M. L. MEARA: The component glycerides of soya bean oil, and of soya bean oil fractions. J. Amer. Oil Chem. Soc. **24**, 321 (1947).

48. HILDITCH, T. P. and M. B. ICHAPORIA: Fatty acids and glycerides of Mowrah fat. J. Soc. chem. Ind. **57**, 44 (1938).

49. HILDITCH, T. P. and (Miss) E. E. JONES: The composition of commercial palm oils. I. The fatty acids and component glycerides of some palm oils of low free acidity. J. Soc. chem. Ind. **49**, 363 T (1930).

50. HILDITCH, T. P. and H. E. LONGENECKER: A further study of the component acids of ox depot fat, with specal reference to certain minor constituents. Biochemic. J. **31**, 1805 (1937).

51. — — Further determination and characterisation of the component acids of butter fat. J. biol. Chemistry **122**, 497 (1938).

52. HILDITCH, T. P. and J. A. LOVERN: The evolution of natural fats: a general survey. Nature [London] 137, 478 (1936).

53. HILDITCH, T. P. and L. MADDISON: The composition of commercial palm oils. V. Partial separation of palm oils by crystallisation as an acid to the determination of the component glycerides. J. Soc. chem. Ind. 59, 67 (1940).

54. — — The mixed unsaturated glycerides of liquid seed fats. II. Low-temperature crystallisation of cottonseed oil. J. Soc. chem. Ind. 59, 162 (1940).

55. — — The mixed unsaturated glycerides of liquid seed fats. III. Low-temperature crystallisation of olive oil. J. Soc. chem. Ind. 60, 258 (1941).

56. — — The mixed unsaturated glycerides of liquid fats. IV. Low-temperature crystallisation of whale oil. J. Soc. chem. Ind. 61, 169 (1942).

57. — — The component acids and glycerides of whale oil. J. Soc. chem. Ind. 67 (in press) (1948).

58. HILDITCH, T. P. and M. L. MEARA: The approximate computation of mixed glycerides present in natural fats from the proportions of their component fatty acids. J. Soc. chem. Ind. 61, 117 (1942).

59. — — The fatty acids and glycerides of *Lophira Alata* kernel fat (Niam fat). J. Soc. chem. Ind. 63, 114 (1944).

60. — — Human milk fat. I. Component fatty acids. Biochemic. J. 38, 29 (1944).

61. — — Human milk fat. II. Component glycerides. Biochemic. J. 38, 437 (1944).

62. HILDITCH, T. P., M. L. MEARA and W. H. PEDELTY: The fatty acids and glycerides of the seed fat of *Hodgsonia capniocarpa*. J. Soc. chem. Ind. 58, 26 (1939).

63. HILDITCH, T. P., M. L. MEARA and O. A. ROELS: The composition of commercial palm oils. VI. The component acids and glycerides of a Belgian Congo plantation palm oil (studied with the aid of low-temperature crystalliation). J. Soc. chem. Ind. 66, 284 (1947).

64. HILDITCH, T. P., R. A. MORTON and J. P. RILEY: Spectrographic determination of linoleic, linolenic and elaeostearic acids. Analyst 70, 68 (1945).

65. HILDITCH, T. P. and K. S. MURTI: The component acids and glycerides of some Indian ox depot fats. Biochemic. J. 34, 1301 (1940).

66. — — Fatty acids and glycerides of seed fats of *Garcinia morella* and *Garcinia indica*. J. Soc. chem. Ind. 60, 16 (1941).

67. HILDITCH, T. P. and S. P. PATHAK: Use of low-temperature crystallisation in the determination of component acids of liquid fats. IV. Marine animal oils. The component acids and glycerides of a Grey (Atlantic) seal. J. Soc. chem. Ind. 66, 421 (1947).

68. HILDITCH, T. P. and H. PAUL: The occurrence and possible significance of some of the minor component acids of cow milk fat. Biochemic. J. 30, 1905 (1936).

69. HILDITCH, T. P. and S. PAUL: The component glycerides of an ox depot fat. Biochemic. J. 32, 1775 (1938).

70. — — The component glycerides of a typical cow milk fat. J. Soc. chem. Ind. 59, 138 (1940).

71. HILDITCH, T. P. and W. H. PEDELTY: Component glycerides of perinephric and outer back pig fats from the same animal. Biochemic. J. 34, 971 (1940).

72. HILDITCH, T. P. and J. PRIESTMAN: The component glycerides of Borneo (Illipé) tallow. J. Soc. chem. Ind. 49, 197 T (1930).

73. — — The component glycerides of *Stillingia* (Chinese vegetable) tallow. J. Soc. chem. Ind. 49, 397 T (1930).

74. HILDITCH, T. P. and J. P. RILEY: Use of low-temperature crystallisation in the determination of component acids of liquid fats. I. Fats in which oleic and linoleic acids are major components. J. Soc. chem. Ind. 64, 204 (1945).

75. — — Use of low-temperature crystallisation in the determination of component acids of liquid fats. III. Fats which contain elaeostearic as well as linoleic and oleic acids. J. Soc. chem. Ind. 65, 74 (1946).

76. HILDITCH, T. P. and S. A. SALETORE: Fatty acids and glycerides of solid seed fats. I. Composition of the seed fats of *Allanblackia Stuhlmannii, Pentadesma butyracea, Butyrospermum parkii* (shea), and *Vateria indica* (dhupa). J. Soc. chem. Ind. 50, 468 T (1931).

77. HILDITCH, T. P. and A. J. SEAVELL: Component acids and glyceride of linseed, conophor, rubberseed and other linolenic-rich drying oils. (In preparation.) (1948.)

78. HILDITCH, T. P. and R. K. SHRIVASTAVA: The spectrophotometric determination of small proportions of linolenic acid in fats. Analyst 72, 527 (1947).

79. — — The component acids and glycerides of neat's foot oil. J. Soc. chem. Ind. 67, 139 (1948).

80. — — The component glycerides of an Indian sheep body fat. J. Amer. Oil Chem. Soc. 25, (in press.) (1948).

81. HILDITCH, T. P. and J. J. SLEIGHTHOLME: Variations in the component fatty acids of butter due to changes in seasonal and feeding conditions. Biochemic. J. 24, 1098 (1930).

82. — — The glyceride structure of butter fats. Biochemic. J. 25, 507 (1931).

83. HILDITCH, T. P. and W. J. STAINSBY: The component glycerides of cacao butter. J. Soc. chem. Ind. 55, 95 T (1936).

84. HILDITCH, T. P. and H. M. THOMPSON: The effect of certain ingested fatty acids upon the composition of cow milk fat. Biochemic. J. 30, 677 (1936).

85. HILDITCH, T. P. and Y. A. H. ZAKY: Sheep body fats. II. Component glycerides of perinephric and external tissue fats from the same animal. Biochemic. J. 35, 940 (1941).

86. KARTHA, A. R. S. and K. N. MENON: Oil of *Mimusops elangi* LINN. Proc. Indian Acad. Sci., Sect. A, 19, 1 (1944).

87. KASS, J. P. and G. O. BURR: ψ-Elaeostearic acid. J. Amer. chem. Soc. 61, 3292 (1939).

88. KLEM, A.: Studies in the biochemistry of whale oils. 3. The milk fat of the blue whale. Hvalrådets Skrifter (Universitets Biologische Laboratorium, Oslo) No. 11, 56 (1935).

89. LONGENECKER, H. E.: Composition and structural characteristics of glycerides in relation to classification and environment. Chem. Reviews 29, 201 (1941).

90. MEARA, M. L. and Y. A. H. ZAKY: The fatty acids and glycerides of the seed fats of *Allanblackia floribunda* and *Allanblackia parviflora*. J. Soc. chem. Ind. 59, 25 (1940).

91. MITCHELL, J. H. jun., A. R. KRAYBILL and F. P. ZSCHEILE: Quantitative spectral analysis of fats. Ind. Eng. Chem., Analyt. Edit. 15, 1 (1943).

92. MOORE, T.: Spectroscopic changes in fatty acids. The formation of liquid and solid "absorptive" acids from the mixed acids of linseed oil. Biochemic. J. 31, 138 (1937).

93. NORRIS, F. A. and K. F. MATTIL: Inter-esterification reactions of tri-glycerides. Oil and Soap 23, 289 (1946).

94. — — A new approach to the glyceride structure of natural fats. J. Amer. Oil Chem. Soc. 24, 274 (1947).

7*

95. RIEMENSCHNEIDER, R. W., F. E. LUDDY, M. L. SWAIN and W. C. AULT: Fractionation of lard and tallow by systematic crystallisation. Oil and Soap 23, 276 (1946).

96. RUSOFF, I. I., R. T. HOLMAN and G. O. BURR: A table of spectroscopic data on fats, fatty acids and their esters. Oil and Soap 22, 290 (1945).

97. SHAW, J. C.: The effect of ketosis and glucose therapy in ketosis upon milk fat synthesis. J. Dairy Sci. 24, 502 (1941).

98. SHAW, J. C., R. C. POWELL and C. B. KNODT: The fat metabolism of mammary glands of the normal cow and of the cow in ketosis. J. Dairy Sci. 25, 909 (1942).

99. SMEDLEY, I.: The fatty acids of butter. Biochemic. J. 6, 451 (1912).

100. SMITH, J. A. B. and W. N. DASTUR: Studies in the secretion of milk fats. II. The effect of inanition on the yield and composition of milk fat. Biochemic. J. 32, 1868 (1938).

101. SMITH, J. A. B.: Biennial reviews of the progress of dairy science. J. Dairy Res. 12, 94 (1941).

101a. THOMAS, K. und G. WEITZEL: Fett und Fettstoffwechsel. Fiat Rev. deutsch, Wiss., Biochemie, I, 1 (1947).

102. TSUJIMOTO, M.: A new method for the separation of the highly unsaturated fatty acids in fish oils. J. Soc. chem. Ind. Japan, 23, 272 (1920).

103. TWITCHELL, E.: Precipitation of solid fatty acids with lead acetate in alcoholic solution. J. Ind. Eng. Chem. 13, 806 (1921).

104. VIDYARTHI, N. L. and C. J. D. RAO: Fatty acids and glycerides of the fat from the seeds of Garcinia Indica (kokum butter). J. Indian chem. Soc. 16, 437 (1939).

105. WALKER, F. T. and M. R. MILLS: The component glycerides of linseed oil. Segregation by chromatography. J. Soc. chem. Ind. 61, 125 (1942).

106. — — The component glycerides of linseed oil. Segregation by chromatography. II. J. Soc. chem. Ind. 62, 106 (1943).

(Received, April 20, 1948.)

Enzymatically Synthesized Polysaccharides and Disaccharides.

By **W. Z. Hassid** and **M. Doudoroff**, Berkeley, California.

With 8 Figures.

Introduction.

It is known that enzymes in living cells are capable of synthesizing complex carbohydrates from monosaccharides such as glucose or fructose. Numerous polysaccharides among which cellulose, pectin or gums and oligosaccharides which include sucrose, trehalose, raffinose are thus produced by plants; glycogen, mucoproteins and lactose are synthesized by animals. A great number and variety of diverse polysaccharides are also synthesized by bacteria and fungi. It is of interest that the bacterium, *Acetobacter xylinum*, when grown with hexoses, sugar alcohols, or glycerol produces a polysaccharide consisting of D-glucose joined by $1,4$-β-glucosidic linkages identical with plant cellulose (2). Various species of *Leuconostoc* also produce polysaccharides consisting of D-glucose units, but these differ structurally from cellulose in that the glucose units are mutually joined through $1,6$-α-glucosidic linkages (26, 73, 20, 37).

Levans, which are condensation products of D-fructose are produced by the common aerobic spore-formers of the *B. mesentericus* and *B. subtilis* types (56, 70). These polysaccharides, in which the fructofuranose units are joined through $2,6$-glycosidic linkages, appear to be identical in structure with a water-soluble levan found in grass (49, 7).

The polysaccharides of most bacteria which have been studied are composed of more than one carbohydrate. In addition to D-glucose, D-galactose, D-mannose, pentoses, amino sugars and acetylated sugars are among the constituents of the molecules. An interesting example of a mold polysaccharide produced by *Penicillium varians* is varianose (50, 78). This polysaccharide, in addition to D-galactose and D-glucose, contains the rare sugar, L-altrose.

Comparatively little is known about the mechanism of formation of these polysaccharides and of the naturally occurring disaccharides, such as lactose, trehalose and other oligosaccharides (raffinose, melizitose). Only within the last decade have some fundamental advances been made towards elucidating the problem. The synthesis of glycogen, starch, bacterial dextran and bacterial levan, sucrose, and several other disaccharides have now been achieved through enzymatic action *in vitro*.

The scope of this survey will be limited only to those carbohydrates of which the mechanism of formation is known or partially known.

Formation of Starch and Glycogen through the Phosphorolytic Condensation.

Although some of the minor details of the structure of starch and glycogen have not yet been worked out, the salient features of the structure of these polysaccharides are well known. It has now been established that naturally occurring starch is not homogeneous but can be separated into at least two fractions, amylose and amylopectin, possessing different physical properties and differing in their behavior toward the enzymes β-amylase and phosphorylase (*69, 45*). The chief distinguishing chemical difference between amylose and amylopectin lies in the fact that the former consists of unbranched chains of glucose residues joined only through α-glucosidic-1,4-linkages, while the latter is a branched molecule which contains, in addition to 1,4-linkages, α-glucosidic-1,6-linkages, at the points of branching. Glycogen resembles the amylopectin fraction of starch in its chemical constitution and mechanism of formation. Like amylopectin, the glycogen molecules are branched, the branches being attached through α-glucosidic-1,6-linkages (*48, 68*).

The most important structural feature common to starch and glycogen is the α-1,4-glucosidic linkage. This feature is responsible for the similarity in specific rotation which starches and glycogen or their derivatives exhibit when dissolved in the same solvents. The similarities in biochemical and physiological behavior of starch and glycogen may also be attributed to the fact that these polysaccharides possess this common structural characteristic. Glycogen is attacked by the same plant and animal amylases that attack starch, and, like starch, it is degraded to maltose and dextrins. Both glycogen and starch are broken down by animal and plant phosphorylases in the presence of inorganic phosphate with the production of glucose-1-phosphate.

The *mechanism of starch and glycogen formation* has been the subject of interest to biochemists for many years. Since the starch splitting amylases were known to occur in many different plants, it was at first believed that these enzymes might be responsible for the interconversion of starch into sugar in the living plant. A similar mechanism for glycogen breakdown and synthesis was assumed to occur in animal cells. The fact that maltose and dextrins, the main products of amylase activity in vitro, are not found in plants, necessitated the assumption that the kinetics of the amylase in the cell are of a different nature from the kinetics in vitro.

CORI and CORI's (9, 10) discovery of glucose-1-phosphate and the isolation of the enzyme phosphorylase paved the way for the study of the mechanism of glycogen and starch formation. These authors demonstrated that phosphorylase, an enzyme of an entirely different nature than amylase, is involved in the breakdown of glycogen (29). The phosphorylase found in animal tissues acts upon glycogen in the presence of inorganic phosphate to produce α-D-glucopyranose-1-phosphate (CORI ester). Due to the reversible nature of the phosphorolytic decomposition, the phosphorylases present in yeast, muscle and liver act on glucose-1-phosphate, producing polysaccharide (62, 15, 8). HANES (31, 32) has found that higher plants such as peas and potatoes contain phosphorylase which catalyzes the reversible reaction:

$$\text{starch} + \text{inorganic phosphate} \rightleftharpoons \text{glucose-1-phosphate.}$$

A phosphorylase capable of acting on glucose-1-phosphate to produce a starch-like polysaccharide which stains blue with iodine was recently shown by HEHRE et al. (54) to exist in bacteria, C. diphtheriae and some streptococci. This indicates that some bacterial polysaccharides are formed through phosphorolytic condensation similar to that found in animals and higher plants.

The formation of glucose-1-phosphate from starch and inorganic phosphate is considered to occur as the result of phosphorolytic cleavage of successive terminal glucose units from the chains of the starch molecule, the starch molecule being decomposed without the entrance of water into the reaction. The reverse reaction, viz. the formation of polysaccharide from the glucose-1-phosphate, is the result of successive "de-phosphorolytic" condensations and may be written as follows:

$$(n) \text{ glucose-1-phosphate}^{--} \rightarrow (n) \text{ } HPO_4^{--} + \text{polysaccharide.}$$

The process whereby starch is broken down to glucose with the simultaneous entrance of inorganic phosphate at the maltosidic linkages is described by PARNAS (72) as "phosphorolysis" in contradistinction to "hydrolysis" where water is involved. A diagrammatic representation showing the entrance of inorganic phosphate at the α-glucosidic linkages resulting in formation of the glucose-1-phosphate from glycogen, as formulated by CORI, COLOWICK, and CORI (14), is given in Fig. 1. This diagram is also applicable to starch (31).

Glycogen and starch are synthesized from glucose through the utilization of energy-rich phosphate bonds from compounds, such as adenosine triphosphate. In this reaction energy has to be expended because the glucosidic residue in glycogen is on a higher energy level than free glucose; and the glucosidic bond energy is approximately equal to the difference in energy between free and ester-bound phosphate (65, 69a). The mechanism apparently used by the cell for the synthesis of polysaccharide

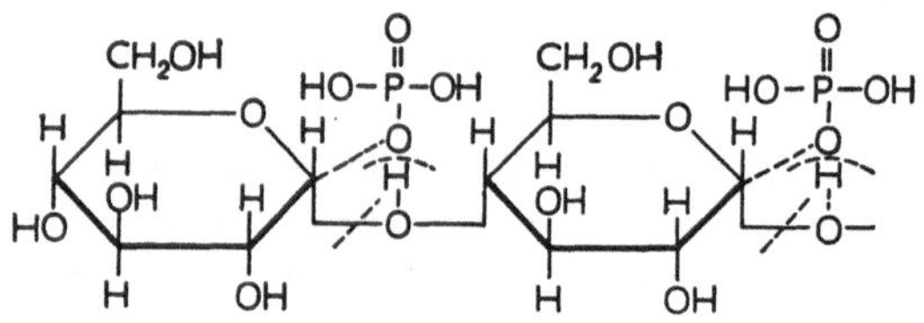

Fig. 1. Phosphorolysis of starch or glycogen.

is, first to combine glucose with phosphate, forming glucose-1-phosphate, by using the energy drop from *adenyl*~*ph* to glucose ester-ph. (Symbol ~*ph* is assigned to the energy-rich bond containing approximately 11 000 calories and −*ph* to the ordinary ester bond having an average energy of 400 calories.) The ester bond of the glucose-1-phosphate is subsequently exchanged for a glucosidic bond on the same carbon atom, thus building up the polysaccharide, as shown in the following sequence of reactions:

Adenyl~ph + glucose
↓
Glucose-6-ph
↓↑
Glucose-1-ph
↓↑
Glycogen or starch + inorganic phosphate

The first step is accompanied with a loss of 8000 calories, because of the conversion of energy-rich phosphate into the ordinary ester phosphate; and it is, therefore, essentially irreversible. The subsequent steps are all readily reversible, involving only small energy changes. The reaction which requires the greatest amount of energy is the phosphorylation of glucose to glucose-1-phosphate. When glucose-1-phosphate is formed, it is transformed into polysaccharide practically without any expenditure of energy (65, 60).

Conditions Affecting Polysaccharide Formation.

The equation for the formation of starch or glycogen from glucose-1-phosphate suggests that two or more glucose-1-phosphate molecules react with the enzyme, starting an α-glucosidic chain which grows by further additions of glucose-1-phosphate. Actually, no reaction takes place when highly purified enzyme (crystalline muscle *phosphorylase* or pu rified potato *phosphorylase*) and synthetic glucose-1-phosphate are used, unless a priming agent, such as starch, glycogen, or dextrin, is added. Crude potato phosphorylase and glucose-1-phosphate prepared enzymatically from starch usually both contain sufficient dextrin impurities to start the reaction of polysaccharide formation. The rate of polysaccharide synthesis depends on the nature as well as the amount of

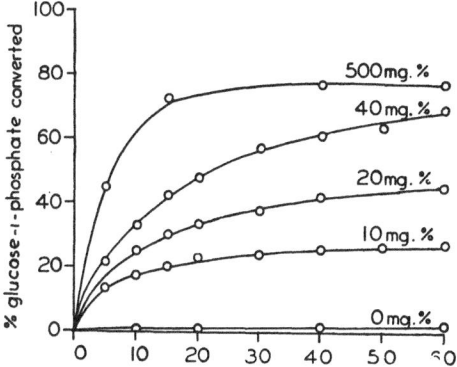

Fig. 2. Effect of glycogen concentration on the rate of reaction of glucose-1-phosphate.

activating polysaccharide added to the reaction mixture of glucose-1-phosphate and muscle phosphorylase.

The effect of glycogen concentration on the rate of reaction of glucose-1-phosphate is shown by CORI, SWANSON, and CORI (*19*) in Fig. 2. In the absence of priming agent no reaction occurs; when 10 mg. per cent. of glycogen is added, some glucose-1-phosphate is converted to polysaccharide but the reaction fails to reach equilibrium; with 40 mg. per cent. the reaction reaches equilibrium, but its rate falls off more rapidly than that of a first order reaction; when 500 mg. per cent. is added, the reaction proceeds at maximum rate and is kinetically of the first order. This kinetic behavior of polysaccharide synthesis is explained by CORI *et al.* as follows:

The catalytic polysaccharide (dextrin, glycogen or starch) is pictured as a central core or nucleus with a large number of branches or side chains, each averaging from 6 to 25 glucose units in length. These branches are extended by the addition of glucose units through repetition of the following process:

$$(C_6H_{10}O_5)_x + C_6H_{11}O_5 \cdot O \cdot PO_3H_2 \rightarrow (C_6H_{10}O_5)_{x+1} + H_3PO_4$$

activating   glucose-1-phosphate
polysaccharide

$$(C_6H_{10}O_5)_{x+1} + C_6H_{11}O_5 \cdot O \cdot PO_3H_2 \rightarrow (C_6H_{10}O_5)_{x+2} + H_3PO_4$$

$$(C_6H_{10}O_5)_{x+2} + C_6H_{11}O_5 \cdot O \cdot PO_3H_2 \rightarrow \ldots$$

When the chain length becomes a limiting factor the above reaction can no longer proceed. This explains why starch and glycogen, having many branches of relatively short chain length, possess strong activating power, while the linear polysaccharide, amylose, has little effect. It also accounts for the fact that the linear polysaccharide formed by the enzyme, phosphorylase, has no activating effect during the course of the reaction. In view of the evidence indicating that single chains of amylose or synthetic polysaccharide represent whole molecules (45, 47) it must be assumed that the large, synthetic polysaccharide macromolecule formed through the lengthening of the branched chain primary agent, breaks up into smaller single chain molecules.

The reversibility of the starch—glucose-1-phosphate reaction is shown by the fact that the ratio of inorganic phosphate to ester phosphate reaches the same value in either direction.

Experiments with potato phosphorylase show that the equilibrium state of the reaction, defined by the values of the ratio, inorganic phosphate/ester phosphate, is not significantly affected by wide alterations of the starch or the inorganic phosphate concentration. The position of the equilibrium is, however, markedly affected by a change in hydrogen ion concentration. This may be explained by the fact that the inorganic phosphate and the ester phosphate have different acid strengths, the latter being stronger. The H^+ concentration, therefore, enters into the true equilibrium equation. Under conditions of greater acidity, a larger proportion of starch and inorganic phosphate is formed. Thus, when the pH value is varied from 5,0 to 7,0, the values of the ratio, inorganic phosphate/ester phosphate decrease progressively from about 10,8 to 3,1 (33).

A similar hydrogen ion concentration effect was found by Cori and Cori (15) in the case of glycogen and animal phosphorylase. The ratio of concentration of inorganic phosphate to glucose-1-phosphate, reached from either side, is 5,7 at pH 6,0 and 2,7 at pH 7,6.

Synthetic Polysaccharides.

Potato phosphorylase and muscle phosphorylase both synthesize polysaccharides *in vitro* which are similar to the amylose fractions of starch (47, 44, 38). Like amylose, the synthetic polysaccharides are slightly soluble in water, rapidly retrograde from solution on standing, and produce a sharp X-ray V-diffraction pattern. They are completely converted to maltose by β-amylase, give a very intense blue color with iodine, and have a low activating power on muscle phosphorylase. Data obtained by the end group method show that, like amylose, the synthetic starches are made up of unbranched chains of glucopyranose units. In contrast to muscle phosphorylase, the heart and liver phosphorylases form a polysaccharide (3) *in vitro* which resembles natural glycogen in solubility, amorphous nature of its X-ray diffraction pattern, reddish-brown coloration with iodine, and strong activating power for phosphorylase. The relatively impure enzyme preparations of liver and heart apparently contain, in addition to the phosphorylase which produces straight chain polysaccharide, another factor capable of forming branched chains. There is reason to believe that two such synthesizing enzymes also exist in the grain of waxy maize, inasmuch as a phosphorylase containing extract from the grain synthesizes amylose only, and yet the starch in this plant appears to consist entirely of amylopectin (*16*). Thus, the combined action of another factor and of phosphorylase, such as found in muscle extract, is required to produce a branched type of polysaccharide. Different branched types of polysaccharides, such as glycogen, amylopectin and corn glycogen (*43*), are presumably formed, depending on the ratio of the two factors.

The enzymatic nature of the branched chain factor is shown by the fact that when crystalline muscle phosphorylase is combined with a heat-labile protein fraction of liver or heart and allowed to act on glucose-1-phosphate, a polysaccharide having the properties of glycogen is formed. This polysaccharide gives a brown color with iodine and activates phosphorylase.

CORI and CORI (*16*) regard the synthesis of the branched glycogen from glucose-1-phosphate as the result of collaboration of two enzymes, one of which produces chains having α-1,4-glucosidic linkages, while the other presumably forms 1,6-glucosidic linkages at the points of branching.

The existence of such two enzymes was demonstrated by HAWORTH, PEAT, and BOURNE (*5*). These workers isolated from the potato a 1,6-linking enzyme which presumably, in conjunction with potato phosphorylase, catalyses the synthesis of amylopectin from glucose-1-phosphate. The 1,6-linking enzyme is referred to as the *Q*-enzyme, while the potato phosphorylase as purified by HANES' method is referred to as the *P*-enzyme. A combination of *P*- and *Q*-enzymes acting upon glucose-1-phosphate yields a branched polysaccharide resembling amylopectin in all its properties.

Arsenolysis and Phosphorolysis of Amylose and Amylopectin.

It was found [KATZ, HASSID and DOUDOROFF (*61*)] that the addition of inorganic arsenate to potato phosphorylase will catalyze the decomposition of starch to glucose according to the reaction:

$$\text{Starch} + H_2O \xrightarrow[\text{arsenate}]{\text{potato phosphorylase}} \text{glucose.}$$

Apparently, an intermediate glucose-1-arsenate is formed which does not accumulate but which is decomposed rapidly to glucose and arsenate.

There is, however, a distinct difference in the limit to which arsenolysis of amylose and of amylopectin proceeds when potato phosphorylase is used. Whereas amylose is completely decomposed to glucose by this enzyme in the presence of arsenate, amylopectin is degraded in this process only to approximately 54%.

It has been observed that potato phosphorylase acts similarly towards amylose and amylopectin in the process of phosphorolysis. Under conditions in which the amount of amylose is insufficient to allow equilibrium to be reached between glucose-1-phosphate and inorganic phosphate, the amylose is almost completely phosphorolyzed to glucose-1-phosphate, but only about 57% of the amylopectin is converted to this ester.

Such behavior is characteristic also of β-amylase. In the well-known hydrolytic reaction, amylose is broken down by β-amylase almost completely to maltose, while amylopectin is degraded by this enzyme only to the extent of about 55%. It can, therefore, be postulated that, like β-amylase, potato phosphorylase acts upon the branched amylopectin through the process of arsenolysis or phosphorolysis, attacking the non-reducing ends, splitting off successive terminal glucose fragments (β-amylase splits off maltose), until it encounters an obstruction. The obstruction is considered to be a modified structure, which is a 1,6-glucosidic linkage at or near the point of branching. With amylose, which has an unbranched structure and therefore no such linkages, arsenolysis or phosphorolysis continues until the whole molecule is degraded. This difference in behavior of potato phosphorylase towards the amylose and amylopectin components of starch constitutes further evidence that the former component is linear while the latter is branched.

Phosphorylase.

Phosphorylases are found in animal, plant and bacterial cells. Only the phosphorylase from animal sources has been well characterized. Cori, Cori, and Green (28, 18, 12, 27, 17, 13, 11) have made an exhaustive study of this enzyme in skeletal muscle and succeeded in preparing it in highly purified crystalline state. It can be obtained in two forms, a and b, both of which are crystalline. The former is a euglobulin of 340,000 to 400,000 molecular weight and has 60 to 70% of its maximum activity without the addition of adenylic acid. The more soluble b form is inactive without the addition of adenylic acid, while both a and b are equally active in its presence. Glucose competitively inhibits the activity of the phosphorylase, while cysteine increases both its activity and solubility.

Extracts of muscle and spleen contain an enzyme ("PR") which, like trypsin, is capable of removing the prosthetic group from phosphorylase a, converting it into form b. This enzyme which catalyzes the reaction, phosphorylase $a \rightarrow$ phosphorylase b, has been separated and purified by fractionation with ammonium sulfate. The activity of this enzyme is increased by the addition of small amounts of cysteine and Mn^{++} ions. The nature of the prosthetic group which is split off by "PR" or trypsin from phosphorylase a has not been elucidated. Attempts to demonstrate free adenylic acid among the products of hydrolysis gave negative results (17). The question as to whether or not phosphorylase a contains adenylic acid is therefore still unanswered in spite of the fact that the activity of phosphorylase b can be restored by the addition of adenylic acid.

Resting muscle of rabbits contains chiefly phosphorylase a. During strong muscular contraction, produced by strychnine or by electrical stimulation, most of the phosphorylase a is converted to phosphorylase b by the action of the "PR" enzyme *in vivo*. This apparently accounts for the fact that crystalline phosphorylase cannot be prepared from extracts of muscles in the state of fatigue. Experiments on frogs indicated that after the phosphorylase a content of the hind legs has been reduced by electrical stimulation, it is completely restored during a short period of rest. CORI (13) considers that the temporary inactivation of phosphorylase a by enzymatic removal of its prosthetic group represents a regulatory mechanism whereby the supply of glycogen is conserved during fatigue.

Phosphorylase from plants has not been obtained in crystalline form, although GREEN and STUMPF (30) prepared purified potato phosphorylase by a series of ammonium sulfate fractionations.

The Mechanism of Dextran and Levan Formation.

With the discovery of the reversible phosphorolytic reaction of starch and glycogen, we have gained an insight into the mechanism of formation and breakdown of these polysaccharides in nature. In this reaction the phosphoric acid groups of a number of glucose-1-phosphate molecules are exchanged for glycosidic linkages of an equal number of monosaccharide units, thus forming a polysaccharide. However, this mechanism is by no means universal in the formation of complex carbohydrates existing in nature. Some polysaccharides are known to be formed by a mechanism which does not involve phosphoric acid esters, but in which an already existing glycosidic linkage is exchanged for a new glycosidic linkage. The formation of polysaccharides through what appears to be such a reaction has been observed in the *bacterial synthesis* of dextran and levan from sucrose.

Dextran is a polymer of glucose in which the hexose units are joined through 1,6-glycosidic linkages. It is produced in large quantities by *L. mesenteroides* when the latter is grown with sucrose, but not with invert sugar (*26*, *73*, *20*). An immunologically identical compound is found in the polysaccharide capsule of some of the *Pneumococci* (*55*, *51*). Levan, a fructose polymer having a 2,6-linkage, is known to be formed abundantly by many species of *Bacillus* (*4*), one of *Aerobacter* (*56*), and two of *Streptococcus* (*70*) in the decomposition of sucrose and raffinose, but not of hexoses. It is interesting that a very similar or identical substance appears in large amounts in barley and in certain wild grasses (*49*, *7*). In the bacteria, dextran and levan syntheses may be regarded as being analogous to capsule formation, although the abundance and solubility of the material causes it to accumulate in the medium instead of remaining in close association with the cell.

While in some cases the enzymes responsible for the polysaccharide production appear mainly within the cell, in the *Bacilli*, and in at least some strains of *Leuconostoc*, they may diffuse into the medium. HEHRE's (*55*, *51*, *52*, *53*) as well as HESTRIN and AVINIERI-SHAPIRO's experiments (*56*) with cell-free enzyme preparations have shown that the formation of neither dextran nor levan requires any organic or in- organic phosphate. In both cases, the polymerization of one component of the sucrose molecule is accompanied by the accumulation of an equi- valent amount of the other component in the form of free hexose, *i. e.*, fructose in dextran production, and glucose in levan production.

The formation of the two polysaccharides from sucrose can be written by the following equations:

$$(n) \text{ glucose-1-fructoside} \rightarrow (\text{glucosan})_n + (n) \text{ fructose,}$$
$$\quad\quad\text{(sucrose)} \quad\quad\quad\quad\quad\quad\text{(dextran)}$$

$$(n) \text{ glucose-1-fructoside} \rightleftarrows (\text{fructosan})_n + (n) \text{ glucose.}$$
$$\quad\quad\text{(sucrose)} \quad\quad\quad\quad\quad\quad\text{(levan)}$$

These reactions are analogous to the reaction representing glycogen or starch formation from glucose-1-phosphate:

$$(n) \text{ glucose-1-phosphate} \rightleftarrows (\text{glucosan})_n + (n) \text{ phosphate.}$$
$$\quad\quad\text{(Cori ester)} \quad\quad\quad\quad\text{(glycogen or starch)}$$

The cell-free enzyme preparations which act on sucrose with the production of polysaccharides do not act on glucose-1-phosphate. Neither do the animal or plant phosphorylases act on sucrose to produce polysaccharide.

The reason that glucosan and fructosan are synthesized in the presence of the proper enzymes from sucrose but not from a mixture of glucose and fructose is that the glycosidic bond in sucrose, as in glucose-1-phosphate, is on a higher energy level than free glucose or fructose. In glycogen or starch this level is equal to the difference in energy between free and carbonyl bond phosphate, which is approximately 2000 calories (69a). The energy of the glycosidic bond in sucrose (glucosido-1-fructose) is approximately the same as that of the C—O—P bond in glucose-1-phosphate; therefore, sucrose, like glucose-1-phosphate can be used for polysaccharide production.

Although HESTRIN and AVINIERI-SHAPIRO (56) showed that the presence of glucose retards the synthesis of levan, they were unable to demonstrate the reversible nature of levan formation. Indirect evidence of the reversibility of the process has been obtained recently with cell-free concentrates and precipitated preparations of *Bacillus subtilis* cultures by DOUDOROFF and O'NEAL (25).

It was assumed that the principal difficulties in showing the reaction from right to left lie in the high molecular weight and therefore relatively low concentration of the polysaccharide, and in the presence of insufficient invertase in the preparations to remove sucrose as it is formed.

To displace this equilibrium to the left, yeast invertase was added to the bacterial enzyme. In such a system, the addition of glucose markedly increased the decomposition of levan to reducing sugar. No effect of phosphate on the reaction in either direction could be observed. This indirect evidence that the formation of levan depends on a reversible reaction, in which the glycosidic bond of sucrose is exchanged directly for the glycosidic bond of fructosan, indicates a new mechanism of sucrose synthesis through the reaction from right to left. Presumably, sucrose would also be formed under the influence of the appropriate enzyme from fructose and dextran, although the reversibility of dextran synthesis has not yet been tested experimentally.

From these considerations and from the discussion that will follow later, it appears that some glycosides can serve a function similar to that of the phosphoric ester group of glucose-1-phosphate in the formation of complex carbohydrates. The exchange of glycosidic bonds may play an important role in the synthesis of a variety of polysaccharides.

Mechanism of Sucrose Formation.

Just as amylase was at first thought to be responsible for both the breakdown and the synthesis of starch, so the widely distributed plant invertase was considered for a long time as having the double role of an hydrolytic and synthetic enzyme. Formation of sucrose was regarded as a result of reversed inversion. More recently OPARIN and KURSSANOV (71) claimed to have synthesized sucrose from invert sugar under the influence of invertase and phosphatase in the presence of inorganic phosphate. However, this claim could not be substantiated by other investigators (64). The fact that leaves of sucrose-producing plants, such as beets and peas, were found to contain considerable amounts of hexose phosphates, suggested that sugar phosphates might enter into the mechanism of sucrose formation and that phosphorylation was an essential step in this process (6, 36, 63).

While this hypothesis has not as yet been supported by the isolation from a plant source of an enzyme capable of sucrose synthesis, DOUDOROFF, KAPLAN and HASSID (24) succeeded in preparing such an enzyme from a bacterial source. It was found that dried cells of *Pseudomonas saccharophila* contain an enzyme system which catalyzes the breakdown of sucrose in the presence of inorganic phosphate with the formation of glucose-1-phosphate (CORI ester) and fructose. Furthermore, the sucrose phosphorylase catalyzes the reverse reaction, i. e. the synthesis of sucrose from glucose-1-phosphate and fructose. The dried bacteria also contain invertase and this hydrolytic enzyme competes with the sucrose phos-

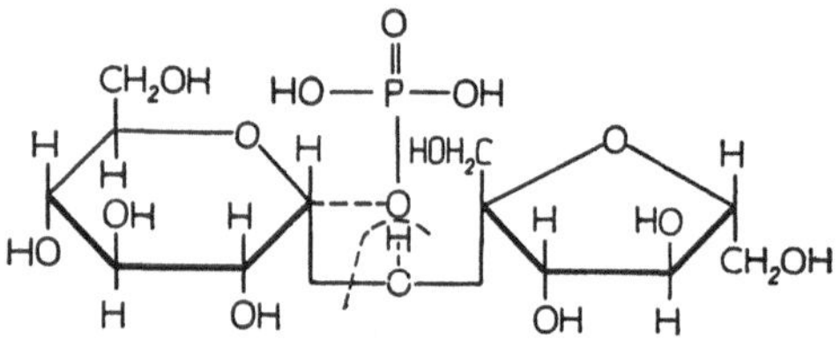

Fig. 3. Phosphorolysis of sucrose.

phorylase for the sucrose. It is, however, possible to eliminate most of the invertase from the bacterial preparations by repeated precipitations with ammonium sulfate. Using a partially purified sucrose phosphorylase preparation and a mixture of glucose-1-phosphate and fructose, HASSID, DOUDOROFF and BARKER (39) succeeded in crystallizing a non-reducing disaccharide which was indistinguishable from natural sucrose. Thus, the first laboratory synthesis of sucrose was achieved.

The empirical formula of the synthetic disaccharide on the basis of analysis is $C_{12}H_{22}O_{11}$. The compound does not reduce FEHLING solution before hydrolysis. After acid or enzymatic hydrolysis, the reducing value and the yield of glucose and fructose are theoretical for invert sugar. The specific rotation, $[\alpha]_D + 65,5°$, is changed by inversion to $- 20°$. The synthetic product gives an identical X-ray diffraction pattern and is hydrolyzed with acid at the same rate as natural sucrose. The optical properties of the crystals are also the same as for sucrose.

As in the phosphorolysis of starch and glycogen, the formation of glucose-1-phosphate and fructose from sucrose and inorganic phosphate may be considered to occur as the result of phosphorolytic cleavage of glucose from the sucrose molecule, the sucrose being disrupted without water entering into the reaction. The reverse reaction, the formation of sucrose from glucose-1-phosphate and fructose is the result of "de-phosphorolytic" condensation of the two monosaccharides. The reversible phosphorolysis of sucrose can be represented as shown in Fig. 3.

In this connection it is noteworthy that the addition of inorganic arsenate to sucrose phosphorylase will catalyze the hydrolytic decomposition of sucrose (*22*) according to the reaction:

$$\text{sucrose} + H_2O \xrightarrow[\text{arsenate}]{\text{sucrose phosphorylase}} \text{glucose} + \text{fructose}.$$

Similarly, arsenate in the presence of sucrose phosphorylase will decompose glucose-1-phosphate to glucose and inorganic phosphate.

As with starch and potato phosphorylase (*61*), apparently an intermediate, unstable glucose-1-arsenate, is formed which does not accumulate but is decomposed to glucose and inorganic arsenate. The system, enzyme + arsenate becomes a catalyst which behaves both as an invertase and as a phosphatase.

Owing to the successful *in vitro* synthesis of sucrose we have gained further knowledge of the *chemical constitution* of this carbohydrate. The type of glycosidic linkages which combine glucose and fructose in the sucrose molecule has hitherto not been known with certainty. Although the work of ARMSTRONG (*1*) and that of PURVES and HUDSON (*76*) indicated that the structure of sucrose is α-D-glucopyranosido-β-D-fructofuranoside, the question as to whether the α- or β-form of either hexose was involved in joining the monosaccharide units to form the disaccharide had not been settled conclusively. By showing that the enzymatically produced sucrose is identical with the naturally occurring compound and that it is synthesized through "de-phosphorolytic" condensation of the α-form of glucose-1-phosphate and fructose, the *α-configuration of glucose in the sucrose molecule* has been confirmed. Evidence regarding this type of linkage is adduced from the observation that when the α-linkage of the phosphoric acid in glucose-1-phosphate is exchanged for a glycosidic linkage with another monosaccharide (as in the process of starch or glycogen formation) the type of linkage is not altered.

The enzymatic synthesis of sucrose also throws light on the formation of the *furanose form of fructose* in the sucrose molecule. The fact that sucrose is directly formed from glucose-1-phosphate and fructose supports the view that the latter monosaccharide occurs in solution in an equilibrium mixture of furanose and pyranose forms. This makes it unnecessary to postulate a special mechanism of stabilization of a five-membered ring before the formation of compound sugars containing the fructose molecule (*36*).

Many unsuccessful attempts to effect the chemical synthesis of sucrose have been made in the past, but Pictet and Vogel (74) claim to have accomplished the synthesis of sucrose by coupling a tetraacetyl-γ-fructose with tetraacetylglucose in .the presence of a dehydrating agent. However, other investigators (79) were not able to reproduce their results. The difficulties encountered in the synthesis of sucrose by chemical means are apparently due to the fact that glucose and fructose, the constituents of sucrose, may each exist in more than one form. Since either of these monosaccharides can exist in the α- or β-form, at least four different disaccharide configurations are possible when the two hexoses are combined. Thus, in attempting to condense tetraacetyl-γ-fructose with tetraacetylglucose, Irvine et al. (58, 57) were unable to obtain sucrose octaacetate, but did obtain a disaccharide derivative with a different glycosidic linkage, the so-called isosucrose octaacetate.

In considering the process of glycogen and starch synthesis, it is observed that these polysaccharides are formed by a general mechanism which operates in various living cells. Thus, regardless of the sources of the enzyme phosphorylase, whether it be derived from animal, yeast or higher plant cells, it will act on glucose-1-phosphate to form a similar polysaccharide. It can, therefore, be assumed that, except for minor variations, the mechanism for glycogen and starch formation is similar in all living cells. Thus far, this generalization cannot be made with regard to the process of sucrose formation. The mechanism for sucrose synthesis in which the enzyme system derived from the bacterial organism, P. saccharophila is involved does not appear to apply to higher plants.

A similar enzyme system that would combine glucose-1-phosphate and fructose to form sucrose and inorganic phosphate has not yet been isolated from the tissues of higher plants, despite various attempts to do so. However, biochemical studies on various species of plants support the view that the synthesis of sucrose may involve chemical reactions in which phosphate esters of glucose, fructose or both hexoses serve as substrates, although the mechanism is probably not identical with that of the bacterial enzyme system. It is also significant that the experimental evidence now available shows that for the synthesis of sucrose from glucose and fructose in the plant, aerobic metabolism is indispensable (35, 66). Possibly, aerobic oxidations are essential to the phosphorylation of one of the substrates involved in the synthesis of sucrose. The problem is complicated by the observation that various substrates other than glucose and fructose may result in sucrose formation by plant tissues. For example, in experiments with barley shoots, the infiltration of galactose as well as of various other carbohydrates was shown to lead to sucrose formation (66).

Specificity of Sucrose Phosphorylase.

Sucrose phosphorylase from *P. saccharophila*, like potato and muscle phosphorylase, is specific with regard to the α-*D*-glucose portion of its substrates. Potato and muscle phosphorylase do not form poly-saccharide from α-maltose-1-phosphate, α-*D*-galactose-1-phosphate, α-*D*-mannose-1-phosphate, α-*D*-xylose-1-phosphate or α-*L*-glucose-1-phosphate (*67, 75*). Similarly, these hexose phosphates cannot be substituted for α-*D*-glucose-1-phosphate when use is made of bacterial sucrose phosphorylase, which has the ability of combining α-*D*-glucose-1-phosphate with fructose to form sucrose. This enzyme is, however, quite versatile with regard to substituents for the second sucrose component, fructose. The sucrose phosphorylase will combine the same ester with other ketose monosaccharides, such as *D*-xyloketose, *L*-araboketose and *L*-sorbose to form the corresponding non-reducing disaccharides, *D*-glucosido-*D*-xyloketoside (*41*), *D*-glucosido-*L*-araboketoside (*23*), and *D*-glucosido-*L*-sorboside (*40*).

It appeared at first as though this enzyme were only capable of catalyzing the reaction between ketose monosaccharides and glucose-1-phosphate. Since the disaccharides produced are non-reducing and their ketose constituents exist in the furanose form, they can be considered as analogues of sucrose. However, it was later discovered that the same enzyme catalyzes a reaction between glucose-1-phosphate and *L*-arabinose, to form a reducing disaccharide (*23, 42*) having no obvious structural relation to sucrose or to any of the previously prepared sucrose analogues.

It seemed unique for a single enzyme to catalyze two such diverse reactions as that involving the carbonyl group on the second carbon atom of a ketose, and the alcohol group on the third carbon atom of an aldose. The question as to whether or not more than one phosphorolytic enzyme is involved in these reactions therefore had to be settled.

That sucrose phosphorylase is involved in the reaction of *L*-arabinose was established beyond reasonable doubt (*23*) from the following observation: When *L*-arabinose was added to a mixture containing the enzyme, glucose-1-phosphate and *D*-fructose, of which the last was present in insufficient concentration to give the maximum rate of sucrose formation, an increase in the rate of utilization of glucose-1-phosphate was observed. However, when the same amount of *L*-arabinose was added to a similar mixture (with the exception that the *D*-fructose concentration was so selected as to give the maximum rate of sucrose formation) the total rate of glucose-1-phosphate utilization was decreased. Since it is known that the rate of reducing disaccharide formation from glucose-1-phosphate and *L*-arabinose is considerably slower than that of sucrose formation from the same ester and *D*-fructose, a decrease in the rate of glucose-1-phosphate utilization would be expected if the two sugars were competing for the enzyme. These results are presented in Table 1.

Table 1. Initial Rate of Inorganic Phosphate Production from Glucose-1-Phosphate in the Presence of Sucrose Phosphorylase upon the Addition of Fructose and Arabinose.

Dilute enzyme preparation incubated with 0,1 M glucose-1-phosphate with additions indicated below at 30°.

Additions	Rate of total evolution of phosphate as μ moles per ml. per hour
$M/4$ L-arabinose	2,1 ($\pm$ 0,2)
$M/64$ D-fructose	3,1 ($\pm$ 0,2)
$M/64$ D-fructose + $M/4$ L-arabinose	4,1 ($\pm$ 0,2)
$M/4$ D-fructose	9,7 ($\pm$ 0,2)
$M/4$ D-fructose + $M/4$ L-arabinose	8,6 ($\pm$ 0,2)

Following is additional evidence that the same enzyme is involved in reactions with both D-fructose and L-arabinose:

(a) The enzyme catalyzing both reactions is produced to a marked extent when sucrose is used as substrate for the growth of the organisms, but not when D-glucose or L-arabinose is used.

(b) The relative rates of reaction with D-fructose and L-arabinose, respectively, remain constant after partial inactivation by heat.

(c) In the preparation and partial purification of phosphorylase from the crude extract of dried cells, the relative rates of reaction with the fructose and L-arabinose remain constant, although the total activity toward both is considerably decreased and many other enzymes are destroyed or removed.

(d) On fractionation of the enzyme preparation with various concentrations of ammonium sulfate, the relative activities of the fractions remains the same for both sugars.

The above considerations indicate not only that the same enzyme is involved in both reactions, but also that no additional enzyme is required for the formation of D-glucosido-L-arabinose.

D-Glucosido-D-xyloketoside, D-Glucosido-L-araboketoside and D-Glucosido-L-sorboside.

The enzymatically synthesized D-glucosido-D-xyloketoside (41), D-glucosido-L-araboketoside (23), and the D-glucosido-L-sorboside (40) are non-reducing, showing that, as in sucrose, the glucose and ketose monosaccharides are linked through the carbonyl groups. In proving the *structure* of these disaccharides, the technique of HUDSON and co-workers (59) for oxidizing carbohydrates with sodium periodate was used. In a disaccharide consisting of glucopyranose and ketofuranose united through positions 1 and 2, the glucose residue would possess three adjacent free hydroxyls, on carbon atoms 2, 3 and 4, and the ketose residue would possess two free hydroxyls, on carbon atoms 3 and 4. On oxidation of such a disaccharide with periodate a total of 3 moles of periodate would be consumed and one mole of formic acid would be formed per mole of disaccharide. Actually, the three disaccharides gave experimental data which agreed with this expectation. Any other ring structure for either the glucose or the ketose component would have given different results.

The fact that these disaccharides, like sucrose, are formed as a result of "dephosphorolytic" condensation involving α-D-glucose-1-phosphate

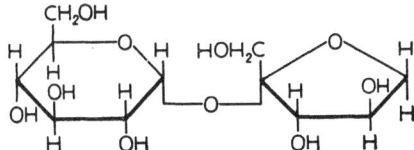

Fig. 4.   α-D-Glucopyranosido-β-D-xyloketofuranoside.

supports the view that glucose also exists in these disaccharides as the α-*form*.

All of the non-reducing disaccharides give a blue-green color with diazouracil. This reaction has been shown by RAYBIN (77) to be specific for sucrose and other compounds containing the same type of glycosidic glucose-fructose linkage, such as raffinose, gentianose, and stachyose. The analogy of the synthetic, non-reducing disaccharides to sucrose in their reaction with diazouracil and with the bacterial sucrose phosphorylase indicates that the local structure about the glycosidic linkage is the same as that of sucrose in all cases.

This evidence makes it possible to postulate the structural formula for glucosido-xyloketoside as shown in Fig. 4.

This compound and sucrose would be structurally identical, except that sucrose has an additional carbon atom with a primary alcohol attached to the ring.

The structure for glucosido-arabofuranoside is α-D-glucopyranosido-β-L-araboketofuranoside as shown in Fig. 5.

The striking similarity between sucrose and glucosido-L-sorboside is immediately evident when the structural formula of the latter is constructed on the same assumption (Fig. 6).

The sorbose component of this disaccharide is an *L*-sugar in contrast to the *D*-fructose unit existing in sucrose. Since β-*D*-fructose and α-*L*-sorbose have the same configuration about their second carbon atoms (Fig. 7), it is necessary to designate the ketose portion of this disaccharide as α-*L*-sorboside.

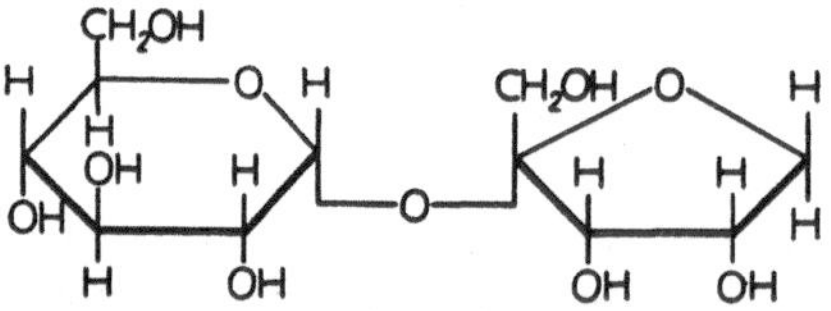

Fig. 5. α-*D*-Glucopyranosido-β-*L*-araboketofuranoside.

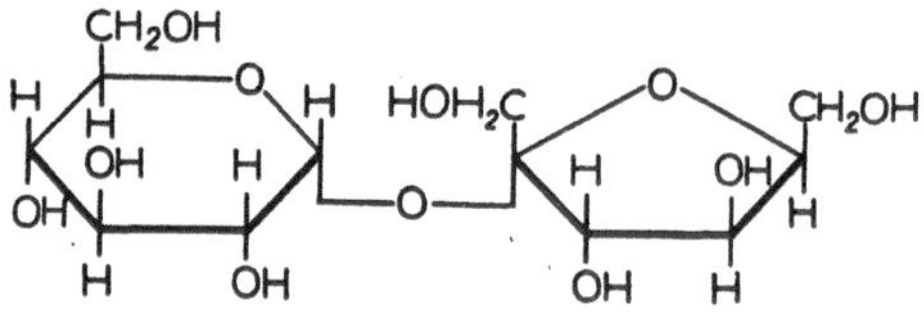

Fig. 6. α-*D*-Glucopyranosido-α-*L*-sorbofuranoside.

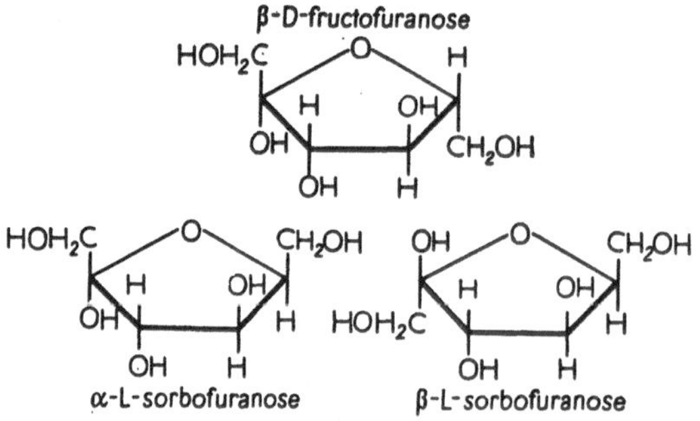

Fig. 7. β-*D*-Fructofuranose, α-*L*-sorbofuranose, and β-*L*-sorbofuranose.

It is of interest to note that fructose, which has a pyranose structure when existing in the free state, assumes a furanose configuration whenever it combines with another sugar to form an oligosaccharide or poly-saccharide. Apparently, the ketohexose, *L*-sorbose, shows the same behavior.

α-*D*-glucosido-*L*-arabinose.

The disaccharide (*42*) formed from α-*D*-glucose-1-phosphate and *L*-arabinose under the influence of the enzyme from *P. saccharophila* reduces Fehling and alkaline ferricyanide solutions. It contains two molecules of water of crystallization and has a specific rotation, $[\alpha]_D = +156°$ (in water). Unlike sucrose and its analogues, this disaccharide is difficultly hydrolyzable with acid. Upon hydrolysis it yields one mole of D-glucose and one mole of L-arabinose. The phenylosotriazole derivative of the disaccharide prepared according to HUDSON *et al.* (*34*) is readily hydrolyzed with acid to D-glucose and L-arabinose phenyloso-triazole, showing that L-arabinose constitutes the free reducing unit in the disaccharide. Since α-D-glucose-1-phosphate is involved in the enzymatic synthesis of the disaccharide it may be assumed that the glucose component exists in the α-form.

On oxidation of the phenylosotriazole derivative of this disaccharide with sodium periodate, three moles of periodate are consumed with the formation of one mole each of formic acid and formaldehyde, per mole of phenylosotriazole derivative. The structure of this compound is, therefore, 3-[α-D-glucopyranosido]-L-arabinose phenylosotriazole in which the D-glucose unit is attached through carbon atom 1 to carbon atom 3 of L-arabinose as shown by Formula (I).

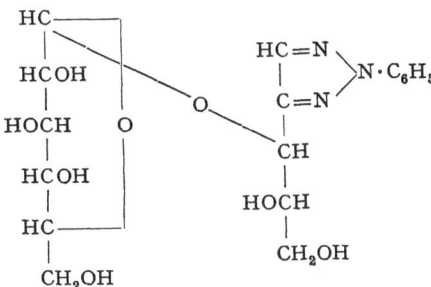

(I.) 3-[α-D-glucopyranosido]-L-arabinose phenylosotriazole.

If the D-glucose in the D-glucopyranosido-L-arabinose phenylosotriazole were attached to carbon atom 4 of the L-arabinose derivative, oxidation of this compound with sodium periodate would require two moles of periodate and would liberate one mole of formic acid without formaldehyde production. Junction of D-glucose to carbon atom 5 of the L-arabinose phenylosotriazole would require three moles of periodate whereby one mole of formic acid would be produced, and no formaldehyde formed.

On methylation of the disaccharide with dimethyl sulfate and sodium hydroxide, a hexamethyl methyl derivative of the compound was obtained. When this fully methylated derivative (II) was hydrolyzed with acid, 2,3,4,6-tetramethyl-D-glucose (III) and dimethyl-L-arabinose (IV) were produced. Since position 3 in the L-arabinose component of (I) was shown to be occupied in glycosidic linkage with D-glucose, the dimethyl-L-arabinose could be either the 2,5- or 2,4-dimethyl derivative (IV), depending on whether the L-arabinose unit originally exists in the disaccharide in the furanose or pyranose form. The ring type of the L-arabinose was ascertained by subjecting the dimethyl-*L*-arabinose to oxidation with sodium periodate, after it had been oxidized with hypoiodite to the corresponding lactone (V) compound and subsequently hydrolyzed to the straight chain compound, dimethyl-*L*-arabonic acid (VI).

If the dimethyl derivative were the 2,5-dimethyl-*L*-arabonic acid, it would possess a pair of adjacent hydroxyls, in positions 3 and 4, which on oxidation with sodium periodate would consume one mole of periodate in the reaction. On the other hand, the 2,4-dimethyl-*L*-arabonic acid (VI), lacking a pair of adjacent hydroxyls, cannot be oxidized. Actually, no periodate was consumed when the dimethyl-*L*-arabonic acid was treated with the reagent. This shows that the dimethyl derivative is 2,4-dimethyl-*L*-arabonic acid. The free hydroxyl in position 3 is obviously restored in the dimethyl-*L*-arabinose when the methylated disaccharide is hydrolyzed; the hydroxyl in position 5 is formed when its internal ring is broken in the process of hydrolysis of the lactone to dimethyl-*L*-arabonic acid which is, as mentioned, a straight chain compound.

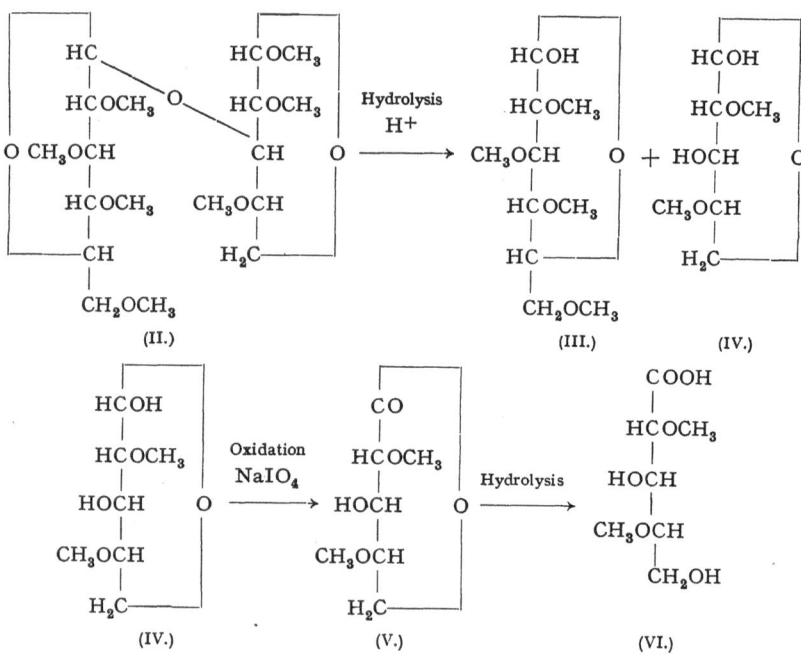

A more direct confirmation that the dimethyl-*L*-arabinose possesses a pyranose configuration was obtained from the study of the rate with which its lactone derivative is hydrolyzed to the corresponding open chain acid (*46*).

When the dimethyl-*L*-arabono-lactone was dissolved in water, it was found to be almost completely hydrolyzed within four hours. This was indicated by a change of its rotation from $[\alpha]_D$ $+60°$ to $+24°$. A constant value of $[\alpha]_D$ $+17°$ was reached within less than twenty-four hours. Since the rate of change in rotation of this methylated lactone due to hydrolysis is high, it strongly indicates that the lactone possesses a pyranose configuration. This observation confirms the periodate oxidation data, showing that the dimethyl derivative is 2,4-dimethyl-*L*-arabonic acid.

On the basis of these results this reducing disaccharide may be designated as *3-[α-D-glucopyranosido]-L-arabopyranose* and its structural formula written as in Fig. 8.

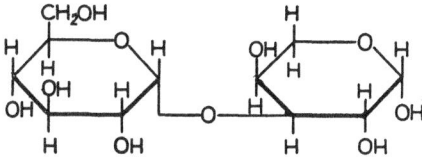

Fig. 8. 3-[α-*D*-Glucopyranosido]-*L*-arabopyranose.

Formation of Sucrose and other Disaccharides through Exchange of Glycosidic Linkages.

Further information has recently been obtained concerning the mechanism of formation and breakdown of the synthetic disaccharides by the use of radioactive phosphate [DOUDOROFF, BARKER, and HASSID (*21*)]. It was observed that when glucose-1-phosphate and radioactive inorganic phosphate were added to sucrose phosphorylase preparations in the absence of ketose sugars, a rapid redistribution of the isotope occurred between the organic and inorganic fractions. This is illustrated by the results listed in Table 2.

Table 2. Exchange of P^{32} between Inorganic Phosphate and Glucose-1-phosphate.

Experiment No.	Reaction Mixture	Radioactivity found in fractions as counts per minute per mole after 60 min. at 30⁰	
		Inorganic phosphate	Glucose-1-phosphate
1	0,1 *M* glucose-1-phosphate + 0,033 *M* inorganic phosphate	1098 (± 40)	0 (± 1)
2	Same as 1 but with enzyme preparation.	859 (± 40)	119 (± 3)
3	Same as 2 but with 0,06 *M* fructose	886 (± 40)	99 (± 3)
4	Same as 2 but with 0,12 *M* glucose	1096 (± 40)	7 (± 1)

The P^{32} appeared rapidly in the glucose-1-phosphate fraction. In the presence of fructose, less P^{32} was found in glucose-1-phosphate. This is due, at least in part, to a dilution of the radioactive phosphate with inactive phosphate liberated as a result of the synthesis of sucrose. The chief reason, however, must be the competition of fructose with phosphate for combination with glucose.

The presence of glucose almost completely stops the exchange of P^{32}. This is consistent with the observation that glucose inhibits phosphorylase activity and supports the view that it competes with glucose-1-phosphate for combination with the enzyme.

The rapid redistribution of the isotopes that occurred between the organic and inorganic fractions when glucose-1-phosphate and inorganic phosphate were added to sucrose phosphorylase preparations indicated a temporary bond between the liberated glucose and the enzyme, the glucose-enzyme combination existing in equilibrium with the glucose-1-phosphate. The following reaction was thus postulated:

Glucose-1-phosphate + enzyme $\rightleftarrows$ glucose-enzyme + phosphate.

This observation suggested that an analogous reaction might occur between the enzyme and sucrose with fructose being liberated instead of phosphate:

Glucose-1-fructoside + enzyme $\rightleftarrows$ glucose-enzyme + fructose.
(sucrose)

The latter scheme was supported by experiments in which phosphate-free enzyme preparations were found capable of synthesizing the sucrose analogue, glucosido-sorboside, directly from sucrose and sorbose:

Glucose-1-fructoside + sorbose $\rightleftarrows$ glucose-1-sorboside + fructose.
(sucrose)

Similarly, sucrose was formed from glucose-1-xyloketoside and fructose in the absence of inorganic phosphate:

Glucose-1-xyloketoside + fructose $\rightleftarrows$ glucose-1-fructoside + xyloketose.
(sucrose)

It may be concluded from these experiments that sucrose phosphorylase is not only capable of substituting glycosidic linkages for an ester linkage but can also exchange equivalent glycosidic linkages. The function of the enzyme is, therefore, to combine reversibly with the glucose residue of glucose-1-phosphate or of those disaccharides which can act as substrates, and to transfer the glucose to suitable acceptors, such as phosphate and various ketoses.

We can, therefore, say that the reversible phosphorolysis of sucrose can consist of the following reactions:

Glucose-1-fructoside + enzyme $\rightleftharpoons$ glucose-enzyme + fructose

(sucrose) $\uparrow\downarrow \pm$ HPO₄⁻⁻

glucose-1-fructoside + enzyme

The direct interconversion of disaccharides, such as the formation of sucrose from glucosido-xyloketoside is illustrated by an analogous reaction as follows:

Glucose-1-xyloketoside + enzyme $\rightleftharpoons$ glucose-enzyme + xyloketose

$\uparrow\downarrow \pm$ fructose

glucose-1-fructoside + enzyme

(sucrose)

The direct exchange of glycosidic linkages has been observed with other bacterial enzymes. As previously pointed out, the enzyme from *Leuconostoc* produces a dextran from sucrose while the enzyme from *B. subtilis* produces a levan from the same sugar. Neither of these polysaccharides requires accumulation of phosphoric esters for their formation. The enzymes which catalyze these reactions simply exchange a glucosido-fructoside linkage for a 1,6-glucosido-glucose bond in one case, and for a 2,6-fructosido-fructose bond in the other.

Since the transfer of glucose or fructose residues may be compared to the well known transmethylation and transamination reactions, it seems appropriate to class the sucrose phosphorylase as well as the enzymes involved in the synthesis of levan and dextran as *"transglycosidases"*.

It is quite possible that production of many disaccharides and polysaccharides in plant and animal tissues may depend on transfers of energy-rich sugar residues without the intermediate accumulation of phosphoric esters. The remarkable versatility of the sucrose phosphorylase which can produce both reducing and non-reducing disaccharides, suggests that one and the same enzyme might, in some cases, account for the formation of a number of different compounds.

References.

1. Armstrong, E. F. and K. F. Armstrong: The Carbohydrates, pp. 181–182. London: Longmans, Green and Co. 1934.

2. Barsha, J. and H. Hibbert: Studies on Reactions Relating to Carbohydrates and Polysaccharides. XLVI. Structure of the Cellulose Synthesized by the Action of *Acetobacter xylinus* on Fructose and Glycerol. Canad. J. Res., Sect. B **10**, 170 (1934).

3. Bear, R. S. and C. F. Cori: X-Ray Diffraction Studies of Synthetic Polysaccharides. J. biol. Chemistry **140**, 111 (1941).

4. Beijerinck, M.: Die durch Bakterien aus Rohrzucker erzeugten schleimigen Wandstoffe. Folia Microbiol. **1**, 377 (1912).

5. Bourne, E. J. and S. Peat: The Enzymic Synthesis and Degradation of Starch. I. The Synthesis of Amylopectin. J. chem. Soc. [London] **1945**, 877.

6. Burkard, J. u. C. Neuberg: Zur Frage nach der Entstehung des Rohrzuckers. Biochem. Z. **270**, 229 (1934).

7. Challinor, S. W., W. N. Haworth and E. L. Hirst: Polysaccharides. XVII. The Constitution and Chain Length of Levan. J. chem. Soc. [London] **1934**, 676; The Carbohydrates of Grass. Isolation of a Polysaccharide of the Levan Type. J. chem. Soc. [London] **1934**, 1560.

8. Cori, C. F.: Glycogen Breakdown and Synthesis in Animal Tissues. Endocrinology **26**, 285 (1940).

9. Cori, C. F. and G. T. Cori: Mechanism of Formation of Hexosemonophosphate in Muscle and Isolation of a New Phosphate Ester. Proc. Soc. exp. Biol. Med. **34**, 702 (1936).

10. — — Formation of Glucose-1-Phosphoric Acid in Muscle Extract. Proc. Soc. exp. Biol. Med. **36**, 119 (1937).

11. — — The Activity and Crystallization of Phosphorylase b. J. biol. Chemistry **158**, 341 (1945).

12. Cori, C. F., G. T. Cori and A. A. Green: Crystalline Muscle Phosphorylase. III. Kinetics. J. biol. Chemistry **151**, 39 (1943).

13. Cori, G. T.: The Effect of Stimulation and Recovery on the Phosphorylase a Content of Muscle. J. biol. Chemistry **158**, 333 (1945).

14. Cori, G. T., S. P. Colowick and C. F. Cori: The Formation of Glucose-1-phosphoric Acid in Extracts of Mammalian Tissues and of Yeast. J. biol. Chemistry **123**, 375 (1938).

15. Cori, G. T. and C. F. Cori: The Kinetics of the Enzymatic Synthesis of Glycogen from Glucose-1-phosphate. J. biol. Chemistry **135**, 733 (1940).

16. — — Crystalline Muscle Phosphorylase. IV. Formation of Glycogen. J. biol. Chemistry **151**, 57 (1943).

17. — — The Enzymatic Conversion of Phosphorylase a to b. J. biol. Chemistry **158**, 321 (1945).

18. Cori, G. T. and A. A. Green: Crystalline Muscle Phosphorylase. II. Prosthetic Group. J. biol. Chemistry **151**, 31 (1943).

19. Cori, G. T., M. A. Swanson and C. F. Cori: The Mechanism of Formation of Starch and Glycogen. Federation Proc. **4**, 234 (1945).

20. Daker, W. D. and M. Stacey: Polysaccharides. XXX. The Polysaccharide produced from Sucrose by *Betabacterium Vermiforme* (Ward-Mayer). J. chem. Soc. [London] **1939**, 585.

21. Doudoroff, M., H. A. Barker and W. Z. Hassid: Studies with Bacterial Sucrose Phosphorylase. I. The Mechanism of Action of Sucrose Phosphorylase as a Glucose-transferring Enzyme (Transglucosidase). J. biol. Chemistry **168**, 725 (1947).

22. — — — Studies with Bacterial Sucrose Phosphorylase. III. Arsenolytic Decomposition of Sucrose and of Glucose-1-phosphate. J. biol. Chemistry **170**, 147 (1947).

23. Doudoroff, M., W. Z. Hassid and H. A. Barker: Studies with Bacterial Sucrose Phosphorylase. II. Enzymatic Synthesis of a New Reducing and a New Non-Reducing Disaccharide. J. biol. Chemistry **168**, 733 (1947).

24. Doudoroff, M., N. Kaplan and W. Z. Hassid: Phosphorolysis and Synthesis of Sucrose with a Bacterial Preparation. J. biol. Chemistry **148**, 67 (1943).

25. DOUDOROFF, M. and R. O'NEAL: On the Reversibility of Levulan Synthesis by *Bacillus subtilis*. J. biol. Chemistry **159**, 585 (1945).

26. FOWLER, F. L., I. K. BUCKLAND, F. BRAUNS and H. HIBBERT: Studies of Reactions Relating to Carbohydrates and Polysaccharides. LIII. Structure of the Dextran Synthesized by the Action of *Leuconostoc mesenterioides* on Sucrose. Canad J. Res., Sect. B **15**, 486 (1937).

27. GREEN, A. A.: The Diffusion Constant and Electrophoretic Mobility of Phosphorylases *a* and *b*. J. biol. Chemistry **158**, 315 (1945).

28. GREEN, A. A. and G. T. CORI: Crystalline Muscle Phosphorylase. I. Preparation, Properties, and Molecular Weight. J. biol. Chemistry **151**, 21 (1943).

29. GREEN, A. A., G. T. CORI and C. F. CORI: Crystalline Muscle Phosphorylase. J. biol. Chemistry **142**, 447 (1942).

30. GREEN, D. E. and P. K. STUMPF: Starch Phosphorylase of Potato. J. biol. Chemistry **142**, 355 (1942).

31. HANES, C. S.: Breakdown and Synthesis of Starch by an Enzyme System from Pea Seeds. Proc. Roy. Soc. [London], Ser. B **128**, 421 (1940).

32. — Reversible Formation of Starch from Glucose-1-phosphate Catalyzed by Potato Phosphorylase. Proc. Roy. Soc. [London], Ser. B **129**, 174 (1940).

33. HANES, C. S. and E. J. MASKELL: The Influence of Hydrogen-Ion Concentration upon the Equilibrium State in Phosphorylase Systems. Biochemic. J. **36**, 76 (1942).

34. HANN, R. M. and C. S. HUDSON: The Action of Copper Sulfate on Phenylosazones of the Sugars. Phenyl-*D*-Glucosotriazole. J. Amer. chem. Soc. **66**, 735 (1944).

35. HARTT, C. E.: Synthesis of Sucrose in Sugar Cane Plant I. Hawaiian Planter's Record **47**, 113 (1943).

36. HASSID, W. Z.: Isolation of Hexosemonophosphate from Pea Leaves. Plant Physiol. **13**, 641 (1938).

37. HASSID, W. Z. and H. A. BARKER: The Structure of Dextran Synthesized from Sucrose by *Betacoccus Arabinosaceus*, ORLA-JENSEN. J. biol. Chemistry **134**, 163 (1940).

38. HASSID, W. Z., G. T. CORI and R. M. MCCREADY: Constitution of the Polysaccharide Synthesized by the Action of Crystalline Muscle Phosphorylase. J. biol. Chemistry **148**, 89 (1943).

39. HASSID, W. Z., M. DOUDOROFF and H. A. BARKER: Enzymatically Synthesized Crystalline Sucrose. J. Amer. chem. Soc. **66**, 1416 (1944).

40. HASSID, W. Z., M. DOUDOROFF, H. A. BARKER and W. H. DORE: Isolation and Structure of an Enzymatically Synthesized Crystalline Disaccharide, *D*-Glucosido-*L*-Sorboside. J. Amer. chem. Soc. **67**, 1394 (1945).

41. — — — — Isolation and Structure of an Enzymatically Synthesized Crystalline Disaccharide, *D*-Glucosido-*D*-Ketoxyloside. J. Amer. chem. Soc. **68**, 1465 (1946).

42. HASSID, W. Z., M. DOUDOROFF, A. L. POTTER and H. A. BARKER: The Structure of an Enzymatically Synthesized Reducing Disaccharide, *D*-Glucosido-*L*-Arabinose. J. Amer. chem. Soc. **70**, 306 (1948).

43. HASSID, W. Z. and R. M. MCCREADY: The Molecular Constitution of Glycogen and Starch from the Seed of Sweet Corn *(Zea Mays)*. J. Amer. chem. Soc. **63**, 1632 (1941).

44. — — The Molecular Constitution of Enzymatically Synthesized Starch. J. Amer. chem. Soc. **63**, 2171 (1941).

45. — — The Molecular Constitution of Amylose and Amylopectin of Potato Starch. J. Amer. chem. Soc. **65**, 1157 (1943).

46. HAWORTH, W. N.: The Constitution of Sugars, p. 24. London: E. Arnold and Co. 1929.

47. Haworth, W. N., R. L. Heath and S. Peat: Constitution of the Starch Synthesised in vitro by the Agency of Potato Phosphorylase. J. chem. Soc. [London] 1942, 55.

48. Haworth, W. N., E. L. Hirst and F. A. Isherwood: Polysaccharides. XXIII. Determination of the Chain Length of Glycogen. J. chem. Soc. [London] 1937, 577.

49. Haworth, W. N., E. L. Hirst and R. R. Lyne: A Water-soluble Polysaccharide from Barley Leaves. Biochemic. J. 31, 786 (1937).

50. Haworth, W. N., H. Raistrick and M. Stacey: Polysaccharides Synthesized by Microorganisms. II. The Molecular Structure of Varianose Produced from Glucose by Penicillium varians G. Smith. Biochemic. J. 29, 2668 (1935).

51. Hehre, E. J.: Serological Properties of Products Synthesized from Sucrose by Enzymes from Different Strains of Leuconostoc Bacteria. Proc. Soc. exp. Biol. Med. 54, 18 (1943).

52. — Phenomenon of Precipitation on Mixing Hemoglobin with Tissue Extract Antigen. Proc. Soc. exp. Biol. Med. 54, 240 (1943).

53. — Serological Reactions of Levans Synthesized from Sucrose and Raffinose by Bacterial Enzymes. Proc. Soc. exp. Biol. Med. 58, 219 (1945).

54. Hehre, E. J., A. S. Carlson and J. M. Neil: Production of Starch-like Material from Glucose-1-phosphate by Diphtheria Bacilli. Science [New York] 106, 523 (1948).

55. Hehre, E. J. and J. Y. Sugg: Serologically Reactive Polysaccharides Produced through the Action of Bacterial Enzymes. I. Dextran of Leuconostoc mesenteroides from Sucrose. J. exp. Medicine 75, 339 (1942).

56. Hestrin, S. and S. Avineri-Shapiro: The Mechanism of Polysaccharide Production from Sucrose. Biochemic. J. 38, 2 (1944).

57. Irvine, J. C. and J. W. H. Oldham: The Coupling of Glucose and γ-Fructose. Conversion of Sucrose. J. Amer. chem. Soc. 51, 3609 (1929).

58. Irvine, J. C., J. W. H. Oldham and A. F. Skinner: Condensation of Glucose and Fructose. Synthesis of an Iso-sucrose. J. Amer. chem. Soc. 51, 1279 (1929).

59. Jackson, E. L. and C. S. Hudson: Studies on the Cleavage of the Carbon Chain of Glycosides by Oxidation. A new Method for Determining Ring Structures and Alpha and Beta Configurations of Glycosides. J. Amer. chem. Soc. 59, 994 (1937); The Cleavage of the Carbon Chain of Levoglucosan by Oxidation with Periodic Acid. J. Amer. chem. Soc. 62, 958 (1940). — Hann, R. M., W. D. Maclay and C. S. Hudson: The Structures of the Diacetone Dulcitols. J. Amer. chem. Soc. 61, 2432 (1939).

60. Kalckar, H. M.: The Nature of Energetic Coupling in Biological Syntheses. Chem. Reviews 28, 71 (1941).

61. Katz, J., W. Z. Hassid and M. Doudoroff: Arsenolysis and Phosphorolysis of the Amylose and Amylopectin Fractions of Starch. Nature [London] 161, 96 (1948).

62. Kiessling, W.: Über den das Glykogen phosphorylierende Fermentprotein-komplex und eine enzymatische, reversible Glykogensynthese. Biochem. Z. 302, 50 (1939).

63. Kursanov, A. and N. Kriukova: Participation of Phosphatase in the Synthesis of Sucrose. Biokhimiya 4, 229 (1939); Chem. Abstr. 34, 1710 (1940).

64. Lebedew, A. u. A. Dikanowa: Über die enzymatische Rohrzuckersynthese. Hoppe-Seyler's Z. physiol. Chem. 231, 271 (1935).

65. Lipmann, F.: Metabolic Generation and Utilization of Phosphate Bond Energy. Advances in Enzymology 1, 99 (1941).

66. McCready, R. M. and W. Z. Hassid: Transformation of Sugars in Excised Barley Shoots. Plant Physiol. 16, 599 (1941).

67. Meagher, W. R. and W. Z. Hassid: Synthesis of Maltose-1-phosphate and D-Xylose-1-phosphate. J. Amer. chem. Soc. 68, 2135 (1946).

68. Meyer, K. H.: The Chemistry of Glycogen. Advances in Enzymology 3, 109 (1943).

69. Meyer, K. H., M. Wertheim et P. Bernfeld: Reserches sur l'amidon. IV. Méthylation et détermination des groupes terminaux d'amylose et d'amylopéctine de maïs. Helv. chim. Acta 23, 865 (1940). Recherches sur l'amidon. XIII. Contribution à l'étude de l'amidon de pommes de terre. Helv. chim. Acta 24, 378 (1941).

69a. Meyerhof, O. and P. Oesper: The Free Energy of Phosphorylation. Federation Proc. 7, 174 (1948).

70. Niven, C. F. Jr., K. L. Smiley and J. M. Sherman: The Polysaccharides Synthesized by Streptococcus salivarius and Streptococcus Bovis. J. biol. Chemistry 140, 105 (1941).

71. Oparin, A. u. A. Kursanov: Über die enzymatische Synthese des Rohrzuckers. Biochem. Z. 239, 1 (1931).

72. Parnas, J. K.: Der Mechanismus der Glykogenolyse im Muskel. Ergebn. Enzymforsch. 6, 57 (1937).

73. Peat, S., E. Schlüchterer and M. Stacey: Polysaccharides. XXIX. Constitution of the Dextran produced from Sucrose by Leuconostoc Dextranicum (Betacoccus Arabinosaceous Haemolyticus). J. chem. Soc. [London] 1939, 581.

74. Pictet, A. et H. Vogel: Synthèse du saccharose. Helv. chim. Acta 11, 436 (1928). Zur Synthese des Rohrzuckers. Ber. dtsch. chem. Ges. 62, 1418 (1929).

75. Potter, A. L., J. C. Sowden, W. Z. Hassid and M. Doudoroff: α-L-Glycose-1-phosphate. J. Amer. chem. Soc. 70, 1751 (1948).

76. Purves, C. B. and C. S. Hudson: The Analysis of Fructoside Mixtures by Means of Invertase. VI. Methylated and Acetylated Derivatives of Crystalline β-Benzylfructopyranoside. J. Amer. chem. Soc. 59, 1170 (1937).

77. Raybin, H. W.: A New Color Reaction with Sucrose. J. Amer. chem. Soc. 55, 2603 (1933). The Direct Demonstration of the Sucrose Linkage in Oligosaccharides. J. Amer. chem. Soc. 59, 1402 (1937).

78. Stacey, M.: Macromolecules synthesized by Microorganisms. (Tilden Lecture). J. chem. Soc. [London] 1947, 853.

79. Zemplén, G. u. Á. Gerecs: Notiz zur Synthese des Rohrzuckers. Ber. dtsch. chem. Ges. 62, 984 (1929).

(Received, April 23, 1948.)

Recent Developments in the Structural Problem of Cellulose.

By E. Pacsu, Princeton, New Jersey.

I. The Linear Structure of Cellulose.

With 2 Figures.

Introduction.

Although a proper definition of "pure cellulose" has not yet been found, the term "cellulose", as commonly used by chemists, is restricted to a product which may be obtained from raw cotton by removal of fat, wax, pectin and other impurities. It is tacitly assumed that the necessary operations of purification involving treatment with hot dilute alkali, often followed by bleaching with hypochlorite, do not cause deep-rooted changes in the chemical structure of native cellulose. The resulting product, termed "standard cellulose", is now usually employed as the starting material in structural investigations. During the last one hundred years an enormous mass of experimental data has accumulated from which conflicting views and opposing conceptions have emerged, each of them offering evidence as to the true chemical constitution of cellulose. Many of these suggestions, however, have been based on data interpreted by several workers in support of their favorite theories which reveal only part of the picture and fail to allow a comprehensive generalization.

Fundamentally, all competing formulae may be classified into two groups according to the underlying hypotheses. In the first group there belong the symbols which are based on the view that cellulose is best represented by "association" or "aggregation" of anhydroglucose or anhydrocellobiose *molecules*. The second group comprises the formulae advanced by those investigators who regard cellulose as a "linear macro-molecule" in which uniform and equivalent, covalent glycosidic linkages unite a large number of glucose-anhydride *residues*. In the light of the modern electronic concept of valence the first hypothesis cannot be readily understood since it calls into play "residual forces of affinity" and some similar, ill-defined conceptions such as auxiliary or secondary valencies, association forces, etc. The rise and decline of the association theory has been recently discussed by PURVES (75), and the reader is referred to his article and to other accounts of the historical development of the chemistry of cellulose (2).

At present, we are mainly concerned with the second school of thought, almost universally accepted today, which represents the modern period of cellulose chemistry since 1921 when the molecular chain concept was put into focus by FREUDENBERG (20) and by HAWORTH and HIRST (35). The ground for this development was well prepared by STAUDINGER's studies on the macromolecular structure of natural and synthetic high polymers (5). Soon the theory gained powerful support from SPONSLER and DORE's (86) pioneering work of X-ray investigation. The successful isolation by ZECHMEISTER and TÓTH (101) of higher oligosaccharides of the cellobiose-cellulose series from acid-catalyzed degradation products of cellulose contributed significantly to the rise of the molecular chain theory which appeared firmly entrenched in 1935 when FREUDENBERG and BLOMQVIST (21) submitted their comprehensive summary of *chemical, polarimetric, kinetic* and *static* proofs for the linear structure (I) of cellulose. Indeed, in the last decade there was scarcely anything left for the cellulose chemist but to attempt to ascertain the precise number of glucose-anhydride units in a rather long chain molecule which was characterized by its property of breaking readily and in a *random* fashion into fragments,

under certain experimental conditions. However, if this simple picture of a linear macromolecule with its "uniform and equivalent, covalent glycosidic linkages" were truly representative of the chemical structure of the cellulose molecule, then it should account for all the facts known about the chemical and physical properties of this polysaccharide.

Obviously, it is of paramount importance to ascertain that the arguments put forth in support of the linear formula (I) stand up in the light of a critical examination. It will be shown in the following that this is not the case. Indeed, there are numerous experimental facts which point directly against the conception of a simple linear cellulose structure.

Critical Review of the Proofs Supporting the Linear Structure of Cellulose.

1. Nature and uniformity of the building units.

Evidently, one of the most important questions centers around the nature and uniformity of the building unit, $C_6H_7(OH)_3O_2$. Although no evidence is known that would indicate the presence of any sugar other than *D*-glucose in cellulose, it is equally true that it has not yet been possible to isolate this sugar in 100 per cent. yield of the theoretical quantity. Monier-Williams (*61*) obtained 90,7 per cent. of crystalline glucose and could detect no other products of hydrolysis. Karrer and Illing (*53*) found that the enzyme lichenase, present in the intestinal juice of *Helix pomatia*, converted regenerated cellulose into glucose to the extent of 95 per cent. of the theoretical quantity, although it had no effect on native cellulose. Applying an acetylation method Irvine and Hirst (*49*) prepared cellulose triacetate in more than 99 per cent. yield, which on methanolysis gave the crystalline mixture of the α- and β-glucosides accounting for 95,5 per cent. of the cotton originally employed. The same authors also made special efforts to detect pentoses if they were present in the acid hydrolysate. However, the furfural test gave a negative result. Neither was any other compound than methyl glucoside found. Therefore, "this work was generally accepted as decisive proof that pure cellulose consisted exclusively of glucose residues" (*76*).

In view of the great volume of literature involving the detection, quantitative estimation, and possible role of the carboxyl group in pure cellulose, one should expect that the material unaccounted for in the above experiments includes some "-onic" or "-uronic" acids which are known to be present in a very small, though definite amount in every cellulose sample. It is, however, not known whether these acidic units constitute an integral part of the cellulose molecule or whether they have been formed only as a result of the treatment with dilute alkali at the boil of the raw cotton. Recent data indicate that the carboxyl content of "standard cellulose" amounts to about 0,052 per cent. (*99*) or to 0,04 to 0,05 per cent. (*67*) for cotton cellulose carefully freed of pectin. Since native cellulose always has been and probably will be treated with chemicals in order to remove the various non-cellulosic constituents, the problem of origin of these carboxyl groups can perhaps not be solved with certainty.

2. Location and number of the free hydroxyl groups.

Extensive studies by many investigators showed that, on the average, cellulose contains three hydroxyl groups for each glucose residue, and that these groups are located at the second, third and sixth carbon atoms of the glucose residues. These results were obtained from experiments yielding in one case (*50*) 91,5 per cent. of the theoretical amount of methyl trimethylglucoside from methanolyzed methylcellulose, and about 99 per cent. of the same product in another case (*42*), in which alkali-soluble "cellulose A" was used as starting material. It should be noted with interest that, in spite of numerous and special attempts, it had not yet been possible to reach the theoretical methoxyl content (as calculated from the linear formula I) of trimethylcellulose (45,58 per cent. methoxyl), unless the starting material consisted of distinctly degraded cellulose and, possibly, unless it suffered additional degradation while it was being methylated. The cause of this difficulty in reaching the theoretical value was sought either in "steric hindrance" (*4*) involving some of the hydroxyl groups in the glucose residues or in the presence of "blocked" hydroxyl groups as postulated by KARRER and ESCHER (*52*). This blocking effect was thought to be due to elimination of water molecules from the hydroxyl groups in the 6-position at every fourth to fifth glucose residue, presumably between adjacent chain molecules, thus giving rise to cross linkages and causing the observed deficiency in the number of available hydroxyl groups. It is significant to note that this explanation casts serious doubts on the correctness of the linear formula (I).

3. The chemical proof of the constitution of cellulose.

The allocation of the hydroxyl groups in cellulose to the second, third and sixth positions of the glucose residues did not answer the question whether the $C_6H_7(OH)_3O_2$ units possessed an open-chain, or a 1,5- or 1,4-ring structure; furthermore, whether the assumed glycosidic bonds were attached to the fourth or fifth position in the adjacent units; and whether the glycosidic bond had an α- or a β-configuration. Since the partial breakdown of cellulose in acetic anhydride containing concentrated sulfuric acid yielded cellobiose octaacetate (*84*), the presence of this disaccharide among the degradation products appeared to lend support for the hypothesis that cellobiose is *preformed* in cellulose and that the latter possesses a straight chain structure. The molecular structure of cellobiose was definitely proved by HAWORTH and collaborators (*37*) to be 4-(β-D-glucopyranosido)-D-glucopyranose. Later, cleavage products like cellobiose, cellotriose, cellotetraose and cellohexaose were isolated by ZECHMEISTER and TÓTH (*101*) from cellulose degraded by 42 per cent. hydrochloric acid at 15° C. Since it was found that in the first three of these oligosaccharides the glucose residues were united through 1,4-β-glycosidic linkages, it appeared *probable* that all glucose residues in cellulose were linked through such 1,4-β-glycosidic bonds. These experimental results were regarded by Freudenberg and Blomqvist as representing the *chemical proof* of the constitution of cellulose.

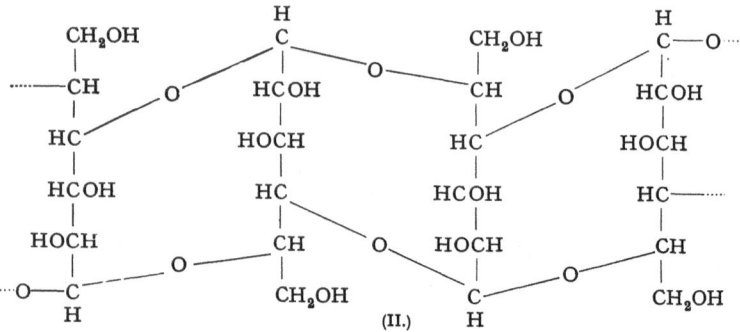

(II.)

However, formula (II), a variation of which was proposed by Pictet (73) for starch, but which is equally applicable to cellulose, should not be ignored just because it postulates a *dicyclic acetal condensation* between adjacent open-chain units, which type of structure has not yet been proved to be present in the oligosaccharides. If cellulose were truly represented by formula (II), then the formation of cellobiose, cellotriose, etc., during acidic degradation could be just as readily interpreted as it is explained on the basis of formula (I). In any case, it seems doubtful whether the formation of these relatively simple oligosaccharides under the drastic conditions of acetolytic or acidic degradation constitutes a valid argument either for or against any cellulose symbol which is based on the *exclusive* (or *partial*) occurrence of either divalent or tetravalent $C_6H_7(OH)_3O_2$ units.

4. The polarimetric proof of the constitution of cellulose.

According to Freudenberg and his collaborators (21, 23), if the cellulose chain were entirely uniform, its molecular rotation would be an additive function of the rotations contributed by its individual constituent units. By the aid of a formula based on this deduction, these authors calculated the molecular rotations of cellotriose and cellotetraose. On the other hand, the rotations of the latter made it possible to calculate that of cellulose. Alteration of one bond in every hundred from the β- to the α-configuration in the fully methylated chain would have caused a deviation of 2,5° between the observed and the calculated specific rotations. The agreement was so close that Freudenberg concluded that the cellulose chain must possess the uniformity assumed in the calculations (β-glycosidic bonds only); and he considered this work as the *polarimetric proof* of the cellulose structure. However, it should be pointed out that while Freudenberg's results failed to supply any evidence for the presence of α-linkages *within* one hundred glucose-anhydride units, they left the problem of the nature and uniformity of the bonds open beyond this limit of chain length.

5. The kinetic proof of the constitution of cellulose.

FREUDENBERG, KUHN, and co-workers (*21, 24*) measured the rate of hydrolysis of cellulose in 51 per cent. ($d_{18}^{18} = 1,415$) sulfuric acid solution at 18° C. and 30° C. The hydrolysis was followed optically and by hypoiodite titrations.

For calculation of the form of the rate plots the assumption was made that when the infinite chain disintegrated, all glycosidic bonds adjacent to the reducing ends of the various-sized fragments broke at the rate given for cellobiose ($K_1 = 1,07 \times 10^{-4}$ at 18° C.; $6,94 \times 10^{-4}$ at 30° C.), and that all other bonds in the fragments hydrolyzed at the rate ($K_i = 0,305 \times 10^{-4}$ at 18° C.; $2,34 \times 10^{-4}$ at 30° C.) determined for the initial stage of hydrolysis by extrapolation.

A mathematical formula which expressed the degree of hydrolysis, $(1 - x)$, as an exponential function of K_2, K_i, and t was derived by KUHN (*22, 54*). The agreement between the calculated and experimental $(1 - x)$ values at any time t was so close that FREUDENBERG regarded this work as a *kinetic proof* of the constitution of cellulose (formula I). The underlying assumption was that all the hydrolyzable bonds, except the ones adjacent to the reducing ends, were equivalent and of the $1,4$-β-glucosidic type which was proved to be present in cellobiose or cellotriose.

However, it should be clearly understood that while this result leaves very little doubt about the nature and uniformity of the glycosidic bonds *in cellulose which is dissolved in strong sulfuric acid*, it gives no information as to the nature, type, and location of any other hydrolyzable bonds which may conceivably be present in the fibrous material to the extent of one per cent. or less.

The crucial point in this argument is that FREUDENBERG and BLOMQVIST (*21*) prepared their solution by "quickly" dissolving cotton in 65 per cent. sulfuric acid and then diluting the solution to 51 per cent. "as soon as it was possible to do so without precipitating dextrins". It is, therefore, evident that the solution actually used for these important rate measurements did not any longer contain native cellulose but a product which (having D. P.* 50), was even more degraded than dextrin. Consequently, if in the cotton fibers there occurred a few acid-sensitive bonds which hydrolyzed, say, one thousand times as fast as ordinary β-glycosidic linkages, then their presence could not be detected in these kinetic measurements, since such bonds would have been cleaved long

* D. P. = *degree of polymerization*, indicating the number of glucose-anhydride units.

before the regular glycosidic bonds began to split. Therefore, an extrapolation to cellulose *fibers* of Freudenberg's kinetic proof of the molecules
dissolved in strong mineral acid is not permissible. It will be shown
below that the non-recognition of this limitation is mainly responsible
for the almost universal acceptance of formula (I).

6. The static proof of the constitution of cellulose.

The partial degradation of cellulose upon acetolysis in strongly acidic
media gives rise to the oligosaccharides, cellobiose (35 to 51 per cent.),
cellotriose (9,9 per cent.), cellotetraose (1,2 per cent.), and cellohexaose
(0,5 per cent.) (*101*), the latter sugar by some authors believed to be
cellopentaose (*90*).

Mathematical treatment of the problem showed (*54*) that the random
breakdown of a long chain with uniform glycosidic bonds could not,
possibly, produce more than 18,7, 13,8, and 9,1 per cent. of cellotriose,
cellotetraose, and cellohexaose, respectively. Calculations also showed
that, if in the acetolysis experiments cellobiose octaacetate were completely
stable in the reaction mixture, no more than 70 per cent. would be formed.
The actual yield of 51 per cent. (*26*) appears, therefore, to be in agreement
with the molecular chain concept. Since all the yields actually obtained
were very far from the calculated values, this *static proof* of the constitution
of cellulose seems to be rather unconvincing. All that may be said is
that the experimental results are not in obvious contradiction to a linear
structure of cellulose.

7. The X-ray pattern of cellulose.

The chain structure of cellulose is in agreement with the results
of X-ray analysis. After the fundamental contribution by Sponsler
and Dore (*86*) and establishment of the chemical structure of cellobiose
by Haworth and collaborators (*37*), the best molecular models for the
interpretation of the X-ray pattern of cellulose were developed by
Mark and Meyer (*55, 59*) and by Meyer and Misch (*58, 60*).

In these models the length of the unit cell, 10,3 Å., is identical with the length
of two glycopyranose residues or one cellobiose unit in the direction parallel to the
fiber axis b. The dimension of the basic cell along the *a* axis is 8,35 Å, whereas
that along the *c* axis, which is perpendicular to the *a*, *b* plane and forms an angle,
$\beta = 84°$, with *a*, is 7,95 Å. From the volume and mass of the $C_6H_7(OH)_3O_2$ unit
and the density of cellulose it was calculated that four cellobiose residues form
the vertical edges of the unit cell, the center of which is occupied by another, but
inverted cellobiose residue. Since the cellobiose residues at each vertical edge
are shared by the four neighboring unit cells, this leaves one cellobiose residue
per unit cell for the four edges. The latter cellobiose residue, together with the
inverted one in the center, gives two cellobiose residues or four glucose residues
per unit cell, and this geometrical pattern repeats itself in all directions. Along
the *a* axis the glucopyranose rings are separated only by a distance of 2,5 Å. It

is, therefore, reasonable to assume that hydrogen bonds are effective between two oxygen atoms belonging to adjacent chains. Along the c axis the nearest distance of atomic centers is about 3,1 Å. This corresponds closely to the distance to be expected if VAN DER WAALS' forces hold the lattice together in this direction.

According to the current views (83), in a cellulose fiber the three-dimensional arrangement of the glucose anhydride units is repeated in all directions to build up crystalline areas (crystallites or micelles) which are separated by amorphous or intercrystalline areas. An approximate idea of the size and form of the crystallites has been derived from the breadth of the layer lines of the X-ray diagram. Accordingly, the micelle of ramie fiber is calculated to be a rhombus which measures at least 600 Å. along the fiber axis and about 50-by-60 Å. across this direction. Since ultracentrifugal, viscosity, and some other physico-chemical methods all indicate that the "molecular cellulose chains" are considerably longer (15400 Å. for a cellulose with D. P. 3000), it is now assumed that a substantial part of the fibrous system is amorphous—that is, the long-chain molecules in certain regions are not strictly parallel as they are in the crystalline areas, or they are not so nearly parallel as in the "mesomorphous" portions (expression for the area neither fully crystalline nor quite amorphous). Obviously, the exact lengths of the chains along the b axis of the fiber, that is the number (D. P.) of glucose-anhydride units present in the "average chain molecules" of cellulose will determine the molecular weight of cellulose. This problem is discussed in the following section.

Molecular Weight Determination of Cellulose.

a) Chemical Methods.

All these chemical methods are based upon the fundamental *assumption* that the chains in cellulose are open and that either the reducing or the non-reducing end-groups of the finite chains undergo certain reactions which can be carried out quantitatively. The ratio of end-groups to "normal" groups is then usually given as the "number-average D. P." or the weight of a "cellulose molecule". This treatment of the problem is, of course, permissible only if the basic assumption is correct. It is well known, however, that similar estimations when carried out on the so-called "amylopectin component" of starch led to the absurdly low value of D. P. 20 to 30, whereas molecular weight determinations by physico-chemical methods resulted in a wide array of values of D. P. (500, 2000, 4300, 20000 and even 70000 were reported by different laboratories). Evidently, a chemical end-group analysis may give "equivalent weights" or the weights of the "repeating sections" only; in contrast, the physico-chemical methods will furnish the average weights of the particles as present in the solution.

Another interesting problem is, just how much importance should be placed on the analytical data obtained by *any* chemical method of molecular weight estimation in cellulose chemistry. First of all, the heterogeneous nature of the reactions usually employed makes it difficult to obtain reproducible results. Secondly, the previous history of the samples studied is not always known with certainty; and variations in the methods of "purification" as applied by the respective investigators prevent a fair comparison of the recorded data.

Methods for isolating cellulose without any degradation are still unknown. Consequently, all the analytical procedures refer to more or less degraded products. The major difficulty, however, lies in the lack of reliable analytical methods that would permit the accurate estimation of a single glucose unit which carries some particular reacting group, in the presence of two or three thousand "indifferent" units. Thus, a native cellulose sample may contain only one carboxyl group per three thousand glucose anhydride units, that is, a 4,86 gram sample of such a cellulose would consume only 0,1 cc. of 0,1 normal-alkali. In general, analytical precision and accurate figures can only be attained on severely degraded products.

Of the various chemical methods which have been used in determination of end-groups, the estimations based upon the potential reducing groups are frequently employed. It is important to note that the reducing units customarily, but *for no compelling reason*, are written in the pyranoid form. The interpretation of the analytical results rests on the following assumptions: (*a*) each chain contains one glucose molecule which is substituted at its fourth position with the rest of the chain; (*b*) all of the reducing power is contributed by such terminal units only; (*c*) the reagents used do not produce new reducing groups. In most cases there is no way of telling whether or not these assumptions are justified with respect to a particular sample.

For determination of the reducing power the *"copper number"* method of Schwalbe (*82*) or one of its numerous variations is being traditionally employed. It has been the target of much criticism and, at the present time, the erratic results it gives have caused it to be relegated to the role of a rough estimate at best (*97*).

Concerning the well-known *hypoiodite* method, a recent investigation (*69*) indicated that the procedure is inapplicable to the reducing type of oxycellulose, although Harris and coworkers (*56*) claimed to have made the method usable for hydrocelluloses. Because of the instability of the reagent and the dependence of the analytical results upon the conditions, this method has been adversely criticized from many sides. It gives reliable results only with simple sugars and oligosaccharides, but in the case of polysaccharides like cellulose or starch the consumption of iodine does not stop at any given time, and "over-oxidation" with the production of iodoform readily occurs.

In order to avoid the use of alkaline reagents in end-group analysis, ice-cold *potassium permanganate* in dilute sulfuric acid solution was employed recently (43). The reaction was found to be sufficiently selective and reasonably complete in that fairly constant values were obtained after about forty minutes of reaction time. However, all these oxidation methods, just as several others which depend on the formation of condensation products with various reagents, give reasonably accurate results only on cellulosic materials when these are considerably degraded to show any measurable reaction at all.

The end-group assay based upon the quantitative formation of *tetramethylglucose* from the non-reducing end of the assumed chain molecule is well known. The consensus of many investigators is expressed by SOOKNE and HARRIS (85) in stating that the method suffers from two obvious disadvantages: (a) degradation during the long series of rather severe treatments; and (b) the requirement of isolation of a material which is present in only minute proportions in the hydrolysate. This opinion is borne out of the large and consistent differences that exist between the analytical results of the various schools of cellulose experts. In spite of its questionable value as an accurate procedure for chain-length determination, the method—through the conflicting results it furnished—contributed substantially to the newer developments in the cellulose problem, as it will be discussed below.

Chain-length estimations based upon the *carboxyl* content of cellulose suffer from the uncertainty regarding the structural arrangement or position of the carboxyl groups in the sample. If it were true that the carboxyl groups of all the chains in the fiber originated from the reducing end-groups and that none of the hydroxyls of the chains became oxidized to carboxyl groups during isolation and purification of the cellulose, then the values of acidity could be used for chain-length estimation. Mostly, neither of these requirements is fulfilled.

A second difficulty is encountered in the selection of a suitable method of carboxyl estimation. Various attempts have been made to develop a sufficiently exact procedure by the use of direct or indirect titration with alkali but the results are not altogether satisfactory.

Recently, DAVIDSON and NEVELL (17) have made a comparative study of the following methods: (a) the methylene blue absorption method, (b) a modified form of the silver absorption method of SOOKNE and HARRIS, (c) the calcium acetate method of LÜDTKE, and (d) the alkali-titration method of NEALE and STRINGFELLOW. The materials used were: a purified cotton cellulose and five series of oxycelluloses prepared from it by oxidation with metaperiodate, chlorous acid, alkaline hypobromite, neutral hypochlorite, and dichromate in the presence of dilute sulfuric acid. The range of the carboxyl content studied did not extend much above a stage of cellulose oxidation corresponding to one carboxyl group in every sixty glucose units (10 m.moles/100 g.).

According to Davidson and Nevell (*17*) in the range of relatively low carboxyl content, the methylene blue absorption method is the most generally useful. The silver absorption method gives satisfactory results provided that a correction can be made for the reduction of silver ion by the cellulose material. The calcium acetate procedure as usually carried out is likely to give too low values for the carboxyl content and a modified procedure had to be devised. The Neale and Stringfellow analysis gives satisfactory results with acidic oxycelluloses of low reducing power, but high and fictitious values with reducing oxycelluloses, owing to the formation of acidic substances when they are treated with alkali. In the authors' opinion this method is unsuitable for general use.

Nevell (*17*) also investigated the applicability to cotton and different types of oxycelluloses of the Lefèvre-Tollens method for the determination of uronic acids, based on the measurements of the rate at which carbon dioxide is evolved upon the action of boiling 12 per cent. hydrochloric acid (*99*). The results show that the evolution of carbon dioxide from the oxycelluloses examined does not come to an end within ten hours, nor does its rate fall to the constant value characteristic of purified cotton cellulose. This method may, nevertheless, be used to obtain a rough estimate of the proportion of the uronic acid groups present; the amount of carbon dioxide evolved in ten hours, corrected for the amount evolved from unmodified cotton in the same period, is suggested as an approximate measure of the uronic acid content. The results thus obtained indicate that almost all the acidic groups in dichromate oxycelluloses, but not more than 40 per cent. of those in hypobromite oxycelluloses, are of the uronic type. Determinations on periodate and periodate-chlorite oxycelluloses indicate that these materials contain few, if any, uronic acid groups.

A recent method by Hirst et al. (*9*) is claimed to effect quantitative liberation of *formic acid* from the terminal sugar residues of cellulose, starch, etc., in about 180 hours at 15° C., by the use of periodic acid under special conditions. According to these authors, over-oxidation does not take place and the formic acid can be titrated by 0,01 *N*-baryta either in the presence of the oxidation products (dialdehydes) or after ether extraction.

b) Physical Methods.

Almost all physical methods in use for determination of molecular weight were developed on true (molecular disperse) solutions. Once such a solution is secured, the molecular weight can be determined by various procedures such as the viscosity method, the osmotic method, the sedimentation and diffusion methods, the light scattering etc. Although some cellulose and starch chemists still believe that such high-polymers are "merely in a state of colloidal dispersion or in a state of transition from colloidal to molecular dispersion" [(*2*), p. 587], all evidence supports the theory that molecular dispersions are possible and that complete dispersion does take place. In a clear presentation Spurlin (*87*) convincingly points out the arguments which favor the theory, originally supported mainly by Staudinger (*5*), that each cellulose molecule is molecularly dispersed and is only occasionally associated with one of its neighbors. Any such association, if formed, is soon broken up by thermal agitation.

The main argument against the existence of permanent micelles in the solution lies in the easy chemical conversion of cellulose into tri-substituted derivatives of the same degree of polymerization. For instance, in cellulose triacetate where all the chains must have reacted, a micelle structure could not possibly continue to exist any longer. Since the same degree of polymerization is found after de-acetylation as before, it would be necessary to assume that the micelles actually "remembered" their original state and would accordingly reform with the same size (21). From the results of the X-ray investigation it can be concluded that the micellar strands of cellulose probably consist of a hundred chain molecules. These are arranged in such a manner that eight chain mole-cules, connected by a network of hydrogen bonds, form the width (7 × 8,35 Å. = 58,45 Å.) of the crystallite along the a axis. Some thirteen layers (6 × 7,95 Å. = 47,7 Å.) of the resulting laminae, held together by the weaker VAN DER WAALS' forces along the c axis, will then constitute a crystallite or micelle of indefinite length along the fiber axis, b. In an appropriate solvent this structure should disintegrate first into laminae, then — owing to the cleavage of the hydrogen bonds in the latter — into the individual, electrically almost neutral chain molecules.

Thus, any dilute solution of cellulose or its trisubstituted derivatives should indeed consist of molecularly dispersed chain molecules possessing indefinite length. The "molecular weight" will then largely depend on the fate of these indefinitely long chains in the solution. Unless most careful precautions have been taken for the elimination of the last traces of air, cellulose is subject to oxidative degradation in cuprammonium solution or in other alkaline media. This has been known since early days and extensively utilized by the industry for breaking down the cellulose molecules to a suitable level by the so-called "ageing" process. Today it appears to be certain that the wide variations in molecular weights of cellulose as obtained by physico-chemical measurements in alkaline solutions are caused solely by the cracking action of oxygen—first, during the purification which usually ends with an alkali boil; and, second, because of the accidental inclusion of air in the cuprammonium solution. According to GOLOVA (31) when oxygen is removed by helium instead of nitrogen in a special apparatus, a molecular weight of D. P. 10000 is usually obtained for native cotton.

In view of the widely scattered values reported from various laboratories, it appears to be a justified *a priori* conclusion that the molecular weight of native cellulose is indefinite and that the results of various measurements invariably refer to a degraded material. Native cellulose is unintentionally degraded to about D. P. 3000, which is then reported as the average molecular weight. Clearly, the latter figure is a measure only of the extent and gravity of the chemical accidents which the original fiber has suffered from the time it was removed from the boll until it became the object of scrutiny in an apparatus for determination of its "molecular weight".

It is also evident that the correctness of the theory of molecular dispersion in the last analysis rests upon the linear chain concept of cellulose. If it becomes necessary to take recourse to assumptions such as "aggregation", "association", or "interplay of secondary valence forces" in order to explain phenomena that do not fit into the simple picture of molecular dispersion, then a re-examination of the linear chain concept of cellulose structure is strongly indicated.

All physical methods when applied to cellulose itself suffer from the inescapable disadvantage that cellulose is soluble only in concentrated mineral acids or in strong bases, such as cuprammonium hydroxide. The former are reaction solvents which immediately attack the cellulose molecule, and in due course only D-glucose is found in the solution. Although alkali alone cannot be regarded as a reaction solvent, in combination with oxygen it becomes one of the most dangerous media when employed for cellulosic materials. Since application of either alkaline or acidic reagents is unavoidable in cellulose chemistry, it is of paramount importance to ascertain the chemical nature of the degraded materials, no matter how small the damage appears to be.

In the next two sections some outstanding problems of both oxycellulose and hydrocellulose will be discussed.

Oxycellulose.

Much confusion exists concerning the properties of oxidized celluloses since early investigators seldom followed any particular method of preparation. Unruh and Kenyon (97) have recently published an excellent summary of our present knowledge regarding the structural problems of oxidized celluloses. A comprehensive survey is also presented by Rutherford and Harris (79).

It is now generally recognized that "oxycellulose" is not a chemical entity since many varieties of materials can be produced by oxidation of cellulose, which can be classified into two main groups: (a) the reducing type, which is characterized by a high copper number and general behavior of aldehydes, and (b) the acidic type, which possesses low reducing power but a high carboxyl content. The first type is well represented by the "periodate oxycellulose", which can be prepared according to Jackson and Hudson (51) by application of periodic acid to cellulose. The structure of the product approximates that demanded by the ideal equation, i. e. a poly-(2,3'-erythro-glyoxal monoacetal) (III).

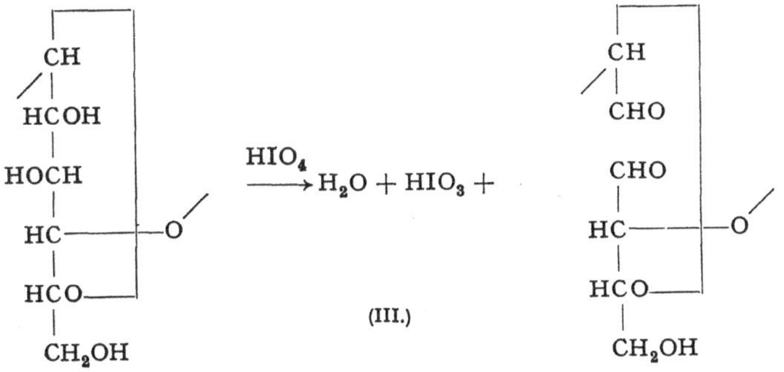

$$
\begin{array}{c}
\text{CH} \\
\diagup \;| \\
\text{HCOH} \\
| \\
\text{HOCH} \\
| \\
\text{HC}\!-\!\!-\!\!-\!\text{O} \\
| \\
\text{HCO}\!-\!\!| \\
| \\
\text{CH}_2\text{OH}
\end{array}
\quad
\xrightarrow[\text{}]{\text{HIO}_4}
\text{H}_2\text{O} + \text{HIO}_3 +
\quad
\begin{array}{c}
\text{CH} \\
\diagup \;| \\
\text{CHO} \\
| \\
\text{CHO} \\
| \\
\text{HC}\!-\!\!-\!\!-\!\text{O} \\
| \\
\text{HCO}\!-\!\!| \\
| \\
\text{CH}_2\text{OH}
\end{array}
$$

(III.)

The second type of oxycellulose is obtained by application of nitrogen dioxide to cellulose as shown by UNRUH, KENYON, and YACKEL (96, 100). The reaction essentially stops at one carboxyl group for each glucose unit, giving rise to a poly-anhydroglucuronic acid or "celluronic acid" (IV) with a carboxyl content of 25,57 per cent., according to the ideal equation.

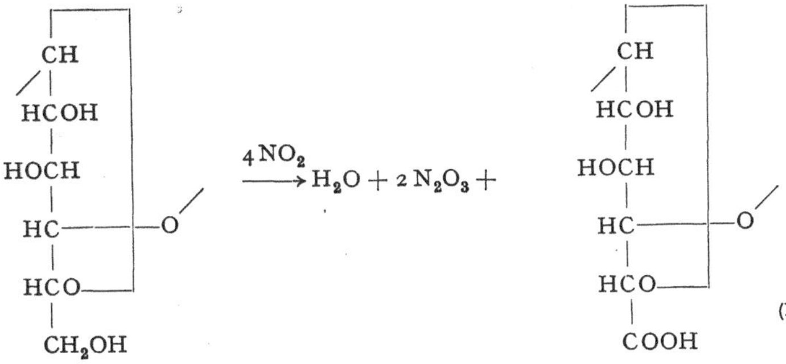

$$
\begin{array}{c}
\text{CH} \\
\diagup \;| \\
\text{HCOH} \\
| \\
\text{HOCH} \\
| \\
\text{HC}\!-\!\!-\!\!-\!\text{O} \\
| \\
\text{HCO}\!-\!\!| \\
| \\
\text{CH}_2\text{OH}
\end{array}
\quad
\xrightarrow[\text{}]{4\,\text{NO}_2}
\text{H}_2\text{O} + 2\,\text{N}_2\text{O}_3 +
\quad
\begin{array}{c}
\text{CH} \\
\diagup \;| \\
\text{HCOH} \\
| \\
\text{HOCH} \\
| \\
\text{HC}\!-\!\!-\!\!-\!\text{O} \\
| \\
\text{HCO}\!-\!\!| \\
| \\
\text{COOH}
\end{array}
$$

(IV.)

Substances of lower carboxyl content are to be regarded as copolymers of anhydroglucose and anhydroglucuronic acid units. They are insoluble in alkali if the carboxyl contents fall below about 12 per cent. Both the aldehyde and acidic types of oxycelluloses may occur in the numerous modifications that cellulose will undergo in a great variety of less specific oxidizing media, such as hypochlorite, hypobromite, chromic acid, permanganate, etc. solutions. Since in the various commercial processes the partial formation of oxidized cellulose mostly represents an undesirable feature, it is of great importance clearly to understand the properties and chemical reactions of such oxycelluloses as can be regarded to be definite chemical entities. Once the nature of the complicated reactions displayed by such oxycelluloses toward various chemical agents, particularly acid and alkali, are understood, it should be possible to apply this knowledge to the interpretation of the reactivity of the more or less severely oxidized celluloses.

While acidic reagents in general do not seem to initiate unduly great degradation in oxidized celluloses of either type, the latter are characterized by the striking change in properties that takes place when they become exposed to alkaline media. Particularly, the aldehyde type of oxycelluloses suffer profound changes on alkaline treatment which otherwise may be regarded as comparatively mild. Most of our present knowledge regarding the degradation of oxidized celluloses originates from the careful investigations by DAVIDSON (*10, 12–15*). For explanation of the remarkable sensitivity of the reducing type oxycelluloses toward alkali, DAVIDSON suggested that this behavior was due to the presence in the chain molecules of such glycosidic linkages which had been rendered unstable to alkali by the oxidation process. In view of the characteristic stability in alkali and lability in acid of the acetals, a class to which cellulose belongs, it is difficult to see why some of the C—O—C linkages of an oxidized cellulose should hydrolyze with extraordinary facility, even in a scarcely alkaline solution (*68*).

The paradoxical behavior of the oxycelluloses can be understood only by the assumption that during the reaction with alkali the substance changes its structure in such a manner that the C—O—C "glycosidic" or "acetal" linkages become susceptible to hydrolysis. This leads to a consideration of the prototropic changes which may conceivably take place in alkaline solution. According to this view the enolic form of periodate oxycellulose (V) belongs to the negatively substituted ketene

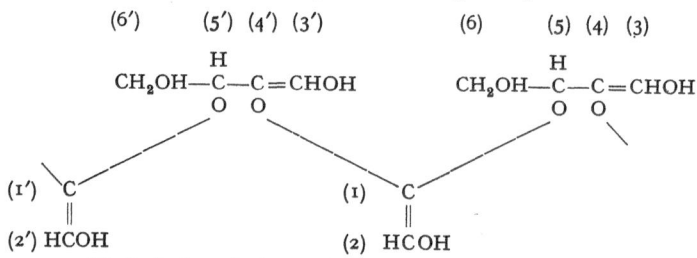

(V.) Enolic form of glucose-anhydride units oxidized by periodate.

acetals (VI) which are known (*28, 29, 78*) to hydrolyze rapidly to the corresponding esters in water or even in alkaline media. The first step (VII, p. 143)

$$\begin{array}{ccc} \mathrm{H} & & \mathrm{OR} \\ \diagdown & & \diagup \\ & \mathrm{C}\!=\!\mathrm{C} & \\ \diagup & & \diagdown \\ \mathrm{HO} & & \mathrm{OR} \end{array}$$

(VI.)

in the reaction of the glyoxal part of periodate oxycellulose probably consists of the removal of the α-proton by the hydroxyl ion of the alkali

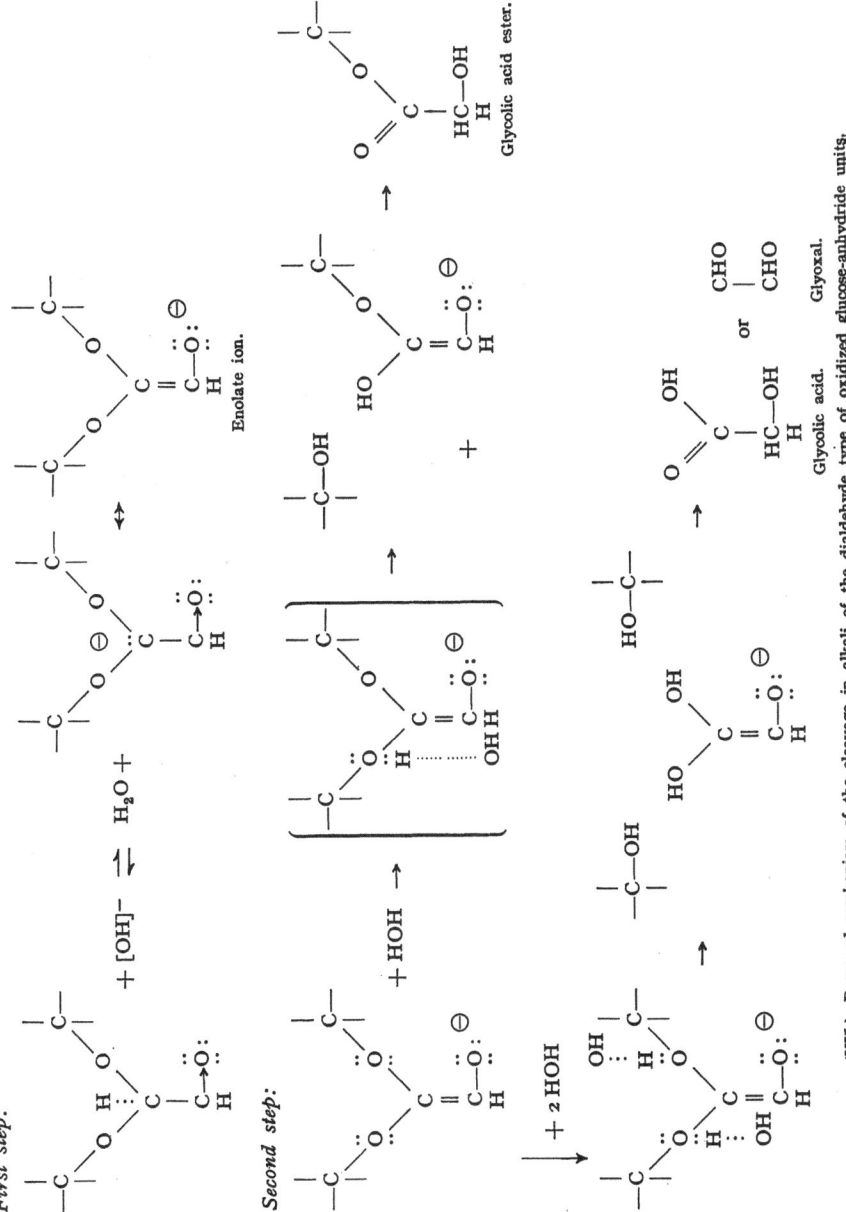

(VII.) Proposed mechanism of the cleavage in alkali of the dialdehyde type of oxidized glucose-anhydride units,

with the production of the mesomeric oxycellulosate ion. In the second step (cf. VII), the enolate ion, being essentially of a ketene acetal type (VI), is cleaved by an electrophilic attack of the water molecules, which (by a displacement mechanism) break one or both of the C—O—C linkages to yield, respectively, the enol-form of glycolic acid ester or of glyoxal. This process gives rise to either of these compounds and to D-erythrose, which then is subject to numerous reversible and irreversible changes that the reducing sugars undergo in alkaline solution. Since glyoxal, in form of its phenylosazone and m-nitrobenzoylosazone, has actually been isolated from alkali-treated periodate oxycellulose HEAD (38) concludes that the sensitivity of the periodate oxycelluloses in weakly alkaline solutions is due to straightforward alkaline hydrolysis of the acetal linkages. This conclusion apparently does not apply to a similarly constructed dialdehyde obtained from β-methyl glucoside, in which instance no glyoxal has been found.

The formation of periodate type of oxycellulose has been postulated (68) to occur by the action on cellulose of air oxygen in alkaline media. On the basis of this theory the ageing of alkali cellulose can be readily interpreted. The attack of oxygen produces a number of dialdehyde type glucose-anhydride units which become immediately cleaved in alkali through the mechanism just outlined. The result is a degraded cellulose with shorter chains and more end-groups, and hence with higher copper numbers and lower viscosity. This theory has found experimental confirmation in the work by HOLLIHAN (47) who has shown that passing oxygen through a cuprammonium solution of cellulose does not result in increased carboxyl content of the precipitated degradation product.

Concerning the exact location of the oxidized glucose-anhydride units in oxycellulose, at present there is no evidence available. Although no compelling reason appears to be in its favor, a random distribution of such oxidized units throughout the chains is assumed by most investigators.

Hydrocellulose.

It is well known that native cellulose is readily degraded by hydrolytic processes when exposed to the action of mineral or organic acids even at low temperatures. The degraded product exists as fibers or as a friable powder with low degree of polymerization depending upon conditions in the treatment. Since the early work by GIRARD in 1875 hydrocellulose has been the subject of extensive study by numerous investigators who prepared a great variety of degradation products by using different experimental conditions. Of these products, all termed "hydrocellulose", only those formed in the early stages of degradation are of commercial interest and have therefore received more attention than the more severely degraded materials. From the structural chemists' point of view hydrocellulose represents an important key-material in the investigation of cellulose structure. HESS and WITTELSBACH (39) first believed hydrocellulose to be a chemical individual and proposed the erroneous formula of a pentaglucosyl-glucose, $(C_6H_{10}O_5)_6 + H_2O$.

With the rise of the linear chain concept of cellulose structure hydrocellulose has been generally regarded as a degradation product consisting of a homologous series of more or less shortened chain molecules which are believed to possess the original macromolecular structure. The shortening of the original cellulose chain is supposed to be caused exclusively by hydrolysis in a completely random fashion of a number of the 1,4-glycosidic linkages with the formation of fragments having lower average D. P.'s than that of the untreated cellulose. The cleavage of each glycosidic link results in the formation of two terminal units, one being a reducing, the other a non-reducing end group. Among the several characteristic changes which accompany the formation of hydrocelluloses, the most important are: loss of tensile strength, decreased viscosity in cuprammonium solution, increase in reducing properties, and solubility increase particularly in alkali. In a brief review of the preparation and properties of hydrocellulose RUTHERFORD and HARRIS (80) summarized the results of recent important contributions in this field.

Most of the investigations on hydrocellulose were carried out on materials which had been obtained by heterogeneous acidic degradation under comparatively mild conditions. The results were generally interpreted in the light of the linear chain concept which, as was pointed out above, derived its most powerful support from FREUDENBERG's kinetic studies. Indeed, so firm was the grip of this concept over the reasoning of many investigators that certain experimental facts which did not seem to fit into this favored theory were either not given sufficient weight or were interpreted on the basis of some auxiliary hypotheses. Thus, general recognition of the inadequacy of the theory has been retarded in spite of early publications which now appear to be important links in more recent developments.

Conclusions Regarding the Linear Structure of Cellulose.

As a summary of the preceeding analysis of the proofs offered in favor of the linear chain structure (I) the following conclusions may be drawn.

1. Native cellulose appears to consist of D-glucose [isolated (53) 95 per cent.] and a small amount of -onic or -uronic acid units.

2. The glucose residues possess unoccupied hydroxyl groups at the second, third and sixth carbon atoms [isolated (50) 91,5 per cent. of methyl 2,3,6-trimethyl glucoside]. The theoretical methoxyl content, 45,58 per cent., calculated from the linear formula (I) has not yet been reached for undegraded cellulose.

3. Cellulose may contain one per cent. of links which are not of the β-glycosidic type.

4. Since Freudenberg's kinetic measurements in strong sulfuric acid were made on a severely degraded product the inferred uniformity of bonds does not necessarily apply to native cellulose.

5. The results of partial degradation of native cellulose upon acetolysis or in strongly acidic media are not in obvious contradiction to the linear structure (I), but they do not prove it either.

6. The unit cell of cellulose consists of two cellobiose-anhydride units. The fiber consists of crystallites or micelles. A micell is calculated to be a rhombus which measures at least 600 Å. along the fiber axis and about 50-by-60 Å. across this direction.

7. The molecular weight of native cellulose cannot be determined with any measure of accuracy by chemical methods.

8. Molecular weight determination of native cellulose by physico-chemical methods leads to indefinite values which depend upon the extent and gravity of accidental degradation.

In view of this unsatisfactory situation one is confronted with the possibility that the true cellulose structure is entirely different from formula (I). In the following section a brief review is given of certain conflicting conceptions which appear to be in opposition to the linear macromolecular theory. In the presentation of the material similarity in the outgrowing ideas rather than chronological order is being observed.

II. The Laminated Chain Structure of Cellulose.

Theory of Ester Linkages.

It is commonly known that acetylation by acetic anhydride and sulfuric acid or zinc chloride of cellulose results in degradation of the material, in that the D. P. of the product is smaller than that of the starting material. Similar degradations were frequently observed under various conditions, for instance when cellulose acetate was de-acetylated in the presence of atmospheric oxygen.

The opposite phenomenon, an apparent increase in the D. P.'s of the reaction products was first observed by Staudinger (89) who found by viscosity measurements that the degree of polymerization in acetone solution of cellulose nitrate (prepared from cotton by means of a mixture of nitric and phosphoric acids) was considerably higher than that of the original starting material in cuprammonium solution. This difference was explained on the basis that cellulose possessed the structure of a high polymer ester, which suffered rapid hydrolysis in cuprammonium solution but its ester linkages remained practically unbroken during the process of nitration. Later Staudinger and Sohn (91) found that a great variety of cellulosic materials, such as bleached linters and ramie, celluloses from spruce, straw, beech wood, etc., showed similar behavior provided that they were not "purified" prior to nitration by precipitation from cuprammonium solution. When the samples had been first submitted to such a pre-treatment, then the nitrated products showed practically the same degree of polymerization as did the "purified" materials. The difference between the "cupra D. P." and the "nitrate D. P." was expressed as a per cent. chain length difference ("Kettenlängen-differenz"):

$$\text{Chain length difference (\%)} = \frac{D.\,P.\,\text{nitrate} - D.\,P.\,\text{cell.} \times 100}{D.\,P.\,\text{cell.}}$$

Likewise, the number of ester linkages, E. L., was determined from the expression

$$E.\,L. = \frac{D.\,P.\,\text{nitrate}}{D.\,P.\,\text{cell.}} - 1,$$

the numerical value being one hundredth of that of the chain length difference.

For instance, the authors found for a spruce cellulose sample, D. P. 720 from the viscosity in cuprammonium solution; however, after nitration, it had D. P. 2400 as determined from the viscosity in acetone (chain length difference: 230%; number of ester linkages: 2,3). Pre-treatment of the cellulose sample by precipitation from cuprammonium solution resulted in D. P. 640 in that solvent and in D. P. 650 for the nitrate in acetone.

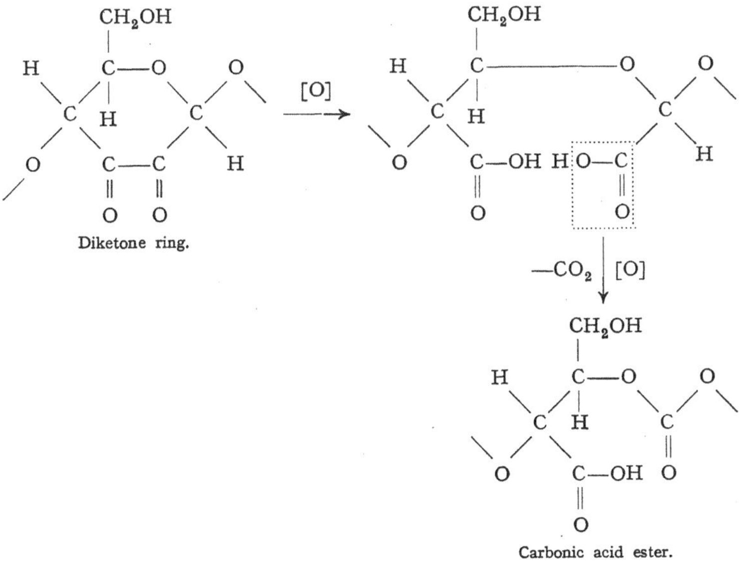

(VIII.) Formation of ester linkages by oxidation according to STAUDINGER and SOHN (92).

If the E. L. value is greater than, then 1 each macromolecule must contain at least one ester linkage. An E. L. value less than 1 indicates that the cellulose sample consists of a mixture of "normal cellulose" and "ester cellulose".

Staudinger and Sohn (92) envisaged the formation of ester linkages from normal glucose-anhydride units as an oxidation process which was supposed mainly to take place under acidic conditions, in various ways. One of the alternatives consists of the oxidation of the hydroxyl groups in the 2- and 3-positions to carbonyl groups and subsequent oxidation of the diketone ring to a carbonic acid ester unit (VIII).

According to these authors, carbonic esters are generally stable toward nitration acid mixtures; for example, diphenylcarbonate can easily be nitrated to yield dinitro-diphenylcarbonate. However, in alkaline media, such as cuprammonium solution in which the viscosity measurements were made, the ester would suffer instantaneous hydrolysis to give rise to two fragments: one with a normal non-reducing end-group, and the other terminating in a presumably erythronic acid residue.

As an alternative way of formation of ester linkages, the rupture of a pyranose ring and subsequent oxidation of the resulting hemiacetal to an ester was postulated by Staudinger and Sohn (IX).

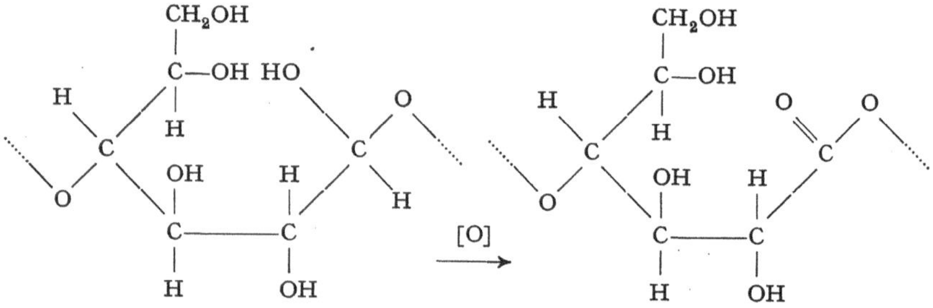

Hydrolyzed pyranose ring.   Gluconic acid ester.

(IX.) Formation of ester linkages by oxidation according to Staudinger and Sohn (92).

The study of this group of "ester celluloses" was greatly facilitated by the previous investigation of Davidson (12) who had found that, after treatment by dilute chromic acid, the modified cellulose showed a lower D. P. value than the nitrate prepared from it and dissolved in acetone. Clearly, the structural problem of cellulose was thus shifted to the constitution of *oxycellulose*. As a result, Staudinger's original theory on the structures of "normal cellulose" and "defective cellulose" ("Fehlerhafte Cellulose"), which postulated the pre-existence in native cellulose of chain molecules united end to end by ester linkages, lost its initial interest. The breakdown in alkaline media of the aldehyde type of oxycelluloses, into which group Staudinger and Sohn's chromic acid pre-treated samples belong, is to be expected, and the mechanism most likely follows the path previously discussed under the section "Oxycellulose".

While the apparent increase of the degree of polymerization during nitration of certain oxycelluloses has thus become understandable, further studies are required for a general explanation of similar phenomena which take place on nitration of *hydrocellulose*. It was first observed by STAUDINGER and SOHN (*92*) that the nitrates of hydrocelluloses also possessed a higher degree of polymerization than the hydrocelluloses themselves.

Thus, a purified cotton with D. P. 2800, after acidic degradation at 100° C. for ten minutes by 0.5 per cent. sulfuric or hydrochloric acid, gave hydrocelluloses with D. P. 910 and 620 respectively, in cuprammonium solutions. After nitration, the corresponding figures were found to be D. P. 1350 (chain length difference: 48 per cent.) and D. P. 930 (chain length difference: 50 per cent.) in acetone solution.

On the other hand, strongly degraded hydrocelluloses with D. P. 165 or 155 in cuprammonium solutions gave rise to "polymer-analogous" nitrates with no increase in their D. P. values in acetone. Since ester linkages are certainly not created during the preparation of hydrocelluloses, STAUDINGER and SOHN (*92*) made the assumption that, under the influence of the nitration acid, the shortened chains undergo condensation between the reducing group of one fragment and an appropriate hydroxyl in the non-reducing end-group of the other fragment. It is interesting to note that these authors also considered the possibility of an opening during nitration, of the pyranoid ring in the reducing unit, and envisaged a condensation occurring in the resulting open-chain glucose unit. In any case, such a condensation was believed to take place only on lesser degraded hydrocelluloses (D. P. 910 and 620) where the crystal lattice of the original, thread-like molecules was supposedly still undisturbed. On the other hand, the chain molecules in a severely degraded hydrocellulose were thought to be in a state of disarrangement as evidenced by the increased solubility of such products in various solvents. Consequently, condensation between two chains of such products was unlikely to occur upon the action of the nitration acid; hence, the products with D. P. 155 and 165 did not show the phenomenon of re-combination.

It is difficult to reconcile this view with the fact that the sharpness of the crystalline pattern usually increases with acidic degradation (*74*).

Although the theory of ester linkages in native cellulose suffered a serious set-back when DAVIDSON's work on the nitration of oxycelluloses became known, until recently it has not been possible to replace it by another theory that satisfactorily would have explained the results obtained by AF EKENSTAM (*18*, *19*) in his study on the behavior of cellulose in strong phosphoric acid solution. He found that native cellulose exhibited a much lower degree of polymerization in cuprammonium solution (D. P. 150) than it did in phosphoric acid (D. P. 230). Actually, af Ekenstam used such cotton which was first dissolved "for a few minutes" in hydrochloric acid, then precipitated by water and washed acid-free. For interpretation of these results he assumed the existence in native cellulose of alkali-sensitive bonds which are comparatively resistant to acid hydrolysis.

On the other hand, from the kinetic measurements of the acidic degradation of cellulosic materials in concentrated mineral acids af Ekenstam concluded that native cellulose decomposed at a much higher initial rate than did hydrocellulose. This behavior he attributed to the presence in native cellulose of "ester-like" or "native" links which evidently hydrolyzed much more readily than did the ordinary β-glycosidic links. Since all these observations purporting to prove the occurrence of ester linkages in native cellulose can be more readily interpreted in the light of some recent developments, the conception of such linkages has only historical interest.

Theory of Blocked Hydroxyl Groups.

In their work on the acetylation and methylation of cellulose, Karrer and Escher (52) pointed out that until 1925 all attempts had been unsuccessful in reaching the 45,58 per cent. theoretical methoxyl content of methylated cellulose, calculated from formula (I). By that time some investigators began to employ specially pre-treated celluloses as starting materials in order to achieve "complete" methylation, but most of the results indicated a methoxyl content of 42 to 44 per cent. only. Karrer and Escher used acetylated cotton celluloses as starting materials which were produced by (a) the use of acetic anhydride—zinc chloride catalyst on regenerated, ether-moist cellulose, (b) application of a glacial acetic acid—acetic anhydride—sulfuric acid mixture, and (c) glacial acetic acid—chloroacetic anhydride—sulfuric dioxide, according to Haworth and Machemer (36). When acetylation followed by de-acetylation of the samples was repeated three to seven times, a considerable rise (from 1,11 to 12,61) in the copper number indicated that extensive degradation had taken place, particularly during process (b).

Subsequent exhaustive methylation by dimethyl sulfate and sodium hydroxide resulted in the low methoxyl values of 41,25 and 42,77 per cent. for the less degraded samples with the low copper numbers 1,23 and 2,86 respectively, whereas 45,45 and 45,93 (!) per cent. methoxyl were found in the severely degraded samples with copper numbers 12,61 and 27,93.

Karrer and Escher (52) mentioned special efforts to establish the presence of the assumed free hydroxyl groups in the methylated samples containing 42 to 43 per cent. methoxyl groups. However, attempted acetylation of such samples by the pyridine-acetic anhydride method failed to show the presence of free, unmethylated hydroxyls; neither did the

ZEREWITINOFF method reveal the presence of active hydrogen. On the other hand, cleavage experiments by the IRVINE and HIRST (50) process with methanol-hydrogen chloride on a methylated cellulose containing 42 per cent. methoxyl, resulted in the isolation of 2,3-dimethyl-glucose, thus indicating that the "blocked" hydroxyl occurred at the 6-position. From an average methoxyl value of 42,5 to 43 per cent KARRER and ESCHER (52) calculated that approximately at every fourth or fifth glucose unit there must occur one unmethylated hydroxyl group. If the "blocking" effect were due to anhydride formation between the supposedly free hydroxyl groups at the 6-positions, then one water molecule would be split off from every eight or ten glucose residues of two parallel chains, thus leading to a formula, $(C_{48}H_{78}O_{39})_x$, which illustrates one cross link of ether character at each fourth glucose unit (X).

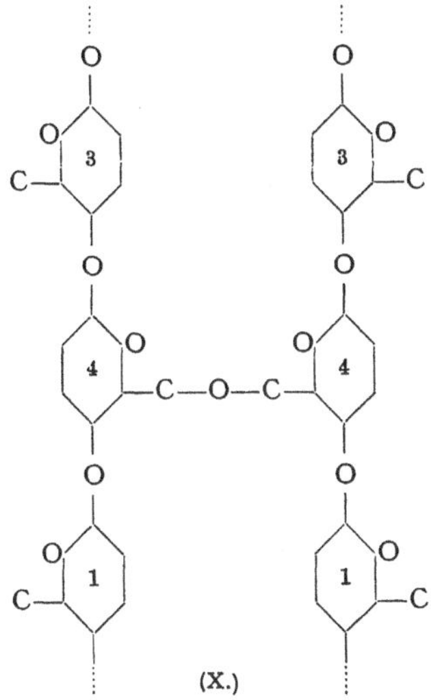

$$
\begin{array}{c}
O \\
\text{| } \quad \text{O} \quad \text{| }\\
\text{C—} \overset{3}{\diamond} \qquad \overset{3}{\diamond} \text{—C} \\
O \qquad O \\
\overset{4}{\diamond}\text{—C—O—C—}\overset{4}{\diamond} \\
O \qquad O \\
\text{C—}\overset{1}{\diamond} \qquad \overset{1}{\diamond}\text{—C}
\end{array}
$$

(X.)

KARRER and ESCHER have also given consideration to the possibility that "steric hindrance" might prevent the chemical reactivity of some of the hydroxyl groups, or that in native cellulose such hydroxyl groups might be "somehow saturated", for example, by formation of rings. Recourse to the former expedient was also taken by MEYER (58) who thought that the neighboring hydroxyl groups around the postulated "ester bridges" might be hindered in the methylation or even in the salt formation by alkali.

Theory of Looping Bonds.

It is evident from formula (I) that, on hydrolysis of completely methylated cellulose, one of the terminal units would give rise to 2,3,4,6-tetramethyl-glucose whereas the rest of the material would consist of 2,3,6-trimethyl-glucose. Until 1932 it was not possible to establish with certainty the existence of tetramethyl-glucose among the products of hydrolysis of methylated cellulose. Therefore, no definite statement could be made regarding the fate of the terminal groups and hence the correctness of formula (I).

Of the alternative structures the one with a large closed ring (with no end groups) was first rejected by Haworth (33), after he and Machemer (36) had succeeded in isolating tetramethyl-glucose from a two hundred grams-sample of highly methylated cellulose (45 per cent. methoxyl). From the amount of this impotant compound Haworth estimated that their preparation must have consisted of not less than 100 and probably not more than 200 glucose units (mol. wt., 20000 to 40000). He considered this result as evidence for the correctness of formula (I) and also as an answer to the question of how the chains are terminated. In contrast, Hess and Neumann (40) contended that the formation of tetramethyl-glucose as observed by Haworth et al. was a result of the far-reaching oxidative degradation which the cellulose had suffered during methylation. In applying their refined method of end-group determination (62) to cellulose, these authors observed that a mild pre-treatment of the fibers, such as the technical purification, bleaching, extraction with dilute alkali, produces hydrolytic and oxidative changes which, in the analysis, result in the formation of tetramethyl-glucose and probably some methylated hydroxy acids. Similar changes caused by atmospheric oxygen during methylation were then postulated, and the scheme (XI) was offered.

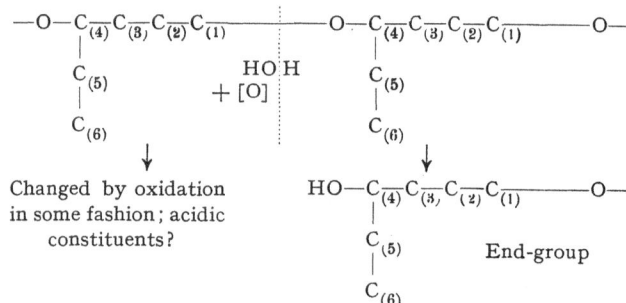

(XI.) Formation of end-groups by hydrolytic oxydation during methylation according to Hess and Neumann (40).

In support of their contention that end groups were created during methylation, Hess and Neumann (40) referred to some experiments in which cellulose samples were methylated in the presence and in the absence of air. In one of the former cases they were able to isolate 0,026 per cent. of methylated product from the non-reducing end-group, thus indicating D. P. 2810 for the methylcellulose employed. However, no tetramethyl-glucose was formed in a nitrogen atmosphere. In general, less degraded materials yielded only small amounts of tetramethyl-glucose, whereas celluloses regenerated from viscose and from a solution in concentrated sulfuric acid gave considerably higher quantities. HESS and NEUMANN interpreted their results to mean that either the cellulose chains are so long as to make the detection of a minute quantity of tetramethyl-glucose impossible, or that cellulose does not consist of chains at all, but it is represented by large loops with no ends to them, the number of the glucose units in the loops still remaining unknown. They accepted the latter possibility as the more probable explanation since they could not see any reason for the existence of a giant chain molecule.

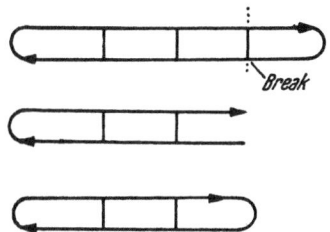

Fig. 1. Looped cellulose chains with cross-linkages [HAWORTH (34)].

When MEYER (58) suggested that the chains in the cellulose fiber were so aligned that their potential reducing groups pointed alternatively in opposite directions, he also envisaged the formation, by each such chain pair, of large rings in that "at the ends of the crystallites the chains were connected through glycosidic linkages".

The observation that no tetramethyl-glucose is obtained from cotton when methylation is carried out in a nitrogen atmosphere was later confirmed by HAWORTH (34). Although the D. P. values of his products (from 1300 to 170) as measured by viscosity and osmotic pressure methods clearly indicated that diminution of the molecular size had occurred, presumably by chain shortening, no end-groups remained after this breakdown (or "disaggregation"). According to HAWORTH, this was only possible if the exposed end-groups of two chains have become involved "in some form of recombination". Since it is theoretically inconceivable that the chains should align themselves under any circumstances, HAWORTH believes that at least two of the chains, and probably more, are held together by "some intermediate bond" which serves to tie together, e. g. a pair of chains at various positions along their length (Fig. 1).

The nature of these "cross-linkages" situated along and between pairs of chains was not revealed by HAWORTH, neither was the precise way of junction indicated. He only stated that "we are concerned here in the polymeric link with a primary valency of the glucosidic type but clearly differentiated from that which join the succeeding glucose members in a single chain". HAWORTH further assumed that these intermediate links pre-existed in cellulose where they joined together a pair or even a greater number of chains laterally, at intervals of 25 to 30 glucose units. Their occurrence may correspond to the isolation of the proportion of dimethyl-glucose which is *always* obtained from the hydrolysis products of methylated cellulose. HAWORTH also suggested that a break in the chain may occur at or near these bonds, thus presumably leading to small fragments (D. P. 170) which still reveal no end-groups on hydrolysis.

This theory of *cross-linked* cellulose structure was further developed by HESS and STEURER (*41*) who gave an exact definition as to the nature of the postulated cross-linkages. They found that equi-viscous cellulose preparations originating from various degradation reactions may show considerable differences in end-group content. Conversely, cellulose preparations with about equal end-groups may differ widely in viscosity. HESS and STEURER assumed that there were two types of bonds joining the glucose units in cellulose: the first type (β-glycosidic bond) occurred along the fiber axis, whereas the second type was represented by lateral *oxygen bridges* between chains and originated in open-chain glucose-anhydride units. Oxygen-bridges of the latter type ("Vernetzungs-brücken") were thought to occur at indefinite intervals between two anti-parallel chains and to give rise to 12-membered rings according to (XII).

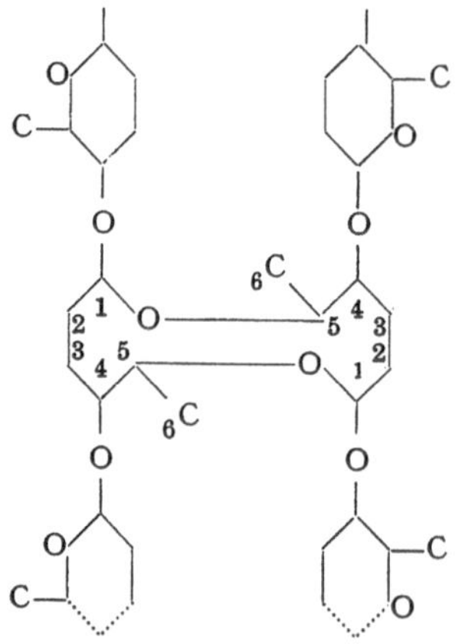

(XII.) Oxygen bridges between two anti-parallel chains in cellulose according to HESS and STEURER (*41*).

In HESS and STEUER's opinion the cleavage of only the normal β-glycosidic oxygen bridges would affect the number of end-groups, whereas scission of the cross-linking O-bridges (resulting in the restoration of the normal pyranose units) would not produce new end-groups; however, it would cause changes in viscosity. These authors also expressed the view that in the various heterogeneous hydrolytic and oxidative reactions of cellulose, an overlapping of the two scission types should be generally expected. However, the oxygen bridges between two lateral open-chain units would be kinetically more favored, and under certain conditions only these bonds may be cleaved. Thus, methylcellulose (40–42% OCH_3) showed (*94*) a marked drop in viscosity, from D. P. 762 to D. P. 190, upon irradiation by ultraviolet light; the methoxyl content remained unchanged, while end-group analysis indicated a much smaller decrease in the degree of polymerization (rupture of 0,8 bonds, in the original molecule with av. D. P. 162). STEURER (*94, 95*) interpreted this deviation by assuming that ultraviolet light splits the cross-linkages at a faster rate than the normal glycosidic bonds.

HESS and STEURER (*41*) also believe that the existence of "inter-meshing" bridges is responsible for the fact that spreading of certain cellulose derivatives on a liquid surface does not produce homogeneous mono-molecular layers but it gives rise to rather thick films which result from the three-dimensional arrangement of the chains.

Theory of Acid-Sensitive Bonds and the Principle of Periodicity.

It was pointed out that at present no experimental evidence is known which would prove beyond doubt the occurrence to the extent of 100 per cent. of β-glycosidic bonds in native cellulose. The best available figure appears to be 99 per cent. which is based on FREUDENBERG's polarimetric proof of the linear chain concept. It is, therefore, obvious that the nature and the role of 1 per cent. of the hydrolyzable bonds at the most, that is some 30 bonds in a cellulose with D. P. 3000, presents a problem of great importance regarding the structure and properties of cellulose. If cellulose were really nothing else but a long chain of glucose-anhydride units held together by "uniform and equivalent, covalent glycosidic linkages", then the statistical laws would apply to its various chemical transformations, for example, to the hydrolytic degradation.

In 1942 SCHULZ and HUSEMANN (*81*) undertook the investigation of the molecular weight distribution both in native cellulose and in the "polymer-homologous" hydrocellulose mixtures obtainable from the former by homogeneous or heterogeneous acidic degradation. Their careful fractionation work indicated that the starting material represented a surprisingly uniform cellulose consisting of about 80 per cent. of molecules with D. P. 3100 ± 100. Furthermore, they found that in the case of rather severely degraded celluloses considerable difference existed between the experimental and the theoretical distribution of molecular weights, in that degradation products possessing less than D. P. 1000 showed a much greater uniformity than would be theoretically expected from formula (I) with equivalent β-glycosidic bonds throughout. Because of these unexpected results these authors were forced to the conclusion that cellulose must contain, at regular intervals, certain links which hydrolyzed at a much faster rate than other bonds; therefore, by dissolution of the sensitive bonds alone, cellulose would disintegrate into fragments with uniform lengths. Since Schulz and Husemann (81) realized that, by chemical methods, it is impossible to identify the nature of a few, perhaps only two or three, special bonds per one thousand "regular" ones, they treated the problem as follows.

First, they have calculated the non-uniformity factor, $U = (K_m/k_m)-1$, from the degree of degradation, $\beta = P_0/P$, by the aid of an exponential equation based on the assumption that (a) all the molecular chains in native cellulose are of uniform lengths, and (b) that the rate constant of the hydrolysis has an identical value for all glycosidic bonds. In the above expression K_m and k_m represent the Staudin-ger constants for cellulose nitrates in acetone solution as determined by osmotic pressure measurements, after nitration of the non-fractionated and sharply fractionated degradation products, respectively. Furthermore, P_0 denotes the number of glucose-anhydride units (D. P. 3100) in the original sample, and P gives the degree of polymerization of the degraded samples as obtained by the osmotic method. By plotting the calculated U values against a series of β values, the authors obtained the theoretical curve which turned out to be of entirely different shape from the curve given by plotting the experimentally determined values of U and β. Schulz and Husemann, therefore, concluded that at least one of the assumptions underlying their theoretical equation was incorrect. Since careful fractionation of the starting material did show a satisfactory uniformity of chain length (about 80 per cent. of D. P. 3100 ± 100), they postulated that their assumption (b) concerning the uniformity of the hydrolytic rate constants was invalid. The experimental curve showed that with increasing degradation the U values pass through a maximum and a minimum.

In an elaborate mathematical treatment, Schulz and Husemann then arrived at an equation that was based on the assumption that a chain of 3000 units was split into six equal fragments, and that the rate of this cleavage was one thousand times as high as that of the "regular" bonds. Since the experimental value fitted the new theoretical curve, the authors regarded this as proof of the correctness of their assumptions.

The presence of some unusually *acid-sensitive bonds* and the principle of their even distribution in cellulose thus appear to be well-established.

Regarding the character of these acid-sensitive bonds Schulz and Husemann expressed the view that they cannot be identical with Staudinger's ester linkages. It seemed to them more probable that these bonds originate from *β-xylose units* which are known in xylan to hydrolyze about one thousand times as fast as the β-glycosidic bonds of glucose units. Moreover, the xylose units need not be present as such in cellulose but may be formed during the acidic degradation, by loss of carbon dioxide from the corresponding glucuronic acid units. It is even conceivable, according to the authors, that decarboxylation is not a necessary step to produce an increase in the rate of hydrolysis, since the latter effect may result merely from the oxidation of the primary hydroxyls of particular glucose units to carboxyl groups. In support of this theory SCHULZ and HUSEMANN pointed out that the carboxyl content of native cellulose—one carboxyl group for every 520 or 530 glucose units (48)—agrees remarkably well with the number of the weak linkages ("Lockerstellen"). Therefore, it appeared that the carboxyl content is of great importance in regard to the chemical resistance of cellulose; the more carboxyl groups are present in a sample, the greater sensitivity the latter will reveal during hydrolysis.

In accordance with their experimental results and deductions, SCHULZ and HUSEMANN advanced the theory that native cellulose consists of a chain of $3 \times 2^{10} = 3072$ glucose-anhydride units. These are linked together by $1,4$-β-glycosidic bonds with the exception of five, equally spaced glucuronic acid residues, which, by loss of carbon dioxide, may have become xylopyranoside units. Since the 1,4-xylopyranosidic linkages hydrolyze one thousand times faster than the normal, glucopyranosidic bonds, the cellulose molecule is rapidly degraded by acidic media to give rise to the six chain molecules, each consisting of $2^9 = 512$ units.

SCHULZ and HUSEMANN envisage the formation of such regularly built chains in nature as a result of enzymic syntheses which would build up stepwise structures consisting of 2,4,8,16 etc., glucose-anhydride units. Furthermore, they postulate the occurrence of the special groups ("Sondergruppen") not only at determined intervals in the chains but also at regularly arranged spatial positions. A regularly built cellulose crystal may then possess two types of arrangements: a basic lattice ("Grundgitter") and a long-period lattice structure. In the first pattern the molecules are supposed to be arranged in a regular way but the side-groups are statistically distributed in the space. In the long-period lattice the molecules are so arranged that the side-groups are close together at comparatively large intervals to give rise to intermeshing planes ("Netzebenen"). Degradation of the fiber is then supposed to take place along the surface of these intermeshing planes "as though the fiber were cut across with a fine knife". Since the dimension along the fiber axis of the assumed long periodicity which is caused by the carboxyl groups of adjacent chains should be about 2600 Å. (corresponding to 512 glucose-anhydride units), in these authors' opinions the length of the cellulose micelles ($\sim$ 600 Å.) seems to be in no way related to the dimension of the postulated long-period of the fiber.

Theory of Acetal Linkages in Equi-Distant Open-Chain Units and the Laminated Chain Structure of Cellulose.

The latest concept of cellulose structure was advanced by the writer as a result of a series of some studies (*43–46, 57, 68–71, 98*) which were published in 1945–1947. Because of war conditions most of the foreign publications, particularly the German ones were then inaccessible, and the new theory, in its original form (*71*), was developed independently from Schulz and Husemann's views (see above). This theory has undergone considerable modifications since, particularly in regard to the principle of periodicity which had been not clearly recognized until Schulz and Husemann's data became available. Although it seems to eliminate many difficulties concerning the interpretation of existing experimental data, it undoubtedly will need further improvements in the details. In the following presentation an account is given of certain experimental facts and of new interpretations of such earlier data which contributed to the development of this theory.

a) Heterogeneous and homogeneous degradation of cellulose in acidic media.

As it was stated above, Freudenberg and co-worker's measurements of the hydrolysis rates of cellulose in strong sulfuric acid led to the conclusion that at least 95 per cent. of the ruptured bonds were of β-glucosidic character. On the other hand, af Ekenstam's (*18, 19*) similar measurements in concentrated phosphoric acid indicated the presence of "some other type" of linkages which were more rapidly cleaved. In 1937 Staudinger and Sorkin (*93*) undertook an investigation dealing with the *heterogeneous* degradation of cellulose in various acidic media. Samples of cotton with D. P. 1650 were kept at 53° C. for 1, 6, 24, 120 and 1226 hours in N inorganic or organic acids, and the degree of polymerization was determined from the viscosities of the degraded products in cuprammonium solution. The number of the cleaved bonds was calculated from the expression (D. P. cell./D. P. hydrocell.) — 1, and the values so obtained were plotted against time.

The resulting curves revealed the striking fact that the degradation at first proceeded with great rapidity, but soon it became extraordinarily slow. These authors were surprised to find that even with strong mineral acids, such as N hydrochloric or nitric acid, the number of ruptured bonds in the solid fibers failed to increase appreciably, once the degree of polymerization (after, say, 120 hours) dropped to 150–200. They also found that during that period of time only about 10 out of the 1650 bonds present in the long chains of solid cellulose became cleaved. Staudinger and Sorkin did not offer any explanation for this remarkable behavior of the fiber except in referring to the "ester-like" linkages which were postulated by AF EKENSTAM in order to explain the initial fast rate of hydrolysis of cellulose *dissolved* in phosphoric acid.

In a systematic investigation of the action of acids on cellulose, DAVIDSON (*16*) treated cotton linters under various conditions with mineral acids and followed the variations of the properties of the resulting hydrocelluloses with the duration of this treatment. The concentration of the acids were fairly high: 10 N and 6 N hydrochloric acids, and 10 N sulfuric acid, all employed at 20° C. However, 0,5 N sulfuric acid was also used at 100° C. The results have indicated that there is initially a progressive change in the properties of the resulting hydrocelluloses, but eventually a stage is reached when little further change in properties occurs; at this stage only a small proportion of the glycosidic linkages in the cellulose have been hydrolyzed. After this point, the hydrolysis does not stop but proceeds extremely slowly.

DAVIDSON has pointed out that these results are very different from the behavior of cellulose in concentrated acid solution, in which *homogeneous system* the hydrolysis to glucose proceeds rapidly to completion. According to him, the difference in the acid hydrolysis of cellulose in homogeneous and heterogeneous systems may be explained in terms of a difference in the accessibility of the glycosidic linkages to acid: "When cellulose is dissolved in an acid medium, the crystalline regions in the cellulose are broken up and the acid acquires access to every linkage. Treatment of cellulose with acid in a heterogeneous system, on the other hand, produces no change in the X-ray diffraction pattern of the cellulose, and it must be, therefore, supposed that the acid does not penetrate the crystalline portions of the cellulose and that its action is confined to the hydrolysis of glycosidic linkages postulated in the amorphous region." Since mercerized cotton is hydrolyzed more rapidly than unmercerized, and the properties of the resulting hydrocelluloses attain constancy at values representing a greater extent of hydrolytic degradation,—e. g. mercerized cellulose, D. P. 95; unmercerized cellulose, D. P. 190—DAVIDSON attributed this behavior to the presence of a greater proportion of the easier accessible amorphous phase.

Although DAVIDSON's explanation regarding the difference in reactivity of the crystalline and amorphous regions may be used with advantage for understanding certain phenomena, it is clearly invalid for two reasons when applied to the *initial* phase of acidic degradation. First, it is an indisputable fact that native cellulose (D. P. ∼ 3000) undergoes a very rapid, heterogeneous degradation to the hydrocellulose stage (D. P. ∼ 250), where the reaction abruptly comes almost to a standstill, *without any appreciable loss of material*. It is evident, therefore, that during this important first phase only about 3 out of 1000 hydrolyzable bonds are affected, whereas the other 997 bonds, now belonging to the shortened and comparatively acid-resistant molecules (average D. P. ~ 250), remain practically intact. *In the light of* Davidson's *definition the "accessible region" in this case becomes identical with a few extremely acid-sensitive bonds.*

The second important reason for the invalidity of Davidson's interpretation is seen in the behavior of native cellulose in concentrated phosphoric acid solution. As we have seen from af Ekenstam's kinetic measurements, native cellulose decomposed at a much higher initial rate than did hydrocellulose. Certainly, in a homogeneous system such as phosphoric acid solution, (as Davidson correctly pointed out), all linkages must be accessible; yet the initial high rate of degradation was so striking that for explanation af Ekenstam was compelled to assume in native cellulose the presence of "ester-like" bonds which were supposed to be particularly sensitive toward acid hydrolysis. To be sure, no reference was made by Davidson to the behavior of native cellulose in phosphoric acid; the comparison between homogeneous and heterogeneous acidic degradation was confined to solutions in concentrated hydrochloric and sulfuric acids in which "the hydrolysis to glucose proceeds rapidly to completion". However, it has already been pointed out that in such solutions only severely degraded products with D. P. ~ 50 are present when the initial measurements are being made.

b) The concept of "limit hydrocellulose".

The phenomenon of rapid formation of hydrocelluloses with D. P. ~ 250 and the problem of the nature of the few bonds involved in this process led the writer to an investigation with the following purposes: (a) to find, if possible, a suitable chemical method of analysis for reducing end-groups in acidic media and then to follow the *initial* phase of the heterogeneous acidic degradation of cellulose fiber; (b) to follow the *initial* homogeneous degradation of cellulose in an appropriate acidic solution by measuring the change of viscosity; (c) to study the *initial* heterogeneous degradation of cellulose in dilute acids by following (in cuprammonium) the changes of the viscosity of the hydrocelluloses thus produced.

(a) In order to eliminate the use of alkali in the oxidation reaction of the reducing end-groups, potassium permanganate in acidic media as a possible oxidizing agent was considered.

Cellulose samples of approximately 1 gram were allowed to react with a measured excess of 0,1 N potassium permanganate in 0,25 N sulfuric acid at 0° C. for different intervals of time. The reaction was terminated by addition of a measured excess of ferrous ammonium sulfate, and the solution was backtitrated with potassium permanganate.

This oxidation method was then applied to various hydrocelluloses. If the destructive effect of dilute acids on cellulose fiber was caused by the same random hydrolysis of the 1,4-glycosidic bonds, which reaction has been established by FREUDENBERG on dissolved cellulose molecules, then a gradual increase in the reducing power would take place. Contrary to this expectation, it was found that the mild acid treatment apparently *decreased* the reducing power of cellulose, and that the more energetic the treatment, within the acidic region employed, the less became the reducing power.

This observation, together with the viscosity readings to be discussed below, led to the assumption that, besides normal 1,4-glycosidic bonds, cellulose must contain another type of covalent bonds which is sensitive to acid-catalyzed hydrolysis and which is somehow correlated with the reducing units of the cellulose molecule. Such a linkage can be represented only by a *hemiacetal* bond of an open-chain glucose unit which could either occur at the reducing end of the chain molecule or originate from open-chain glucose, cellobiose, etc. residues which may function as cross-links between adjacent cellulose chains.

The original reducing power of cotton would then correspond to the total number of hemiacetal linkages, and the gradual decrease in the reducing power would be caused by the loss of some small molecules during acid treatment. Since on leaving the fiber, after hydrolysis of their acid-sensitive hemiacetal linkages, these small molecules do not create new reducing groups, the reducing power of the resulting hydrocellulose would then become less than that of the original material. The amount of material lost in the initial attack by acids is minute by comparison with the weight of the sample. No such loss could be detected by analytical weighing, but volumetric estimation on the acid filtrate of a sample indicated one lost reducing molecule for every 548 glucose anhydride units. In the light of this interpretation the D. P. value, 287, obtained for surgical cotton, demonstrates only the ratio of units with hemiacetal bonds to normal groups and not the true molecular weight of the original substance.

(*b*) FREUDENBERG's polarimetric study has conclusively demonstrated that about 99 per cent. of the hydrolyzable bonds in cellulose are uniformly of the 1,4-β-glycosidic type. If the remaining 1 per cent. of bonds were more acid-sensitive than this main type, one could detect them in a homogeneous degradation only if the acid exerted its hydrolyzing effect at a much lower rate than did sulfuric acid.

Experiments have shown (*46*) that in concentrated *phosphoric acid* the viscosity of surgical cotton rapidly diminishes, reaching a very low value after a comparatively short time. A hydrocellulose with D. P. $\sim$ 300 gave immediately a low viscosity value which, when the solution was allowed to stand, diminished to the same final value as that of the cotton solution. The type of curve obtained in the case of cotton has been observed before by STAMM and COHEN (*88*), who interpreted it as a measure for the rate of hydrolysis of 1,4-glycosidic bonds. However, this interpretation appears to be incorrect for the following reason. The average rate of the reaction, *during the first five hours*, viz. $K = 3.4 \times 10^{-3}$, as determined from the corresponding viscosity data, would indicate a half period of about 90 minutes for the degradation. On this basis, cellulose in 85 per cent. phosphoric acid should change into pure D-glucose in about 15 hours, which is not the case. Indeed, even at the stage of practically zero viscosity, upon dilution, a precipitate of high-polymer material can be obtained. Neither can the presence of D-glucose in the solution be detected by polarimetric readings even after several weeks.

According to the interpretation advanced by the present writer, the chemical reaction which takes place in phosphoric acid solution is at the initial stage identical with the action of weak acids upon native cellulose fiber which produce hydrocellulose—that is, splitting of hemiacetal or acetal linkages of open-chain glucose units. Although in phosphoric acid solution this is a relatively slow and measurable process in comparison with that in a concentrated sulfuric acid solution, its rate, $K = 3.4 \times 10^{-3}$, is about 2000 times as fast as the hydrolysis of the 1,4-glycosidic bonds, for the rate constant of which—*discarding the data for the first 4 to 6 hours*—STAMM and COHEN (*88*) gave the value of 7.05×10^{-6}.

Consequently, when cotton is treated with phosphoric acid, the initial rapid hydrolysis of the acid-sensitive bonds must begin as soon as some of the fiber structure is loosened and (after rupture of hydrogen bonds) large "aggregates" are able to go into solution through solvation of their freed hydroxyl groups. These huge particles are responsible for the tremendous initial viscosities observed. By catalyzing the hydrolysis of the acid-sensitive bonds, the phosphoric acid continues to split the aggregates or "micellar plates" to smaller and smaller sections, and thus the viscosity decreases. After a relatively short time of two or three days, the original cellulose is completely "dispersed", and only true chain molecules are present which contain exclusively "normal" glycosidic bonds. As expected, the rate of hydrolysis of the latter is quite slow, thus confirming the results of AF EKENSTAM (*18, 19*) and of STAMM and COHEN (*88*).

(*c*) It is evident from the foregoing discussion that for the measurement of the hydrolysis rate of the postulated acid-sensitive bonds a mild *heterogeneous* degradation of cellulose by dilute acid might be more advantageous than the usual homogeneous degradation in acid solution. Accordingly, both surgical cotton and viscose rayon samples were degraded (*57*) in the solid state loy the use of (*1*) a mixture of 10 per cent.

acetic acid and 10 per cent. sodium chloride; (2) 10 per cent lactic acid, at the temperature of the boiling water-bath, 87° C. and at 97° C.; and (3) N-hydrochloric acid at 60° C. for viscose rayon. These experiments resulted in the significant observation that the hydrolysis is very rapid in the initial stage but comes almost to a standstill when the D. P. has reached ~ 250 for cotton, and ~ 60 for viscose rayon. Further action of the acids has no appreciable effect on the D. P. values. The velocity constants and energy and entropy of activation for the initial hydrolysis from D. P. 2500 to about 550 have been calculated for surgical cotton to be, $K_{97}° = 3,4 \times 10^{-3}$; $K_{87}° = 1,3 \times 10^{-3}$; $E = 26\,300$ cal.; $\Delta S^{\ddagger} = -11,4$ e. u. Since experiments with cellulose, regenerated from cuprammonium solution and then submitted to a similar heterogeneous treatment, have also indicated that the reaction stops at D. P. ~ 60, *the value of $2^6 = 64$ has now been chosen as the probable degree of polymerization of a "limit hydrocellulose chain molecule" which is devoid of acid-sensitive linkages and which is the building unit of native cellulose fiber.*

c) Proposed structure of cellulose.

From the above results it is concluded that extrapolation to cellulose fibers of FREUDENBERG's "kinetic proof" of the structure of *dissolved* cellulose molecules is not permissible, and that cellulose contains at least 0,3 per cent., but possibly as much as 1,5 per cent. of *covalent* bonds which hydrolyze about 1400 to 2000 times faster than do normal glycosidic bonds. It is suggested that these acid-sensitive, covalent linkages represent *acetal* and possibly also hemiacetal bonds which originate from open-chain glucose-anhydride residues of the "limit hydrocellulose chain molecules".

It is also proposed that *the molecular weight of cellulose is indefinite* in the sense that the visible fiber represents a *three-dimensional network* of glucose-anhydride residues. At certain regular intervals, probably at every 330 Å. (D. P. 64) such residues occur in open-chain form and are connected through acetal linkages with similar units along the *a* axis. These acetal linkages appear to be responsible for the rapid initial degradation of cellulose. (Some variations in the rates of hydrolysis of these special bonds are to be expected and are presumably due to the different locations of the open-chain units from which such bonds originate, just as this is assumed to be the case with the 1,4-glycosidic bonds of the primary chain molecules.)

A rigid structural formula respresenting all features of this concept cannot be written readily. However, symbol (XIII) may illustrate the essential requirements of the theory and it displays the postulate structure of a *micellar lamina*; and a schematic diagram (Fig. 2, p. 164) represents a lamina of a portion of the fiber in the *a*, *b* plane.

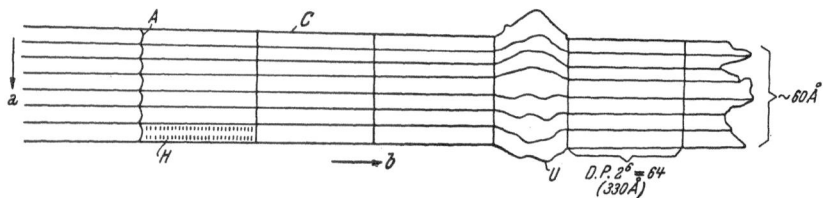

(XIII.) Proposed structure of a cellulose lamina.

Fig. 2. Proposed structure of native cellulose: a lamina in the a, b plane. C = crystalline (stretched) portion (80%); U = accessible (unstretched) region (20%); A = acid-sensitive acetal bonds; and H = hydrogen bonds.

It is perhaps the most characteristic feature of this cellulose model that the eight open-chain glucose units which furnish the acid-sensitive acetal bonds, are joined together along the a axis straight through the eight glucose-anhydride chains which supposedly form the width of a crystallite. Beside these open-chain units the eight chains are also connected by a network of hydrogen bonds operating between the appropriate oxygen atoms of the normal glucose-anhydride units in the *fully stretched* or crystalline ("inaccessible") area. On the other hand, in the accessible region ·very few, if any, hydrogen bonds are present because the *unstretched* state of the eight micellar chains, which results in increased distances between parallel chains, makes the formation of such bonds difficult or impossible.

Some thirteen layers of the long, ribbon-like laminae, held together by the weaker VAN DER WAALS' forces along the *c* axis, will then constitute a three-dimensional *micelle* of indefinite length. Such a micelle consisting of about 100 chains should appear under the electron microscope as a fine fibril with well defined dimensions except along the *b* axis. In a hydrocellulose which has been obtained by heterogeneous acid degradation, the length should correspond to that of 64 glucose-anhydride units or to some multiple of this figure.

Recently, FREY-WYSSLING and MÜHLETHALER (*25*) used supersonic waves to disintegrate fiber samples for observation under the electron microscope. This treatment resulted in the disintegration of the fiber into fibrils which appeared to be split into fine strands, the smallest of which measured 60 A. in diameter. These micellar strands have approximately the same diameter as the crystallized cellulose strands reported by X-ray analysts. In a study of regenerated cellulose films, ROSEVEARE, WALLER and NELSON (*77*) presented an electron micrograph of a gold-shadowed silica replica of the surface of a regenerated film showing an oriented network. Many of the particles appeared to be joined together into fibrils with diffuse spots which were 300 A. to 400 A. apart. Moreover, many of these spots appeared to be joined to similar spots in adjacent fibrils. The periodic distance of 300 Å. to 400 A. agrees in its order of magnitude with the limiting D. P. 64 value mentioned and it is about one-half of the distance estimated by means of X-ray data. The average linear chain length of the cellulose in the sample was found to be about 2000 A., i. e. six times larger than the periodic distance along the chains.

A striking evidence in favor of the proposed principle of periodicity remained apparently unnoticed by SCHULZ and HUSEMANN (*81*) when they fractionated cotton samples which had previously been degraded in concentrated phosphoric acid *solution*. Table 1 shows the result of their fractionation on three nitrated samples with the respective average D. P. values, 1050, 620, and 156.

In the last column of Table 1 (p. 166) the present writer has calculated the D. P. values on the basis of the apparent "limit hydrocellulose" unit, D. P. $2^8 = 256$, which has been obtained for cotton cellulose in heterogeneous acid hydrolysis. It is evident from these data that *the initial homogeneous acid degradation of cotton cannot result from random hydrolysis*. The latter takes place at a later phase of the reaction as is clearly indicated by Schulz and Husemann's fractionation results on the sample (D. P. 156) which was obtained by keeping cotton cellulose at 27° C. in about 14 M phosphoric acid solution for 65 hours. In this case no periodicity in the D. P. values could be detected and the hydrolysis followed a random path.

Table 1. Fractionation of Cellulose Nitrate from Cotton (*81*).

Average D. P.	Fraction	Per cent.	D. P.	
			Observed	Calculated[1]
1050	1	1,95		
	2	9,50	780	768 (3)
	3	20,90	1005	1024 (4)
	4	45,60	1530	1536 (6)
	5	22,05	2460	2560 (10)
620	1	10,3	225	256 (1)
	2	31,5	570	512 (2)
	3	18,5	835	768 (3)
	4	16,2	1180	1024 (4)
	5	23,5	1290	1280 (5)
156	1	1,7	55	
	2	10,0	58,5	
	3	6,6	98	
	4	17,2	144	
	5	18,1	208	
	6	31,1	278	
	7	7,0	394	
	8	8,3	570	

In order to prove the uniformity of their starting material, Schulz and Husemann fractionated cellulose nitrate prepared from native cotton (Table 2).

Table 2. Fractionation of Cellulose
Nitrate from Cotton (*81*).

Fraction	Per cent.	D. P.	
1	1,0	470	} 18%
2	17,0	1200	
3	30,0	2940	
4	35,9	3140	} 80,4%
5	14,5	3200	

In spite of some degradation which is supposed to take place during the nitration process, the cotton samples showed a high degree of uni-

---

[1] Numbers in parentheses indicate factors (times $2^8 = 256$) by which the respective calculated D. P. values were obtained.

formity in their chain lengths. In view of the experimental facts which indicate that native cellulose possesses an indefinitely high molecular weight, such a uniformity can only be understood on the basis of the principle of periodicity.

Convincing evidence for the presence of equally spaced, acid-sensitive bonds in cellulose is seen in GRALÉN's (32) data regarding the molecular weights of various cellulose samples from sedimentation and diffusion experiments in cuprammonium (Table 3). The last column was included by the present writer and shows a regularity in the architecture of cellulose.

Table 3. Molecular Weights of Cellulose in Cuprammonium from Sedimentation and Diffusion (32).

Cellulose	D. P.	Unit, 12 × 256 = 3072
Bleached American linters	3,000	1
Sulfite cellulose......................	3,100	1
Unbleached American linters..........	9,300	3
Georgia cotton	10,800	3
Nettle fibers	11,600	4
Ramie...............................	12,400	4
Flax fiber...........................	36,000	12

These facts strongly support the conclusion that native cellulose contains periodically occurring acid-sensitive bonds which are responsible for its rapid initial (homogeneous or heterogeneous) hydrolysis in acid media. The fast *initial* degradation cannot be explained solely on the basis of the theory of accessibility. The order of accessibility as found by different methods is of entirely different magnitude than the percentage of bonds broken during the formation of hydrocelluloses.

Thus, MARK and co-workers (27) found that cellulose reacts very rapidly with heavy water and concluded that the extent of this reaction is a measure of the accessibility of the cellulose sample. From the accessibility they estimated the amount of amorphous material present in the samples studied to be 21 per cent. for cotton, 46 per cent. for wood pulp and 66 per cent. for rayon. These figures are in fairly good agreement with HERMANS' estimates (1) obtained from density measurements, but they are at variance with the results of the chemical methods such as NICKERSON's (63–66) or CONRAD and SCROGGIE's (11) oxidative hydrolysis to carbon dioxide and water, or GOLDFINGER, SIGGIA and MARK's (30) rate measurements of the periodate oxidation of cellulose. The NICKERSON method gave 5,3 per cent. for cotton linters, 10 per cent. for wood pulp, and 24 per cent. for rayon as the values of the accessible region, whereas the periodate treatment yielded 1 to 2 per cent. for cotton and 7 to 19 per cent. for rayon.

Another method for the determination of the ratio of crystalline and amorphous cellulose has recently been developed by Philipp et al. (72). It is based on a study of the kinetics of the heterogeneous cellulose hydrolysis by 6 N-hydrochloric acid at 100° C. It indicates that there are two distinct rates which have been attributed to the rapid hydrolysis of the loose amorphous regions and to the slow hydrolysis of the dense crystalline portion of the fiber. The raw cottons examined were found to range in degree of crystallinity from 82 to 88 per cent.; ramie, 95 per cent.; and textile rayon, 68 per cent. Of all the various procedures used, only Purves and co-workers' thallation method (6, 7) resulted in a figure, 0,4 per cent., which is close to the numerical value (0,3 to 1,5 per cent.) of the postulated acid-sensitive bonds; however, this agreement is clearly fortuitous since the measurements were made in an anhydrous system on bone-dry cellulose samples.

In any event, it should be clearly understood that the postulated occurrence of crystalline and amorphous regions in cellulose does not constitute a basis for the explanation of the *initial phase* of the heterogeneous degradation, under relatively mild conditions. Indeed, the drastic treatment of the samples in the chemical procedures for the determination of accessibility is such as to exclude any claims that might be made in this respect; the initial phase of hydrolysis is over long before the studied chemical decomposition begins. Conversely, it appears improbable that the initial rapid decrease in the average D. P. values, observed under mild conditions is a measure of the accessibility as proposed by some investigators (8).

The initial action of strong phosphoric acid on the fiber is then envisaged essentially to consist of solvation of the free hydroxyl groups both in the amorphous and the crystalline regions; this process results first in swelling of the fiber, then eventually in passing into solution in the form of extended laminae. Precipitation at this point by dilution with water gives rise to a slightly degraded cellulose with regenerated structure, i. e. one which is practically devoid of hydrogen bonds.

On the other hand, keeping the cellulose in acid solution for a longer period of time causes (at random frequencies in the laminae) the rupture of the acid-sensitive acetal bonds of the open-chain glucose units. This hydrolytic scission is a comparatively slow, catalytic process, since the hydrogen ion activity of the solvent is small. In the first stage of this process probably the 1,4-acetal bonds across the laminae are cleaved as shown in formula (XIV). This action would result in the scission of the chains and the production of shorter laminae ending in units with hemiacetal links (formula XV). Subsequent hydrolysis of the remaining 1,5-acetal bonds of the open-chain units would then completely disintegrate the shortest laminae (330 Å.), thus liberating the individual chain molecules with D. P. 64. Cleavage of the 1,5-acetal links as a first step at certain intervals in the laminae would result not only in the liberation of the individual chains but it would also give rise to longer chains (XVI),

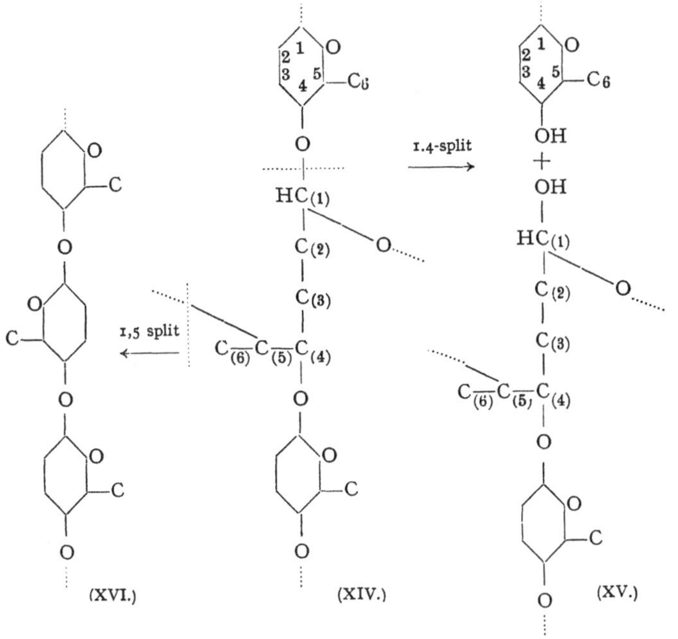

$$
\begin{array}{ccc}
 & \overset{2\ \ 1}{\underset{3\ \ 4\ \ 5}{\bigcirc}}\!\!O\!\!-C_6 & \overset{2\ \ 1}{\underset{3\ \ 4\ \ 5}{\bigcirc}}\!\!O\!\!-C_6 \\
 & O & \text{1.4-split} \quad OH \\
 & \vdots & + \\
 & HC_{(1)} & OH \\
 & C_{(2)}\!\!-\!\!O & HC_{(1)} \\
 & C_{(3)} & C_{(2)}\!\!-\!\!O \\
 \xleftarrow[\text{1,5 split}]{} & C_{\overline{(6)}}C_{\overline{(5)}}C_{(4)} & C_{(3)} \\
 & O & C_{\overline{(6)}}C_{\overline{(5)}}C_{(4)} \\
 & & O
\end{array}
$$

(XVI.)   (XIV.)   (XV.)

with lengths corresponding to an integer multiple of D. P. 64. Once the acid-sensitive bonds are all split, the hydrolysis continues on the 1,4-β-glycosidic bonds with its characteristic slow rate.

A very similar process is visualized regarding the initial phase of the *heterogeneous* acid hydrolysis of the native cellulose fiber. In this case, however, the first attack is supposed to be confined to the 1,4-acetal linkages occurring in the accessible region. Consequently, the acid-sensitive bonds in the resulting, crystalline "hydrocelluloses" would mostly remain intact. Since the glucose-anhydride chains in the terminal sections of these hydrocellulose molecules would stay aligned in their original positions, a recombination by loss of water and restoration of the original covalent bonds will be possible. Thus, STAUDINGER's observation regarding the increase in the D. P. value of degraded cellulose after nitration can be easily understood.

From the proposed concept of cellulose structure it follows that cellulose, hydrocellulose and their derivatives are not really "molecularly dispersed" in solutions but they also contain, beside individual chain molecules, laminae of various frequencies, and, possibly, also such laminae which have been split along the *b* axis. "Associations" of this kind, however, can not be broken up, as earlier supposed, by thermal agitation, since they are held together by covalent bonds. This conclusion is of fundamental importance concerning the determination of the "molecular weight" or the degree of polymerization of cellulose and its derivatives by physico-chemical methods which all require that the substance studied must be present in molecular dispersion.

References.

Books.

1. HERMANS, P. H.: Physics of Cellulose Fibers. New York-Amsterdam: Elsevier Publ. Co., Inc. 1946.
2. HEUSER, E.: The Chemistry of Cellulose. New York: John Wiley and Sons, Inc. 1944.
3. OTT, E. [Editor]: Cellulose and Cellulose Derivatives. New York: Interscience Publ., Inc. 1943.
4. PRINGSHEIM, H.: Die Polysaccharide, 2nd ed. Berlin: Julius Springer. 1923.
5. STAUDINGER, H.: Die hochmolekularen organischen Verbindungen. Berlin: Julius Springer. 1932; Makromolekulare Chemie und Biologie. Basel: Wepf u. Co. 1947.

Articles.

6. ASSAF, A. G., R. H. HAAS and C. B. PURVES: A Study of the Amorphous Portion of Dry, Swollen Cellulose by an Improved Thallous Ethylate Method. J. Amer. chem. Soc. **66**, 59 (1944).
7. — — — A New Interpretation of the Cellulose-Water Adsorption Isotherm and Data Concerning the Effect of Swelling and Drying on the Colloidal Surface of Cellulose. J. Amer. chem. Soc. **66**, 66 (1944).
8. BATTISTA, O. A. and S. COPPICK: Hydrolysis of Native versus Regenerated Cellulose Structures. Text. Res. J. **17**, 419 (1947).
9. BROWN, F., S. DUNSTAN, T. G. HALSALL, E. L. HIRST and J. K. N. JONES: Application of New Methods of End-group Determination to Structural Problems in the Polysaccharides. Nature [London] **156**, 785 (1945).
10. BROWNSETT, T. and G. F. DAVIDSON: The Solution of Chemically Modified Cotton Cellulose in Alkaline Solutions. VI. The Effect of the Method of Modification on the Relation between Fractional Solubility in Sodium Hydroxide Solution and Fluidity in Cuprammonium Solution. J. Textile Inst. **32**, T 25 (1941).
11. CONRAD, C. C. and A. G. SCROGGIE: Chemical Characterization of Rayon Yarns and Cellulosic Raw Materials. Ind. Engng. Chem. **37**, 592 (1945).
12. DAVIDSON, G. F.: The Progressive Oxidation of Cotton Cellulose by Chromic Acid over a Wide Range of Oxygen Consumption. J. Textile Inst. **29**, T 195 (1938).
13. — Properties of Oxycelluloses Formed in the Early Stages of Oxidation of Cotton Cellulose by Periodic Acid and Metaperiodate. J. Textile Inst. **31**, T 81 (1940).
14. — The Progressive Oxidation of Cotton Cellulose by Periodic Acid and Metaperiodate over a Wide Range of Oxygen Consumption. J. Textile Inst. **32**, T 109 (1941).
15. — The Progressive Oxidation of Cotton Cellulose by Chromic Acid over a Wide Range of Oxygen Consumption. J. Textile Inst. **32**, T 132 (1941).

16. DAVIDSON, G. F.: The Rate of Change in the Properties of Cotton Cellulose under the Prolonged Action of Acids. J. Textile Inst. 34, T 87 (1943).

17. DAVIDSON, G. F. and T. P. NEVELL: The Acidic Properties of Cotton Cellulose and Derived Oxycelluloses. Parts I–VI. J. Textile Inst. 39, T 59 (1948).

18. EKENSTAM, AF, A.: Über das Verhalten der Cellulose in Mineralsäure-Lösungen. I. Die Bestimmung des Molekulargewichts in Phosphorsäure-Lösung. Ber. dtsch. chem. Ges. 69, 549 (1936).

19. — Über das Verhalten der Cellulose in Mineralsäure-Lösungen. II. Kinetisches Studium des Abbaus der Cellulose in Säure-Lösungen. Ber. dtsch. chem. Ges. 69, 553 (1936).

20. FREUDENBERG, K.: Zur Kenntnis der Cellulose. Ber. dtsch. chem. Ges. 54, 767 (1921).

21. FREUDENBERG, K. u. G. BLOMQVIST: Die Hydrolyse der Cellulose und ihre Oligosaccharide. Ber. dtsch. chem. Ges. 68, 2070 (1935).

22. FREUDENBERG, K. u. W. KUHN: Die Hydrolyse der Polysaccharide. Ber. dtsch. chem. Ges. 65, 484 (1932).

23. FREUDENBERG, K., K. FRIEDRICH u. I. BUMANN: Über Cellulose und Stärke. Liebigs Ann. Chem. 494, 41 (1932).

24. FREUDENBERG, K., W. KUHN, W. DÜRR, F. BOLZ u. G. STEINBRUNN: Die Hydrolyse der Polysaccharide (14. Mitteil. über Lignin und Cellulose). Ber. dtsch. chem. Ges. 63, 1510 (1930).

25. FREY-WYSSLING, A. and K. MÜHLETHALER: Use of Supersonics in the Preparation of Fiber Samples for Electron-microscope Studies. Text. Res. J. 17, 32 (1947).

26. FRIESE, H. u. K. HESS: Zur Cellobiosebildung. III. Mitteil. über die Acetolyse der Cellulose. Liebigs Ann. Chem. 456, 38 (1927).

27. FRILETTE, V. J., J. HANLE and H. MARK: Rate of Exchange of Cellulose with Heavy Water. J. Amer. chem. Soc. 70, 1107 (1948).

28. FRITSCH, P.: Über die Umwandlung des Pentachloracetons in Trichloracrylsäure und Monochlormalonsäure. Liebigs Ann. Chem. 297, 318 (1897).

29. FRITSCH, P. u. F. FELDMANN: Synthese aromatisch disubstituierter Essigsäuren mittelst Chloral. Liebigs Ann. Chem. 306, 72 (1899).

30. GOLDFINGER, G., H. MARK and S. SIGGIA: Kinetics of Oxidation of Cellulose with Periodic Acid. Ind. Engng. Chem. 35, 1083 (1943).

31. GOLOVA, O. P.: The Molecular Weight of Cellulose. Chem. Abstr. 40, 457 (1946).

32. GRALÉN, N.: Sedimentation and Diffusion Measurements on Cellulose and Cellulose Derivatives. Thesis; Uppsala. 1944.

33. HAWORTH, W. N.: Die Konstitution einiger Kohlenhydrate. Ber. dtsch. chem. Ges. 65, (A) 60 (1932).

34. — The Structure of Cellulose and other Polymers Related to Simple Sugars. J. Soc. chem. Ind. 58, 917 (1939).

35. HAWORTH, W. N. and E. L. HIRST: The Constitution of the Disaccharides. V. Cellobiose (Cellose). J. chem. Soc. [London] 119, 196 (1921).

36. HAWORTH, W. N. and H. MACHEMER: Polysaccharides. X. Molecular Structure of Cellulose. J. chem. Soc. [London] 1932, 2270.

37. HAWORTH, W. N., C. W. LONG and J. H. G. PLANT: The Constitution of the Disaccharides. XVI. Cellobiose. J. chem. Soc. [London] 1927, 2809.

38. HEAD, F. S. H.: The Alkali-Sensitivity of the Aldehydes Obtained by Periodate Oxidation of β-Methyl Glucoside, β-Methyl Cellobioside, and Cellulose. J. Textile Inst. 38, T 389 (1947).

39. HESS, K.: Über die Konstitution der Zellulose I; HESS, K. u. W. WITTELSBACH: Die Acetolyse der Äthyl-Zellulose. Z. Elektrochem. angew. physik. Chem. 26, 232 (1920).

40. Hess, K. u. F. Neumann: Endgruppen-Frage und Konstitution der Cellulose
Ber. dtsch. chem. Ges. **70**, 728 (1937).

41. Hess, K. u. E. Steurer: Vergleich von Endgruppenbestimmung und Viscosität
bei Cellulose. Ber. dtsch. chem. Ges. **73**, 669 (1940).

42. Hess, K. u. W. Weltzien: Über Trimethylcellulose A und ihre Spaltung. Liebigs
Ann. Chem. **442**, 49 (1925).

43. Hiller, L. A., jr. and E. Pacsu: Cellulose Studies. V. Reducing End-Group
Estimation: A New Method Using Potassium Permanganate. Text. Res.
J. **16**, 318 (1946).

44. — — Cellulose Studies. VI. Determination of Carboxyl Groups in Cellulosic
Materials. Text. Res. J. **16**, 390 (1946).

45. — — Cellulose Studies. VII. The Nature of "Hydrocellulose". Text. Res.
J. **16**, 490 (1946).

46. — — Cellulose Studies. VIII. Viscosity and Hydrolytic Degradation of
Cellulose in Phosphoric Acid Solution. Text. Res. J. **16**, 564 (1946).

47. Hollihan, J. P.: Effect of Ageing of the Alkali Cellulose on the Carboxyl
Content of Rayons. Text. Res. J. **16**, 487 (1946).

48. Husemann, E. and O. H. Weber: The Carboxyl Content of Fibers and Wood
Celluloses. J. prakt. Chem. **159**, 334 (1942); Chem. Abstr. **37**, 3931 (1943).

49. Irvine, J. C. and E. L. Hirst: The Constitution of Polysaccharides. V. The
Yield of Glucose from Cotton Cellulose. J. chem. Soc. [London] **121**, 1585
(1922).

50. — — The Constitution of Polysaccharides. VI. The Molecular Structure
of Cotton Cellulose. J. chem. Soc. [London] **123**, 529 (1923).

51. Jackson, E. L. and C. S. Hudson: Application of the Cleavage Type of
Oxidation by Periodic Acid to Starch and Cellulose. J. Amer. chem. Soc. **59**,
2049 (1937).

52. Karrer, P. u. E. Escher: Über Acetylierung und Methylierung der Cellulose.
Ein Beitrag zur Konstitutionsfrage des Kohlenhydrates. Helv. chim. Acta **19**,
1192 (1936).

53. Karrer, P. u. H. Illing: Polysaccharide. XXX. Zur fermentativen Spaltung der Gerüstzellulose. Kolloid-Z. **36**, Suppl. 91 (1925).

54. Kuhn, W.: Über die Kinetik des Abbaues hochmolekularer Ketten. Ber.
dtsch. chem. Ges. **63**, 1503 (1930).

55. Mark, H. u. K. H. Meyer: Über den Bau des kristallisierten Anteils der
Cellulose. II. Z. physik. Chem., Abt. B **2**, 115 (1928).

56. Martin, A. R., L. Smith, R. L. Whistler and M. Harris: Estimation of
Aldehyde Groups in Hydrocellulose from Cotton. J. Res. nat. Bur. Standards
27, 449 (1941).

57. Mehta, P. C. and E. Pacsu: Cellulose Studies. X. Heterogeneous Degradation
of Cellulose and Viscose Rayon in Organic Acid Solutions. Text. Res. J.
18, 387 (1948).

58. Meyer, K. H.: Über den Bau des krystallisierten Anteils der Cellulose. V.
Ber. dtsch. chem. Ges. **70**, 266 (1937).

59. Meyer, K. H. u. H. Mark: Über den Bau des krystallisierten Anteils der
Cellulose. Ber. dtsch. chem. Ges. **61**, 593 (1928).

60. Meyer, K. H. et L. Misch: Positions des atoms dans le nouveau modèle
spatial de la cellulose. Helv. chim. Acta **20**, 232 (1937).

61. Monier-Williams, G. W.: The Hydrolysis of Cotton Cellulose. J. chem.
Soc. [London] **119**, 803 (1921).

62. Neumann, F. u. K. Hess: Nachweis kleinster Mengen endständiger Gruppen
bei Polysacchariden. Ber. dtsch. chem. Ges. **70**, 721 (1937).

63. NICKERSON, R. F.: Hydrolysis and Catalytic Oxidation of Cellulosic Materials. Hydrolysis of Natural, Regenerated, and Substituted Celluloses. Ind. Engng. Chem. **33**, 1022 (1941).

64. — Hydrolysis and Catalytic Oxidation of Cellulosic Materials. Ind. Engng. Chem., Analyt. Edit. **13**, 423 (1941).

65. — Hydrolysis and Catalytic Oxidation of Cellulosic Materials. Hydrolysis of Mercerized Cotton. Ind. Engng. Chem. **34**, 85 (1942).

66. — Hydrolysis and Catalytic Oxidation of Cellulosic Materials. Characterization of Celluloses. Ind. Engng. Chem. **34**, 1480 (1942).

67. NICKERSON, R. F. and C. B. LEAPE: Distribution of Pectic Acid in Cotton Fibers. Ind. Engng. Chem. **33**, 83 (1941).

68. PACSU, E.: Cellulose Studies. I. Reaction of Oxycellulose with Aqueous Acids and Alkalies. Text. Res. J. **15**, 354 (1945).

69. — Cellulose Studies. II. Estimation of Aldehyde Groups in Oxycellulose by the Hypoiodite Method. Text. Res. J. **16**, 105 (1946).

70. — Cellulose Studies. IX. The Molecular Structure of Cellulose and Starch. Text. Res. J. **17**, 405 (1947); J. Polymer Sci. **2**, 565 (1947).

71. PACSU, E. and L. A. HILLER, jr.: Cellulose Studies. IV. The Chemical Structure of Cellulose and Starch. Text. Res. J. **16**, 243 (1946).

72. PHILIPP, H. J., M. L. NELSON and H. M. ZIIFLE: Crystallinity of Cellulose Fibers as Determined by Acid Hydrolysis. Text. Res. J. **17**, 585 (1947).

73. PICTET, A.: L'amidon et ses produits de dégradation. Rapports sur les hydrates de carbon, Xième Conférence de l'Union Internationale de Chimie, p. 122. Liège. 1930.

74. PLÖTZE, E. u. H. PERSON: Die Kristallitorientierung in Fasercellulosen. Z. physik. Chem., Abt. B **45**, 193 (1940).

75. PURVES, C. B.: Historical Survey. Ref. *(3)*, p. 29.

76. — Chain Structure. Ref. *(3)*, p. 54.

77. ROSEVEARE, W. E., R. C. WALLER and J. N. NELSON: Structure and Properties of Regenerated Cellulose. Text. Res. J. **18**, 114 (1948).

78. REITTER, H. u. A. WEINDEL: Versuche zur Darstellung von Orthosäureestern. Ber. dtsch. chem. Ges. **40**, 3358 (1907).

79. RUTHERFORD, H. A. and M. HARRIS: Oxycellulose. Ref. *(3)*, p. 175.

80. — — Hydrocellulose. Ref. *(3)*, p. 164.

81. SCHULZ, G. V. u. E. HUSEMANN: Über die Verteilung der Molekulargewichte in abgebauten Cellulosen und ein periodisches Aufbauprinzip im Cellulosemolekül. Z. physik. Chem., Abt. B **52**, 23 (1942).

82. SCHWALBE, C. G.: Über das Reduktionsvermögen einiger Cellulosearten. Ber. dtsch. chem. Ges. **40**, 1347 (1907).

83. SISSON, W. A.: X-Ray Examination. Ref. *(3)*, p. 203.

84. SKRAUP, H. u. J. KÖNIG: Über Cellose, eine Biose aus Cellulose. Ber. dtsch. chem. Ges. **34**, 1115 (1901).

85. SOOKNE, A. M. and M. HARRIS: End Groups. Ref. *(3)*, p. 77.

86. SPONSLER, O. L. and W. H. DORE: The Structure of Ramie Cellulose as Derived from X-Ray Data. Fourth Colloid Symposium Monograph, p. 174. New York: Chemical Catalog Co. 1926.

87. SPURLIN, H. M.: Solubility. Ref. *(3)*, p. 853.

88. STAMM, A. J. and W. E. COHEN: The Viscosity of Cellulose in Phosphoric Acid Solution. J. physic. Chem. **42**, 921 (1938).

89. STAUDINGER, H.: Über die Konstitution der Cellulose. Cellulosechemie **15**, 53 (1934).

90. Staudinger, H. u. E. O. Leupold: Über das Cellopentaose-acetat und die Konstitution der Cellulose. Ber. dtsch. chem. Ges. **67**, 479 (1934).

91. Staudinger, H. u. A. W. Sohn: Über makromolekulare Verbindungen. 227. Mitteil.: Über normale und fehlerhafte Cellulosen. Ber. dtsch. chem. Ges. **72**, 1709 (1939).

92. — — Über makromolekulare Verbindungen. 242. Mitteil.: Über native und umgefällte Cellulosen und deren Nitrate. J. prakt. Chem. **155**, 177 (1940).

93. Staudinger, H. u. M. Sorkin: Über hochpolymere Verbindungen. 162. Mitteil.: Über Hydro-cellulosen. Ber. dtsch. chem. Ges. **70**, 1565 (1937).

94. Steurer, E.: Über den Einfluß des Lichtes auf Celluloselösungen. Z. physik. Chem., Abt. B **47**, 139 (1940).

95. Steurer, E. u. H.-W. Mertens: Der Sauerstoffeinfluß beim Lichtabbau von Methylcellulosen. Ber. dtsch. chem. Ges. **74**, 790 (1941).

96. Unruh, C. C. and W. O. Kenyon: Investigation of the Properties of Cellulose Oxidized by Nitrogen Dioxide. J. Amer. chem. Soc. **64**, 127 (1942).

97. — — The Formation and Properties of Oxidized Celluloses. Text. Res. J. **16**, 1 (1946).

98. van Fossen, P. and E. Pacsu: Cellulose Studies. III. Hyperoxidation of Cellulose with Concentrated Sodium Hypobromite Solutions. A Simple Method for the Determination of Hypobromite and Bromate Ions in the Presence of Each Other. Text. Res. J. **16**, 163 (1946).

99. Whistler, R. L., A. R. Martin and M. Harris: Determination of Uronic Acids in Cellulosic Materials. J. Res. nat. Bur. Standards **24**, 13 (1940).

100. Yackel, E. C. and W. O. Kenyon: Oxidation of Cellulose by Nitrogen Dioxide. J. Amer. chem. Soc. **64**, 121 (1942).

101. Zechmeister, L. u. G. Tóth: Zur Kenntnis der Hydrolyse von Cellulose und der dabei auftretenden Zwischenprodukte. Ber. dtsch. chem. Ges. **64**, 854 (1931).

(Received, April 27, 1948.)

Lignin.

By F. E. BRAUNS, Appleton, Wisconsin.

With 2 Figures.

The last report on lignin in these "Fortschritte" is that of FREUDEN-
BERG (*102*). Since then a considerable amount of work on lignin has
been carried out, especially by ERDTMAN, FREUDENBERG, HIBBERT, and
WACEK, and their co-workers. Some progress in fundamental knowledge
has been made, although the final solution of the problem of the lignin
structure is still to be found. Brief reviews of special phases of lignin
chemistry, particularly on the structure, were published recently by
BARONI (*27*), BRAUNS (*36*), ERDTMAN (*85*), FREUDENBERG (*103*, *104*),
HIBBERT (*152*), KRÜGER (*193*), LANGE (*208*), McCARTHY (*226*), PERCI-
VAL (*254*), PHILLIPS (*256*), WACEK (*308*), and YORSTON (*332*).

The *utilization of lignin* in general was recently discussed by ARIES (*6, 8*),
BAILEY and WARD (*24*), BRAUNS (*39*), DUNN and SEIBERLICH (*77, 78*),
HARRIS (*135*), HILL (*153*), JAHN (*170, 171*), KEILEN (*181, 182*), KRATZL (*188*),
PEPPER (*252*), REFLE (*268*), SCHWARTZ (*293*), SEIBERLICH (*294, 295*),
and WAGNER (*317*).

I. Occurrence, Formation and Detection.

The State of Lignin in Wood.

The argument started by HILPERT (*154*) in 1935, as to whether lignin
exists as such in wood or whether it is a secondary reaction product
formed by the action of mineral acids or alkalies on certain sensitive
carbohydrates in the plant, is still in progress. In a series of publications,
HILPERT studied the relationship between the pentosans and lignin (*155*),
the action of SCHWEIZER's reagent (*156*), of cuproxide ethylenediamine
solution (*157*), and of nitric acid (*158*) on woods and straws. In the
results of these investigations, HILPERT believed that he had found
further evidence for his theory of the non-existence of "lignin" in plants.

From the results of the action of nitric acid on shells from nuts and
pits, KRÜGER (*194*) agreed with HILPERT that lignin is a secondary

reaction product of sensitive carbohydrates and that the degree of lignification is best measured by the methoxyl content. This, however, is incorrect, because a considerable amount of the methoxyl in wood, particularly in deciduous woods, is combined with carbohydrates [BRAUNS (36); HÄGGLUND (131, p. 198)].

HILPERT's theory is strongly supported by SCHÜTZ (283, 284, 285, 286). He found that, when beechwood is extracted with flowing steam at 100–105°, up to 26,4% of the wood goes into solution without any change in the elementary composition of the residual substance. On boiling the aqueous extract with 1% hydrochloric acid, from 5 to 40% of "lignin-like" products (containing 24,5% methoxyl) are precipitated. In an attempt to prove the presence of lignin in spruce, SCHÜTZ and SARTEN (287, 288) treated spruce groundwood with diazobenzene-4-sulfonic acid in sodium bicarbonate solution. They found that oxidation takes place and that, after three treatments with two intermediate extractions with dilute sodium hydroxide at 60 to 90°, all the wood, with the exception of about 22% of cellulose, was dissolved. No aromatic compounds related to lignin were found in the aqueous extracts. SCHÜTZ and co-workers (290) also succeeded in bringing about an almost complete dissolution of beechwood chips by alternate treatment with 10% hydrogen peroxide at room temperature for one hour and with steam at 100° for 12 to 20 hours. In the aqueous solution (as far as it was investigated), formic, acetic, oxalic, succinic, dl-tartaric, vanillic, and protocatechuic acids, in addition to glycol aldehyde, were found. Because α-cellulose or unbleached wood pulp, when treated under the same conditions, lost only 3% or 18%, respectively, of its weight, these authors concluded that neither cellulose nor lignin occurs in the wood in the same state as that in which it is isolated. Unlike FREUDENBERG (105), SCHÜTZ (289) does not consider the refractive index of 1,61 for lignin as a proof that lignin exists as such in woods.

Accepting HILPERT's theory, ENDERS (80) put forward a purely speculative hypothesis of lignin formation. According to this hypothesis, two molecules of glycerol, formed by biochemical reduction of a triose, and one molecule of glycerylaldehyde (I) condense with elimination of six molecules of water to form an aldol-like product (II) called "ligninogen" of a nonatetraene type similar to that suggested by FREUDEN-BERG (101, p. 119). The condensation is followed by a partial methylation. In this ligninogen the hydroxypolyene chains are arranged in such a manner that, by splitting off water under the conditions used in the isolation of lignin, cyclization takes place with the formation of phenyl-propane building stones (III) which are assumed to be present in isolated ("artificial") lignin. The folded structure of the ligninogen molecule, which is similar in shape to that proposed by ASTBURY and WRINCH (9) for proteins, may contribute, according to ENDERS, to the strength of the lignified fiber.

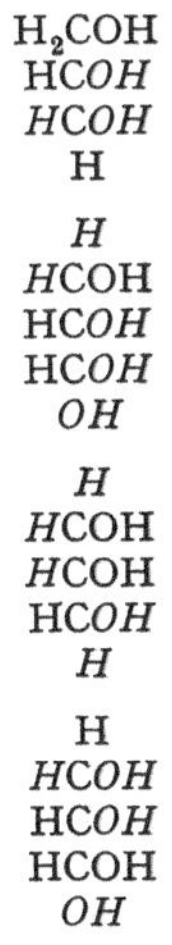

H₂COH
HCOH
HCOH
H

H
HCOH
HCOH
HCOH
OH

H
HCOH
HCOH
HCOH
H

H
HCOH
HCOH
HCOH
OH

(I.)

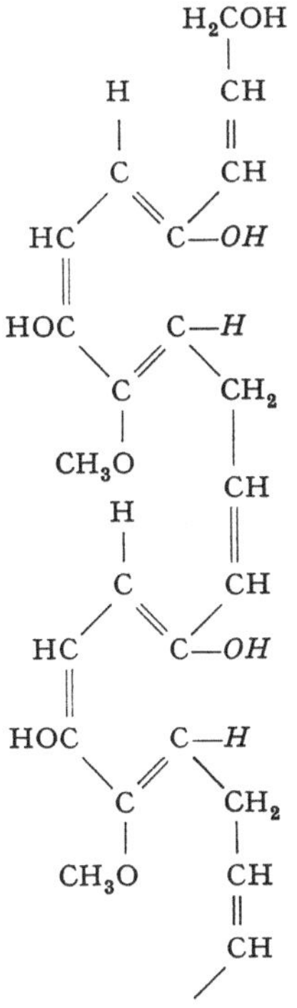

H₂COH
|
H   CH
| ‖
C   CH
/   ‖
HC   C—OH
‖
HOC   C—H
\ ‖
C   CH₂
| |
CH₃O CH
H   ‖
| CH
C   CH
/   ‖
HC   C—OH
‖
HOC   C—H
\ ‖
C   CH₂
| |
CH₃O CH
‖
CH
/

(II.) "Ligninogen".

H₂COH
|
CH
‖
CH

HO—

CH₂
|
CH₃O   CH
‖
CH

HO—

CH₂
CH₃O |
CH
‖
CH
/

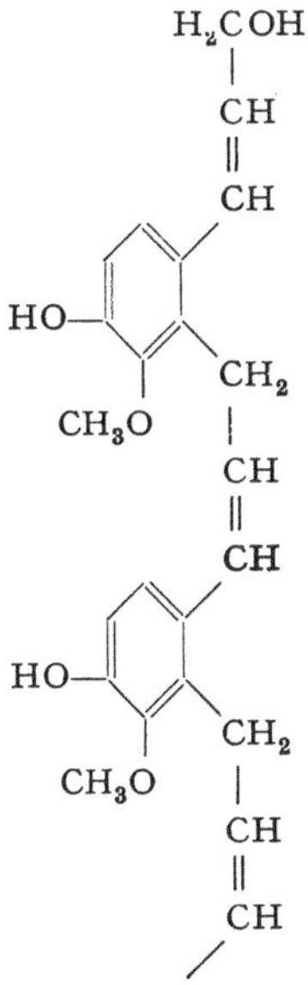

(III.) Phenylpropane building stones.

A compromise between HILPERT's hypothesis and the rather general view that lignin is built up of phenylpropane building stones and exists as an aromatic compound in the plant, was suggested by JAYME (*173*). His investigation (*174*) of the waste liquor obtained in the isolation of "holocellulose" from sprucewood by means of sodium chlorite and acetic acid revealed that, in spite of the fact that holocellulose is supposed to contain the total carbohydrates of the wood, an appreciable amount of carbohydrate material is found in the waste liquor. To explain the presence of this so-called "excess polysaccharide material", JAYME suggested that what is called "lignin" is present in wood as a polysaccharide of a hexose type in which guaiacyl groups are substituted at carbon atoms 1 and 4 of the glucose anhydride unit as shown in scheme (IV). It is assumed that, during the isolation of lignin, guaiacyl-substituted sugar groups undergo condensation reactions that destroy their identity, and that they are instrumental in giving typical phenylpropane structures [as indicated by the broken lines in (IV)] similar to those postulated by FREUDENBERG for lignin. JAYME's views, however, are opposed by FREUDEN-

Berg (*111*), who obtained carbonyl-containing, ether-insoluble products on oxidation of lignin and of catechin with hydrogen peroxide or ozone [cf. Richtzenhain (*272*)].

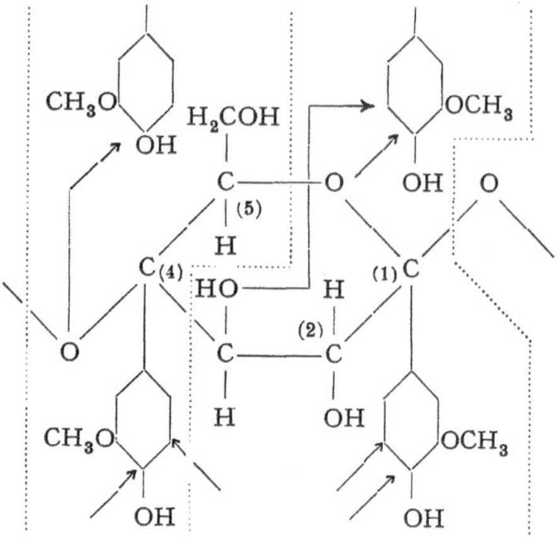

(IV.) Jayme's guaiacyl-carbohydrate complex.
(The arrows indicate the points at which condensation may take place.)

Another hypothesis on the formation of lignin was recently suggested by Kürschner (*197*). According to him a partially methylated pentosan, "prolignan", could occur in wood and would be hydrolyzed on treatment with hot water to a water-soluble (hypothetical) 2-methylpentose termed "prolignose" (V). The latter, through the action of acids or alkali or on heating would condense, with the formation of lignin, by splitting off five molecules of water and one molecule of acetic acid from two molecules prolignose, according to scheme (V).

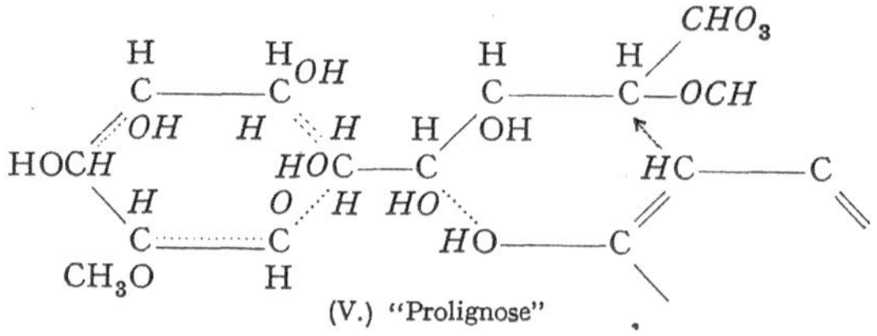

(V.) "Prolignose"

That lignin occurs as such in wood has been demonstrated by the fact that a small amount can be extracted from sprucewood and western hemlock by indifferent solvents, such as dioxane or ethyl alcohol, without the use of a catalyst [Brauns (*35, 36*)]. This so-called "isolated native lignin" behaves chemically very similarly to the remainder of the lignin and shows ultraviolet absorption spectra identical with those given by other lignin preparations [see also Aulin-Erdtman (*12*)].

The occurrence of lignin in wood as an aromatic compound was further shown by Lange (*206, 207*) who, by means of an ultraviolet quartz microscope, studied the absorption of lignin *in situ*. He found an absorption maximum at 2800 Å. for the lignin and, by comparing it with a series of model substances, he estimated that one aromatic ring is present per ten carbon atoms, thus confirming the aromatic nature of "protolignin".

Lignin-Complexes in Wood.

In a series of investigations, FREUDENBERG and PLOETZ (*115*) and, later, PLOETZ (*258, 259, 260, 261*) studied the enzymatic decomposition of polymeric carbohydrates in wood and other lignified tissues and arrived at the conclusion that the major part of lignin is in *chemical combination* with carbohydrates. Carbohydrate-lignin complexes with a ratio of 3 : 1 and 1 : 1 have been isolated, the latter type resulting from the former by further decomposition. When basswood is extracted with cuproxide ethylenediamine, 66% of the wood dissolves. From this solution, 42% (based on wood) is precipitated on acidification and 24% remains in the aqueous solution. Enzymatic decomposition of the precipitated portion with a digestive fluid from *Helix pomatia* results in the formation of a lignin-carbohydrate complex 1 : 1. PLOETZ (*260*) concluded from these results that the lignin in wood is almost entirely combined with carbohydrates.

BAILEY (*14*) digested aspen and jack pine in a buffered butanol-water mixture at 180° and removed all the lignin from aspen and 80% of the lignin from pine. The remaining 20% of the latter was readily removed with the same mixture in the presence of 2% sodium hydroxide. The conclusion drawn that the lignin is entirely uncombined in aspen is not justified, however, because the assumption that the buffered solution would prevent hydrolysis is erroneous.

A linkage very resistant to cleavage by sodium hydroxide or by phenol and hydrogen chloride appears to exist between the cellulose and the "lignin" in redwood bark as was found by BRAUNS and co-workers (*46*). A part of the "lignin" seems to be closely related structurally to spruce and redwood lignin. This seems to be true also for the "lignin" in spruce bark as was shown by WACEK and SCHÖN (*316*). Although the methoxyl content (5,6%) of the lignin was relatively low, the fact that the bark, on alkaline oxidation with nitrobenzene, gave 2,2% vanillin, indicated that the spruce bark lignin contains, at least in part, the same building stones as the lignin in the wood.

As model substances for a possible lignin-carbohydrate linkage in wood, HIBBERT and co-workers (*97*) prepared some β-d-xylosides of ethanolysis and other lignin degradation products; furthermore, a β-cello-bioside of acetovanillone, and the β-d-glucoside of α-hydroxypropio-veratrone. From the rate of hydrolysis of these and other aliphatic and phenolic glycosides, they considered it possible that a phenolic-glycosidic type of linkage combines the lignin with carbohydrates (98).

In order to decide whether or not lignin is combined with other components of the wood and, if this is the case, to find out in what manner, the classical method used for the elucidation of the structure of natural glycosides—i. e., complete methylation followed by hydrolysis—was applied to sprucewood [BRAUNS and YIRAK (48)]. A completely methylated sprucewood (38,5% methoxyl) was subjected to various reactions used for the isolation of lignin (such as 42% hydrochloric acid, 72% sulfuric acid, methanol and hydrochloric acid, phenol and hydrochloric acid, and bromohydrin). In all cases, lignin preparations with lower methoxyl contents than anticipated were obtained, the only exception being the methanol-hydrochloric acid extraction which gave a methanol lignin with 30,6% methoxyl, corresponding to nine methoxyl groups per one lignin building unit. However, no conclusions can be drawn from the latter experiment because methoxyl groups are introduced into the lignin molecule during the reaction and may substitute for carbo-hydrate groups if they are attached to lignin in the wood. Experiments concerning the stability of methoxyls in fully methylated hydrochloric acid spruce lignin [BRAUNS (40)] revealed that two methoxyl groups per unit are split off when the methylated lignin is treated with strong mineral acids or acetic acid and magnesium chloride, or phenol and a trace of hydrogen chloride. With thioglycolic acid only one methoxyl group was removed. Neither the fully methylated sprucewood nor the fully methylated hydrochloric acid spruce lignin is attacked when heated with 5% sodium hydroxide at 175° to 180° C.

Formation, Detection and Estimation of Lignin.

The *formation* of lignin in the oat plant during its growth was studied by PHILLIPS and co-workers (257), who concluded that the plant does not synthesize lignin from cellulose, pectin or pentosans. There was, moreover, no indication that a coniferyl derivative is an intermediate in the biosynthesis of lignin. It seems to be more likely that simple carbohydrates, having firmly bound methoxyl groups, are the precursors of lignin, since the amount of these substances decreases with increasing lignin formation in the wood.

Because lignified tissues in general contain very little pectic substance and considerable amounts of lignin, whereas in unlignified tissues the opposite is true, BENNETT (29) concluded that pectic substances are the precursors of lignin.

NORD and VITTUCI (234, 235) found that a wood-destroying mold, *Lentinus lepideus*, which, when grown on white Scot's pine, produces methyl *p*-methoxycinnamate, also forms this ester from glucose, xylose, glycerol or ethyl alcohol, and that acetaldehyde is an intermediate in the enzymatic degradation of the carbohydrate and the enzymatic ester synthesis. Since methyl *p*-methoxycinnamate is structurally related to the lignin building stones, they concluded that the progress of lignification is governed by the respective rates of the enzymatic carbohydrate degradation and the synthesis of methylated products.

The course of formation of lignin during the growth of sprucewood was studied by HIBBERT and co-workers (*34*). Hydrogenation of 2,5 to 3 weeks' old spruce tips in dioxane in the presence of copper chromite gave no phenylpropyl derivatives. With 3,5 to 4 months' old spruce buds, 4-propylcyclohexanol was obtained in an amount corresponding to about 20% of the lignin present in mature wood.

KRATZL (*186*) found that potato germs grown in the absence of light contain a substance which he considered to be lignin because it gave the color reactions with phloroglucinol-hydrochloric acid, aniline sulfate, and the MÄULE test. On isolation with 50 to 57% sulfuric acid, 14% "lignin" with 4,2 to 4,4% methoxyl and 6% nitrogen was obtained as compared with 13,4% methoxyl and 2% nitrogen in the lignin (15%) from the potato stalks. KRATZL assumed that photo-energy is not required for lignin formation.

COPPICK and FOWLER (*66*) described a method for determining the *location* of lignin in wood fibers by treating them with chlorine water followed by silver nitrate in a 3% solution of ethanolamine (in 95% ethyl alcohol). According to these investigators, reducing groups are formed or liberated from a lignin-carbohydrate complex on chlorination, and these groups reduce the silver nitrate, precipitating silver in the region of the lignin. HARLOW (*134*) studied the location of lignin in woody cell walls by removing the carbohydrates with 72% sulfuric acid, followed by SCHWEIZER's reagent. DADSWELL and ELLIS (*73*) demonstrated the distribution of lignin in wood fiber by removing the carbohydrates with 72% sulfuric acid from microtome wood sections, pretreated with 3% iodine in potassium iodide. A new method for the microscopic detection of protolignin in the cell wall was developed by JAYME and HARDERS-STEINHÄUSER (*175*). Microtome wood sections were treated with nitrogen dioxide, causing the formation of a deep yellow color in the lignified portions of the fiber. On treatment with triethanolamine, the color was deepened. It was shown that lignin is present in the secondary cell wall, as well as in the middle lamella of the beechwood fiber.

According to WIECHERT (*328*) the *color reactions* with phenols and amines generally ascribed to lignin are given not by lignin itself but by accessory components, such as vanillin. This is contradicted by the fact that not only isolated native spruce and aspen lignins which have been thoroughly extracted with ether and benzene, but also woods from which the soluble native lignin has been eliminated give a strong color with WIESNER's reagent. Because the diazomethane-methylated native spruce lignin, in which the enolized carbonyl groups are methylated, no longer gives the color reaction, the latter must involve a free carbonyl group [BRAUNS (35)].

A few new amines as color reagents for lignin have been proposed by MOERKE (230).

CAMPBELL and McGOWAN (59) drew attention to the fact that the MITCHELL color reactions for gallotannin and for hardwood lignin (produced on treatment with chlorine and sodium sulfite) are essentially identical. Both are caused by a pyrogallol grouping.

In a systematic investigation of the *quantitative determination* of lignin by means of sulfuric acid, FREUDENBERG and PLOETZ (116) found that with sprucewood an increase in the concentration of the acid from 66% to 75% did not noticeably affect the results. With basswood, beech, and elder pith, however, this concentration had a noticeable effect on the yield of lignin and its elementary composition. FREUDENBERG suggests that the suitable acid concentration should be determined for each case, to give "the minimum yield of lignin with the maximum methoxyl content". This was further confirmed by PLOETZ (262).

When alkali corncob lignin is treated with 72% sulfuric acid, partial degradation of the lignin occurs with simultaneous loss of methoxyl groups [DRYDEN et al. (76)].

According to GRIFFIOEN (129), pre-extraction of lignified material with benzene-alcohol, water, and dilute acid caused very little effect in lignin estimations. The interference of various types of tannin was studied by SHRIKHANDE (297).

II. Isolation of Lignin.

One of the main difficulties in lignin chemistry is the fact that no one thus far has succeeded in isolating the *total lignin in an unchanged state*. A small amount of what is considered to be "unchanged native lignin" has been isolated by BRAUNS (35, 37) from black spruce and from western hemlock in a yield of about 8 to 10% of the total lignin. These lignin samples have a methoxyl content of 14,8%, in agreement with the value calculated for spruce protolignin by HÄGGLUND (131, p. 199). They give strong colors with WIESNER's reagent and dissolve completely in hot bisulfite solution. In the same way, BUCHANAN (57) prepared from aspen a native lignin with 19,5% methoxyl.

That the lignin obtained by other methods is more or less changed is shown by the fact that, when isolated native spruce lignin is treated under the conditions used for the isolation of lignin with 72% sulfuric acid or 42% hydrochloric acid, it loses about 8% of its weight, with a simultaneous increase in carbon and methoxyl contents [BRAUNS (35)].

Several new methods, or modifications of old methods, for the isolation of lignin have been proposed in recent years. Based upon the discovery by HELFERICH and by SCHLUBACH that anhydrous *hydrofluoric acid* readily hydrolyzes polysaccharides such as cellulose, FREDENHAGEN and CADENBACH (100) and later WIECHERT (327, 329, 330) as well as RÜDIGER (276) worked out methods for the isolation (and quantitative determination) of lignin by means of anhydrous hydrogen fluoride. Since the wood is treated with this reagent at a very low temperature and for a short time, it is claimed that hydrogen fluoride lignin is changed very little. Hydrogen fluoride beech lignin is a light brown powder, still showing the morphological structure of the wood and containing 60,8% carbon, 5,5% hydrogen, and 19,6% methoxyl groups. Since this method requires special apparatus, and hydrogen fluoride is not very pleasant to work with, little research has been done with this kind of lignin. The fact that it is no longer soluble in hot bisulfite solution indicates that a considerable change must have taken place during the isolation.

An entirely new approach to the isolation of lignin was made by PURVES and co-workers (274, 318), who treated pre-extracted wood with a solution of sodium paraperiodate at pH 4 and 20°, thus oxidatively cleaving the carbohydrate chains between carbon atoms 2 and 3. The oxidized polysaccharide was then hydrolyzed by boiling the wood with water at pH 6,5, leaving the *"periodate lignin"* as an insoluble residue. Although this lignin underwent chemical changes, presumably oxidation rather than resinification, it is claimed that it resembles protolignin very closely. Periodate lignin still shows the morphological structure of the wood and gives a strong purple color with phloroglucinol in hydrochloric acid. It dissolves almost completely in hot bisulfite under conditions similar to those used in the sulfite cooking process. To what degree the lignin has been oxidized by the periodic acid is still open to discussion. That lignin in lignosulfonic acid is oxidized by periodic acid was shown by PENNINGTON and RITTER (250), who found that the consumption of periodic acid increased with the methoxyl content and that methoxyl groups were removed during the oxidation process.

Whereas, in the above mentioned isolation methods, insoluble lignin samples are obtained, a number of new *soluble lignin derivatives* have also been prepared. *Acetic acid* in the presence of an acid catalyst, first suggested by PAULY (243), was used recently by SCHÜTZ and KNACKSTEDT (286), furthermore, by FREUDENBERG and PLANKENHORN (114) with magnesium chloride as the catalyst. FREUDENBERG also converted cuproxam spruce lignin into an acetic acid lignin which differed only slightly in its elementary composition from acetic acid lignin isolated directly from sprucewood. PAULY (242) claimed that, in the isolation with acetic acid in the presence of 3% sulfuric acid, all ether linkages in the lignin molecule are cleaved, with the exception of the methoxyl linkages. For this reason he called such "highly phenolic hydrolyzates" "acetic acid lignols" in order to emphasize their phenolic nature. BRAUNS and BUCHANAN (42) found, however, that the number of hydroxyl groups was the same as in isolated native spruce lignin.

The extraction of lignin from cornstalks by means of other organic acids (such as lactic, α-methoxypropionic, propionic, acetic, and formic acids) was studied by FISHER (93). When concentrated acids were used, partial esterification of the lignin took place. Lactic acid seems to combine with the lignin with ether formation:

$$LigOR + CH_3CH(OH)COOH \rightarrow CH_3CH(OLig)COOH + ROH.$$

In an attempt to isolate lignin in a state as unchanged as possible, HIBBERT and co-workers (249) subjected red oakwood to *acetolysis*, using acetic anhydride and glacial acetic acid, with sulfuric acid as the catalyst. On solvent fractionation, no carbohydrate-free lignin preparation was obtained. The isolation of a bisulfite-soluble oak lignin by acetolysis of oakwood was claimed by STEEVES and HIBBERT (299).

The separation of lignin from humic substances by means of acetyl bromide was found to be unsuitable [KÜRSCHNER (199)]. Aqueous chloroacetic acid (80 to 85%) was found to dissolve lignin from spruce, beech, and rye straw without the presence of a catalyst, since some hydrochloric acid is always formed in this reaction [SCHÜTZ (283, 284)]. The yields of lignin were in inverse proportion to the sugar yields. (From this, SCHÜTZ concluded that lignin is a secondary reaction product, formed from sensitive carbohydrates.) Similar results were obtained upon extraction of lignin with *ethylene chlorohydrin* and with *α-mono-chlorohydrin* [SCHÜTZ (285)]. The latter reagent was also used by FREUDENBERG (106). Extraction of lignin by means of *chloral hydrate* [OGAIT (236)] gave chloral spruce lignin containing one chloral group per three lignin building stones. In all the lignin preparations isolated with chlorine-containing solvents, chlorine was found, indicating that these "solvents" reacted with the lignin.

What is erroneously called *"butanol lignin"* was prepared by BAILEY (15, 17, 18) by extracting wood with a 1:1 mixture of water and butanol containing 40% sodium hydroxide (based on the wood) at 160°. A true butanol spruce lignin was prepared by CHARBONNIER (61). It contained two butoxyl groups combined in an acetal-like manner.

The lignin isolated by means of an "alkali-catalyzed" butanol-water mixture is identical with alkali spruce lignin [see also BAILEY (22)].

Lignins obtained from the heart- and sapwood of Douglas fir by various means showed no significant differences in their analytical data from corresponding sprucewood lignin preparations [cf. BROOKBANK and BRAUNS (55)]

Although it was once claimed [WEDEKIND (319)] that lignin can be extracted unchanged from wood by treating it with hydrochloric acid and then with ethyl acetoacetate, no details of this method were ever published. VIRASORO (306) isolated an "ester quebracho lignin" in this manner. It had a methoxyl content of 17,1% whereas quebracho lignin isolated according to KÖNIG contained 18,3% and a quebracho Klason lignin had 18,0% methoxyl [WISE (331)]. The low methoxyl content of the ester quebracho lignin indicated either that the acetoacetic ester had reacted with the lignin or that some methoxyl had been split off.

In the extraction methods just discussed, chemicals having reactive groups capable of undergoing a condensation with the lignin were used. The extraction of lignin with an indifferent solvent, such as *dioxane*, in the presence of a catalyst is therefore of particular interest. Originally suggested by WEDEKIND and ENGEL (319), dioxane has recently been used by JUNKER (178) and PERRENOUD (255). They found that, with increasing amounts of concentrated hydrochloric acid as the catalyst, the yield of lignin also increased. At the same time, the methoxyl content decreased to 11,3% when a yield of 25,3% of lignin was obtained.

To isolate lignin, HESS and HEUMANN (144) treated rye straw, previously ground in an oscillating mill for 24 hours to a particle size of 1μ or less, with 25% *hydrazine* in ethanol at room temperature. In this manner, 96% of the lignin went into solution. The hydrazine straw lignin contained 51,8% carbon, 5,6% hydrogen, 6,1% nitrogen, and 13,5 to 14% methoxyl groups, and had a molecular weight of 2000 to 3000 as determined by the osmotic pressure method in 80% alcohol. The low carbon content was explained by the fact that the lignin did not undergo autocondensation. On treatment with 1 to 2% sulfuric acid in ethanol, hydrazine lignin lost nitrogen. On methylation with diazomethane, 4 to 5% methoxyl was taken up, and with dimethyl sulfate a methoxyl content of 32 to 33% was obtained. This lignin showed the same ultraviolet absorption spectrum as methanol spruce lignin; it contained one phenolic hydroxyl group per two lignin building stones, in addition to an alcoholic hydroxyl in the side-chain. The formation of a condensation product with hydrazine indicated the presence of one carbonyl group per lignin building stone.

On extraction of lignin with *ethanolamine*, considerable demethylation occurs [BRAUNS (41); REID et al. (270)] with the formation of an ethanolamine-lignin condensation product [FISHER and BOWER (94)].

III. Properties and Conversions of Lignin.

Physical Properties.

Molecular weight. GRALÉN (*128*) determined the molecular weight of a number of thioglycolic acid spruce lignin preparations, using the Svedberg ultracentrifuge. With a thioglycolic acid lignin he found a molecular weight of about 9400 or, calculated on the thioglycolic acid-free basis, between 6000 and 7000. Assuming that lignin consists of phenylpropane building stones of a molecular weight of about 180, a degree of polymerization (*D. P.*)[1] of 36 was calculated. When the wood was pretreated with hydrochloric acid, the molecular weight increased to approximately 34000 or D. P. 146. A hypobromite spruce lignin showed a molecular weight of 8000 to 9000, or D. P. 32 to 37. An acetic acid spruce lignin had a molecular weight of 13000 to 15000, whereas an alkali spruce lignin showed only 7000.

SCHWABE and HASNER (*292*) separated lignosulfonic acid by means of fractional electrodialysis into three fractions and found molecular weights of 300 to 400, 1500 to 2000, and about 20000, using the diffusion method. A separation of lignosulfonic acid into fractions of similar molecular weights was achieved by means of isopropyl alcohol (*291*). It is possible that, in the sulfite cooking process, the lignin molecule is partially degraded and partially further condensed [HEDLUND (*143*)].

By determining the diffusion coefficients of lignosulfonic acids isolated from sulfite waste liquor, molecular weights of 3000 to 10000 were found by PENNINGTON and RITTER (*251*). The non-lignin components in sulfite waste liquor seem to have a low molecular weight, indicating that lignosulfonic acids are the only macromolecules present. Similar experiments were carried out by ERNSBERGER and FRANCE (*91*) who separated, by fractional diffusion, lignosulfonic acid from a commercial sulfite waste liquor into three fractions with the respective molecular weights, 250 to 290, 2140, and 9500.

Applying a quantitative thermal method for measuring the high frequency energy-loss factor of solutions containing polar molecules and using the DEBYE dipole theory for both parts of the complex dielectric constant, CONNER (*63*) found that methanol aspen, maple, and spruce lignins are fairly homogeneous, showing a molecular weight of approximately 3900 and a shape factor of 8. His results are in very good agreement with those reported by LOUGHBOROUGH and STAMM (*220, 221*).

Form of particles. Viscosity measurements [ERBRING and PETERS (*81*)] of lignin solutions prepared by dissolving the lignin from sprucewood and wheat straw in glycol or glycerol at 150 to 160°, in the presence of

---

[1] The Degree of Polymerization is obtained by dividing the molecular weight by the weight of a C_6 unit.

small amounts of hydrochloric acid or formic acid, indicated that glycol lignin is not made up of thread-like but of almost spherical particles. Measurements of the interfacial surface forces between alkali lignin solutions and paraffin oil or benzene demonstrated that lignin is of an aromatic nature.

The axial ratio (width : length) of a lignin particle of a molecular weight of 7000 was found by GRALEN (*128*) to be 1 : 5.

Spectrum. Recent determinations [GLADING (*125*)] showed that, in many lignin preparations and their derivatives, the absorption maximum at 280 mμ. was directly proportional to the native lignin content of the derivative. This relationship, however, did not hold for derivatives containing a substituent which, in itself, has a high absorption coefficient at 280 mμ. On the basis of the spectra of compounds with known structure and assumed to be related to lignin, it was concluded that each lignin building unit contains two pyran rings as proposed by FREUDENBERG (*101*, p. 136). It was suggested also that a carbonyl group could be responsible for the band of the phenylhydrazone at 352 mμ. When the carbonyl group is completely enolized, this band disappears. Similar spectra were obtained by VIRASORO (*305*) for quebracho and red willow lignins.

A comparison of the spectra of pyrocatechol, pyrogallol 1,3-dimethyl ether, their methyl ethers, and of various derivatives in which a carbon side chain is substituted in the *p*-position to the hydroxyl group, with the spectrum of ethanol lignins, indicates that the latter are derived from lignin precursors of the types exemplified by hydroxy derivatives of 1-(4-hydroxy-3-methoxyphenyl)-1-propanone and 1-(4-hydroxy-3,5-dimethoxyphenyl)-1-propanone. The bands seem to show the presence of a carbonyl group or an ethylenic double bond in conjugation with the aromatic nucleus [HIBBERT and co-workers (*239, 240*)].

The spectra of the lignans agree quite well with those of monomeric phenylpropanes [AULIN-ERDTMAN (*12*)]. A comparison of the spectra of isolated native spruce lignin and conidendrin indicates a close relationship between the lignans and lignin [BRAUNS (*38*)].

The effect of ultraviolet light *in situ* was studied by FORMAN (*99*). He found that lignin undergoes drastic changes which cause a loss in lignin (as determined by 72% sulfuric acid) and also a loss in methoxyl, and are accompanied by a change in the ultraviolet spectrum. A considerable part of the lignin is rendered alcohol soluble. (Among the decomposition products, vanillin was identified.)

Molecular distillation. Although a solvent fractionation method, by which ethanol lignin was separated into various fractions, was described by HIBBERT and co-workers (*222, 241*), attempts to prove the homogeneity or heterogeneity of so-called "butanol" lignin by solvent extraction were found to be inconclusive [BAILEY (*16*)]. Molecular distillation of various lignin preparations at 220°, on the other hand, indicated their heterogeneity. HECHTMANN (*142*) tried to fractionate isolated native spruce lignin and other lignin preparations by means of the molecular still. At 260° and 1 mμ. pressure he obtained 4% of a distillate which was related to lignin, as shown by its ultraviolet spectrum, and which he believed to be a partially depolymerized lignin. The residue was thought to be more highly polymerized than the original material. How stable lignin is at such high temperatures is, however, still open to discussion.

Colloidal properties. A thorough investigation of some properties of cuoxam spruce lignin was carried out by JODL (*176*). He showed that lignin is submicrocrystalline like humic substances. Although the interference rings were not very clear, a height of 9 Å. and a diameter of about 15 A. with a distance of 3,9 Å. for the layer planes of the crystallite were found, indicating an order of three layers, one above another. With water as the pyknometer liquid, a density of 1,41 was found. Although lignin is organophilic, it combines with 7 g. of water per 100 g. by lyosorption and with 27 g. by capillary condensation, indicating a polycapillary system with capillary diameters of 30 to 2000 Å. From the sorption values, a surface area of 180 square meters per gram of lignin has been calculated. According to JODL, lignin cannot be considered as a uniform chemical compound with a definite molecular weight.

The colloidal properties of dioxane lignin, particularly its behavior in soil as a fertilizer[1] and its relationship to the formation of humus were studied by JUNKER (*178*), by PERRENOUD (*255*), and by PALLMANN (*238*). FULLER (*123*) found that the addition of alkali lignin to a sandy loam soil neutralized by addition of calcium carbonate reduces the formation of nitrate from dried blood or ammonium sulfate. The recovery of nitrogen is much less in the latter case, probably because of the absorption of ammonia by lignin.

KLEINERT (*184*) found that lignin absorbs phenol from aqueous solutions quite strongly.

Various Conversions of Lignin.

The formation of *formaldehyde* on distillation of certain lignin preparations with 12% hydrochloric acid was ascribed by FREUDENBERG (*101*, p. 121) to a piperonal grouping since such groups often occur in nature, for instance in safrol and in egonol. HIBBERT and co-workers (*168*), on the other hand, claimed that the formaldehyde does

---

[1] For the use of lignin as fertilizer see: ARIES (*7*); DUNN *et al.* (*77, 78*); SEIBERLICH (*294*).

not originate from a piperonal group because, thus far, a piperonyl-containing derivative has never been obtained from lignin. Since the methylenedioxy group in piperonyl derivatives is quite stable, they ascribed the formation of formaldehyde to the presence of carbohydrates in the lignin. Since the sassafras tree yields oils high in safrole content, it is possible that the lignin of this tree does contain the piperonyl grouping. However, neither in acetic acid sassafras lignin nor in sassafras ligno-sulfonic acid could any evidence for the presence of a methylenedioxy group be detected [BOND et al. (31)]. Further experiments by FREU-DENBERG and co-workers (113) on the origin of formaldehyde with model substances showed that compounds of the 1-phenyl-3-propanol type, either free or etherified and having two additional hydroxyl groups in the side-chain or a carbonyl group at the carbon atom attached to the benzene ring, as well as hydroxybenzene alcohols and their ethers, gave formaldehyde on heating with 12% hydrochloric acid.

From the formation of about 6,5% *acetic acid* on oxidation of cuoxam spruce lignin with chromic acid, FREUDENBERG (120) concluded that about 2,7% CH_3C groups are present in lignin. HIBBERT and his colleagues (224), on the other hand, did not find any acetic acid on oxidation of spruce- or maplewood, except that obtained on saponification and distillation. They assumed that the protolignin in these woods contains no appreciable quantities of terminal methyl groups. They also suggested that the presence of such groups in isolated lignins is the result of intermolecular changes undergone by very reactive side-chains similar to those present in β-ketodihydroconiferyl alcohol.

The effect of *Grignard reagents* on lignin was studied by HIBBERT and co-workers (218, 219) in order to determine the amount of active hydrogen and carbonyl groups. They found that these values varied with the solvent used. Dioxane gave lower figures than pyridine, which was explained by incomplete reaction rather than by increasing aggregation of the lignin complex as a result of intermolecular dehydration. The lignins recovered from the Grignard analysis showed a loss of 3 to 6% of methoxyl, which was greater in pyridine than in dioxane and increased when the lignin was more highly methylated. The loss of these methoxyl groups, which cannot be replaced by diazomethane, is explained by the change of an ethyl iodide-forming group or the conversion of a carbomethyl linkage into a tertiary carbinol group.

The presence of different types of hydroxyl groups in lignin was further demonstrated by the preparation of a number of lignin ethers and esters [JONES and BRAUNS (177); CLARK and BRAUNS (62)].[1]

---

[1] For the preparation of esters and ethers from commercial lignins see: BRAUNS et al. (47); BROOKBANK, et al. (56); LEWIS et al. (217); and McNAIR and JAHN (227).

The number of phenolic hydroxyl groups in cuoxam, hydrochloric acid and acetic acid spruce lignins, and in spruce lignosulfonic acid was determined by FREUDENBERG and WALCH (*121*) by estimating the amount of toluenesulfinic acid formed on reacting the toluenesulfonic esters of these lignins with *hydrazine*. The phenolic hydroxyl content in cuoxam lignin amounted to one group per 6,3 lignin building stones; in hydrochloric acid spruce lignin to one per 5,3; in technical hydrochloric acid lignin to one per 5; in acetic acid lignin to one per 3,2; and in lignosulfonic acid to one per 3,9. The high phenolic hydroxyl content in the last two lignin preparations is explained by the splitting of phenol ether linkages or the opening of coumaran or flavone rings.

Phenol ether linkages and heterocyclic oxygen rings are also cleaved with simultaneous loss of methoxyl groups when lignin is treated with *potassium in liquid ammonia* at room temperature [FREUDENBERG *et al.* (*109*)]. When cuoxam spruce lignin is treated in this way, then methylated and oxidized with potassium permanganate, 5% veratric acid, 3 to 4% isohemipinic acid, and approximately 4% dehydro-dihydro-veratric acid are obtained. The latter compound is probably formed on oxidation of veratric acid by air before the methylation. With model substances, such as diisoeugenol and egonol, similar results were obtained.

That heterocyclic ring systems of a coumaran type of lactone linkage occur in beech lignin was suggested by SPENCER and WRIGHT (*298*), who believed that they had found a similarity in the behavior of coumaran and acetic acid beech lignin. From the results of the *ozonolysis* of the neutral oils from beechwood tar, WACEK and NITTNER (*315*) concluded that this tar consists partially of substituted coumarones.

Because phenols and organic thio compounds in the presence of a trace of hydrochloric acid react readily with lignin, with the formation of condensation products, the behavior of *thiophenol* towards lignin was studied by BRAUNS and LANE (*45*). When sprucewood was extracted by this reagent, a mixture of lignin-thiophenol condensation products was obtained. A similar mixture is formed from isolated native spruce lignin.

Unlike spruce protolignin which gives a tetrathioglycolic acid derivative, hydrochloric acid lignin, fully methylated glycol spruce lignin, and maple or beech protolignins react with *thioglycolic acid* with the formation of trithioglycolic acid derivatives. *Mercaptans*, such as butyl and benzyl mercaptan, in the presence of a small amount of hydrochloric acid, dissolve lignin from sprucewood with the formation of tetramercaptan condensation products [BRAUNS and BUCHANAN (*43*)]. Hydrochloric acid spruce lignin, when heated with a 10% *sodium sulfide* solution at 170°, is completely dissolved with formation of thiolignin.

To establish whether a slight oxidation of lignin affects its reactivity towards thioglycolic acid, HOLMBERG (*163*) treated sprucewood with *sodium hypobromite* for one hour at 0° and overnight at room temperature. The partially demethylated hypobromite lignin, containing 11 to 15% bromine and 9 to 10% methoxyl, had equivalent weights of 312–334 before and 211–230 after treatment with alkali, indicating that the lignin was partially lactonized. From the analytical results the formula, $C_9H_9O_5$, was calculated for the bromine- and methyl-free building stone. The hypobromite lignin reacted with only half as much thioglycolic acid as did the protolignin in untreated wood. The dimethyl sulfate-methylated hypobromite lignin reacted with thioglycolic acid by replacing methoxyl groups by the mercapto acid group, which indicates that the group reacting with thioglycolic acid is capable of being methylated and directly sulfidized, and that this group is either a phenolic or a tertiary alcoholic hydroxyl.

Recently, HOLMBERG (*160*) continued his interesting work on the *mercaptolysis* of woods. On extraction of sprucewood with thiohydracrylic acid, he found that this acid is an even better reagent for the isolation of lignin than thioglycolic acid. The composition of the spruce ligno-thiohydracrylic acid corresponded with that of the spruce lignothio-glycolic acid and lent further support to the fact that the basic structure of the lignin building unit is that of phenylpropane. As is the case with lignothioglycolic acid [HOLMBERG (*162*)], lignothiohydracrylic acid lost most of its mercapto acid groups on methylation with dimethyl sulfate and sodium hydroxide. A comparison of the lignothioglycolic acids and thiohydracrylic acids isolated from the diploid aspen (*Populus tremula* L.) and the heart- and sapwood of the triploid *Gigas*-aspen showed no difference in their composition, although the yield from the latter was lower. On a mercapto acid and methyl-free basis, a composition of $C_9H_{10}O_4$ was calculated for the basic lignin bulding stone (*161*). This formula differs from that of the sulfuric acid aspen lignin, $C_9H_{8.4}O_{3.45}$, in its higher content of the elements of water. LARSSON (*210*), on the other hand, found that aspen lignin isolated as lignosulfonic acid had a basic lignin building stone of the composition of $C_9H_{11.9}O_{6.3}$ to $C_9H_{15.3}O_{7.9}$, which is considerably higher in hydrogen and oxygen content. According to HOLMBERG, aspen may contain several kinds of lignin.

The reaction between *sodium hydrogen sulfide* or *sodium sulfide* and spruce lignin was studied by AHLM (*4*). In this conversion, a mixture of thiolignin and alkali lignin was formed because of the partial hydrolysis of the reagents into sodium hydroxide. The sulfur in thiolignin was present as a mercapto group which was indicated by the formation of a mercury salt. The sulfur and the mercury content of this salt corresponded to the presence of one mercapto group per lignin building unit of a molecular weight 840. From methylation experiments, it was concluded that the sodium hydrogen sulfide had reacted with the carbonyl group in lignin. Hägglund (*132, 133*) suggests that the reaction is similar to that in the sulfite process, with the rupture of an unstable oxygen linkage or the addition of sodium hydrogen sulfide to a double bond. Hydrogen chloride spruce lignin, when treated with sodium hydrogen sulfide under similar conditions, forms a thiolignin very similar to that obtained by Ahlm [Brauns and Buchanan (*43*)].

Vanillin from Lignin.

Freudenberg and co-workers (*108*) found that when cuproxam or hydrochloric acid spruce lignin, lignosulfonic acid, or even sprucewood is heated for three hours with 2 N sodium hydroxide and nitrobenzene at 160°, *vanillin* (4-hydroxy-3-methoxy-benzaldehyde) is formed in yields of 20 to 24% (based upon the lignin material), in addition to guaiacol, and vanillic, acetic, and oxalic acids. This discovery was applied by Hibbert and co-workers (*69, 70, 71*) to a large number of other lignified plant materials. Whereas gymnosperms yielded vanillin only, angiosperms gave a mixture of vanillin and syringaldehyde (4-hydroxy-3,5-dimethoxy-benzaldehyde) in 25 to 45% yield (based on the Klason lignin of the plant). The ratio, vanillin:syringaldehyde from the dicotyledons was found to be 1 : 3, whereas that of the monocotyledons was 1 : 1. Oxidation of the lignin in corn cobs gave a mixture of 4,5% vanillin, 2,6% syringic aldehyde, and 1,4% *p*-hydroxybenzaldehyde. Hibbert suggested that this reaction may be a possible method for the taxonomic classification of plant materials. The results are in agreement with the color reactions given by the Mäule test, since the plants yielding both vanillin and syringic aldehyde gave a positive color reaction.

According to Iwadare (*169*), the yield of vanillin obtained from various lignins on oxidation with nitrobenzene decreases in the following order: alkali lignin, Klason lignin, degraded lignin, methylated lignin.

By heating lignosulfonic acid with alkali, Kratzl (*187*) obtained acetaldehyde in addition to vanillin. The bromination of lignosulfonic acid (*189, 190*) could be conducted in such a way that dibromoacetaldehyde was obtained instead of bromoform. From this result, he concluded that, in at least a part of lignosulfonic acid, the structure of a polymeric coniferylaldehyde hydrosulfonic acid must be present. This sulfonic acid is formed in a manner similar to that suggested by Freudenberg for the sulfonation of Erdtman's acid (*102*). The results support the hypothesis of the presence of a "masked" carbonyl group [Aulin-Erdtman (*13*)] in connection with a conjugated system and would explain the ease of sulfonation of lignin and the results of the alkaline hydrolysis of the sulfonic acid.

ERDTMAN (85) subjected the lignin recovered from the alkaline hydro-·lysis, after removal of the vanillin according to FREUDENBERG, to a treatment with alkali, to methylation with dimethyl sulfate and diazomethane, and to oxidation with potassium permanganate, and obtained 11% veratric acid, 7,2% isohemipinic acid, and 2,9% dehydrodiveratric acid. This lignin, therefore, gave proportionally more isohemipinic acid and less veratric acid than lignosulfonic acid, although a part of the potential veratryl-producing group had already been removed in form of vanillin.

The formation of vanillin from lignin and lignosulfonic acid on oxidation with nitrophenol in alkaline solution was studied by WACEK and KRATZL (310–313) on model substances. The phenylpropane sulfonic acids used were prepared either by reacting the corresponding halogen derivatives with sodium sulfite [WACEK et al. (314); KRATZL (185)] or by heating the appropriate phenylpropane derivative with sodium bisulfite [KRATZL and DÄUBNER (191)]. WACEK and KRATZL found that, when the benzene nucleus is unsubstituted, the attack on the side chain is negligible, regardless of whether a double bond or a sulfonic acid group is present at the carbon atom attached to the benzene ring. When, however, the double bond is conjugated to a carbonyl group or a sulfonic acid group is present at the carbon atom attached to the benzene nucleus, oxidation of the side chain readily takes place. In the latter case, a double bond may be formed by splitting off sulfurous acid. The introduction into the benzene ring of indifferent groups, such as methoxyl, has almost no effect upon the result. A free hydroxyl group in the p-position to the side chain, on the other hand, strongly increases the yield of aldehyde. Although, in general, the presence of a sulfonic acid group has little effect on the oxidation of the side chain, α-guaiacylacetone-α-sulfonic acid gave almost twice as much vanillin as the corresponding hydroxy derivative. This result is of particular interest since the yield of vanillin from lignosulfonic acid increases with the degree of sulfonation.

Oxidation, Reduction and Halogenation.

On electrolytic *oxidation* of lignins (alkali, so-called "butanol", methylated butanol lignin, and lignosulfonic acid) from Western hemlock with a lead cathode and a mercury anode, BAILEY and BROOKS (23) claimed that they obtained 82% of identified products such as acetone (15,2%), methyl ethyl ketone (23,3%), acetic acid (15,1%), β-resorcylic acid (7,8%), protocatechuic acid (7,6%), p-hydroxybenzoic acid (4,2%), m-toluic acid (3,9%), oxalic acid (3,1%), and isobutyl methyl ketone (1,1%). The large number of compounds containing methyl and ethyl groups on oxidation is incompatible with the present concept of the lignin structure.

On *reduction* of cuproxam spruce lignin with hydrazine and sodium ethoxide in ethanol at 160 to 180°, according to WOLFF-KISHNER, LAUTSCH (*211*) obtained about 10 % ethylguaiacol and traces of isoeugenol.

HARRIS and LOFTAHL (*138*) found that, in the reaction of methyl hypochlorite on maple lignin, two new methoxyl groups are introduced in addition to chlorine. With spruce lignin, two to three new methoxyl groups are formed. From the addition of these methoxyl groups, they concluded that maple lignin contains two and spruce lignin two or three ethylene groups per lignin unit of a molecular weight of 950.

The absorption of bromine, presumably by addition to a double bond in lignin, was utilized by KÜRSCHNER (*201*) to determine the amount of lignin in woods.

Chlorination of so-called "butanol lignin", sulfite waste liquor, and hemlock wood itself gave chlorolignins with almost identical properties, although the sulfonic acid group was not removed during the chlorination [PEDERSEN and BENSON (*247*)]. The scission of sulfonic acid groups, however, depends entirely on the degree of chlorination. After a certain degree has been reached, rapid desulfonation takes place [cf. BRAUNS (*41*); LARSON (*209*)]. A study of the mechanism of the chlorination of lignin [WHITE *et al.* (*326*)], with special emphasis on the bleaching of pulp, seemed to indicate that the chlorine enters the guaiacyl nuclei either in the 5- or 6-position, depending upon whether or not the *p*-hydroxy groups are covered by an ether linkage. With the entrance of the chlorine, the methoxyl groups are rendered unstable and are partially split off. The regenerated hydroxyl groups are then oxidized on continuous chlorination, and quinone groups are formed which are further cleaved to give acidic groups. In this way the lignin is converted into alkali- or even water-soluble compounds.

A comparison of chlorolignin from various plant materials with the chlorinated humic substances from levulose, glucose, furfural, and furfuryl alcohol indicated a close relationship between these two types of compounds [MÜLLER (*233*)].

Bromination of lignosulfonic acid purified through its quinoline salt gives a brominated lignosulfonic acid, with simultaneous loss of about 50% of its sulfur content [KRATZL and BLECKMANN (*189, 190*)]. Bromination of model substances, such as sodium propioguaiacone-α-sulfonate or sodium propioveratrone-α-sulfonate, also causes the loss of the sulfonic acid group with formation of 5-bromo-α,α'-dibromopropioguaiacone or 6-bromo-α,α'-dibromopropioveratrone, which is in agreement with the results found by WHITE *et al.* (*326*).

On oxidation of bromolignin in an alkaline solution with cobaltic hydroxide, LAUTSCH and PIAZOLO (*214*) obtained 8% 6-bromovanillin, indicating that the bromine entered the benzene nuclei of lignin in the 6-position. Because 6-bromovanillin is not obtained by direct bromination of vanillin but only by first blocking the free hydroxyl group, the hydroxyl group in lignin should be covered, probably by etherification. This conclusion, however, is contradicted by AULIN-ERDTMAN (*11*), who found that, on bromination of coerulignol and its methyl ether, the bromine does enter the nucleus in 6-position regardless of the etherified phenolic hydroxyl group of the methyl ether. From spruce iodo lignin (prepared from acetomercuric lignin), about 10% of 5-iodovanillin is obtained, whence it follows that the mercuration occurred, in part at least, at the 5-position [LAUTSCH and PIAZOLO (*214*)].

A small amount of 6-chlorovanillin was isolated [PEARL (*244*)] from a chlorite waste liquor obtained in the preparation of holocellulose from black spruce.

Nitration.

Spruce nitrolignin, prepared by extraction with ethyl alcohol containing nitric acid, was further investigated by KÜRSCHNER (*195, 200*) and its aromatic structure was confirmed. According to KÜRSCHNER (*196*), it is free of carbohydrates. With aromatic amines, spruce nitrolignin gives high-molecular azohydroxy compounds which are reduced by an excess of amine to the corresponding azo hydrocarbons. The statement by MÜLLER (*233*) that, in nitrolignin prepared according to KÜRSCHNER, methoxyl groups are replaced by ethoxyl is disputed [KÜRSCHNER and SCHINDLER (*198*)]. On nitration of pine and red beechwood with nitric acid in glacial acetic acid, only slight nitration (2,3% and 1,9%, respectively) takes place, whereas considerable demethylation of the lignin occurs [FRIESE and LÜDECKE (*122*)].

FREUDENBERG and co-workers (*111*) nitrated an acetylated hydrochloric acid spruce lignin (acetyl content 24%) in chloroform suspension at — 20° with nitrogen pentoxide, and obtained a nitrolignin with 4,7% nitrogen, corresponding to 0,9 nitro group per lignin building stone. Since all the hydroxyl groups were acetylated and no acetyl was lost during the nitration, no nitric ester could have been formed and the nitro groups were certainly present in the aromatic ring only. Nitrolignin derivatives with a nitrogen content of about 4% were also prepared from sulfite waste liquor, so-called "butanol" lignin, and hemlock lignin [CARPENTER and BENSON (*60*)].

Sulfonation.

The sulfonation of lignin is one of its most important reactions which seems to proceed more rapidly with hardwood than with softwood lignin.

MAASS and co-workers (58) de-lignified sprucewood by heating it with bisulfite (10 to 11% free and 1% combined SO_2) at 50° for 1000 to 1500 hours and obtained a kind of "holocellulose" in about 70% yield. They did not investigate the lignosulfonic acid in the waste liquor.

FREUDENBERG, LAUTSCH, and PIAZOLO (110) studied the sulfonation of lignin by preparing a series of lignosulfonic acids by fractional extraction of sprucewood with sodium bisulfite at 70°.

The lignosulfonic acid from the first 4 fractions had a carbon to sulfur ratio of 20 : 1 and a methoxyl to sulfur ratio of 1 : 0,5. In the next 2 fractions, the sulfur content gradually increased and, in fraction 7, reached a carbon to sulfur ratio of 11 : 1 and a methoxyl to sulfur ratio of 1 : 1. The lignosulfonic acid of the last fraction had a sulfur content of 15,3% and a ratio of methoxyl to sulfur, 1 : 1,5. The carbon to methoxyl ratio was 10—12 : 1 in all fractions. The hydroxyl content (8,6%) of the lignosulfonic acid from the first fraction decreased with increasing sulfur content to 5,0% in the lignosulfonic acid from the last fraction. The phenolic hydroxyl content was hardly affected by the degree of sulfonation.

Potentiometric titrations and equivalent weight calculations from the ratio of sodium to sulfur in the sodium salts indicated the presence of one carboxyl group for an equivalent weight of 1500–3000. The elementary composition and the methoxyl and hydroxyl contents of the lignosulfonic acids (calculated on a sulfurous acid-free basis) agree very well with the values found for a carefully prepared cuoxam spruce lignin. These results and the fact that, with increasing sulfonation, the phenolic hydroxyl content did not increase, showed that, in the sulfonation of lignin at 70°, no benzofuran rings were split but that ether linkages were cleaved without the formation of new phenolic hydroxyl groups.

The effect of the concentration of the total sulfur dioxide of sodium, calcium, ammonium, and magnesium cooking liquors on the sulfonation of lignin in western hemlock, Sitka spruce, and Douglas fir was studied by MOTTET (231), who found further support for HÄGGLUND's observation (131, p. 150) that lignin contains groups of different reactivity toward sulfonation. This was found also by ERDTMAN (83) in his studies on the effect of phenols in the sulfite cooking process. When pine heartwood (which is known to be difficult to pulp) or sprucewood, in the presence of phenols, was first treated with a slightly acidic bisulfite liquor, the lignin was readily dissolved because some of its groups were sulfonated which otherwise would have preferentially reacted with the phenols. Those groups, which are inert towards phenol even at high acidity, were then sulfonated by increasing the acidity, and yielded normal, highly sulfonated lignosulfonic acids.

That protolignin can condense under the effect of even weak acids was shown by the fact that, when the sulfite cook was interrupted after two hours, the wood washed, cooked with water or buffered phosphate solution at 120°, and washed and recooked with bisulfite, the residual lignin was much more difficult to remove. The partially sulfonated lignin catalytically condensed the unsulfonated lignin [cf. HEDLUND (*143*)].

DORLAND and HIBBERT (*75*) found that formic and acetic acid birch lignins, which are normally insoluble in hot bisulfite solution, are rendered soluble on treatment with ozone. The lignosulfonic acids thus obtained gave vanillin and acetovanillin on alkaline cleavage. This ozone treatment is best carried out in formic acid (*74*).

In a series of investigations, ERDTMAN (*84*, *86–90*) studied the sulfonation of lignin and the separation of lignosulfonic acids into different fractions by stepwise precipitation with various organic bases such as 4,4'-bis(dimethylaminodiphenyl)methane ("bis"), quinoline, β-naphthyl-amine, strychnine, β-naphtholquinaldine, and 6,6'-diquinoline. The amount precipitated by the respective reagents increased in the order given. Basic lead acetate was found to precipitate lignosulfonic acid almost quantitatively. The lignosulfonic acids most readily precipit-ated—i. e., those with "bis" and quinoline—had high molecular weights, did not pass through a dialyzing membrane, and showed high methoxyl and low sulfur contents. They reduced Fehling solution only slightly. In the order of precipitability, the molecular weight and the methoxyl content of the lignosulfonic acid decreased, whereas the dialyzability, the reducing power towards Fehling solution, and the sulfur content increased. The precipitability and the ratio of the various fractions depend upon the cooking conditions under which the sulfite waste liquor was prepared and on the wood species, since lignosulfonic acids from deciduous woods are in general less completely precipitated than those originating from coniferous woods (*86*).

Contrary to the report by FRIESE (*122*), no evidence of the presence of a carbohydrate-lignosulfonic acid complex was found in sulfite waste liquor from deciduous woods [ERDTMAN (*87*)]. Because the actual reaction taking place between lignin and bisulfite is still unknown, it is difficult to devise an empirical formula for the "native lignin" from lignosulfonic acids. This could be done either by subtracting sulfurous acid (assuming that simple addition took place) or sulfur dioxide (assuming a reaction with the elimination of water and replacing of a hydroxyl group), which results in carbon contents ranging from 61 to 63% for the lignin. These values are in good agreement with those found for isolated native spruce lignin [BRAUNS (*35*)] and with those calculated by HOLMBERG and GRALÉN (*165*) from lignin derivatives of thio acids (on a thio acid-free basis), assuming that the latter react with lignin through elimination of the elements of water. ERDTMAN (*87*) did not find a drastic degradation of the lignin molecule in lignosulfonic acid as claimed by RACKY (*267*). The free lignosulfonic acids were methylated with diazomethane, giving products of various degrees of methylation.

That a lignosulfonic acid in sulfite waste liquor is a mixture of differently sulfonated compounds of a wide range of molecular weights was also shown by SCHENCK (281) by means of fractional precipitation and.dialysis. A fractionation of lignosulfonic acid in sulfite waste liquor on ion-exchangers was described by SAMUELSON (280).

The residual lignin in sulfite pulp is present as a sulfonated lignin (solid lignosulfonic acid) [LARSON (209)] and is rendered partially soluble in a lignin determination with strong acids.

Lignosulfonic acid can be desulfonated by heating with 35 to 40% sodium hydroxide in a ratio of 1 : 1 at 175 to 180° or with a 1 : 1 ratio of lime in 10 to 15% concentration [PEARL, BAILEY and BENSON (245, 246)].

Model substances. To study the sulfonation of lignin on models, RICHTZENHAIN (271) subjected various coumaran and flavan derivatives and flavanone to a bisulfite cook. Of these, only flavan reacted by opening its oxygen ring with formation of β-(2-hydroxybenzoyl)-phenyl-ethyl-α-sulfonic acid (VIa).

(VI.) Flavan.   (VI a.) β-(2-Hydroxybenzoyl)-phenylethyl-
                                               α-sulfonic acid.

AULIN-ERDTMAN *et al.* (12), on the other hand, heated 2,5-dimethyl-3-coumaranone (VII) with bisulfite and obtained 2,5-dimethyl-3-coumara-none-2-sulfonic acid (VIII). The latter showed an ultraviolet absorption

(VII.) 2,5-Dimethyl-3-coumaranone-   (VIII.) 2,5-Dimethyl-3-coumaranone-
            2-sulfonic acid.   2-sulfonic acid.

spectrum almost identical with that of the original coumaranone, indicating that no scission of the furan ring had occurred during the sulfonation.

WACEK (307) and KRATZL and co-workers (191, 192) subjected phenyl- and guaiacylpropane derivatives to a bisulfite cook at elevated tempera-tures. In all cases in which a reaction took place, the sulfonic acid group entered the molecule at the side chain. In ethylphenylcarbinol or benz-hydrol the secondary alcohol group was replaced by the sulfonic acid group. In some aromatic compounds containing carbonyl in conjugation with a double bond, C=C—C=O, addition of the sulfurous acid to a double bond took place while SO_3H was added to the carbon atom attached to the benzene ring.

The utilization of lignosulfonic acid for the preparation of ion exchangers was described by LAUTSCH (*213*).

Ethanolysis of Wood and Lignin.

In the isolation of methanol spruce lignin with methanol and hydrochloric acid [BRAUNS and HIBBERT (*44*)], it was observed that only about half of the lignin actually dissolved was recovered on pouring the concentrated methanol extract into water, whereas the other half remained in solution. A few years later, HIBBERT and co-workers (*51, 52, 68, 166, 167, 202, 266*) studied the ethanolysis of wood with special regard to the water-soluble lignin degradation products which they called "distillable oils". From the aqueous alcoholic mother liquor of ethanol *spruce* lignin they isolated the following compounds: vanillin (formula IX, $R=A$) (*52*), 1-ethoxy-1-(3-methoxy-4-hydroxyphenyl)-2-propanone (X, $R=A$) (*324*), 1-(3-methoxy-4-hydroxyphenyl)-2-propanone (XI, $R=A$) (*324*), 2-ethoxy-1-(3-methoxy-4-hydroxyphenyl)-1-propanone (XII, $R=A$) (*68*), and 1-(3-methoxy-4-hydroxyphenyl)-1,2-propanedione (XIII, $R=A$) (*203*). From the mother liquor of ethanol *maple* lignin, syringic aldehyde (IX, $R=B$) (*266*), 1-(3,5-dimethoxy-4-hydroxyphenyl)-2-propanone (XI, $R=B$) (*204, 323*), α-ethoxypropiosyringone (XII, $R=B$) (*166*), and syringoyl acetyl (XIII, $R=B$) (*203*) were isolated, in addition to the corresponding guaiacyl derivatives.

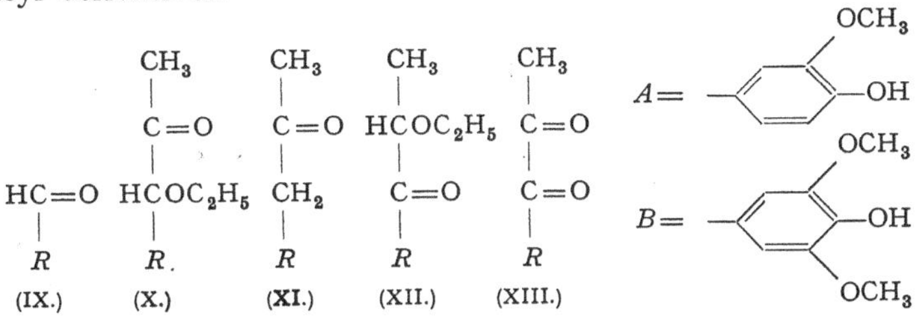

The previously isolated syringoyl acetaldehyde (*266*) was later found to be (XIII, $R=B$) (*52*). The ethoxy groups in (X) and in (XII) are the results of ethylation during the isolation with ethyl alcohol and hydrochloric acid. The yield of these monomeric phenylpropane derivatives is relatively small. From sprucewood it amounts to only 3 to 5% and from maple to about 8%.

HIBBERT and co-workers (*248, 325*) claimed that acetylated oak lignin, isolated by acetolysis of oakwood (*249*), can be converted into monomolecular lignin building stones by ethanolysis. HEWSON and HIBBERT (*145*) subjected mildly prepared ethanol maple and spruce lignins to successive re-ethanolyses and succeeded in further "depolymerization" of the lignin into low molecular units. At the same time, a part of the lignin was converted into alcohol-insoluble, more complex polymerization products. The structures of the ethanolysis products were identified by their syntheses [HIBBERT et al. (67, 95, 96, 124, 205, 320, 322)]. HIBBERT (225) suggested that the ethanolysis of wood may be used as a means for the botanical classification of gymnosperms and angiosperms.

A study of the mechanism of the reaction occuring on treatment with ethanol and hydrogen chloride of the lignin in wood and of ethanol lignin revealed that both depolymerization and polymerization take place [HEWSON, McCARHTY and HIBBERT (146, 147)]. Apparently, the low molecular "distillable oils" are formed by alcoholysis of high molecular lignin molecules. At the same time, polymerization of the lignin takes place with formation of complex, irreversible polymers, insoluble in alcohol. Such a polymerization prevents a complete delignification of wood by methanol or ethanol in the presence of small amounts of hydrogen chloride [BRAUNS and HIBBERT (44)], and at the same time limits the yield of distillable oils. The latter are also obtained from maplewood with ethanol at high temperature (150 to 200°) in the absence of hydrogen chloride (147). According to HIBBERT and co-workers (145), the delignification of maplewood by ethanolysis involves changes of linkages between lignin-lignin or lignin-carbohydrate complexes. This reaction is catalyzed by the presence of hydrogen ions or hydroxyl ions.

Ethanolysis of red cedar, Douglas fir, and western hemlock gave products similar to those obtained by HIBBERT and co-workers from sprucewood [BRAWN et al. (49)].

In contrast to HIBBERT, HOLMBERG (164) was unable to detect α-ethoxypropiovanillone in the aqueous-alcoholic mother liquor of the ethanolysis of sprucewood; however, he found that ethanol lignin can react as a keto compound, since upon treatment of this kind of lignin with thiobenzhydrazide, a compound of the composition $C_{25}H_{28}O_7 = N \cdot NH \cdot CS \cdot C_6H_5$ was obtained. This discrepancy is probably due to the fact that HOLMBERG worked in the presence of water, whereas HIBBERT extracted under anhydrous conditions. The lignin left in the wood after alcoholysis gave thioglycolic acid lignins similar to those obtained by direct extraction of sprucewood.

Hydrogenation.

A survey of earlier experiments on this subject was presented by LAUTSCH (212). A great deal of work has recently been carried out on the hydrogenation of lignin, probably in an attempt to find a commercial use.

HARRIS, D'IANNI, and ADKINS (137) were the first to conduct pressure hydrogenation of methanol aspen lignin in dioxane, at 260°, in the presence of copper chromite as a catalyst. In addition to methanol originating from the methoxyl groups in the lignin, they obtained 1-propyl-4-cyclo-hexanol (XIV), 1-propyl-3,4-cyclohexanediol (XV), and 3-(4-hydroxycyclohexyl)-1-propanol (XVI) in a yield of about 50% of the lignin used. Methanol aspen lignin represents only about 50% of the lignin originally present in the wood. Later, the hydrogenation was carried out in an aqueous solution giving the same reaction products [HARRIS (*136*); HARRIS, SAEMAN, and SHERRARD (*139*)].

ADKINS and co-workers (*3*) hydrogenated Meadol[1] over copper chromite in dioxane at 250° to 300°. Although the amount of hydrogen absorbed per gram of lignin was about the same as with methanol aspen lignin, the reaction products were different. The yield of cyclohexyl-

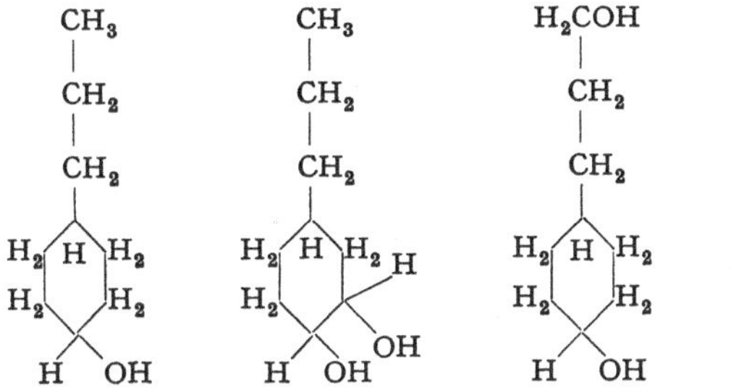

CH_3	CH_3	H_2COH
CH_2	CH_2	CH_2
CH_2	CH_2	CH_2

(XIV.) 1-Propyl-4-cyclohexanol. (XV.) 1-Propyl-3.4-cyclohexanediol. (XVI.) 3-(4-Hydroxy-cylohexyl)-1-propanol.

propane derivatives was lower and the main product consisted of polycyclic hydrocarbons with 20 to 70, or even more, carbon atoms. Whereas the high boiling products from the methanol aspen lignin contained one oxygen for each 6 carbon atoms, those from the alkali lignin had one oxygen for an average of 13,5 carbon atoms. These results are in agreement with the concept that the phenylpropane building stones are arranged in a chainlike manner in methanol lignin, whereas in the formation of alkali lignin, cyclization takes place either during the cooking process or the isolation. In methanol lignin, linkages between the C_6—C_3 building stones are cleaved; in alkali lignin, hydrogenation of unsaturated rings and hydrogenolysis of hydroxyl, methoxyl, and cyclic ether linkages take place, since the rings are stable towards hydrogenolysis.

The hydrogenation of wood, lignin, and lignin degradation products was studied also by HIBBERT and co-workers (*32, 33, 64, 65, 126*). Ethanol maple lignin yielded water, methanol, ethanol, 4-propylcyclohexanol, and 4-propyl-1,2-cyclohexanediol. Spruce- and maplewoods were hydro-

---

[1] "Meadol" is a technical alkali hardwood lignin with a methoxyl content of 21,5% marketed by The Mead Corporation, Chillicothe, Ohio [PLUNGIAN (*263, 264*); BROOKBANK (*53, 54*)]. Other commercially available lignins are "Tomlinite" [POWTER (*265*)], an alkali lignin from coniferous woods, sold by the Howard Smith Paper Co., Cornwall, Ontario, Canada; and "Indulin" [KEILEN (*181, 182*)], a mixture of alkali and thiolignins from a pine sulfate waste liquor, sold by the Industrial Chemical Sales Division, West Virginia Pulp and Paper Co., New York.

genated at 280° under 3500 lb. pressure in the presence of a copper chromite catalyst (*127*) and gave 19,5% of 4-propylcyclohexanol as well as 5,8% of 3-(4-hydroxycyclohexyl)-1-propanol. In a later investigation (*33*), 3-cyclohexyl-1-propanol was also found. Hydrogenation of maple holocellulose yielded no cyclohexylpropane derivatives, indicating that the latter originated from the lignin in the hydrogenation of the wood. The hydrogenation of maplewood in an ethanol-water mixture (1:1) under mild conditions, with RANEY nickel as a catalyst, at 165° to 170° for four hours did not yield simple cyclohexylpropane derivatives [HIBBERT *et al.* (*50*)]. The lignin was separated from the cellulose degradation products by extraction with chloroform in a yield of 70 to 80% (of the original lignin) as so-called *"hydrol lignin"*. It had a "methoxyl" content of 24 to 25%, which is higher than that of isolated lignin and which was later found to include 5,6% ethoxyl [PEPPER and HIBBERT (*253*)]. "Hydrol lignin" is a mixture.

By solvent fractionation and alkali extraction, 8,8% (all yields based upon the original KLASON lignin) of 3-(4-hydroxy-3,5-dimethoxyphenyl)-1-propanol (XVII, $R=B$), 0,84% of 3-(4-hydroxy-3-methoxyphenyl)-1-propanol (XVII, $R=A$), and 0,83% of 3-(4-hydroxy-3,5-dimethoxyphenyl)-propane (XVIII, $R=B$) were isolated (for A and B of p. 199).

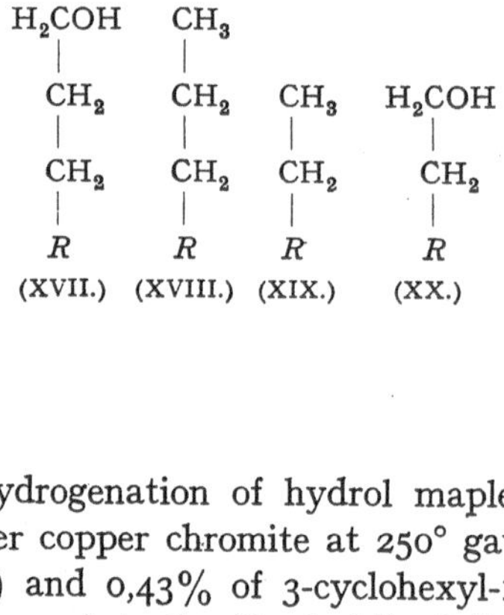

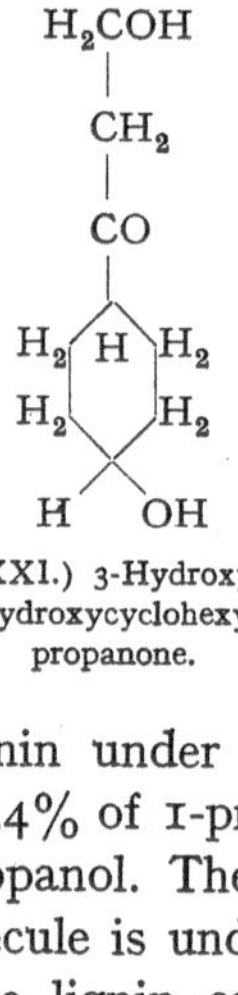

(XXI.) 3-Hydroxy-1-
(4-hydroxycyclohexyl)-1-
propanone.

Further hydrogenation of hydrol maple lignin under more drastic conditions over copper chromite at 250° gave 0,4% of 1-propyl-4-cyclohexanol (XIV) and 0,43% of 3-cyclohexyl-1-propanol. The entrance of the ethoxyl group into the "hydrol lignin" molecule is undoubtedly the result of the reaction of the solvent with the lignin caused by the formation of acid.

In order to prevent this ethoxylation, the hydrogenation was carried out in dioxane-water (1 : 1) in the presence of 30% sodium hydroxide, based on the wood (*253*). At 173° and 3000 lb. pressure for six hours, over RANEY nickel, 0,38 gram-molecule of hydrogen per 20,6 grams of lignin was absorbed. The latter was rendered completely soluble in chloroform. From this hydrogenation mixture, 2,2% 4-hydroxy-3-methoxyphenylethane (XIX, $R=A$), 15,4% 4-hydroxy-3,5-dimethoxyphenylethane (XIX, $R=B$), and 6,2% 2-(4-hydroxy-3,5-dimethoxyphenyl)-ethanol (XX, $R=B$) were obtained. The isolation of the last compound was considered by HIBBERT to be evidence of a carbon-oxygen linkage through the β-carbon atom of the side chain and of a close relationship of lignin to the lignans. It must be remembered, however, that all lignans are dimers of phenylpropane derivatives linked together through a carbon-carbon bond of the β-carbon atoms, and not through a carbon-oxygen linkage [ERDTMAN (85)].

To study the behavior of model ethanolysis products on hydrogenation, α-ethoxypropiovanillone was hydrogenated by HIBBERT et al. (64) over copper chromite at 250° and 250 atm. and gave 4-propylcyclohexanol, which is obtained also on the hydrogenation of lignin. Under the same conditions, hydrogenation of 3-(4-hydroxycyclohexyl)-1-propanol yielded 60% 4-propylcyclohexanol, indicating that in lignin phenylpropane groups with a terminal oxygen linkage may exist to a greater proportion than is indicated by the amount found in its hydrogenation products. Hydrogenation of 3-hydroxy-1-(4-hydroxycyclohexyl)-1-propanone (XXI), synthesized by WEST and HIBBERT (322), as a possible lignin progenitor, gave a mixture of 4-propylcyclohexanol and 3-cyclohexyl-1-propanol in a ratio 1 : 3. In these results, HIBBERT saw further support for his hypothesis that the 3-carbon side chains attached to the aromatic ring in lignin contain a terminal hydroxyl group or an ether grouping.

Because dimeric compounds of substituted phenylpropane derivatives may be formed during the various degradation processes of lignin, the following compounds were synthesized by HIBBERT and his colleagues (25, 322) in order to study their behavior in reactions previously applied to

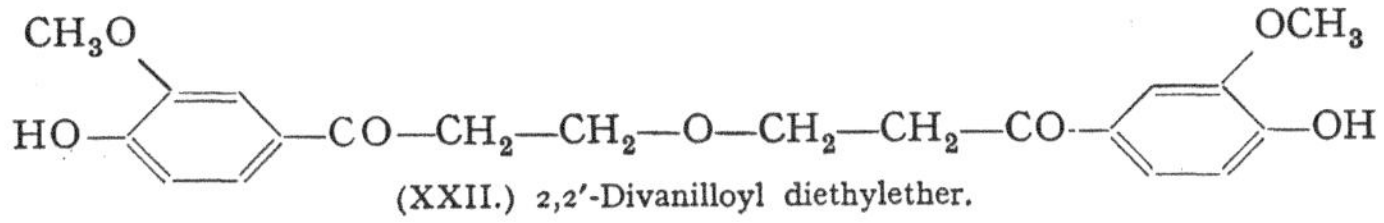

(XXII.) 2,2'-Divanilloyl diethylether.

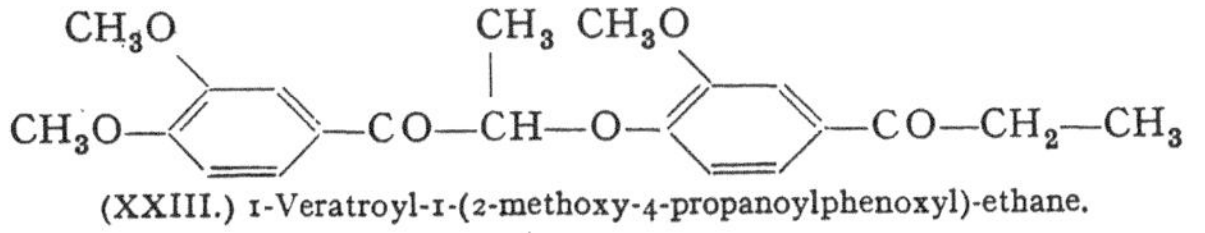

(XXIII.) 1-Veratroyl-1-(2-methoxy-4-propanoylphenoxyl)-ethane.

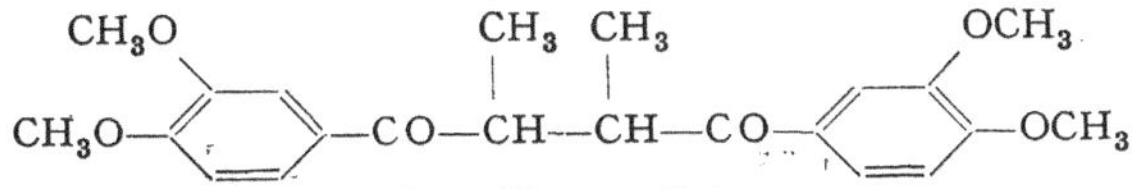

(XXIV.) 2,3-Diveratroylbutane.

lignin: 2,2'-divanilloyl diethyl ether (XXII), 1-veratroyl-1-(2-methoxy-4-propanoylphenoxyl)-ethane (XXIII), and 2,3-diveratroylbutane (XXIV). None of these derivatives, when subjected to an ethanolysis as applied to wood, gave any monomolecular fission products similar to those obtained from lignin [WEST, HAWKINS, and HIBBERT (321)]. Hydrogenation of (XXII) in aqueous ethanol containing 3% sodium hydroxide at 165° gave 60% scission products consisting of almost equal parts of 3-methyl-4-hydroxyphenylpropane, 3-(4-hydroxy-3-methoxyphenyl)-1-propanol, and 2,2'-diveratroyl dimethyl ether (26). Under the same conditions, (XXIII) yielded equal amounts of 3-hydroxy-3-methoxyphenylpropane, 3,4-dimethoxyphenylpropane, and 1-veratryl-1-(3-methoxy-4-propyl-phenoxy)-ethane. At 185°, 85% of 4-propylcyclohexanol was formed. On hydrogenation of (XXIV), ring hydrogenation occurred, but there was no scission of the carbon-carbon linkage of the dimer.

By hydrogenating hydrochloric acid lignin (16,3% methoxyl) from *Picea gezoensis* at 260° to 270° and 230 atm. with some milder catalysts (such as nickel oxide) in dioxane for 35 to 55 hours, HACHIHAMA, ZYODAI, and UMEZU (130) found that one molecule of hydrogen was taken up by 40 to 44 grams of lignin, yielding 46,7% (of the weight of the lignin) of an ether-soluble fraction which contained dehydroeugenol, pyrocatechol, protocatechuic acid, and p-hydroxybenzoic acid. The methylated hydrochloric acid lignin (30,9% methoxyl) did not react with hydrogen.

Hydrogenation or low temperature carbonization in a hydrogen atmosphere of hydrochloric acid spruce lignin on the surface of which nickel has been precipitated as a catalyst was carried out by FREUDENBERG and ADAM (107). In this process, 47% of the lignin was converted into ether-soluble, distillable products which were separated into phenols, acids, and neutral compounds. The phenols, amounting to 35% of the lignin, consisted of phenol, p-ethylphenol, guaiacol, p-cresol, o- and p-ethylguaiacol, isoeugenol, pyrocatechol, p-propylpyrocatechol, homopyrocatechol, and some high boiling compounds. From the neutral fraction, toluene, o-ethylanisol, p-homoveratrol, methanol, ethanol, methylcylopentanol, cyclohexanediol, and some high boiling products were isolated. Similar results were obtained from hydrochloric acid beech lignin.

Because the *waste liquors* of the pulp industry are the main source of isolated lignins, FREUDENBERG and co-workers (112), particularly LAUTSCH (212, 215), turned their efforts to the hydrogenation of these products. Preliminary experiments with hydrochloric acid spruce lignin, cuproxam spruce lignin, and spruce lignosulfonic acid in aqueous alkaline solution indicated that temperature ranges of 250° to 260° in the presence of catalysts, and 340° to 350° with or without catalysts, were especially advantageous. At 250° to 260° the results depended to a great extent upon the catalyst used. With Raney nickel or a nickel catalyst according to RUPE (277), ring-hydrogenated compounds of the cyclohexanol type were the princip1 products, whereas with less active catalysts, such as nickel-copper-aluminium oxide or cobalt catalysts, phenolic compounds were obtained. It was far more difficult to stop the hydrogenation at the phenol stage than to further hydrogenate the intermediate lignin degradation products to cyclohexyl derivatives. The principal action of the catalyst is the hydrogenolytic degradation of the lignin structure and, for this purpose, a specific catalyst may be required.

In order to obtain phenols, FREUDENBERG and co-workers (211) hydrogenated spruce lignin in 6% sodium hydroxide for several hours in the presence of copper-nickel or cobalt-nickel at 260° and 80 to 100 atm., whereby 1 to 1,3 molecules of hydrogen for each lignin building stone (molecular weight, about 180) was absorbed, an amount required for the hydrogenolysis of the side chain. About 45 to 50% of the reaction products were phenols, half of which were distillable and consisted of guaiacol, cresol, ethylguaiacol, and the corresponding pyrocatechol derivatives. Experiments on the hydrogenation of lignosulfonic acid under the same conditions failed, even with the use of the sulfur-resistant catalysts, nickel sulfide or cobalt sulfide.

With more reactive catalysts such as Raney nickel but under otherwise un-changed conditions, five molecules of hydrogen were consumed per lignin building stone with the formation of 36 to 40% ether-soluble products, from which 1-methyl-4-cyclohexanol was isolated from the low-boiling fraction. At higher temperatures (340° to 350°), not only isolated lignin but also salts of lignosulfonic acid were hydrogenated, even without the use of a catalyst. In all cases, about 40% of the lignin were obtained as neutral decomposition products, in addition to about 15% of polymeric phenols of a tar-like nature. In the presence of Raney nickel, the phenolic fraction dropped to about 4%. Of the neutral hydrogenation products, 60% were distillable. From the low boiling fraction, cyclopentanol, 2-methyl-cyclopentanol, and 1-ethyl- or dimethylcyclopentanols were isolated. The middle fraction consisted chiefly of alkyl-cyclohexanols and alkyl-cyclohexanediols. The high boiling fraction contained 30 to 40% hydrocarbons. By the same process, fermented and unfermented sulfite waste liquors were directly hydrogenated, giving 45% (based upon the lignin portion of the waste liquor) of hydrocarbons, in addition to a phenolic fraction. A spruce soda black liquor yielded as much as 57% (based on the lignin precipitated by hydrochloric acid) of cyclic alcohols.

Instead of hydrogen, *carbon monoxide* or a mixture of carbon monoxide and hydrogen (water gas) may be used. Carbon monoxide and sodium hydroxide form sodium formate which decomposes with liberation of hydrogen according to the equation:

$$2\ HCOONa + H_2O \rightarrow Na_2CO_3 + 4\ H + CO_2.$$

Hydrogenation of guaiacol, cresol, p-methylphenol, and isoeugenol as model substances gave cyclohexanol, 4-methylcyclohexanol, 4-methyl-cyclohexanol, and 4-propylcyclohexanol, respectively.

An attempt [LOVIN and FRIEDMAN (223)] to hydrogenate a purified *sodium lignosulfonate* in alkaline solution with nickel sulfide, molybdenum oxide, or copper chromite as catalyst or with tetraline as a hydrogen donor at 400° to 410° and 300 atm. resulted in the absorption of one molecule of hydrogen by about 75 grams of lignosulfonate. At 260° to 270° and 230 to 240 atm., first over copper chromite and then twice over Raney nickel, one mole of hydrogen was taken up by 40 to 50 grams of sodium lignosulfonate.

Lignosulfonic acid, in the form of the residue left on evaporation of sulfite waste liquor, was hydrogenated by BOBROV and KOLOTOVA (30). It was suspended in heavy coal t r oil and hydrogenated at 350° to 450° and 40 atm. The lignin was converted to the extent of 55% into oils, of which about 50% distilled below 150°.

Hydrogenation experiments, under conditions similar to those used by LAUTSCH, were carried out with methanol aspen lignin by SAEMAN and HARRIS (279). At 250° and 400 atm. within four to five hours, one mole of hydrogen was taken up per 33 grams of the lignin. When, large catalyst-to-acceptor ratios were used and the reaction was interrupted before the pressure drop ceased, 24% of volatile products were obtained. Ethylene glycol, diethylene glycol, 4-ethylcyclohexanol, 4-propylcyclohexanol, 2-methoxy-4-ethylcyclohexanol, and 3-(4-hydroxycyclohexyl)-1-propanol were isolated.

LAUTSCH and PIAZOLO (215) hydrogenated various lignin preparations by using *hydrogen-donating compounds* in place of hydrogen. According to REID, WORTHINGTON and LARCHER (269), anhydrous alcohols are converted quantitatively by alkali at 330° to 360° into the corresponding fatty acids and hydrogen:

$$R \cdot CH_2OH + NaOH \rightarrow R \cdot COONa + 4 \, H.$$

When hydrochloric acid spruce lignin was heated with 20% aqueous alcohol containing 10% sodium hydroxide for several hours at 350°, 77–80% of the lignin was converted into ether-soluble oils, of which 60% was distillable. The various fractions obtained contained 80 to 84% carbon, 13 to 11% hydrogen, and 0,3 to 0,5% active hydrogen, and had molecular weights varying from 108 to 158. The oxygen was present in hydroxyls. Hydrogenation of lignosulfonates and sulfite waste liquor in the same way gave similar products. Hydrogenation of phenol and isoeugenol as model substances showed that the latter, in the absence of catalysts, gave a mixture of cyclohexyls, apparently by cracking the side chain at the double bond, whereas phenol was hydrogenated only to a small extent in the absence of catalysts, but in good yield in the presence of Raney nickel.

LAUTSCH (215) explained the mechanism of the hydrogenation of lignin in an aqueous alkaline medium at 250° to 260° as follows: The alkali hydrolyzes the lignin building units at their ether linkages (XXVa) and, by hydrogenolytic degradation of the side chain of the lignin building stones at *a*, *b* or *c*, guaiacol, methyl- or ethylguaiacol is formed. Depending upon the activity of the catalyst and the time of the reaction, the hydrogenation may stop at this stage or may continue with demethylation followed by ring hydrogenation. Lignin building units with a furan or pyran structure are similarly hydrolyzed (XXVb) followed by a cracking reaction at *a*, *b*, *c* or *d*, whereby fragments of type (XXVc) and (XXVd) can be formed. The hydrogenation at 340° to 350° differs from that at 250° to 260° in that the presence of a catalyst is not required; this hydrogenation always yields neutral products, and the latter contain, in addition to six-membered cyclic alcohols and cycloparaffins, alcohols

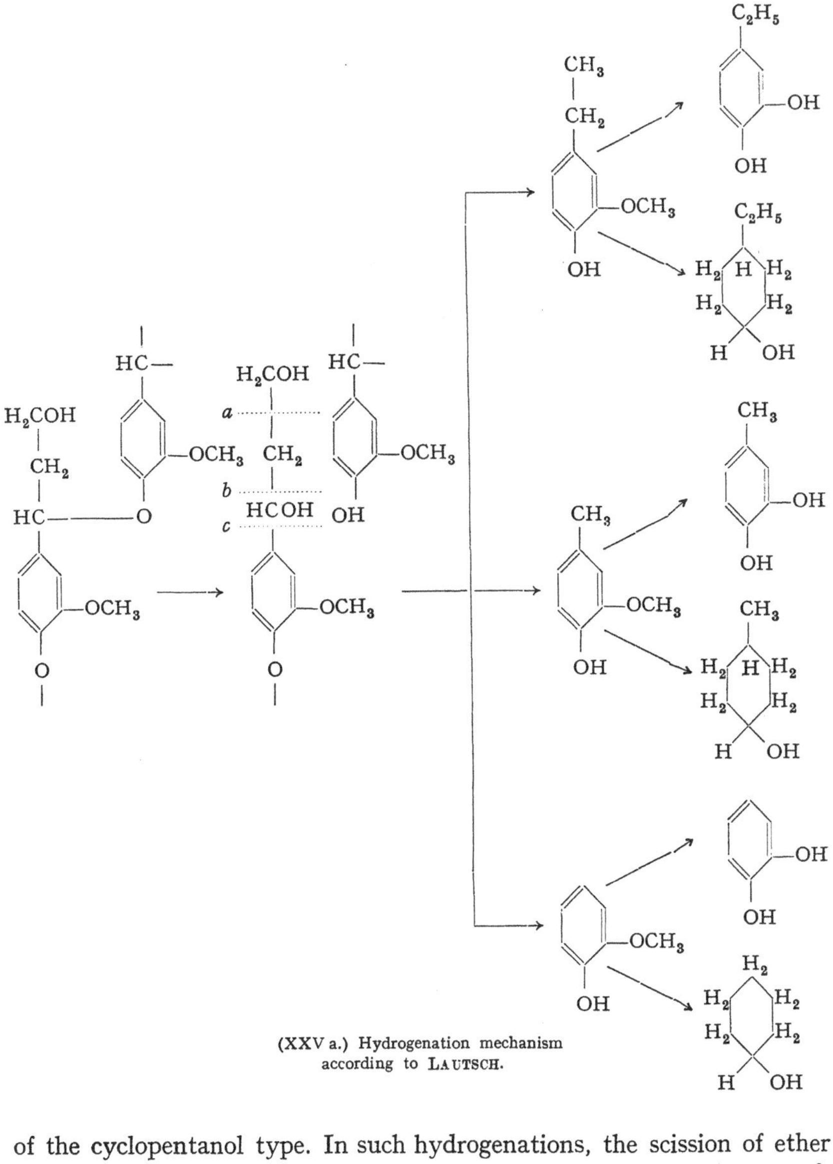

(XXV a.) Hydrogenation mechanism
according to LAUTSCH.

of the cyclopentanol type. In such hydrogenations, the scission of ether groups is followed by a much more drastic cracking. Up to about 310°, cracking of the side chain with low hydrogen consumption takes place, whereas at 320° hydrogenation of the benzene rings sets in with high hydrogen consumption and formation of cyclopentane derivatives. The presence of catalysts does not affect the reaction if the temperature of 260° is passed rapidly. Lower pressures can be applied in this case.

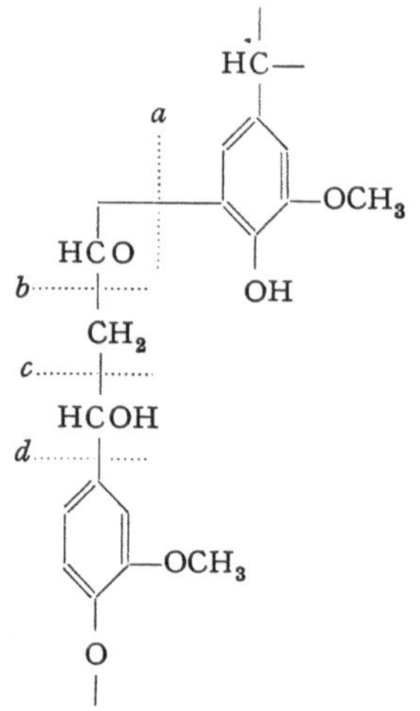

(XXV b.)   (XXV d.)

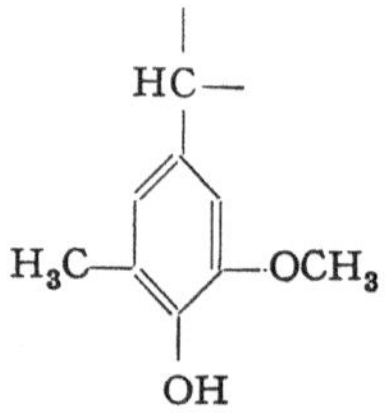

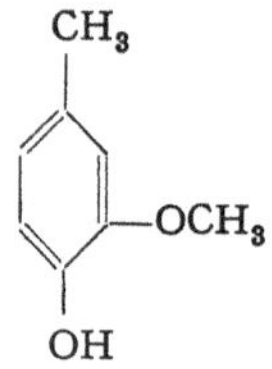

Experiments on the Synthesis of Lignin and Similar Products.

Based on HILPERT's opinion that lignin does not exist as such in wood but is formed during the isolation by condensation of carbohydrate decomposition products, SCHÜTZ and co-workers (290) synthesized what they considered to be "artificial lignins". Because the addition of xylose and other sugars, on one hand, or of certain phenols, on the other, to wood increased the yield in a lignin determination, they theorized that lignin might be a condensation product of sugars with phenols.

By the condensation of xylose with guaiacol in the presence of strong mineral acids they obtained an "artificial lignin" with 67,5% carbon, 5,6% hydrogen, and 14,2% methoxyl, values very similar to those frequently obtained for spruce lignin. With pyrogallol 1,3-dimethylether in lieu of guaiacol, a "deciduous lignin" with 63,8% carbon, 5,38% hydrogen, and 20,15% methoxyl was prepared. Sublimation of the guaiacyl xylose lignin according to KÜRSCHNER gave small amounts of vanillic acid with a strong odor of vanillin.

In spite of contentions to the contrary, there seems to be no doubt that lignin is preformed in wood as a partially *aromatic* compound [FREUDENBERG (105)]. The high yield of vanillin and guaiacyl derivatives from spruce lignin indicates that it is built up of hydroxylated benzene derivatives. The question arises: by what biochemical process involving the phenolic hydroxyl groups can condensation products of various degrees of polymerization be formed from such phenolic substances? Condensation of phenylpropane building stones through the sidechains, as is the case in the formation of the lignans, is improbable because condensation products of this type still possess the free phenolic hydroxyl group and, on oxidative degradation by alkali treatment, methylation, and oxidation, only veratric acid and no iso-hemipinic açid (XXXII) is formed. The discovery by ERDTMAN (*82*) and AULIN-ERDTMAN (*9*) that the dehydrogenation of isoeugenol by the action of ferric chloride or mushroom ferment leads to the formation of a coumaran derivative, dehydro-dihydroisoeugenol, in which one hydroxyl group has been etherified, suggests that lignin might be formed in a similar way. Experiments on model substances along this line were carried out by FREUDENBERG and RICHTZENHAIN (*117, 273*). With vanillin, syringic, ferulic and, perticularly, dihydroferulic acids, condensation products having ether linkages which involve the phenolic hydroxyl group were formed. This result supports FREUDENBERG's hypothesis that lignin may be biosynthesized by enzymatic dehydrogenation of hydroxyphenylpropane compounds.

The synthesis of a gymnosperm lignin was claimed by RUSSELL (*278*). Since most of the decomposition products of spruce lignin contain the guaiacyl group, RUSSELL thought that lignin might be composed of a series of phenylpropane units as proposed by KLASON, FREUDENBERG, and HIBBERT. From this and the analogy with other plant products, such as flavones, flavanones, and anthocyanes, RUSSELL concluded that lignin in gymnosperms is a poly-8-methoxydihydrobenzopyrone (XXX).

This compound is synthesized as follows: Vanillin (XXVI) is acetylated to give acetylvanillin (XXVII); the latter is subjected to a FRIES rearrangement by treating it in nitrobenzene with aluminium chloride to give 5-acetovanillin (XXVIII).

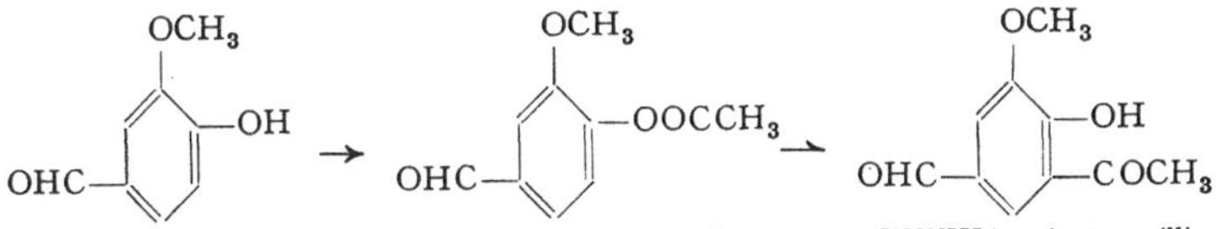

(XXVI.) Vanillin.   (XXVII.) Acetylvanillin.   (XXVIII.) 5-Acetovanillin.

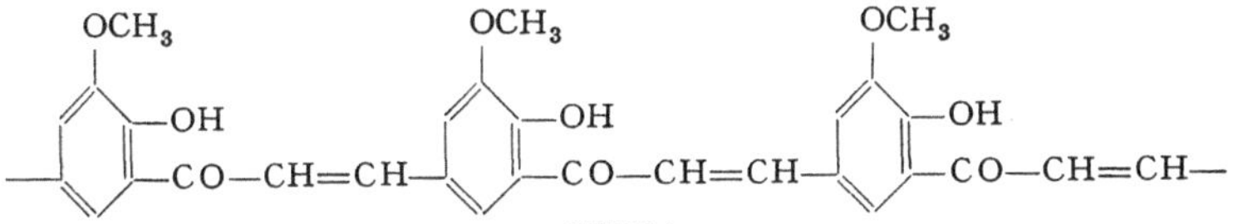

(XXIX.)

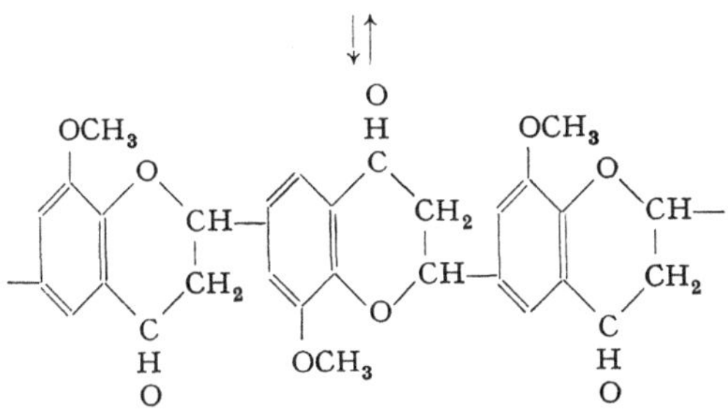

(XXX.) Polymeric 8-methoxy-dihydrobenzopyrone.

Under the reaction conditions used, the latter immediately undergoes polymerization through CLAISEN condensation with the formation of a linear chain molecule (XXIX); this compound cyclizes to the structure (XXX). According to RUSSELL, in a gymnosperm lignin there exists an equilibrium, 20% being present as (XXIX) and 80% as (XXX). He claims that the properties of this "synthetic lignin" do not differ from those recorded for spruce lignin. An identical structural formula for gymnosperm lignin was recently discussed by RITTER and co-workers (275).

Coniferyl alcohol (XXXI), which KLASON suggested as early as 1893 as the precursor of lignin and from which he obtained a sulfonic acid with properties very similar to those of lignosulfonic acid, also served FREUDENBERG (119) as starting material for a lignin model. This mono-meric alcohol, when treated with a trace of mineral acid, gave an amorphous polymer with an unchanged methoxyl content. On methylation with diazomethane, followed by treatment with 70% sodium hydroxide at 170°, methylation, and oxidation with potassium permanganate, 22 to 26% veratric acid but no isohemipinic acid (XXXII) was obtained.

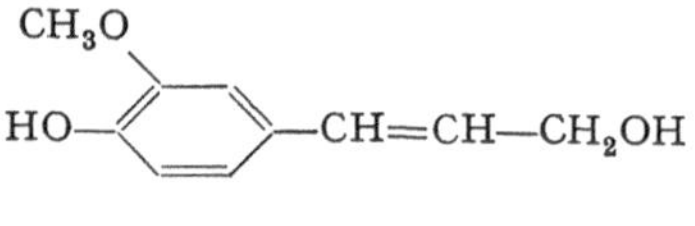

$$CH_3O$$
$$HO-\langle\ \rangle-CH=CH-CH_2OH \qquad CH_3O-\langle\ \rangle-COOH$$
$$HOOC$$
$$CH_3O$$

(XXXI.) Coniferyl alcohol. 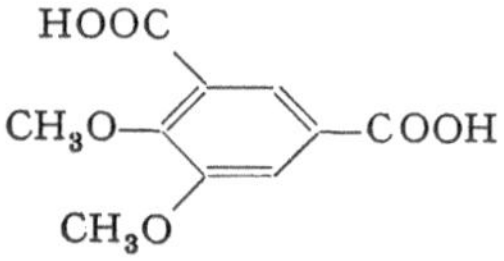   (XXXII.) Isohemipinic acid.

When the polymer was treated with strong mineral acids, it lost water but remained soluble in dilute alkali. The dehydrated polymer dissolved in bisulfite at 135° and also reacted with thioglycolic acid but, on methyla-tion, alkali treatment, remethylation, and oxidation, only very little veratric acid and no isohemipinic acid was formed. Polymerization and dehydration, therefore, apparently took place not by a ring condensa-tion but in the side chain, probably through the double bonds.

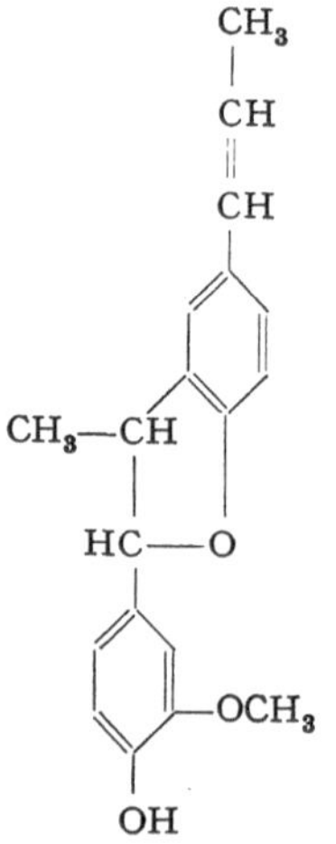

$$CH_3$$
$$|$$
$$CH$$
$$||$$
$$CH$$

$$CH_3-CH$$
$$|$$
$$HC-O$$

$$-OCH_3$$

$$OH$$

(XXXIII.) Dehydro-diisoeugenol.

When coniferyl alcohol was oxidized with ferric chloride, a product was obtained which behaved similarly to dehydro-diisoeugenol (XXXIII). On treatment with strong alkali, followed by methylation and oxidation, it gave veratric and isohemipinic acids. Since FREUDENBERG considers the formation of isohemipinic acid to be the most significant criterion for lignin, he concludes that lignin may be considered as a dehydrogenation product of coniferyl alcohol or a similar compound.

In model experiments with α-ethoxypropiovanillone (XII, $R=A$) and with condensation products of coniferyl aldehyde and coniferin alcohol oxide, in which these compounds were treated with 2% sulfuric acid or 20% hydrochloric acid, FREUDENBERG and co-workers (*119*) obtained amorphous products which were soluble in organic solvents but differed in their structure from lignin because, when heated with 70% potassium hydroxide, followed by methylation and oxidation, they gave no isohemipinic acid.

Decomposition.

Thermal decomposition. Based on the assumption that lignin is a condensation-polymerization product, SUIDA and PREY (*301*, *302*) tried to convert it to low molecular scission products by heating it in a solvent having hydrolyzing properties. With methanol at 300°, low molecular phenols, acids, and neutral compounds were obtained but still about 50% of the lignin remained undissolved. Addition of small amounts of acid or base did not improve the yield. In the presence of large amounts of alkali (50 to 100% based on the lignin), the lignin was rendered completely soluble but about 30% was lost as gaseous products. When a coniferous Scholler lignin was heated with 50% calcium hydrate in ethanol at 300°, up to 85% crude tar or 65% pure tar was obtained. In the latter, guaiacol, *p*-creosol, *p*-ethylguaiacol, propylpyrocatechol, *o*-ethylguaiacol, and isoeugenol were identified. Treatment of the tar with decalin or tetralin at 430° to 440° resulted in further degradation and simultaneous hydrogenation of the low molecular products.

Western hemlock lignin (with a methoxyl content of 15,2%), isolated by heating wood chips with a 1 : 1 mixture of butanol and water at 160°, was hydrolyzed by heating it for three hours at 160° with a mixture of 550 cc. of butanol, 550 cc. of water, and 24,4 cc. of 12 N hydrochloric acid [BAILEY (*19*)]. The reaction mixture contained 27,5% of the lignin as volatile products, which were separated by fractional distillation into 1,9% of acetone, 1,6% of butyraldehyde, 2,5% of methanol, 2,5% of allyl alcohol, 4,8% of propyl alcohol, 11,4% of formic acid, and 2,8% of β-ethyl-γ-methylacrolein [BAILEY (*20*)]. Although the methoxyl groups were all hydrolyzed, only 2,5% were recovered in form of methanol. (The high yield of formic acid was thought to have originated from the oxidation of methyl alcohol.) From the high boiling fraction, 7,3% resorcinol monomethyl ether, 6,7% creosol, 1% butyric acid, 0,3% guaiacol, and 0,2% vanillin were isolated, corresponding to a total yield of 43%, of which about $1/3$ consisted of phenolic and $2/3$ of aliphatic compounds (*21*).

The effect of the prehydrolysis of spruce- and beechwood upon the lignin in these woods was studied in detail by OVERBECK and MÜLLER (237). On heating spruce, beech, or the stems of annual plants with water under pressure, not only are the hemicelluloses dissolved but the lignin also undergoes some changes. A part of it is rendered soluble in alcohol at low temperatures; another part, together with some polysaccharides, becomes soluble in dilute alkali, and the residual lignin is partially rendered insoluble in bisulfite. By heating beechwood for a short time under pressure, a small quantity of a labile lignin-carbohydrate compound goes into solution. Because larger amounts of polysaccharides cannot be dissolved without degrading the lignin, it was concluded that lignin is combined with carbohydrates in wood. Upon cleavage of this linkage, a part of the lignin is converted into a readily soluble state and another part is transformed into a high-molecular state and becomes insoluble in bisulfite.

Decomposition by sea-water and ageing. The resistance of pine lignin against sea-water and attack by teredos was shown by an investigation by SCHMIDT-NIELSEN and HÖYE (282). They extracted lignin with thioglycolic acid according to HOLMBERG (162) from pinewood which had been exposed for 800 years to the attack of sea-water and then for periods varying from six to seventeen months to the attack of teredos. The thioglycolic acid lignin obtained gave elementary analyses very similar to those given by HOLMBERG for thioglycolic acid pine lignin (isolated from fresh pine). It is particularly interesting that this lignin had not lost any of its methoxyl content (12,45%). The same holds true for a lignin sample isolated from an ancient sequoia wood [CUNDY (72)].

An analysis of ancient beech stakes from a fish weir showed that they consisted of 75% lignin. The stakes gave a positive MÄULE reaction and had a methoxyl content of 12,2%. On the assumption that all the methoxyl was combined with the lignin, the latter would have had a methoxyl content of about 15%. The methoxyl content of this lignin was not determined, but the calculated value indicated either that methoxyl groups had been split off during the exposure to the water or that the lignin was contaminated by decomposed carbohydrates [JAHN and HARLOW (172)].

Biochemical decomposition. Corn stalks and other annual plants, when aerobically decomposed for six months, lost $2/3$ of the plant tissue, including $1/3$ of the lignin. The KLASON lignins showed a loss of 27 to 40% of their methoxyl contents [BARTLETT (28)].

ADAMS and LEDINGHAM (1, 2) studied the growth and the lignin-decomposing power of various fungi. Certain species of *Fusarium* and *Alternaria* were found to convert from 12 to 18% of the lignin in lignosulfonic acid. For estimating the amount of sodium lignosulfonate in the mixture, these authors developed an ultraviolet spectrograpich method (*216*). The enzymatic fermentation of lignin was studied by FERNÁNDEZ and REGUEIRO (*92*). The ferment of *Auricularia mesenterica* acts upon lignin with formation of vanillin and thus indicates the presence of lignase. Acacia wood *(Sophora japonica)*, when treated with *Polyporus hispidus*, gives small quantities of vanillin. The elm tree *(Populus nigra)*, when attacked by mold, yielded considerable amounts of uronic acid and a little vanillin.

IV. The Structure of Lignin.

ERDTMAN's Concept of the Lignin Structure.

HAWORTH (*140*) drew attention to the fact that the occurrence of the C_6—C_3 group in natural phenolic resins (lignans) and the present view that lignin is built up of phenylpropane building stones indicates a close relationship of the lignans to lignin. The low content of phenolic hydroxyl groups in lignin suggests, however, an important difference in the type of condensation of the phenylpropane groups in lignans and lignin. VANZETTI (*304*) suggested that lignans, such as olivil, constitute a primary biological product in the formation of lignin. Under normal growth conditions, lignin is biosynthesized whereas, under pathological conditions, resins would appear.

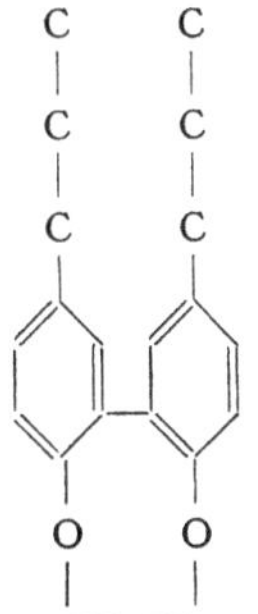

(XXXIV.) 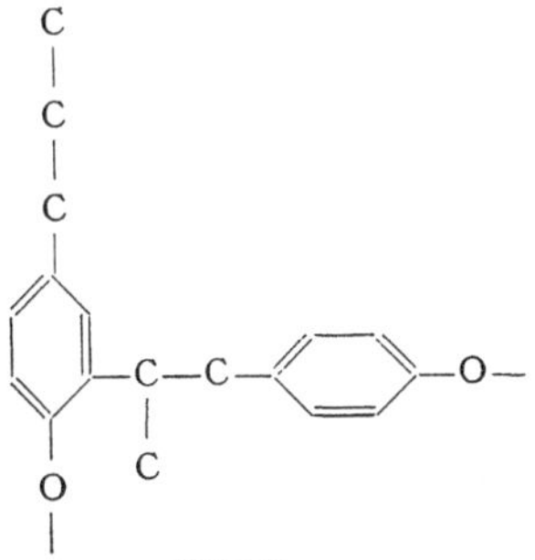 (XXXV.)

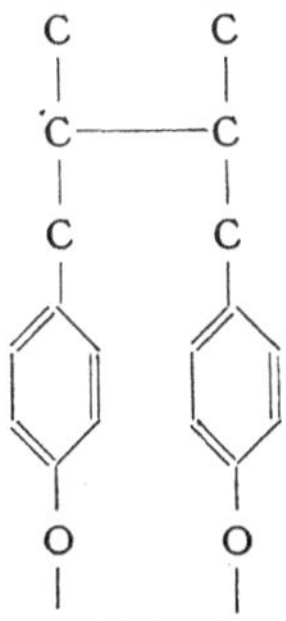

(XXXVI.) 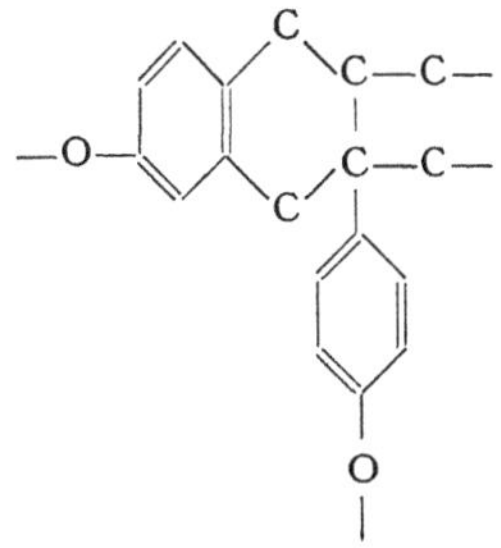 (XXXVII.)

As early as 1933 ERDTMAN (*82*) suggested that lignin might be a *high molecular dehydrogenation product of phenylpropane derivatives.* More recently, he pointed out (*85*) that, in nature, there exists a large number of compounds of dimeric forms of simple phenylpropane building stones. In these compounds the monomers are coupled according to one of the three structures (XXXIV–XXXVI). Of these structures, (XXXIV) is found in magnolol (XXXVIII) (*300*) and in dehydroeugenol (XXXIX) (*82*), whereas (XXXV) is present in egonol (XL) (*179*). Magnolol is

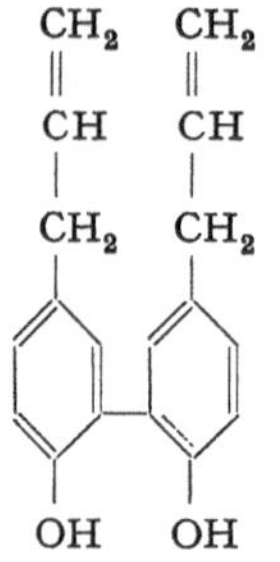

(XXXVIII.) Magnolol. 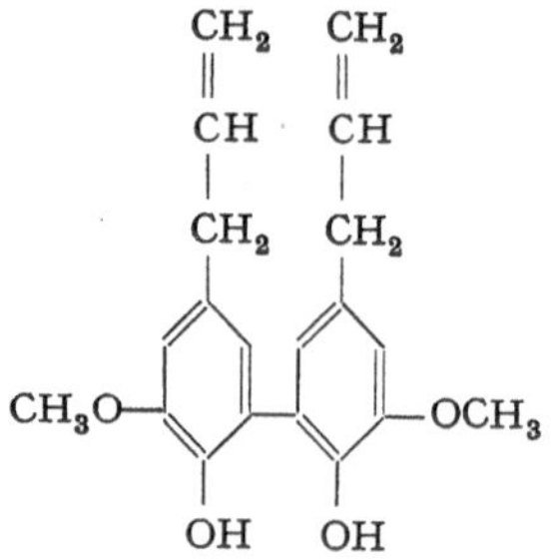   (XXXIX.) Dehydroeugenol.

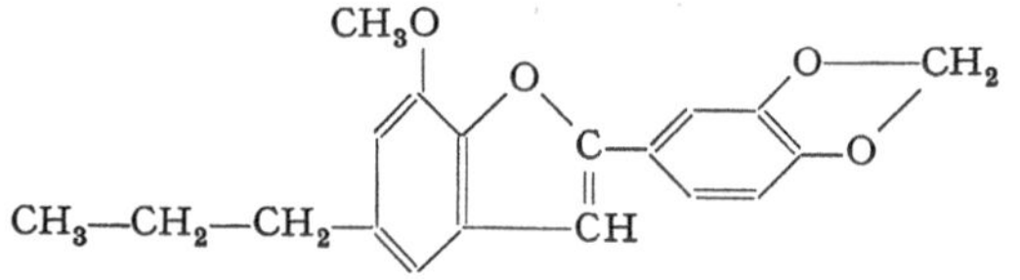

(XL.) Egonol.

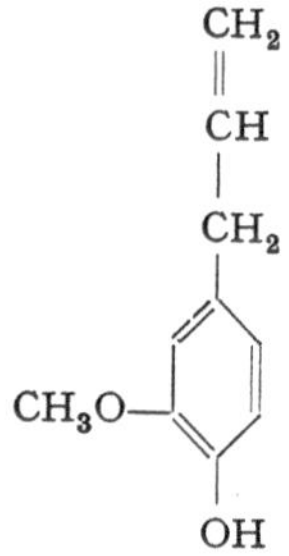

(XLI.) Chavicol.

formed by enzymatic dehydrogenation of chavicol (XLI) by a type of condensation similar to the formation of dehydroeugenol (*82*) from eugenol by mild oxidation with ferric chloride. Egonol is a ring condensation product of a phenylethane with a phenylpropane building stone. It is possible that the phenylethane group is a degradation product of an original phenylpropane group. A compound with a similar carbon structure and a furan ring, dehydro-diisoeugenol (XXXIII) is obtained on biochemical dehydrogenation of isoeugenol [AULIN-ERDTMAN (*10*); ERDTMAN (*82*); FREUDENBERG and PLÖTZ (*118*); see also MÜLLER and HOR-

VATH (*232*)]. The basic carbon structure of the naturally occurring egonol can readily be recognized in FREUDENBERG's formula for lignin (XLII).

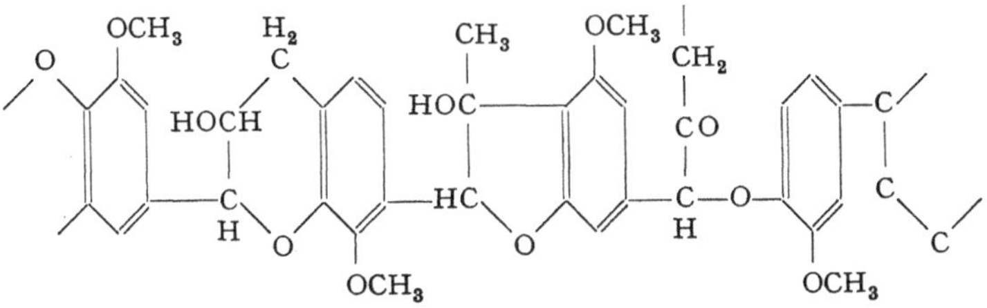

(XLII.) FREUDENBERG's formula for lignin.

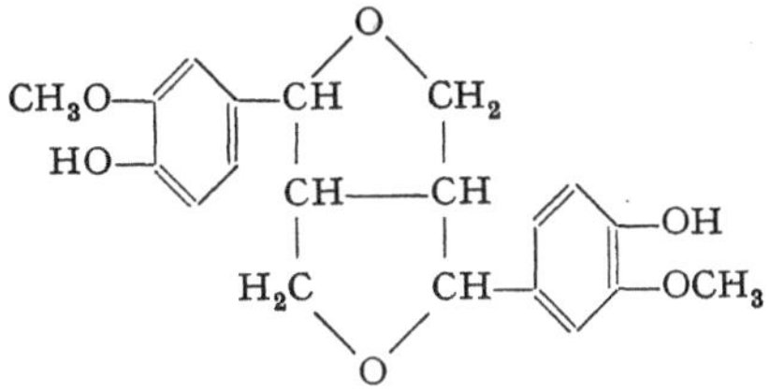

(XLIII.) Pinoresinol.

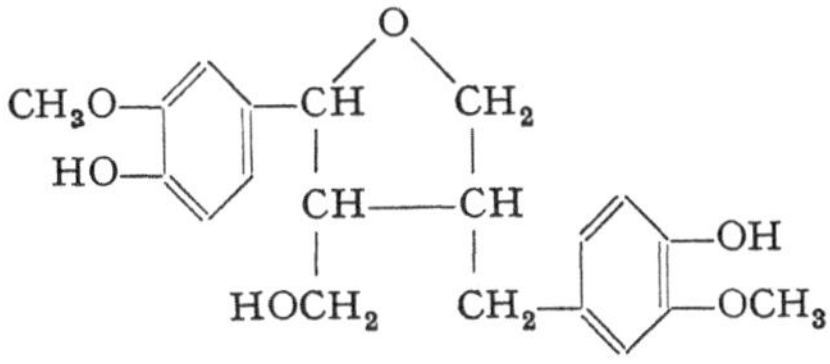

(XLIV.) Lariciresinol.

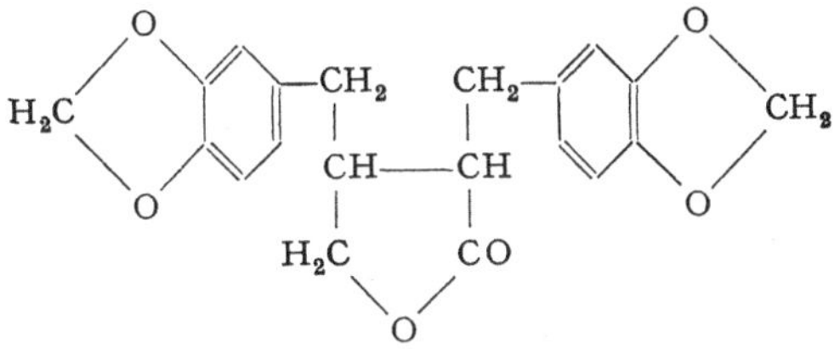

(XLV.) Hinokinen.

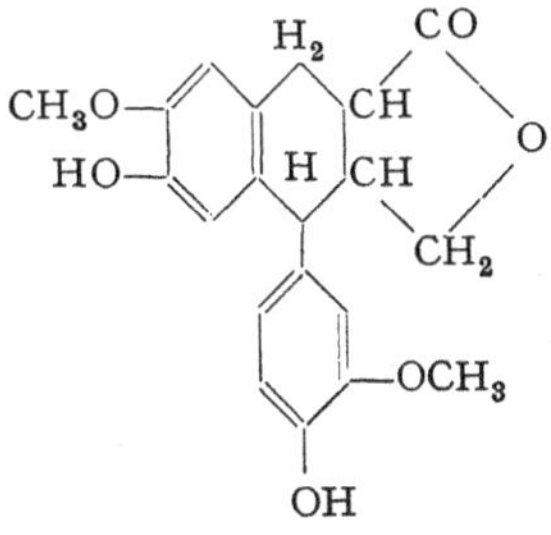

(XLVI.) Conidendrin.

The carbon structure (XXXVI) or its cyclized form (XXXVII) is present in the lignans which play an important role in wood chemistry. Of these, pinoresinol (XLIII) is found in spruce resin [ERDTMAN (83)], lariciresinol (XLIV) in larch resin [HAWORTH and WOODCOCK (141)], hinokinen (XLV) in the heartwood of *Chameocyparis obtusa* [KEIMATSU et al. (183)], and conidendrin (XLVI), long known as "sulfite liquor lactone", in spruce [HOLMBERG (159)], western hemlock [BRAUNS (38)], *Tsuga sieboldii* CARR [KAWAMURA (180)]; and in the resin of *Pinus sylvestris* [HOLMBERG (159)]. In the lignans, the middle or β-carbon atom of the side chain is involved in the condensation of the phenyl-propane building stones.

In the formulas thus far proposed for lignin, FREUDENBERG's scheme of a lignin structure (XLII) is similar to (XXXV) or dehydro-diisoeugenol, whereas a structure having a pyran ring is formed in substances of the catechin (XLVII), flavone (XLVIII), and anthocyanidin type (XLIX).

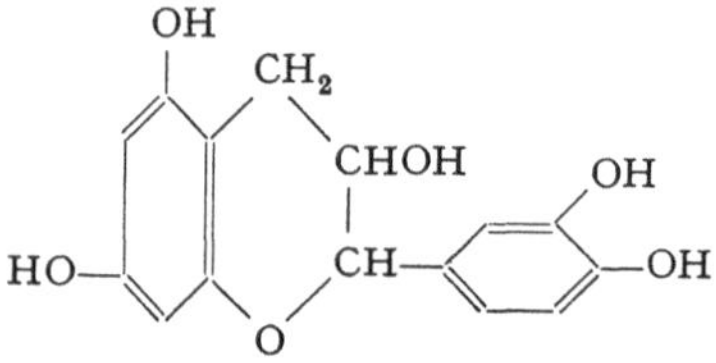

(XLVII.) Catechin type.

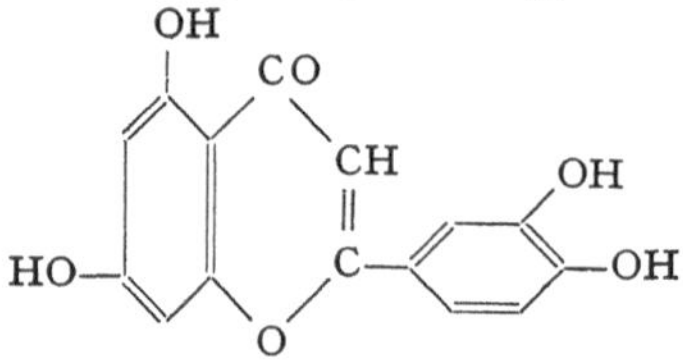

(XLVIII.) Flavone type.

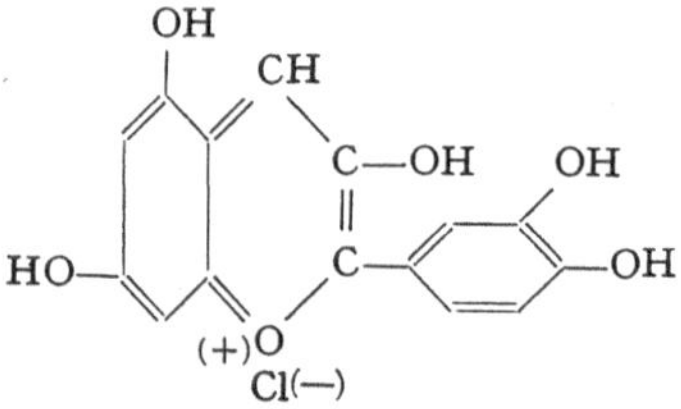

(XLIX.) Anthocyanidin type.

These compounds, however, are composed of only *one* phenylpropane building stone which is condensed with the phenolic grouping. According to ERDTMAN, in the condensation of phenylpropane derivatives in nature, the β-carbon atom of the side chain always seems to be involved. In "comparative anatomical" considerations, ERDTMAN arrived at the concept that condensation by biochemical dehydrogenation of single molecules to more complex ones plays a much greater role in the synthesis of natural products than hitherto realized. On the basis of the dehydro-diisoeugenol structure, ERDTMAN expressed the opinion that lignin may be a *high molecular dehydrogenation product of phenylpropane building stones*. The structure of dehydro diisoeugenol (XXXIII) agrees very well with FREUDENBERG's formula (XLII). Such a concept also explains the formation of isohemipinic acid by hydrolytic opening of the oxygen ring, followed by methylation of the phenolic hydroxyl groups thus formed, and oxidation with potassium permanganate.

FREUDENBERG's Concept of the Lignin Structure.

This was discussed in detail in the last report on lignin appearing in *Fortschritte* [FREUDENBERG (*102*)]. During the last decade, FREUDEN-BERG and his school have carried out a great deal of work much of which was discussed above. They have found further support for the view (*104*) that spruce lignin is a *high molecular product of bifunctionally combined phenylpropane building stones*. In spruce lignin, open ether linkages and condensed ring systems of the benzofuran and benzopyran type which contain secondary hydroxyl groups may be present. A structure of this type is shown in formula (XLII), p. 215.

HIBBERT's Concept of the Lignin Structure.

In 1938, HIBBERT (*148*) presented his "guaiacol theory of lignin formation", according to which lignin is built up from guaiacyl derivatives having a five-carbon side chain. Because no phenylpentane derivative has ever been isolated, HIBBERT soon abandoned this theory in favor of FREUDENBERG's concept that lignin is built up of guaiacylpropane units in which the side chain is oxidized to the same extent as in α- and β-hydroxy-dihydroconiferyl aldehyde. HIBBERT (*151, 152*) completed FREUDENBERG's series of 3,4-dihydroxyphenylpropane derivatives by

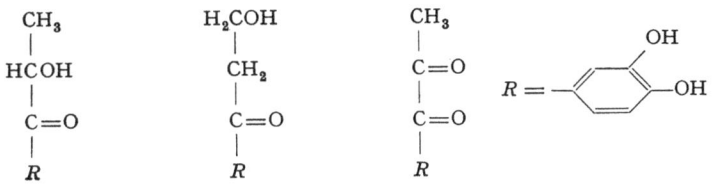

(L.) 3,4-Dihydroxy-benzoylmethyl-carbinol. (LI.) β-(3,4-Dihydroxy-benzoyl)-ethanol. (LII.) 3,4-Dihydroxybenzoyl-acetyl.

adding 3,4-dihydroxybenzoylmethyl-carbinol (L), β-(3,4-dihydroxy-benzoyl)-ethanol (LI), and 3,4-dihydroxybenzoyl acetyl (LII), the latter having a side chain with a somewhat higher degree of oxidation. The guaiacyl and 3,5-dimethoxy-4-hydroxyphenyl derivatives of (L) and

(LII) were isolated by HIBBERT and co-workers on alcoholysis of soft- and hardwood protolignins.

The discovery of these degradation products inspired HIBBERT (*151*) to propose the following theory of the biosynthesis of lignin. Formic acid, acetaldehyde, and glycolaldehyde may be in equilibrium with

$$HCOOH \rightleftarrows \underset{|}{\overset{|}{C}}(OH)_2 \qquad CH_3-CHO \rightleftarrows -CH_2-CH(OH)-$$

(LIII.)   (LIV.)

$$CH_2OH-CHO \rightleftarrows -CH(OH)-CH(OH)-$$

(LV.)

their respective dismutation isomers, viz., (LIII), (LIV), and (LV). By trimerisation of (LV), inositol (LVI) can be formed. The condensation

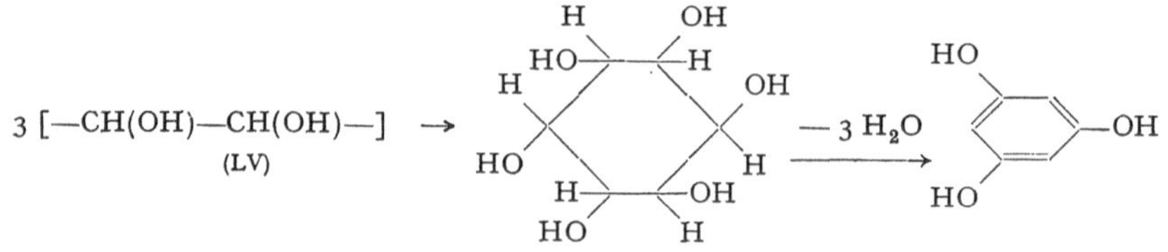

$$3 [-CH(OH)-CH(OH)-] \rightarrow \qquad \xrightarrow{-3 H_2O}$$

(LV)

(LVI.) Inositol.

of two molecules of (LIV) and two molecules of (LIII) would yield 1,3,4-trihydroxybenzene (LVII). Furthermore, 3-methoxy-4-hydroxybenzoyl-

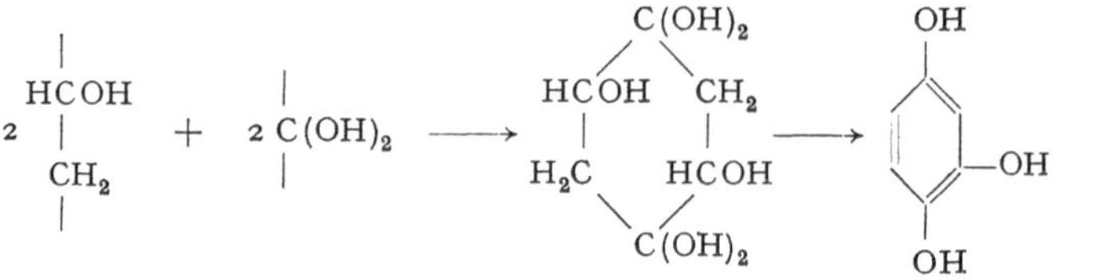

$$2 \begin{matrix} | \\ HCOH \\ | \\ CH_2 \\ | \end{matrix} + 2 \begin{matrix} | \\ C(OH)_2 \\ | \end{matrix} \longrightarrow \longrightarrow$$

(LIV.)   (LIII.)   (LVII.) 1,3,4-
Trihydroxybenzene.

methylcarbinol (LVIII) can change into its dismutation isomers, viz. (LIX) and (LX). According to HIBBERT's concept (*148*, *149*), methylglyoxal,

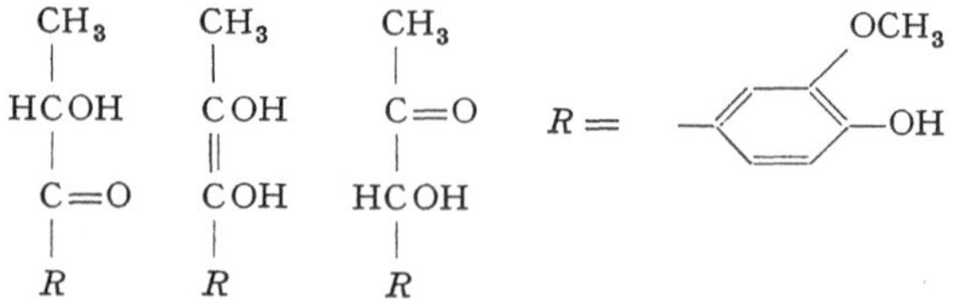

$$\begin{matrix} CH_3 \\ | \\ HCOH \\ | \\ C=O \\ | \\ R \end{matrix} \qquad \begin{matrix} CH_3 \\ | \\ COH \\ || \\ COH \\ | \\ R \end{matrix} \qquad \begin{matrix} CH_3 \\ | \\ C=O \\ | \\ HCOH \\ | \\ R \end{matrix} \qquad R =$$

(LVIII.) 3-Methoxy-   (LIX.)   (LX.)
4-hydroxy-benzoyl-
methylcarbinol.

$CH_3-CO-CHO$, which presumably occurs as an intermediate in carbohydrate metabolism, could assume its cyclic dimer form (LXI) which with two molecules of guaiacol (or 1,3-dimethylpyrogallol) will give (LXII).

$$CH_2\text{———}CH\,(OH)$$

$$OC \diagup \qquad\qquad \diagdown CO$$

$$\diagdown CH\,(OH)\text{—}CH_2 \diagup$$

(LXI.) Dimer of methylglyoxal.

$$
\begin{array}{l}
CH_3O \quad ^{c}\text{----} \\
\qquad\qquad \diagdown H \quad CH_2\text{———}HCOH\,\text{---}^{f} \qquad OCH_3 \\
HO\text{—} \qquad\text{—}C \qquad\qquad\qquad C\diagup\text{——}\text{—OH} \\
CH_3O \qquad ^{e}\text{--} HCOH\text{———}CH_2 H\diagdown\text{--}_{d} \quad OCH_3 \\
\qquad\qquad\qquad ^{a}
\end{array}
$$

(LXII.) Reaction product of dimeric methylglyoxal and 2 molecules of guaiacol.

The formation of the various degradation products from such a compound in the alcoholysis of protolignin is explained as follows: by ring scission along *ab* (LXII), two molecules of α-(3,5-dimethoxy-4-hydroxyphenyl)-lactic aldehyde are formed; along *cd*, two molecules of methyl-syringoyl-carbinol; and along *ef*, two molecules of β-(3,5-dimethoxy-4-hydroxyphenyl)-β-hydroxypropionaldehyde or one molecule of syringoyl-acetaldehyde and one β-hydroxy-β-(3,5-dimethoxy-4-hydroxyphenyl)-propanol.

The idea that the side chains in lignin are present as a cyclohexyl ring, was, however, given up by HIBBERT (*151*) in 1941 in favor of FREUDENBERG's hypothesis that, in lignin, the phenylpropyl building stones condense with each other and form a chain molecule. The discovery by ERDTMAN (*82*) that, in dehydro-diisoeugenol, one molecule of isoeugenol is condensed through the β-carbon atom of its side chain with the *m*-carbon atom of a second molecule, led HIBBERT (*152*) to suggest that the polymerization of lignin progenitors is due to the following factors: "(a) the presence of a phenolic hydroxyl group in *p*-position with respect to a 3-carbon side chain; (b) the presence in the side chain of a propenyl group conjugated with an aromatic ring; (c) the pronounced reactivity of the hydrogen atom in the phenol group and of that attached to the nuclear carbon atom in the position *o* to the phenol hydroxyl group; (d) the reactivity of the terminal carbinol grouping; (e) the tendency of the side chain to undergo an allyl shift; and (f) the labile character of the methylene group in 1-guaiacyl-3-hydroxy-2-propanone (the keto form of oxyconiferyl alcohol) situated, as it is, between a phenyl and a carbonyl group."

By applying the mode of condensation of isoeugenol to 1-guaiacyl-3-hydroxy-2-propanone, dimers of type (LXIII) or (LXIV) are formed, which could react further with the building up a trimer, tetramer, etc. This condensation is of the same type as that suggested earlier by

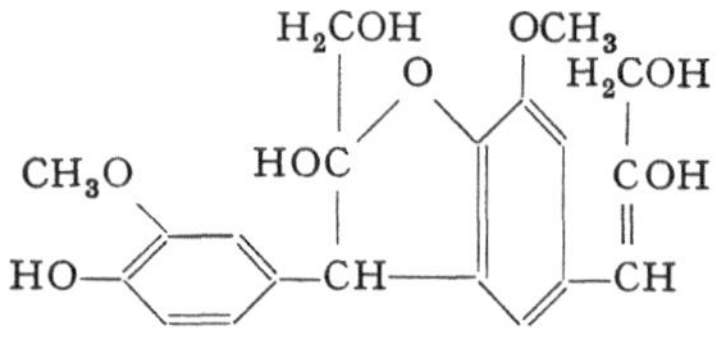

$$\text{(LXIII.)}$$

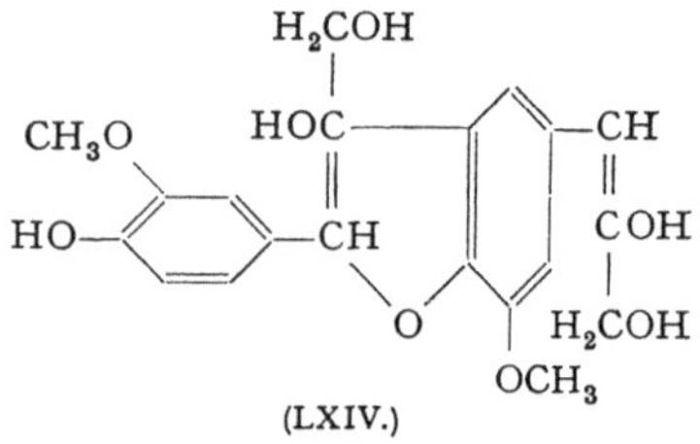

$$\text{(LXIV.)}$$

FREUDENBERG (*101*, p. 136). By splitting off water, benzofuran polymers such as (LXV) and (LXVI) could be formed, whereas ring scission could

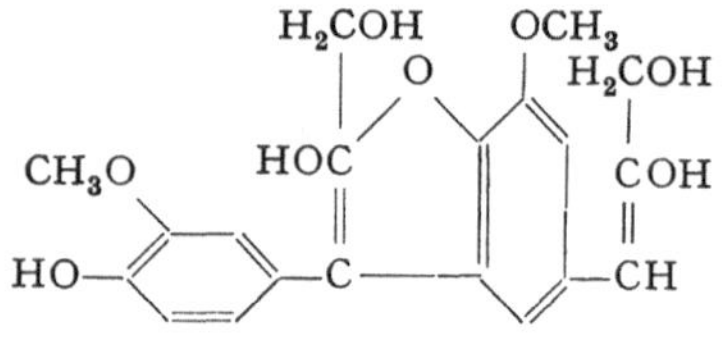

$$\text{(LXV.)}$$

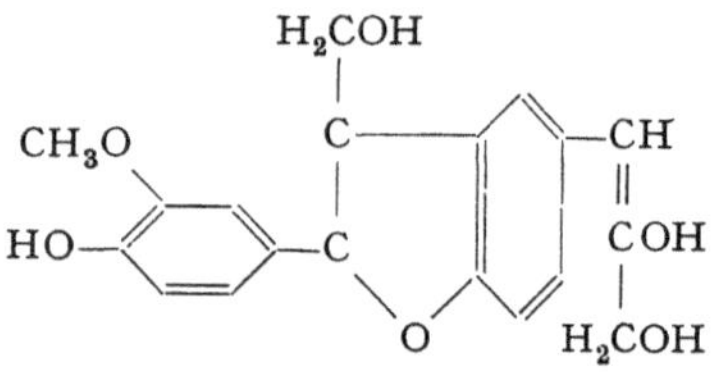

$$\text{(LXVI.)}$$

give the open chain form (LXVII). By double allyl rearrangement, a compound (LXVIII) could be formed which, on alcoholysis (as indicated by the dotted line), could give a diketone or a hydroxyketone.

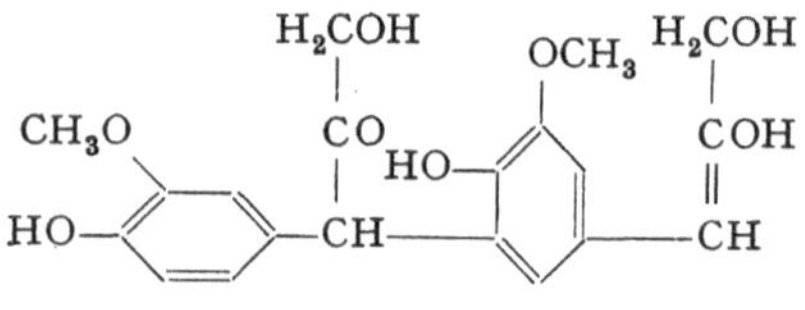

$$\text{(LXVII.)}$$

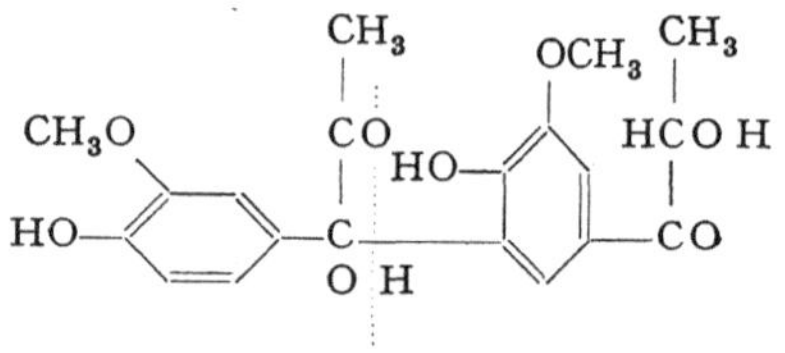

$$\text{(LXVIII.)}$$

HIBBERT did not consider the dehydro-diisoeugenol type to be the only one capable of explaining the experimental facts. For example, two molecules of 1-guaiacyl-3-hydroxy-2-propanone (HIBBERT's oxy-coniferyl alcohol) R—CH_2—CO—CH_2OH (LXIX, R = guaiacyl) could condense to form a polymer of type (LXX) which would presumably undergo preferential dehydration to yield (LXXI) rather than a benzo-furan derivative.

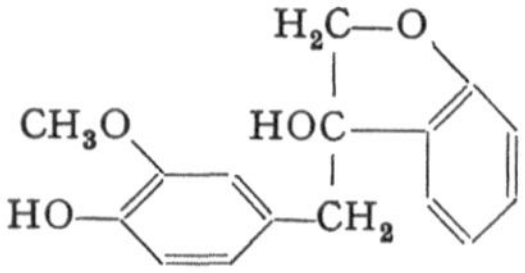

$$\text{(LXX.)}$$

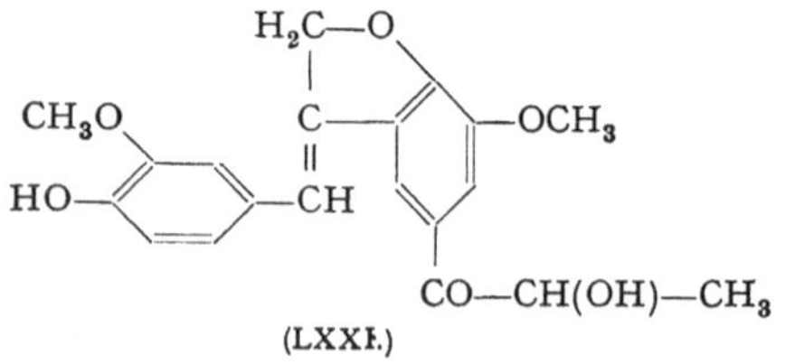

$$\text{(LXXI.)}$$

The systems (LXX) and (LXXI) would explain the following properties of spruce lignin: (a) absence of phenolic hydroxyl groups; (b) presence of aliphatic hydroxyl groups; (c) formation of stable and labile sulfonic acids; (d) formation of vanillin on oxidation; and (e) the presence of terminal methyl groups. They would, however, not explain the formation of 1,2-diketones.

The isolation of the various dismutation products from the ethanolysis mixture of wood induced HIBBERT (150) to suggest a *plant cell respiration system* similar to that put forward by SZENT-GYÖRGYI (303) for animal cell respiration. According to the latter investigator, the dehydrogenase catalysts consist of a series of dicarboxylic acids with four carbon atoms as shown in Figure 1. These catalysts serve as transport media for the

SZENT-GYÖRGYI'S Animal Cell Respiration System

HIBBERT'S Plant Cell Respiration System

$$HO_2C—CO—CH_2—CO_2H$$
$$+2H \downarrow \uparrow -2H$$
$$HO_2C—CH(OH)—CH_2—CO_2H$$
$$\downarrow -H_2O$$
$$HO_2C—CH=CH—CO_2H$$
$$+2H \downarrow \uparrow -2H$$
$$HO_2C—CH_2—CH_2—CO_2H$$

$$3,4\text{-}(CH_3O)\,(HO)\,C_6H_3—CO—CH_2—CH_2OH$$
$$+2H \downarrow \uparrow -2H$$
$$3,4\text{-}(CH_3O)(HO)C_6H_3—CH(OH)—CH_2—CH_2OH$$
$$\downarrow -H_2O$$
$$3,4\text{-}(CH_3O)\,(HO)\,C_6H_3—CH=CH—CH_2OH$$
$$+2H \downarrow \uparrow -2H$$
$$3,4\text{-}(CH_3O)\,(HO)\,C_6H_3—CH_2—CH_2—CH_2OH$$

Fig. 1. Respiration systems.

hydrogen supplied by the carbohydrates. SZENT-GYÖRGYI was able to show that catechol (LXXII), ascorbic acid (LXXIII), and dihydroxy-maleic acid (LXXIV), may act in oxidation-reduction systems. In

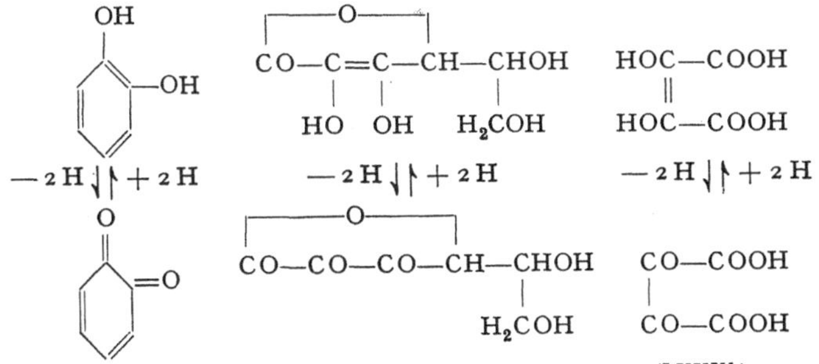

(LXXII.)   (LXXIII.)   (LXXIV.)

HIBBERT's plant respiratory system (Figure 1), the lignin progenitors act in a way very similar to those in the animal cell system. Coniferyl alcohol, which corresponds to fumaric acid in the animal cell system, has been found to occur in practically all plants in the early stage of growth. Through an allylic rearrangement of α-ketodihydroconiferyl alcohol (HIBBERT's "oxyconiferyl alcohol"), the following C_6—C_3 enediol-diketone system can be obtained:

$$R—CH_2—CO—CH_2OH \rightleftharpoons R—CH=C(OH)—CH_2OH \rightleftharpoons$$
$$\rightleftharpoons R—CH(OH)—C(OH)=CH_2 \rightleftharpoons R—CH(OH)—CO—CH_3.$$

The formation of the various monomer guaiacylpropane derivatives cannot be explained on the basis of the present concept of the structure of lignin.

This induced HIBBERT and co-workers (79) to study the rearrangement reactions involved in the transformation of 3-bromo-1-(3,4-dimethoxyphenyl)-2-propanone

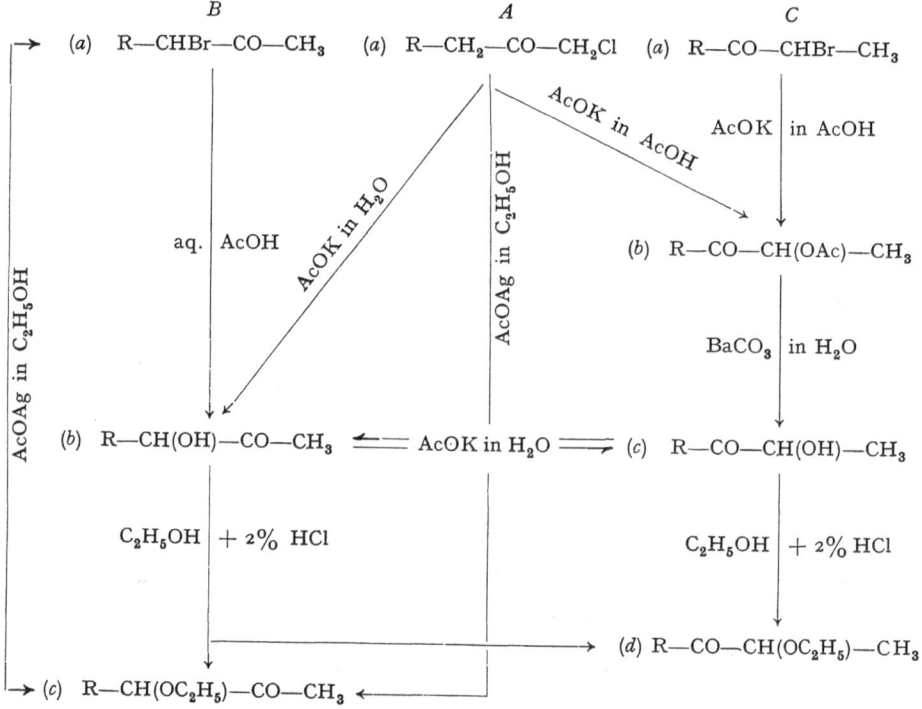

Fig. 2. Rearrangement reactions of 1-(3,4-dimethoxyphenyl)-3-bromo-2-propane, — 1-chloro-2-propanone and — 2-bromo-1-propanone.

(Figure 2, B a), 1-chloro-1-(3,4-dimethoxyphenyl)-2-propanone (A a), and 2-bromo-1-(3,4-dimethoxyphenyl)-1-propanone (C a) into their corresponding hydroxy-, ethoxy-, and acetoxy-derivatives.[1] The results are shown in Figure 2. As can be seen from this scheme, treatment of A a with ethanol in the presence of silver acetate or with potassium acetate in acetic acid results in a complete rearrangement with formation of 1-ethoxy-1-(3,4-dimethoxyphenyl)-2-propanone (B c), and

---

[1] The rearrangement of hydroxypropioguaiacone was recently studied by WACEK and HORAK (309).

1-(3,4-dimethoxyphenyl)-2-acetoxy-1-propanone (*Cb*), respectively. On treatment of *Aa* with potassium acetate in water, double rearrangement takes place with formation of an equilibrium mixture of 1-(3,4-dimethoxyphenyl)-1-hydroxy-2-propanone (*Bb*) and 1-(3,4-dimethoxyphenyl)-2-hydroxy-1-propanone (*Cc*), in addition to a small amount of 1-(3,4-dimethoxyphenyl)-1,2-propanedione (LXXV) as the result of a dismutation rearrangement. Treatment of *Ba* with aqueous

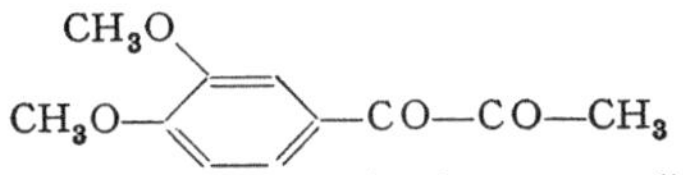

$$CH_3O\diagdown$$
$$CH_3O-\!\!\!\!\langle\quad\rangle\!\!-CO-CO-CH_3$$

(LXXV.) 1-(3,4-Dimethoxyphenyl)-1,2-propanedione.

potassium acetate or with silver acetate in ethanol gives the normal reaction products: 1-hydroxy-1-(3,4-dimethoxyphenyl)-2-propanone (*Bb*) and 1-ethoxy-1-(3,4-dimethoxyphenyl)-2-propanone (*Bc*). But when the former is refluxed with ethanol containing 2% hydrogen chloride, 30% of the normal ethoxy derivative (*Bc*) and 60% of 1-(3,4-dimethoxyphenyl)-2-ethoxy-1-propanone (*Cd*) are formed. The ease with which *Aa* rearranges to give derivatives of *Ba* and *Ca* was considered by HIBBERT to be of great significance for lignin chemistry. All the rearrangement products formed from *Aa* have a terminal methyl group. Whereas on treatment of *Bb* with ethanol and hydrogen chloride, partial rearrangement to *Cd* takes place, the hydroxy derivative, *Cc*, does not undergo rearrangement when treated with this reagent. From this, HIBBERT concluded that, if coniferous protolignin contains the structure R—CH_2—CO—CH_2—O— (Type *A*, Figure 2), this may readily rearrange during the ethanolysis with formation of products of type *Bb* or *Cb* and *Cc*.

Hydrogenation of isolated lignins and of protolignin in wood at high pressure has shown the presence in native lignin of large numbers of building stones-containing a terminal hydroxymethyl group (—CH_2OH) or ether linkages of the type, CH_2—O—C. Because protolignin in sprucewood does not contain terminal methyls [MITCHELL and HIBBERT (*228, 229*)], the formation of *Bc* and *Cd* on ethanolysis permits the conclusion that these compounds represent stabilized end products derived from a more reactive lignin progenitor—i. e., α-keto-dihydroconiferyl alcohol. The fact that, on hydrolysis of *Aa* with aqueous barium carbonate or potassium acetate, a small amount of 1-(3,4-dimethoxyphenyl)-1,2-propanedione is formed, is of marked significance.

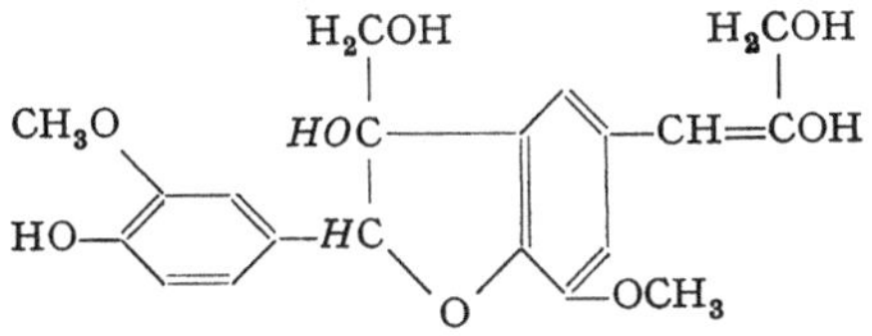

(LXXVI)

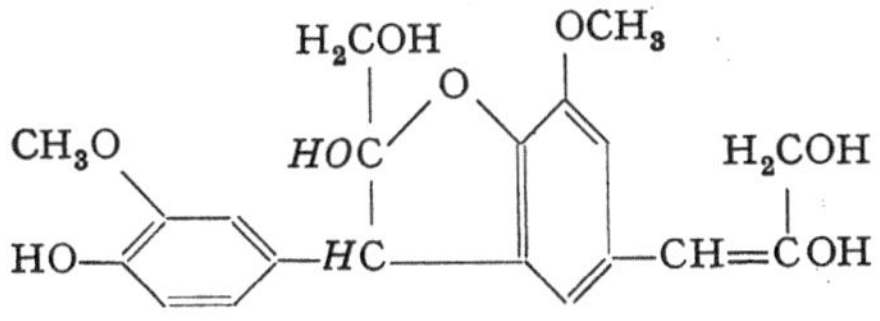

(LXXVII)

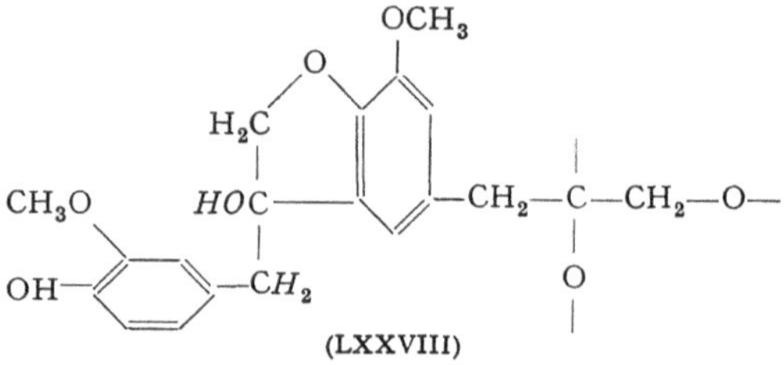

(LXXVIII)

According to HIBBERT (79), spruce protolignin is a mixture of polymers derived by condensation of α-k. todihydroconiferyl alcohol. This condensation and polymerization may occur throug one or more of the following structures: (LXXVI), (LXXVII), and (LXXVIII).

HIBBERT assumed that the corresponding syringyl lignin building stone, viz. 1-(3,5-dimethoxy-4-hydroxyphenyl)-2-keto-3-propanol (LXIX, $R = 3,5$-dimethoxy-4-hydroxyphenyl) is present in the protolignin of all angiosperms, probably as the initial monomeric building unit, because the blocking by methoxyl groups of both positions o to the phenolic hydroxyl group prevents ring condensation. Water may be split off involving the italicized hydroxyls and hydrogens in (LXXVI-LXXVIII), thus forming unsaturated linkages which are capable of giving rise to cross linkages.

Although HIBBERT's basic assumption is identical with that of FREUDENBERG, they differ in the following point: whereas FREUDENBERG believes that *several* types of oxygenated and substituted phenylpropane derivatives take part in lignin formation, HIBBERT considers protolignin as a mixture of polymers derived by condensation of only *one* type of propane derivative—i. e., 1-guaiacyl-(or 3,5-dimethoxy-4-hydroxyphenyl)-3-hydroxy-2-propanone. Although FREUDENBERG's structural formula contains one terminal methyl group for each four to five building stones, and possesses no primary hydroxyl groups, a HIBBERT type condensation product does not include any terminal methyls. Two of the types contain one primary hydroxyl group per building stone, although no definite proof for the presence of such a group has yet been given. HIBBERT himself considered these structural formulas as still being speculative.

Conclusions.

As mentioned in the introduction, a great deal of work has been carried out within the last decade and some progress has been made but the final solution of the lignin problem—i. e., the elucidation of its structure—is still to be found. There seems to be a general agreement that *lignin is built up of phenylpropyl building stones.* Whether these are uniform or whether they differ in their degree of oxidation in the side chain, is still open to discussion. There is no doubt that the aromatic rings of lignin from angiosperms vary in their methoxyl content since there are both vanillyl and syringyl groups present. This difference may consequently lead to various combinations of the building stones. But whether even in gymnosperm lignin, which is supposed to be composed of 3-methoxy-4-hydroxyphenylpropane groups only, these units are linked together according to a single pattern or whether they are condensed in various ways as suggested by FREUDENBERG, is still unsettled. The next goal of lignin research should be the attempt to isolate dimeric or oligo-phenylpropane polymers and to establish the mode of condensation of their monomeric building stones, similar to the isolation of cellobiose [H. SKRAUP (296)] and of some oligosaccharides from cellulose (333).

References.

1. ADAMS, G. A. and G. A. LEDINGHAM: Biological Decomposition of Chemical Lignin. I. Sulfite waste liquor. Canad. J. Res., Sect. C 20, 1 (1942).
2. — — Biological Decomposition of Chemical Lignin. III. Application of a new Ultraviolet Spectrographic Method to the Estimation of Sodium Ligno-sulfonate in Culture Media. Canad. J. Res., Sect. C 20, 101 (1942).
3. ADKINS, H., R. L. FRANK and E. S. BLOOM: The Products of the Hydrogenation of Lignin. J. Amer. chem. Soc. 63, 549 (1941).
4. AHLM, C. E.: An Investigation of Spruce Thiolignin. Paper Trade J. 113, no. 13, 115 (1941).
5. ALLEN, C. F. H. and J. R. BYERS: A Synthesis of Coniferyl alcohol. Science [New York] 107, 269 (1948).
6. ARIES, R. S.: Plastics from Lignin. Chem. Industries 56, 226 and 416 (1945).
7. — Research on Lignin as a Soil Builder. Northeast. Wood Util. Council, Bull. no. 7, 56 (1945).
8. — Lignin as a Fertilizer Material. Paper Trade J. 123, no. 21, 47 (1946).
9. ASTBURY, W. T. and D. M. WRINCH: Intermolecular Folding of Proteins by Keto-enol Interchange. Nature [London] 139, 798 (1937).
10. AULIN-ERDTMAN, G.: Die Struktur des Dehydrodi-isoeugenols. Svensk kem. Tidskr. 54, 168 (1942).
11. — Einige Modellversuche zur Chemie des Lignins. Svensk kem. Tidskr. 55, 116 (1943).
12. — Spektrographische Beiträge zur Ligninchemie. Svensk Papperstidn. 47, 91 (1944).
13. AULIN-ERDTMAN, G., A. BJÖRKMAN, H. ERDTMAN u. S. E. HÄGGLUND: Einige Überlegungen und Modellversuche zur Sulfitierung des Lignins. Svensk Papper-stidn. 50, no. 11 B, 81 (1947).
14. BAILEY, A. J.: The Chemistry of Lignin. I. The Existence of a Chemical Bond between Lignin and Cellulose. Paper Trade J. 110, no. 1, 29 (1940).
15. — The Chemistry of Lignin. II. The Butanolysis of Wood. Paper Trade J. 110, no. 2, 29 (1940).
16. — The Heterogeneity of Lignin. Paper Trade J. 111, no. 7, 27 (1940).
17. — Preparation and Properties of Butanol Lignin. Paper Trade J. 111, no. 6, 27 (1940).
18. — The Chemistry of Butanol Lignin. Paper Trade J. 111, no. 9, 86 (1940).
19. — Hydrolytic Derivatives of Lignin Volatile Compounds. J. Amer. chem. Soc. 64, 22 (1942).

20. Aulin-Erdtman, G.: Volatile Hydrogenation Derivatives of Lignin. J. Amer. chem. Soc. 65, 1165 (1943).
21. — High-boiling Hydrolytic Derivatives of Lignin. J. Amer. chem. Soc. 69, 575 (1947).
22. — The Lignin-alcohol Condensation. The Reaction of Lignin with Amino- and Nitro-butanol. Paper Ind. Paper Wld. 29, 1606 (1948).
23. Bailey, A. J. and H. M. Brooks: Electrolytic Oxidation of Lignin. J. Amer. chem. Soc. 68, 445 (1946).
24. Bailey, A. J. and O. W. Ward: Synthetic Lignin Resin and Plastic. Ind. Engng. Chem. 37, 1199 (1945).
25. Baker, S. B., T. H. Evans and H. Hibbert: Studies on Lignin and Related Compounds. LXXXV. Synthesis and Properties of Dimers Related to Lignin. J. Amer. chem. Soc. 70, 60 (1948).
26. Baker, S. B. and H. Hibbert. Studies on Lignin and Related Compounds. LXXXVI. Hydrogenation of Dimers Related to Lignin. J. Amer. chem. Soc. 70, 63 (1948).
27. Baroni, E.: Struktur des Lignins. I. Neuere Ergebnisse. Papier, Pappe, Zellulose, Holzstoff 58, 1 (1940).
28. Bartlett, J. B.: The Effect of Decomposition of the Lignin of Plant Materials. Iowa State Coll. J. Sci. 14, 11 (1939).
29. Bennett, E.: Are Pectic Substances Precursors to Lignin? Science [New York] 91, 95 (1940).
30. Bobrov, P. A. and L. I. Kolotova: Hydrogenation of Waste Sulfite Liquors. C. R. [Doklady] Acad. Sci. URSS 24, 46 (1939).
31. Bond, W. J., I. G. Goddard and G. F. Wright: The Isolation of Sassafras Lignin. Canad. J. Res., Sect. B 25, 535 (1947).
32. Bower, J. R., Jr., J. L. McCarthy and H. Hibbert: Studies on Lignin and Related Compounds. LXIII. Hydrogenation of Wood (part 2). J. Amer. chem. Soc. 63, 3066 (1941).
33. Bower, J. R., Jr., L. M. Cooke, and H. Hibbert: Studies on Lignin and Related Compounds. LXX. Hydrogenolysis and Hydrogenation of Maple Wood. J. Amer. chem. Soc. 65, 1192 (1943).
34. — — Studies on Lignin and Related Compounds. LXXI. The Course of Formation of Native Lignin in Spruce Buds. J. Amer. chem. Soc. 65, 1195 (1943).
35. Brauns, F. E.: Native Lignin. I. The Isolation and Methylation. J. Amer. chem. Soc. 61, 2120 (1939).
36. — Recent Advances in the Chemistry of Lignin. Paper Trade J. 111, no. 14, 33 (1940).
37. — The Nature of Lignin from Western Hemlock (Tsuga heterophylla). J. org. Chemistry 10, 211 (1945).
38. — The Occurrence of Conidendrin in Western Hemlock (Tsuga heterophylla). J. org. Chemistry 10, 216 (1945).
39. — The Utilization of Lignin. T. A. P. P. I. Bull. no. 84, 3 pp. (1946).
40. — The Stability of the Methoxyl Groups in Methylated Hydrochloric Acid Spruce Lignin. J. Amer. chem. Soc. 68, 1721 (1946).
41. — (unpublished).
42. Brauns, F. E. and M. A. Buchanan: Acetic Acid Spruce Lignin and Acetic Acid Willstätter Spruce Lignin. J. Amer. chem. Soc. 67, 645 (1945).
43. — — The Reaction between Thio Compounds and Lignin. Paper Trade J. 122, no. 21, 49 (1946).

44. BRAUNS, F. E. and H. HIBBERT: Methanol Lignin. Canad. J. Res., Sect. B 13, 28 (1935).

45. BRAUNS, F. E. and W. H. LANE: Thiophenol Spruce Lignin. Paper Trade J. 122, no. 8, 38 (1946).

46. BRAUNS, F. E. and H. F. LEWIS: The Nature of the Lignin in Redwood Bark. Paper Trade J. 119, no. 22, 34 (1944)

47. BRAUNS, F. E., H. F. LEWIS and E. B. BROOKBANK: Lignin Ethers and Esters; Preparation from Lead and other Metallic Derivatives of Lignin. Ind. Engng. Chem. 37, 70 (1945).

48. BRAUNS, F. E. and J. J. YIRAK: The Decomposition of Methylated Spruce Wood. Paper Trade J. 125, no. 12, 55 (1947).

49. BRAWN, J. S., R. D. HEDDLE and J. A. F. GARDNER: The Ethanolysis of Western Red Cedar, DOUGLAS Fir and Western Hemlock. J. Amer. chem. Soc. 62, 3251 (1940).

50. BREWER, C. P., L. M. COOKE and H. HIBBERT: Studies on Lignin and Related Compounds. LXXXIV. The High Pressure Hydrogenation of Maple Wood: Hydrol Lignin. J. Amer. chem. Soc. 70, 57 (1948).

51. BRICKMAN, L., J. J. PYLE and H. HIBBERT: The Aldehydic Constituents from the Ethanolysis of Spruce and Maple Woods. J. Amer. chem. Soc. 61, 523 (1939).

52. BRICKMAN, L., J. J. PYLE, W. L. HAWKINS and H. HIBBERT: Structure of the Ethanolysis Products from Spruce and Maple Wood. J. Amer. chem. Soc. 62. 986 (1940).

53. BROOKBANK, E. B.: Recovery and Uses of Byproduct Soda Lignin. Chemurgic Digest 2, 97 (1943).

54. — Industrial Uses of Alkali Lignin. Paper Trade J. 122, no. 13, 44 (1946).

55. BROOKBANK, E. B. and F. E. BRAUNS: The Chemical Nature of DOUGLAS Fir Lignin. Paper Trade J. 110, no. 5, 33 (1940).

56. BROOKBANK, E. B., F. E. BRAUNS, H. F. LEWIS and M. A. BUCHANAN: New Derivatives of Lignin. Paper Trade J. 116, no. 13, 27 (1943).

57. BUCHANAN, M. A.: (unpublished).

58. CALHOUN, J. M., F. H. YORSTON and O. MAASS: A Study of the Mechanism and Kinetics of the Sulfite Process. Canad. J. Res., Sect. B 17, 121 (1939).

59. CAMPBELL, W. G. and J. C. MC GOWAN: Color Reactions of Lignin and Tannins. Nature [London] 143, 1022 (1939).

60. CARPENTER, J. S. and H. K. BENSON: Chemical Derivatives of Lignin: Nitrolignin. Pacific Pulp Paper Ind. 14, no. 12, 17 (1940).

61. CHARBONNIER, H. Y.: Lignins Isolated in the Presence of Butanol. Paper Trade J. 114, no. 11, 31 (1942).

62. CLARK, J. C. and F. E. BRAUNS: Esters of Certain Lignin Derivatives. Paper Trade J. 119, no. 6 33 (1944).

63. CONNER, W. P.: High Frequency Energy Losses in Solutions Containing Macromolecules. J. chem. Physics 9, 591 (1941).

64. COOKE, L. M., J. L. McCARTHY and H. HIBBERT: Studies on Lignin and Related Compounds. LX. Hydrogenation Studies on Maple Ethanolysis Products. I. J. Amer. chem. Soc. 63, 3052 (1941).

65. — — — Studies on Lignin and Related Compounds. LXI. Hydrogenation of Ethanolysis Fractions from Maple Wood. II. J. Amer. chem. Soc. 63, 3056 (1941).

66. COPPICK, S. and W. F. FOWLER, Jr.: The Location of Potential Reducing Substances in Woody Tissues. Paper Trade J. 109, no. 11, 81 (1939).

67. CRAMER, A. B. and H. HIBBERT: Studies on Lignin and Related Compounds. XLV. Synthesis and Properties of α-Hydroxypropiovanillone. J. Amer. chem. Soc. **61**, 2204 (1939).

68. CRAMER, A. B., J. M. HUNTER and H. HIBBERT: Studies on Lignin and Related Compounds. XXXV. The Ethanolysis of Spruce Wood. J. Amer. chem. Soc. **61**, 509 (1939).

69. CREIGHTON, R. H. J., J. L. McCARTHY and H. HIBBERT: Studies on Lignin and Related Compounds. LIX. Aromatic Aldehydes from Plant Materials. J. Amer. chem. Soc. **63**, 3049 (1941).

70. CREIGHTON, R. H. J., R. D. GIBBS and H. HIBBERT: Studies on Lignin and Related Compounds. LXXV. Alkaline Nitrobenzene Oxidation of Plant Materials and Application to Taxonomic Classification. J. Amer. chem. Soc. **66**, 32 (1944).

71. CREIGHTON, R. H. J. and H. HIBBERT: Studies on Lignin and Related Compounds. LXXVI. Alkaline Nitrobenzene Oxidation of Corn Stalks. Isolation of p-Hydroxybenzaldehyde. J. Amer. chem. Soc. **66**, 37 (1944).

72. CUNDY, P. F.: A Comparison of Ancient and Modern Sequoia Wood. Madrono **7**, 145 (1946).

73. DADSWELL, H. E. and D. J. ELLIS: Study of the Cell Wall. I. Methods of Demonstrating Lignin Distribution in Wood. J. Council sci. ind. Res. **13**, 44 (1940).

74. DORLAND, R. M. and H. HIBBERT: Formic Acid as a Solvent for Ozonization Investigations. Canad. J. Res., Sect. B **18**, 30 (1940).

75. DORLAND, R. M., W. L. HAWKINS and H. HIBBERT: Studies on Lignin and Related Compounds. XLVI. The Action of Ozone on Isolated Lignin. J. Amer. chem. Soc. **61**, 2689 (1939).

76. DRYDEN, E. C., J. D. REID and S. I. ARONOVSKY: A Note on the Effect of Repeated Treatment of Corncob Lignin by the 72% Sulfuric Acid Method. Paper Trade J. **119**, no. 11, 119 (1944).

77. DUNN, S. and J. SEIBERLICH: Uses of Lignin in Agriculture. Mechan. Engng. **69**, 197 and 212 (1947).

78. DUNN, S., J. SEIBERLICH and D. S. EPPELSHEIMER: The Use of Lignin in Potato Fertilizer. Northeast. Wood Util. Council, Bull. **7**, 21 (1945).

79. EASTHAM, A. M., H. E. FISHER, M. KULKA and H. HIBBERT: Studies on Lignin and Related Compounds. LXXIV. Relation of Wood Ethanolysis Products to the HIBBERT Series of Plant Respiratory Catalysts. Allylic and Dismutation Rearrangements of 3-Chloro-1-(3,4-dimethoxyphenyl)-2-propanone and 1-Bromo-1-(3,4-dimethoxyphenyl)-2-propanone. J. Amer. chem. Soc. **66**, 26 (1944).

80. ENDERS, C.: Wie entsteht der Humus in der Natur? Die Chemie **56**, 281 (1943).

81. ERBRING, H. and H. PETER: Zur Kenntnis des Lignins. Kolloid-Z. **96**, 47 (1941).

82. ERDTMAN, H.: Dehydrierungen in der Coniferylreihe. II. Dehydrodi-isoeugenol. Liebigs Ann. Chem. **503**, 283 (1933).

83. — Über die Wirkung von Phenolen beim Sulfitkochprozeß. Svensk Papperstidn. **43**, 255 (1940); Cellulosechemie **18**, 83 (1940).

84. — Untersuchungen über Sulfitablaugen. I. Die Fällbarkeit der Ligninsulfonsäuren durch organische Basen. Svensk kem. Tidskr. **53**, 201 (1941).

85. — Developments in Lignin Chemistry in Recent Years. Svensk Papperstidn. **44**, 243 (1941).

86. — Investigations of Sulfite Waste Liquor. II. The Methodical Separation of the Constituents of Sulfonic Acids from Sulfite Waste Liquor with Organic Bases. Svensk Papperstidn. **45**, 315 (1942).

87. ERDTMAN, H.: Untersuchungen über Sulfitablaugen. III. Zusammensetzung und Eigenschaften verschiedener Ligninsulfonsäurefraktionen aus technischen Sulfitablaugen. Svensk Papperstidn. 45, 374 (1942).

88. — Untersuchungen über Sulfitablaugen. IV. Methylierte Ligninsulfonsäuren aus bei der Herstellung von Kunstseidenzellstoff und von starkem Zellstoff anfallenden Ablaugen. Svensk Papperstidn. 45, 392 (1942).

89. — Untersuchungen über schwefelarme Ligninsulfonsäuren. Svensk Papperstidn. 48, 75 (1945).

90. ERDTMAN, H., G. AULIN-ERDTMAN and B. LINDGREN: The Sulfonation of Spruce Lignin. Svensk Papperstidn. 49, 199 (1946).

91. ERNSBERGER, F. M. and W. G. FRANCE: Some Physical and Chemical Properties of Weight-fractionated Lignosulfonic Acids, Including the Dissociation of Lignosulfonates. J. Phys. Colloid Chem. 52, 267 (1948).

92. FERNÁNDEZ, O. and B. REGUEIRO: Enzymatic Degradation of Lignin. Farm. nueva 11, 57, 111, 169 and 223 (1946); Revista de la real Academia de Ciencias 39, 331 (1945).

93. FISHER, E.: The Action of Organic Acids on Corn Stalk Lignin. Iowa State Coll. J. Sci. 17, 241 (1943).

94. FISHER, E. and R. S. BOWER: The Action of Organic Nitrogen Bases on Corn Stalk Lignin. J. Amer. chem. Soc. 63, 1881 (1941).

95. FISHER, H. E. and H. HIBBERT: Studies on Lignin and Related Compounds. LXXXIII. Synthesis of 3-Hydroxy-1-(4-hydroxy-3-methoxyphenyl)-2-propanone. J. Amer. chem. Soc. 69, 1208 (1947).

96. FISHER, H. E., M. KULKA and H. HIBBERT: Studies on Lignin and Related Compounds. LXXIX. Synthesis and Properties of 3-Hydroxy-1-(3,4-dimethoxy-phenyl)-2-propanone. J. Amer. chem. Soc. 66, 598 (1944).

97. FISHER, J. H., W. L. HAWKINS and H. HIBBERT: Studies on Lignin and Related Compounds. XLVII. The Synthesis of Xylosides Related to Lignin Plant Constituents. J. Amer. chem. Soc. 62, 1412 (1940).

98. — — — Studies on Lignin and Related Compounds. LIV. Synthesis and Properties of Glycosides Related to Lignin. J. Amer. chem. Soc. 63, 3031 (1941).

99. FORMAN, L. V.: Action of Ultraviolet Light on Lignin. Paper Trade J. 111, no. 21, 34 (1940).

100. FREDENHAGEN, K. u. J. CADENBACH: Der Abbau der Zellulose durch Fluor-wasserstoff und ein neues Verfahren der Holzverzuckerung durch hoch-konzentrierten Fluorwasserstoff. Angew. Chem. 46, 113 (1933).

101. FREUDENBERG, K.: Tannin, Cellulose, Tannin. Berlin: J. Springer. 1933.

102. — Lignin. Fortschr. Chem. organ. Naturstoffe 2, 1 (1939).

103. — Polysaccharides and Lignin. Ann. Rev. Biochem. 8, 81 (1939).

104. — Die Grundzüge der Ligninchemie. Chemiker-Ztg. 68, 39 (1944).

105. — Die aromatische Natur des Lignins im Holze. Cellulosechemie 22, 117 (1944).

106. FREUDENBERG, K. u. L. ACKER: Über die Einwirkung von Glykolchlorhydrin auf Fichtenlignin. Ber. dtsch. chem. Ges. 74, 1400 (1941).

107. FREUDENBERG, K. u. K. ADAM: Die Verschwelung des Lignins im Wasser-stoffstrom. Ber. dtsch. chem. Ges. 74, 387 (1941).

108. FREUDENBERG, K., W. LAUTSCH u. K. ENGLER: Die Bildung von Vanillin aus Fichtenholz. Ber. dtsch. chem. Ges. 73, 167 (1940).

109. FREUDENBERG, K., W. LAUTSCH u. G. PIAZOLO: Die Einwirkung von Kalium in Ammoniak auf das Lignin und Holz der Fichte und Buche. Ber. dtsch. chem. Ges. 74, 1879 (1941).

110. FREUDENBERG K., W. LAUTSCH u. G. PIAZOLO: Die Einwirkung von Sulfit auf Fichtenholz bei 70°. Cellulosechemie 22, 97 (1944).

111. — — — Die Nitrierung des Fichtenlignins. Die aromatische Natur des Lignins im Holze. Cellulosechemie 21, 95 (1943).

112. FREUDENBERG, K., W. LAUTSCH, G. PIAZOLO u. A. SCHEFFER: Die Druckhydrierung des Lignins und der ligninhaltigen Ablaugen der Fichte. Ber. dtsch. chem. Ges. 74, 171 (1941).

113. FREUDENBERG, K. u. K. PLANKENHORN: Die Herkunft des Formaldehyds aus dem Lignin. Ber. dtsch. chem. Ges. 80, 149 (1947).

114. — — Über Essigsäure-Lignin. Ber. dtsch. chem. Ges. 75, 857 (1942).

115. FREUDENBERG, K. u. TH. PLÖTZ: Über enzymatische Abbauversuche an Holz. Holz als Roh- u. Werkstoff 3, 105 (1940).

116. — — Die quantitative Bestimmung des Lignins. Ber. dtsch. chem. Ges. 73, 754 (1940).

117. FREUDENBERG, K. u. H. RICHTZENHAIN: Enzymatische Versuche zur Entstehung des Lignins. Ber. dtsch. chem. Ges. 79, 997 (1943).

118. — — Die Konstitution des Dehydro-di-isoeugenols und seine Bedeutung für die Chemie des Lignins. Liebigs Ann. Chem. 522, 126 (1942).

119. FREUDENBERG, K., H. RICHTZENHAIN, E. FLICKINGER u. K. ENGLER: Modellversuche zur Ligninfrage. Ber. dtsch. chem. Ges. 72, 1805 (1939).

120. FREUDENBERG, K. u. F. SOHNS: Zur Kenntnis des Lignins. Ber. dtsch. chem. Ges. 66, 262 (1933).

121. FREUDENBERG, K. u. H. WALSCH: Die Phenolgruppen im Lignin. Ber. dtsch. chem. Ges. 76, 305 (1943).

122. FRIESE, H. u. W. LÜDECKE: Lignin. XIII. Über eine Spaltung des Holzes mittels Nitrierung. Ber. dtsch. chem. Ges. 74, 308 (1941).

123. FULLER, J. E.: Influence of Purified Lignin on Nitrification in Soil. Science [New York] 104, 313 (1946).

124. GARDNER, J. A. F. and H. HIBBERT: Studies on Lignin and Related Compounds. LXXXII. Synthesis and Properties of 1,3-Diacetoxy-1-(4-acetoxy-3-methoxyphenyl)-2-propanone and 1-Acetoxy-3-chloro-1-(4-acetoxy-3-methoxyphenyl)-2-propanone and their Relation to Lignin Structure. J. Amer. chem. Soc. 66, 607 (1944).

125. GLADING, R. E.: The Ultraviolet Absorption Spectra of Lignin and Related Compounds. Paper Trade J. 111, no. 23, 32 (1940).

126. GODARD, H. P., J. L. McCARTHY and H. HIBBERT: Hydrogenation of Wood. J. Amer. chem. Soc. 62, 988 (1940).

127. — — — Studies on Lignin and Related Compounds. LXII. High Pressure Hydrogenation of Wood Using Copper Chromite Catalyst. I. J. Amer. chem. Soc. 63, 3061 (1941).

128. GRALÉN, N.: The Molecular Weight of Lignin. J. Colloid Sci. 1, 453 (1946).

129. GRIFFIOEN, K.: Presence of Combined Lignins in Plant Material. Chem. Weekbl. 36, 81 (1939).

130. HACHIHAMA, Y., S. ZYODAI and M. UMEZU: Lignin and Related Compounds. I. Hydrogenation of Softwood Lignin. J. Soc. chem. Ind. Japan, suppl. Bind. 43, 127 (1940).

131. HÄGGLUND, E.: Holzchemie. Leipzig: Akad. Verlagsges. 1939.

132. — Über Schwefellignin und seine Bedeutung bei dem Sulfatkochprozeß. Holz als Roh- u. Werkstoff 4, 236 (1941).

133. — Thio (sulfide) Lignin and its Role in the Sulfate Process. Svensk Papperstidn. 44, 183 (1941).

134. HARLOW, W. M.: Contribution to the Chemistry of the Plant Cell Wall. IX. Further Studies on the Location of Lignin, Cellulose and other Components in Woody Cell Walls. Paper Trade J. **109**, no. 18, 38 (1939).

135. HARRIS, E. E.: Utilization of Waste Lignin; Current Chemical Research. Ind. Engng. Chem. **32**, 1049 (1940).

136. — Hydrogenation of Lignin. Paper Trade J. **111**, no. 24, 27 (1940).

137. HARRIS, E. E., J. D'IANNI and H. ADKINS: Reaction of Hardwood Lignin with Hydrogen. J. Amer. chem. Soc. **60**, 1467 (1938).

138. HARRIS, E. E. and L. J. LOFDAHL: The Reaction of Methyl Hypochlorite with Lignin. J. Amer. chem. Soc. **63**, 112 (1941).

139. HARRIS, E. E., J. SAEMAN and E. C. SHERRARD: Hydrogenation of Lignin in Aqueous Solutions. Ind. Engng. Chem. **32**, 440 (1940).

140. HAWORTH, R. D.: Constitution of Natural Phenolic Resins. Nature [London] **147**, 255 (1941).

141. HAWORTH, R. D. and D. WOODCOCK: The Constituents of Natural Phenolic Resins. XV. The Stereochemical Relationship of Lariciresinol and Pinoresinol. J. chem. Soc. [London] **1939**, 1054.

142. HECHTMAN, J. F.: The Behavior of Some Lignin Preparations in the Molecular Still. Paper Trade J. **114**, no. 22, 45 (1942).

143. HEDLUND, I.: Über die Ligninkondensation beim Sulfitkochprozeß. Svensk Papperstidn. **50**, no. 11 B, 109 (1947).

144. HESS, K. u. K. E. HEUMANN: Über Feinstvermahlung verholzter Zellwände und die Reaktionsfähigkeit des Lignins mit Hydrazin. Ber. dtsch. chem. Ges. **75**, 1802 (1942).

145. HEWSON, W. B. and H. HIBBERT: Studies on Lignin and Related Compounds. LXV. Re-ethanolysis of Isolated Lignin. J. Amer. chem. Soc. **65**, 1173 (1943).

146. HEWSON, W. B., J. L. MCCARTHY and H. HIBBERT: Studies on Lignin and Related Compounds. LVII. Mechanism of the Ethanolysis Reaction. J. Amer. chem. Soc. **63**, 3041 (1941).

147. — — — Studies on Lignin and Related Compounds. LVIII. The Mechanism of the Ethanolysis of Maple Wood at High Temperatures. J. Amer. chem. Soc. **63**, 3045 (1941).

148. HIBBERT, H.: The Structure of Lignin. Canad. J. Res., Sect. B **16**, 69 (1938).

149. — Studies on Lignin and Related Compounds. XXXVII. The Structure of Lignin and the Nature of Plant Synthesis. J. Amer. chem. Soc. **61**, 725 (1939).

150. — The Mechanism of Plant Respiration. J. Amer. chem. Soc. **62**, 984 (1940).

151. — Status of the Lignin Problem. Paper Trade J. **113**, no. 4, 35 (1941).

152. — Lignin. Ann. Rev. Biochem. **11**, 183 (1942).

153. HILL, A. C.: Recent Advances in the Utilization of Lignin in Waste Sulfite Liquor. Pulp Paper. Mag. Canada **41**, 148 (1940).

154. HILPERT, S. R. u. H. HELLWAGE: Buchenholz-Lignin, ein Reaktionsprodukt der Kohlehydrate bei der Ligninbestimmung. Ber. dtsch. chem. Ges. **68**, 380 (1935).

155. HILPERT, S. R. u. H. MEYBIER: Zusammenhänge zwischen den Bestimmungen der Pentosane und des Lignins. Ber. dtsch. chem. Ges. **71**, 1962 (1938).

156. HILPERT, S. R. u. J. PFÜTZENREUTER: Die Charakterisierung der pflanzlichen Zellwand durch Behandeln mit Kupferoxyd-Ammoniak-Lösung. Ber. dtsch. chem. Ges. **71**, 2220 (1938).

157. — — Die Einwirkung von Äthylendiaminkupferoxyd-Lösung auf Holz und Stroh. Ber. dtsch. chem. Ges. **72**, 607 (1939).

158. HILPERT, S. R., W. KRÜGER u. G. HECHLER: Die Einwirkung von Salpetersäure auf Hölzer. Ein Beitrag zur Chemie des Lignins. Ber. dtsch. chem. Ges. **72**, 1075 (1939).

159. HOLMBERG, B.: Ligninuntersuchungen. I. Über das Sulfitlaugenlacton. Ber. dtsch. chem. Ges. **54**, 2389 (1921).

160. — Ligninuntersuchungen. XVI. Fichtenholz und Thiohydracrylsäure. Ark. Kemi, Mineral. Geol. **21** (1945).

161. — Ligninuntersuchungen. XVII. Espenhölzer und Mercaptosäuren. Ark. Kemi, Mineral. Geol. **24** (1947).

162. — Lignin und Thioglykolsäure. Österr. Chemiker-Ztg. **43**, 152 (1940).

163. — Lignin. XV. Über Bromlaugenlignine. Ber. dtsch. chem. Ges. **75**, 1760 (1942).

164. — Lignin. XVIII. Die Alkoholyseprodukte des Fichtenholzes. Svensk Papperstidn. **50**, no. 11 B, 111 (1947).

165. HOLMBERG, G. u. N. GRALÉN: Die Stöchiometrie des Fichtenlignins. Ing. Vetensk. Akad., Handl. **162**, 29 pp. (1942).

166. HUNTER, M. J., A. B. CRAMER and H. HIBBERT: Studies on Lignin and Related Compounds. XXXVI. Ethanolysis of Maple Wood. J. Amer. chem. Soc. **61**, 516 (1939).

167. HUNTER, M. J. and H. HIBBERT: Studies on Lignin and Related Compounds. XLI. The Detection, Isolation and Estimation of the Syringyl Radical in Plant Products. J. Amer. chem. Soc. **61**, 2190 (1939).

168. — — Studies on Lignin and Related Compounds. XLIII. The Absence of the Piperonyl Group in the Lignin Structure. J. Amer. chem. Soc. **61**, 2196 (1939).

169. IWADARE, K.: Lignin. IV. Oxidation with Nitrobenzene. J. chem. Soc. Japan **62**, 1095 (1941).

170. JAHN, E. C.: General Utilization of Lignin and Wood Wastes. Paper Mill Wood Pulp News **63**, no. 23, 18 (1940).

171. — Utilization of Lignin. News Edit. (Amer. chem. Soc.) **18**, 993 (1940).

172. JAHN, E. C. and W. M. HARLOW: Chemistry of Ancient Beech Stakes from the Fishweir. The Boylston Street Fishweir: Papers of the Report of the Robert S. Peabody Foundation for Archaeology **2**, 90 (1942).

173. JAYME, G., L. ESER u. G. HANKE: Über die Entstehung von „Überschuß-substanz" beim Chloritaufschluß von Hölzern und ihre Bedeutung für die Chemie des Holzes und des „Lignins". Naturwiss. **31**, 274 (1943).

174. JAYME, G. u. G. HANKE: Über die Natriumchlorit-Oxydationsprodukte und die Konstitution des nativen Fichtenholzlignins. Cellulosechemie **21**, 127 (1943).

175. JAYME, G. u. M. HARDERS-STEINHÄUSER: Über eine Methode zum mikroskopischen Nachweis des Protolignins in pflanzlichen Zellwänden und ihre Anwendung auf Buchenholz. Holzforsch. **1**, 33 (1947).

176. JODL, R.: Feinstrukturuntersuchungen über Lignine. Brennstoff-Chem. **23**, 163 (1942).

177. JONES, G. M. and F. E. BRAUNS: Ethers of Certain Lignin Derivatives. Paper Trade J. **119**, no. 11, 108 (1944).

178. JUNKER, E.: Kolloidchemische Eigenschaften des Humus. Zur Dispersions-chemie des Lignins. Kolloid-Z. **95**, 213 (1941).

179. KAWAI, S. u. N. SUGIYAMA: Untersuchungen über Egonol. VII. Synthese der beiden Egonol-Abbauprodukte Dihydro-coniferylalkohol und Styraxinol-aldehyde. Zum Reaktionsmechanismus der Flavyliumsalz-Synthese. Ber. dtsch. chem. Ges. **72**, 369 (1939).

180. KAWAMURA, J.: A new Constituent of Tsuga Resin. Bull. Imp. Forestry Exp. Stat. Tokyo **31**, 73 (1932).

181. KEILEN, J. J.: Lignin as Detergent Ingredient. Soap Sanit. Chemicals **21**, 40, 146 (1945).

182. KEILEN, J. J. and A. POLLAK: Lignin for Reinforcing Rubber. Ind. Engng. Chem. **39**, 480 (1947).

183. KEIMATSU, S., T. ISHIGURO and G. YAMAMOTO: The Constituents of Resins. V. The Constitution of Tsuga Lacton (Tsuga resinols). J. pharmac. Soc. Japan **55**, 226 (1935).

184. KLEINERT, TH.: Über die Aufnahme von Phenol durch die Holzsubstanz und ihre Hauptbestandteile. Cellulosechemie **18**, 115 (1940).

185. KRATZL, K.: Über die Synthese von Modellsubstanzen für Ligninsulfosäuren. II. Über einige in der Seitenkette substituierte Propiophenonderivate. Ber. dtsch. chem. Ges. **76**, 895 (1943).

186. — Zur Biogenese des Lignins. Über das Lignin etiolierter Kartoffelkeimlinge. Experientia **4**, 110 (1948).

187. — Über die Konstitution der Ligninsulfosäure. Mh. Chem. **78**, 173 (1948).

188. — Die Ligninsulfosäure und ihre technische Verwertung. Experientia **2**, 469 (1946).

189. KRATZL, K. u. CH. BLECKMANN: Über die Bromierung von Ligninsulfosäure und deren Modellsubstanzen. Experientia **2**, 24 (1946).

190. — — Über die Bromierung der Ligninsulfosäure und deren Modellsubstanzen. Mh. Chem. **76**, 185 (1947).

191. KRATZL, K. u. H. DÄUBNER: Über die Sulfitkochung von Phenylpropanderivaten und Chalkonen. Ber. dtsch. chem. Ges. **77**, 519 (1944).

192. KRATZL, K., H. DÄUBNER u. U. SIEGENS: Über die Sulfitkochung von Phenylpropanderivaten. II. Mh. Chem. **77**, 146 (1947).

193. KRÜGER, D.: Die Gewinnung von Lignin und Ligninderivaten bei verschiedenen Verfahren des Holzaufschlusses. Zellstoff u. Papier **21**, 299 (1941); Holz als Roh- u. Werkstoff **5**, 36 (1942).

194. KRÜGER, W.: Die Einwirkung von Salpetersäure auf pflanzliche Samenschalen. Ber. dtsch. chem. Ges. **73**, 493 (1940).

195. KÜRSCHNER, K.: Azokohlenwasserstoffe aus Nitroligninen (Nitrohumussäuren) und aromatischen Aminen. Cellulosechemie **18**, 70 (1940).

196. — Über Kohlehydratverunreinigungen in isolierten Ligninen und Rohcellulosen aus Holz. Cellulosechemie **18**, 64 (1940).

197. — Schwankende Grundlagen der Ligninchemie? Cellulosechemie **21**, 141 (1943).

198. KÜRSCHNER, K. u. F. SCHINDLER: Lignin und Huminsubstanzen. Papierfabrikant **38**, 254 (1940).

199. — — Bemerkungen über die Trennung von Lignin und Humusstoffen. Papierfabrikant **38**, 34 (1940).

200. — — Neue Bemerkungen über Nitrolignine. Cellulosechemie **18**, 12 (1940).

201. KÜRSCHNER, K. u. K. WITTENBERGER: Nitrolignine verschiedener Holzarten bei der Halogenaufnahme in Vakuum. Cellulosechemie **18**, 21 (1940).

202. KULKA, M., H. E. FISHER, S. B. BAKER and H. HIBBERT: Studies on Lignin and Related Compounds. LXXVII. Re-investigation of the Ethanolysis Products of Maple Wood. J. Amer. chem. Soc. **66**, 39 (1944).

203. KULKA, M., W. L. HAWKINS and H. HIBBERT: Studies on Lignin and Related Compounds. LIII. Isolation of Vanilloyl and Syringoyl Methyl Ketones from Ethanolysis Products of Maple Wood. J. Amer. chem. Soc. **63**, 2371 (1941).

204. Kulka, M. and H. Hibbert: Studies on Lignin and Related Compounds. LXVII. Isolation and Identification of 1-(4-Hydroxy-3,5-dimethoxyphenyl)-2-propanone and 1-(4-Hydroxy-3-methoxyphenyl)-2-propanone from Maple Wood Ethanolysis Products. Metabolic Changes in Lower and Higher Plants. J. Amer. chem. Soc. **65**, 1180 (1943).

205. — — Studies on Lignin and Related Compounds. LXVIII. Synthesis and Properties of 1-Ethoxy-1-(4-hydroxy-3-methoxyphenyl)-2-propanone, 3-Ethoxy-1-(4-hydroxy-3-methoxyphenyl)-2-propanone, and their Methyl Ethers. J. Amer. chem. Soc. **65**, 1185 (1943).

206. Lange, P. W.: The Nature and Distribution of Lignin in Spruce Wood. Svensk Papperstidn. **47**, 262 (1944).

207. — The Ultraviolet Absorption of Solid Lignin. Svensk Papperstidn. **48**, 241 (1945).

208. — Some Views on the Lignin in the Woody Fiber during the Sulfite Cook. Svensk Papperstidn. **50**, no. 11 B, 130 (1947).

209. Larson, L. L.: Nature of Lignin Residues in Unbleached and Partially Bleached Sulfite Pulp. Paper Trade J. **113**, no. 21, 25 (1941).

210. Larsson, A.: Ligninsulfonsäuren aus Ablaugen von Untersuchungen über Sulfitablaugen. V. Espenholz-Sulfit-Kochungen. Svensk Papperstidn. **46**, 93 (1943).

211. Lautsch, W.: Neuere Ergebnisse der Ligninforschung. Brennstoff-Chem. **22**, 265 (1941).

212. — Über die oxydative und hydrierende Destruktion des Holzes, des Lignins und der schwefelhaltigen Ablaugen der Fichte. Cellulosechemie **19**, 69 (1941).

213. — Über Ionenaustauscher auf Lignin-Basis. Die Chemie **57**, 149 (1944).

214. Lautsch, W. u. G. Piazolo: Über den oxydativen Abbau halogensubstituierter Fichtenlignine. Ber. dtsch chem Ges. **73**, 317 (1940).

215. — — Über die Hydrierung von Lignin und ligninhaltigen Stoffen mit wasserstoffabgebenden Mitteln, insbesondere Alkoholen. Ber. dtsch. chem. Ges. **76**, 486 (1943).

216. Ledingham, G. A. and G. A. Adams: Biological Decomposition of Chemical Lignin. II. Studies on the Decomposition of Calcium Lignosulfonate by Wood Destroying and Soil Fungi. Canad. J. Res., Sect. C **20**, 13 (1942).

217. Lewis, H. F., F. E. Brauns, M. A. Buchanan and E. B. Brookbank: Lignin Esters of Mono- and Di-basic Aliphatic Acids. Ind. Engng. Chem. **35**, 1113 (1943).

218. Lieff, M., F. G. Wright and H. Hibbert: Studies on Lignin and Related Compounds. XXXVIII. The Effect of Solvents in the Grignard Analysis for Active Hydrogen and Carbonyl. J. Amer. chem. Soc. **61**, 865 (1939).

219. — — — Studies on Lignin and Related Compounds. XL. The Extraction of Birch Lignin with Formic Acid. J. Amer. chem. Soc. **61**, 1477 (1939).

220. Loughborough, D. L. and A. J. Stamm: Molecular properties of Lignin Solutions from Viscosity, Osmotic Pressure, Boiling Point Raising, Diffussion and Spreading Measurements. J. physic. Chem. **40**, 1113 (1936).

221. — — The Molecular Properties of Lignin Solutions. J. physic. Chem. **45**, 1137 (1941).

222. Lovell, E. L. and H. Hibbert: Studies on Lignin and Related Compounds. LII. New Method for the Fractionation of Lignin and Other Polymers. J. Amer. chem. Soc. **63**, 2070 (1941).

223. Lovin, R. J. and L. Friedman: The Catalytic Hydrogenation of Sulfonated Lignin. Pacific Pulp Paper Ind. **16**, no. 7, 23 (1942).

224. MAC GREGOR, W. S., T. H. EVANS and H. HIBBERT: Studies on Lignin and Related Compounds. LXXVIII. Chromic Acid Oxidation of Lignin-type Substances. Wood Ethanolysis Products and Wood. J. Amer. chem. Soc. 66, 41 (1944).

225. MAC INNES, A. S., E. WEST, J. L. MCCARTHY and H. HIBBERT: Studies on Lignin and Related Compounds. XLIX. Occurrence of the Guaiacyl and Syringyl Groupings in the Ethanolysis Products from Various Plants. J. Amer. chem. Soc. 62, 2803 (1940).

226. MCCARTHY, J. L.: The Chemical Structure of Lignin and of Wood; Current Knowledge. TAPPI, Empire State Section, 1939/40.

227. MCNAIR, J. J. and E. C. JAHN: Properties of Lignin Esters. Paper Trade J. 117, no. 8, 29 (1943).

228. MITCHELL, L., T. H. EVANS and H. HIBBERT: Studies on Lignin and Related Compounds. LXXXI. Properties of 1-Bromo-1-(4-acetoxy-3-methoxyphenyl)-2-propanone and Relation to Lignin Structure. J. Amer. chem. Soc. 66, 604 (1944).

229. MITCHELL, L. and H. HIBBERT: Studies on Lignin and Related Compounds. LXXX. The Ethanolysis of 1-Acetoxy-1-(4-acetoxy-3-methoxyphenyl)-2-propanone and its Relation to Lignin Structure. J. Amer. chem. Soc. 66, 602 (1944).

230. MOERKE, G. A.: Lignin Color Reactions with Amino Compounds. J. org. Chemistry 10, 42 (1945).

231. MOTTET, A. L.: The Sulfonation of Western Hemlock Lignin. Pacific Pulp Paper Ind. 13, no. 10, 22 (1939).

232. MÜLLER, A. u. H. HORVATH: Die Phenylhydrindenstruktur des Diisoeugenol und Diisohomogenol (Bis-[propenylphenoläther]). III. Ber. dtsch. chem. Ges. 76, 855 (1943).

233. MÜLLER, O. A.: Lignin und Humin. Papierfabrikant 37, 237 (1939).

234. NORD, F. F. and J. C. VITUCCI: On the Mechanism of Enzyme Action. XXX. The Formation of Methyl *p*-Methoxycinnamate by the Action of *Lentinus lepideus* on Glucose and Xylose. Arch. Biochem. 14, 243 (1947).

235. — — On the Mechanism of Enzyme Action. XXXI. The Mechanism of Methyl *p*-Methoxycinnamate Formation by *Lentinus lepideus* and its Significance in Lignification. Arch. Biochem. 15, 465 (1947).

236. OGAIT, A.: Über den Aufschluß von Fichtenholz mit Chloralhydrat und eine Chloralverbindung des Lignins. Cellulosechemie 22, 15 (1944).

237. OVERBECK, W. u. H. F. MÜLLER: Über die Hydrolyse verschiedener Hölzer mit Wasser unter Druck und die damit verbundene Veränderung der Holzbestandteile, insbesondere des Lignins. Ber. dtsch. chem. Ges. 75, 547 (1947).

238. PALLMANN, H.: Dispersoidchemische Probleme in der Humusforschung. Kolloid-Z. 101, 72 (1942).

239. PATTERSON, R. F. and H. HIBBERT: Studies on Lignin and Related Compounds. LXXII. The Ultraviolet Absorption Spectra of Compounds Related to Lignin. J. Amer. chem. Soc. 65, 1862 (1943).

240. — — Studies on Lignin and Related Compounds. LXXIII. The Ultraviolet Absorption Spectra of Ethanol Lignins. J. Amer. chem. Soc. 65, 1869 (1943).

241. PATTERSON, R. F., K. A. WEST, E. L. LOVELL, W. L. HAWKINS and H. HIBBERT: Studies on Lignin and Related Compounds. LI. The Solvent Fractionation of Maple Ethanol Lignin. J. Amer. chem. Soc. 63, 2065 (1941).

242. PAULY, H.: Gewinnung huminfreier Essigsäure-Lignole. Ber. dtsch. chem. Ges. 76, 864 (1943).

243. — Scheidung von Lignin-Komponenten. Ber. dtsch. chem. Ges. 67, 1188 (1934).

244. Pearl, I. A.: 6-Chlorovanillin from the Chlorite Oxidation of Lignin. J. Amer. chem. Soc. **68**, 916 (1946).

245. Pearl, I. A., A. Bailey and H. K. Benson: Studies on the Desulfonation of Calcium Lignosulfonate with Lime. Paper Trade J. **113**, no. 17, 47 (1941).

246. Pearl, I. A. and H. K. Benson: Preparation of Desulfonated Lignin from Calcium Lignosulfonate. Paper Trade J. **111**, no. 19, 29 (1940).

247. Pedersen, J. H. and H. K. Benson: Chemical Derivatives of Lignin. Chlorolignin. Pacific Pulp Paper Ind. **14**, 48 (1940).

248. Peniston, Q. P., J. L. McCarthy and H. Hibbert: The Reconversion of an "Extracted" Lignin into its Primary Building Units. J. Amer. chem. Soc. **61**, 530 (1939).

249. — — — Studies on Lignin and Related Compounds. L. Fractionation of Acetylated Cell Wall Constituents of Red Oak Wood. J. Amer. chem. Soc. **62**, 2284 (1940).

250. Pennington, D. and E. M. Ritter: Oxidation of Lignosulfonic Acids by Periodic Acid. J. Amer. chem. Soc. **68**, 1391 (1946).

251. — — A Diffusion Study of Lignin Sulfonic Acids in Sulfite Waste Liquor. J. Amer. chem. Soc. **69**, 665 (1947).

252. Pepper, J. M.: Isolation and Utilization of Lignosulfonic Acids. Pulp Paper Mag. Canada **46**, 83 (1945).

253. Pepper, J. M. and H. Hibbert: Studies on Lignin and Related Compounds. LXXXVII. High Pressure Hydrogenation of Maple Wood. J. Amer. chem. Soc. **70**, 67 (1948).

254. Percival, E. G. V.: The Lignin Problem. Ann. Rep. Progr. Chem. **39**, 142 (1943).

255. Perrenoud, H.: Zur Kenntnis der kolloidchemischen Eigenschaften des Humus. Dioxanextraktion und Dispersitätschemie des Fichtenlignins. Kolloid-Z. **107**, 16 (1944).

256. Phillips, M.: Lignin as Constituent of Nitrogen-free Extract. J. Assoc. off. agric. Chemists **23**, 108 (1940).

257. Phillips, M., M. J. Goss, B. L. Davis and H. Stevens: Composition of the Various Parts of the Oat Plant at Successive Stages of Growth, with Special Reference to the Formation of Lignin. J. agric. Res. **59**, 319 (1939).

258. Ploetz, Th.: Über den enzymatischen Abbau polymerer Kohlehydrate. IV. Vergleichender enzymatischer Abbau einiger isolierter Holzbestandteile mit Schneckenferment. Ber. dtsch. chem. Ges. **73**, 57 (1940).

259. — Über den enzymatischen Abbau polymerer Kohlehydrate. V. Enzymatischer Abbau einiger nativer ligninhaltiger Materialien. Ber. dtsch. chem. Ges. **73**, 61 (1940).

260. — Über den Abbau enzymatischer polymerer Kohlehydrate. VI. Über den Bindungszustand des Lignins im Holz. Ber. dtsch. chem. Ges. **73**, 74 (1940).

261. — Über den enzymatischen Abbau polymerer Kohlehydrate. VII. Weitere Fraktionierungsversuche an Lindenholz und enzymatischer Abbau der Fraktionen. Ber. dtsch. chem. Ges. **73**, 790 (1940).

262. — Beiträge zur Ligninbestimmung mit starker Schwefelsäure. Cellulosechemie **18**, 49 (1940).

263. Plungian, M.: Meadol Lignin and its Use in Plastics. Sixth Annual Natl. Farm Chem. Conference, Chicago. 1940.

264. — Preparation and Properties of Meadol, a Pure Alkali Lignin from Soda Black Liquor. Ind. Engng. Chem. **32**, 1399 (1940).

265. Powter, N. B.: Canada Develops a New Form of Lignin. Mod. Plastics **24**, 106 (1947); Paper Ind. Paper Wld. **28**, 1744 (1947); Brit. Plast. mould. Prod. Trader **19**, 215 (1947).

266. PYLE, J. J., L. BRICKMAN and H. HIBBERT: Studies on Lignin and Related Compounds. XLIV. The Ethanolysis of Maple Wood; Separation and Identification of the Water-soluble Aldehyde Constituents. J. Amer. chem. Soc. 61, 2198 (1939).

267. RACKY, G.: Zur Kenntnis der Sulfitablauge. I. Papierfabrikant 39, 121 (1941); II. Cellulosechemie 20, 22 (1942).

268. REFLE, K.: Lignin als Rohmaterial. I. u. II. Chemiker-Ztg. 65, 267, 276 (1941).

269. REID, E. E., H. WORTHINGTON and A. W. LARCHER: The Action of Caustic Alkali and of Alkaline Salts on Alcohols. J. Amer. chem. Soc. 61, 99 (1934).

270. REID, J. D., E. C. DRYDEN and S. I. ARONOVSKY: Effect of Ethanolamine upon Corncob Lignin. A Preliminary Investigation. Paper Trade J. 113, no. 7, 27 (1941).

271. RICHTZENHAIN, H.: Die Spaltung von Ätherbindungen mit Bisulfit und Thioglykolsäure. Modelle zur Chemie des Lignins. Ber. dtsch. chem. Ges. 72, 2152 (1942).

272. — Vergleichende Oxydationsversuche an Vanillin und Lignin. Ber. dtsch. chem. Ges. 75, 269 (1942).

273. — Enzymatische Versuche zur Entstehung des Lignins. II. Die Dehydrierung des 5-Methyl-pyrogallol-1,3-dimethyläthers. Ber. dtsch. chem. Ges. 77, 409 (1944).

274. RITCHIE, P. F. and C. B. PURVES: Periodate Lignins; their Preparation and Properties. Pulp Paper Mag. Canada 48, no. 12, 74 (1947).

275. RITTER, D. M., D. E. PENNINGTON, E. D. OLLEMAN, K. A. WRIGHT and T. F. EVANS: Constitution of Gymnosperm Lignin. Science [New York] 107, 20 (1948).

276. RÜDIGER, W.: Der Bau verholzter Membranen und ihr Verhalten in flüssigem Fluorwasserstoff. Papierfabrikant 38, 9 (1940).

277. RUPE, H., A. ACKERMANN u. H. TAKAGI: Die Reduktionsprodukte des Oxymethylencamphers. Helv. chim. Acta 1, 453 (1918).

278. RUSSELL, A.: Interpretation of Lignin: the Synthesis of Gymnosperm Lignin. Science [New York] 106, 372 (1947).

279. SAEMAN, J. F. and E. E. HARRIS: Hydrogenation of Lignin over RANEY Nickel. J. Amer. chem. Soc. 68, 2507 (1946).

280. SAMUELSON, O.: The Fractionation of Sulfite Waste Liquor. Svensk Papperstidn. 45, 516 (1942).

281. SCHENCK, A.: The Nature of Lignosulfonic Acids Fractionated by Chemical and Physical Methods. Paper Trade J. 117, no. 14, 97 (1943).

282. SCHMIDT-NIELSEN, S. u. J. HÖYE: Zur Kenntnis der Resistenz des Kiefern-Lignins. Svensk kem. Tidskr. 53, 287 (1941).

283. SCHÜTZ, FR.: Über den Aufschluß von Faserrohstoffen mit wasserhaltiger Monochloressigsäure. Cellulosechemie 18, 76 (1940).

284. — Untersuchung der beim Holzaufschluß mit wasserhaltiger Chloressigsäure entstehenden Lignins. Cellulosechemie 19, 87 (1941).

285. — Über den Aufschluß von Faserrohstoffen mit Chlorhydrinen. Cellulosechemie 19, 33 (1941).

286. SCHÜTZ, FR. u. W. KNACKSTEDT: Holzaufschluß mit Salzsäure oder Chloriden als Katalysator in essigsaurer Lösung. Cellulosechemie 20, 15 (1942).

287. SCHÜTZ, FR. u. P. SARTEN: Beiträge zur Holzchemie. Über die Bildung von Lignin aus Holz und Holzextrakten und die Oxydation der Zucker, der Cellulose, des Holzes und der Holzextrakte mit Diazoverbindungen zu Säuren der Zuckergruppe. Cellulosechemie 21, 35 (1943).

288. Schütz, Fr. u. P. Sarten: Beiträge zur Holzchemie. II. Cellulosechemie **22**, 1 (1944).

289. — — Beiträge zur Holzchemie. II. (Nachtrag.) Über das optische Verhalten des Holzes und des Lignins. Cellulosechemie **22**, 114 (1944).

290. Schütz, Fr., P. Sarten u. H. Meyer: Beiträge zur Holzchemie. III. Weitere Versuche zur Ligninfrage und vollständigen Auflösung des Holzes. Holzforsch. **1**, 2 (1947).

291. Schwabe, K. u. E. Hahn: Über ein neues Verfahren zur Fraktionierung von Ligninsulfonsäuren aus Sulfitablauge nach ihrer Molekülgröße. Holzforsch. **1**, 42 (1947).

292. Schwabe, K. u. L. Hasner: Molekularbestimmung an Ligninsulfonsäuren durch Dialyse. Cellulosechemie **20**, 61 (1942).

293. Schwartz, H.: The Utilization of Lignin in Plastics. Pulp Paper Mag. Canada **45**, 675 (1944).

294. Seiberlich, J.: Fundamentals of Lignin Chemistry as Applied to Fertilizers. Northeast. Wood Util. Council, Bull. no. 7, 94 (1945).

295. — Utilization of Lignin by Zinc Salt Treatment. Chem. Industries **56**, 53 (1945).

296. Skraup, Z. H. u. J. König: Über Cellose, eine Biose aus Cellulose. Ber. dtsch. chem. Ges. **34**, 1115 (1901).

297. Shrikhande, J. G.: Estimation of Lignin in Tannin Materials. Biochemic. J. **34**, 783 (1940).

298. Spencer, E. Y. and G. F. Wright: The Action of Diazomethane on Lactones and on Lignins. J. Amer. chem. Soc. **63**, 2017 (1941).

299. Steeves, W. H. and H. Hibbert: Studies on Lignin and Related Compounds. XLII. The Isolation of a Bisulfite Soluble "Extracted Lignin". J. Amer. chem. Soc. **61**, 2194 (1939).

300. Sugii, Y.: The Constituents of the Bark of *Mongolia officinalis* Rhed. et Wils. and *Mongolia obovata* Thumb. J. pharmac. Soc. Japan **50**, 183 (1930).

301. Suida, H. u. Y. Prey: Über den Aufschluß von Säure-Lignin. Ber. dtsch. chem. Ges. **74**, 1916 (1941).

302. — — Über den Aufschluß von Säure-Lignin. Ber. dtsch. chem. Ges. **75**, 1580 (1942).

303. Szent-Györgyi, A.: Über Zellatmung. Ber. dtsch. chem. Ges. **72 A**, 53 (1939).

304. Vanzetti, B. L.: Lignin and Resins. Atti V. Congr. Chim. appl. (II) **5**, 932 (1936).

305. Virasoro, E.: Absorption Spectra of Lignin in the Ultraviolet. An. Asoc. quím. argent. **30**, 54 (1942).

306. — Extraction of Lignin from White Quebracho Wood by Ethyl Acetoacetate and Phenol. An. Asoc. quím. argent. **30**, 54 (1942).

307. Wacek, A. V.: Über die Synthese von Modellsubstanzen für die Ligninsulfonsäuren. III. Synthese von Veratrylaceton (3,4-Dimethoxy-phenylaceton) und Guajacylaceton (4-Oxy-3-methoxy-phenylaceton) und deren α-Sulfonsäuren. Ber. dtsch. chem. Ges. **77**, 85 (1944).

308. — Der chemische Aufbau des Holzes. Experientia **2**, 171 (1946).

309. Wacek, A. V. u. I. Horak: Über die Konstitution der synthetischen und durch Äthanolyse aus dem Holz isolierbaren Ketole vom Typus des Oxypropioguaiacons. Mh. Chem. **77**, 18 (1947).

310. Wacek, A. V. u. K. Kratzl: Über Oxydation verschieden substituierter aliphatischer Seitenketten mit Natronlauge und Nitrobenzol. Äthanolysenversuche an synthetischen Sulfonsäuren. Cellulosechemie **20**, 108 (1942).

311. WACEK, A. V. u. K. KRATZL: Über die Oxydation verschieden substituierter aliphatischer Seitenketten in Modellsubstanzen für die Ligninbausteine mit Natronlauge und Nitrobenzol. II. Ber. dtsch. chem. Ges. **76**, 891 (1943).

312. — — Über die Oxydation verschieden substituierter aliphatischer Seitenketten in Modellsubstanzen für die Ligninbausteine mit Natronlauge und Nitrobenzol. III. Ber. dtsch. chem. Ges. **77**, 516 (1944).

313. — — Modellversuche zum Ligninproblem. Österr. Chemiker-Ztg. **48**, 36 (1947).

314. WACEK, A. V., K. KRATZL u. A. V. BÉZARD: Über die Synthese von Modellsubstanzen für die Ligninsulfonsäuren. Synthese von α-Phenylaceton-α-sulfonsäure und Propioveratron-α-sulfonsäure. Ber. dtsch. chem. Ges. **75**, 1348 (1942).

315. WACEK, A. V. u. E. NITTNER: Über das Vorkommen von substituierten Cumaronen im Buchenholzteer und deren Beziehung zum Lignin. Cellulosechemie **18**, 29 (1940).

316. WACEK, A. V. u. A. SCHÖN: Untersuchungen zur Frage der Zusammensetzung von Baumrinden. Holz als Roh- u. Werkstoff **4**, 18 (1941).

317. WAGNER, K.: Die Verwertung des Lignins. Vierjahresplan **5**, 924 (1941); Wbl. Papierfabrikat. **73**, 341 (1942).

318. WALD, W. J., P. F. RITCHIE and C. B. PURVES: The Elementary Composition of Lignin in Northern Pine and Black Spruce Woods and of the Isolated KLASON and Periodate Lignins. J. Amer. chem. Soc. **69**, 1371 (1947).

319. WEDEKIND, E.: Aufschluß der Holzarten durch organische Lösungsmittel: die Lignindarstellung als Vorlesungsexperiment. Forstarch. **11**, 53 (1935).

320. WEST, K. A., W. L. HAWKINS and H. HIBBERT: Studies on Lignin and Related Compounds. LV. Synthesis and Properties of β-Hydroxypropioveratrone. J. Amer. chem. Soc. **63**, 3035 (1941).

321. — — — Studies on Lignin and Related Compounds. LVI. Stability of Lignin Building Units and Ethanol Lignin Fractions toward Ethanolic Hydrogen Chloride. J. Amer. chem. Soc. **63**, 3038 (1941).

322. WEST, K. A. and H. HIBBERT: Studies on Lignin and Related Compounds. LXIV. Synthesis and Properties of 3-Hydroxy-1-(4-hydroxy-3-methoxyphenyl)-1-propanone. J. Amer. chem. Soc. **65**, 1170 (1943).

323. WEST, E., W. S. MAC GREGOR, T. H. EVANS, I. LEVI and H. HIBBERT: Studies on Lignin and Related Compounds. LXVI. The Ethanolysis of Maple Wood. J. Amer. chem. Soc. **65**, 1176 (1943).

324. WEST, E., A. S. MAC INNES and H. HIBBERT: Studies on Lignin and Related Compounds. LXIX. Isolation of 1-(4-Hydroxy-3-methoxyphenyl)-2-propanone and 1-Ethoxy-1-(4-hydroxy-3-methoxyphenyl)-2-propanone from the Ethanolysis of Spruce Wood. J. Amer. chem. Soc. **65**, 1187 (1943).

325. WEST, E., A. S. MAC INNES, J. L. McCARTHY and H. HIBBERT: Occurrence of the Syringyl Radical in Plant Products. J. Amer. chem. Soc. **61**, 2556 (1939).

326. WHITE, E. V., J. N. SWARTZ, Q. P. PENISTON, H. SCHWARTZ, J. L. McCARTHY and H. HIBBERT: Mechanism of the Chlorination of Lignin. Techn. Assoc. Pap. **24**, 179 (1941).

327. WIECHERT, K.: Darstellung und Eigenschaften des Rotbuchenlignins. Papierfabrikant **37**, 325, 339 (1939).

328. — Über die Ligninfarbreaktionen des Rotbuchenholzes. Papierfabrikant **37**, 17, 30 (1939).

329. — Eine neue Methode zur Bestimmung des Ligningehaltes von Holz und Papier mittels wasserfreiem Fluorwasserstoff. Cellulosechemie **18**, 57 (1940).

330. — Neuere Methoden der präparativen organischen Chemie. II. Verwendung von Fluorwasserstoff für organisch-chemische Reaktionen. Die Chemie **56**, 333 (1943).

331. Wise, L. E. and E. K. Ratliff: Summary Analysis of Quebracho Wood. Trop. Woods **91**, 40 (1947).
332. Yorston, F. H.: Recent Advances in Chemistry of Wood. Pulp Paper. Mag. Canada **46**, 276, 278, 280 (1945).
333. Zechmeister, L. u. G. Tóth: Zur Kenntnis der Hydrolyse von Cellulose und der dabei auftretenden Zwischenprodukte. Ber. dtsch. chem. Ges. **64**, 854 (1931).

(Received, May 3, 1948.)

The Chemistry of the Constituents of Toad Venoms.

By V. DEULOFEU, Buenos Aires.

Introduction.

Toads have two glands along their neck known as "parotid" glands, the size of which varies with the species. Small similar glands are distributed in the skin. When the animal under certain conditions is excited, these glands secrete a viscous liquid, termed toad poison or toad venom. It can also be obtained by pressing the glands. The amount of venom produced varies with the size of the glands. CHEN and CHEN (2) have prepared from 15 mg. to 580 mg. per animal in twelve different species of toads. From *Bufo paracnemis*, one of the largest species known, we have obtained 1,2–1,4 g. WIELAND and BEHRINGER (67) have reported a difference in the amount of venom yielded by the animal depending on the sex. While from *B. vulgaris* males 31 mg. of fresh secretion was obtained, 64 mg. was produced by the females.

The fresh secretion contains 50–59% water. Besides water the venom consists of an inert, protein-like mass which does not seem to have been studied by modern methods, and a small amount of various substances, some being specific for the secretion and most of them having pharmacological effects.

These substances are: neutral products named *bufagins*, similar to the cardiac aglucones occurring in some plants; the nitrogenous components named *bufotoxins*, yielding on hydrolysis a bufagin-like substance which can be considered a genin; furthermore, arginine, and suberic acid; adrenaline and other bases with indolic structure, specific for the toad poison; finally, sterols.

The pharmacological activity of the venoms derives principally from its content of bufagins, bufotoxine, adrenaline and the indolic bases mentioned.

For chemical studies the dried secretion has been employed in many instances; however, in others the extract of whole toad skins has been used. Some difference may be found not only in the quantity but also in the quality of the respective products, depending on the source.

Ch'an Su (Senso). Ch'an Su is the name given in China to the dried secretion of the regional toad. The name *Senso* is employed by Japanese workers to designate the same material. *Ch'an Su* is employed in the popular Chinese medicine for the treatment of canker sores, local inflammations, toothache, sinusitis and hemorrhages of the gums. This secretion is obtained by expression of the glands or by introducing garlic or pepper into the mouth of the toad, thus, causing secretion which can be easily collected (9).

Chen and Chen (6) assumed that *Ch'an Su* may be prepared from the venom of *B. bufo gargarizans*, one of the common species in China, since the principles isolated from *Ch'an Su* and from fresh secretions of this species were found to be identical. Many bufagins have since been isolated from *Ch'an Su*. One was identified as gamabufagin which was also isolated from *B. vulgaris formosus*. Others are closely related to cinobufagin, and the remaining ones are amorphous substances of which only some derivatives are known in crystalline form.

I. The Bufagins.

The name "bufagin" is used in this review, in a broad sense, to designate neutral, non-nitrogenous constituents of toad venoms. The term "genin" will also be used, thus extending to the toad poisons a name which has been accepted for a particular group of substances obtained by hydrolysis of some natural products. Bufagins are considered as "genins" because the bufotoxins yield on hydrolysis compounds of the bufagin type, and since we know that in certain cases a bufagin is present in combined form in the bufotoxin molecule.

"Bufagin" was the name originally given by Abel and Macht (1) to the crystalline neutral product isolated from the venom of *B. marinus*. It is now usual to prefix the name of a bufagin with the name of the sp cies of the toad from which the venom originates and, thus, the "bufagin" just mentioned should be termed "marinobufagin". Some bufagins have older, special names, and it seems preferable to keep them. Table 1 includes all bufagins which have been characterized to date, although some of them are of doubtful purity.

The bufagins are soluble in chloroform, methanol and ethanol, less so in ethyl acetate or ether but insoluble in petroleum ether. When crystalline, they are insoluble in water. Wieland, Hesse and Hüttel (69) have obtained a mixture of bufagins in a yield of 2 mg. per toad; and Wieland and Behringer (67) isolated 0,55 mg. pure bufotalin per male toad, and 1,23 mg. per female. One hundred and fifty mg. of a bufagin mixture was obtained per gram of dried secretion (69). From

Table 1. Bufagins.

Name	Toad species	Formula	M. p.	References	Acetylated bufagin, M. p.	References
Bufotalin	B. vulgaris, B. vulgaris formosus	$C_{26}H_{36}O_6$ (a)	223°	(68, 18)	255–257°	(74, 48)
Bufotalidin	B. vulgaris	$C_{24}H_{32}O_6$	228°	(68)	248°	(69)
Bufotalinin	B. vulgaris	$C_{24}H_{30}O_6$	233°	(68)	247°	(69)
Marinobufagin	B. marinus, B. paracnemis	$C_{24}H_{32}O_5$	212–213°	(1, 20, 58)	203–204°	(19, 37)
Gamabufagin	B. vulgaris formosus, Senso	$C_{24}H_{34}O_5$	261–265°	(12, 47, 73)	251–252°	(47, 49)
F₃-Bufotalin	B. vulgaris formosus	$C_{24}H_{32}O_5$	243–245°	(44)		
F₁-Bufotalin	B. vulgaris formosus	Amorphous (b)	—	(43)		
F₂-Bufotalin	B. vulgaris formosus	Amorphous (c)	—	(43)		
Cinobufagin ("new")	B. bufo gargarizans, Senso	$C_{26}H_{34}O_6$ (a)	212–213°	(6, 51)	196–197°	(51)
Cinobufotalin	Senso	$C_{26}H_{36}O_7$ (a)	248–249°	(51)	219–220°	(51)
Cinobufotalidin	Senso	$C_{24}H_{34}O_6$	217°	(45)		
Bufalin	Senso	$C_{24}H_{34}O_4$	235–236°	(53)	229–231°	(53)
Pseudo-desacetyl-bufotalin	Senso	Amorphous (d)	—	(43)		
Arenobufagin (arenobufogenin)	B. arenarum	$C_{24}H_{32}O_6$	252°	(67)		
Arenobufagin	B. arenarum	$C_{24}H_{32}O_6$	231–232°	(19)	246–248°	(21, 24)
Arenobufagin	B. arenarum	$C_{25}H_{34}O_6$ (a, e)	220°	(11)	162–163°	(32)
Regularobufagin	B. regularis	$C_{25}H_{34}O_6$ (e)	235–236°	(32)	224–225°	(32)
Vallicepobufagin	B. valliceps	$C_{26}H_{38}O_6$ (e)	212–213°	(5)		
Viridobufagin	B. viridis viridis	$C_{23}H_{34}O_5$ (e)	255°	(10)	253–254°	(10)
Fowlerobufagin	B. fowleris	$C_{23}H_{33}O_6$ (b, e)	153°	(5)		
Quercicobufagin	B. quercicus	$C_{23}H_{34}O_5$ (e)	258–259°	(5)		
Melanosticobufagin A	B. melanostictus	$C_{23}H_{30}O_6$ (e)	243–245°	(27)		
Melanosticobufagin B	B. melanostictus	$C_{29}H_{42}O_7$ (e)	173–175°	(27)		

(a) Contains one acetyl group.
(b) p-nitrobenzoate, m. p. 171–178°.
(c) p-nitrobenzoate, m. p. 286–287°.
(d) See p. 252.
(e) Bufagins with a number of carbon atoms different from 24 or 26 (with an acetyl group) are of doubtful purity.

16*

one gram of dried venom of *B. marinus*, Deulofeu and Mendive (20) isolated 64 mg. of pure marinobufagin.

All the bufagins so far studied in detail are derived from a typical cyclopentano-phenanthrene skeleton (I) whose spatial configuration is unknown; a six-membered unsaturated lactone ring is linked to carbon atom no. 17.

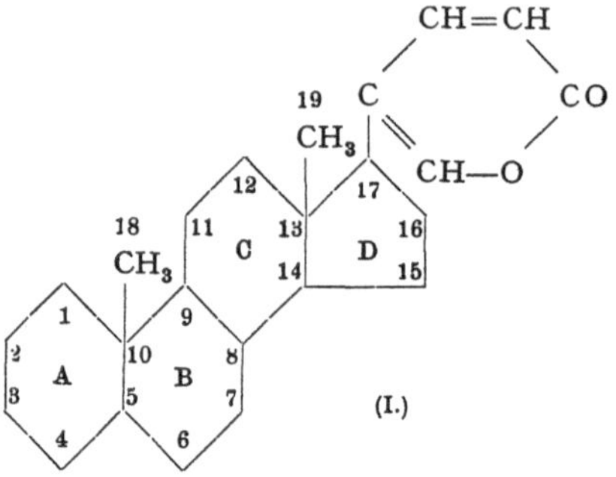

(I.)

In the cyclopentano-phenanthrene structure hydroxyls, and in many cases double bonds are found. In some bufagins one hydroxyl is acetylated. The non-acetylated bufagins belong to the C_{24}-series, and the mono-acetylated ones to the C_{26}-series.

The existence of a cholane structure in the bufagins was experimentally supported by Wieland, Hesse, and Meyer (70); it will be discussed later. They found that bufotalin can be converted into an iso-bufo-cholanic acid, $C_{24}H_{30}O_2$, which is isomeric with the cholanic acids. On dehydrogenation with selenium, bufagins and their derivatives produce methylcyclopentano-phenanthrene. Tschesche and Offe (64) as well as Jensen (33) obtained the latter compound from cinobufagin (m. p. 223°) and Ikawa (31) prepared it starting from pseudobufotalin. These observations were confirmed by Jensen (34) employing marinobufagin. Only impure chrysene was obtained from bufotalin by Wieland and Hesse (68).

The existence of a lactone ring in bufagins was postulated in their first paper by Wieland and Weil (74) who observed that an acid was produced by treating bufotalin with alkali. That this ring constitutes an enolic lactone with two conjugated double bonds which are also in conjugation to a carboxyl group was suggested almost simultaneously by Wieland, Hesse, and Hüttel (69), and by Tschesche and Offe (65). Both groups found that bufagins have an absorption spectrum with a maximum at 300 mμ., which is expected for a doubly unsaturated, conjugated lactone. As Tschesche pointed out, the methyl ester of coumalinic acid (Ia) with a similar structure also shows a maximum at 290–300 mμ.

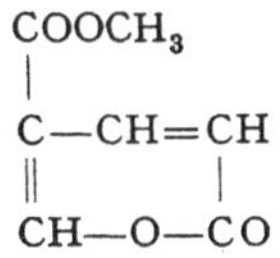

$$COOCH_3$$
$$|$$
$$C—CH=CH$$
$$||\qquad\quad|$$
$$CH—O—CO$$

(I a.) Coumalinic acid methyl ester.

A similar type of lactone had been found in scillaren A by STOLL and his collaborators (61–63), and some of the reactions which these investigators described were eventually applied to the bufagins.

Treatment with methanolic alkali opens the lactone ring (II) giving the methyl ester of an enolic potassium salt (III) which on acidification yields the free ester (IV).

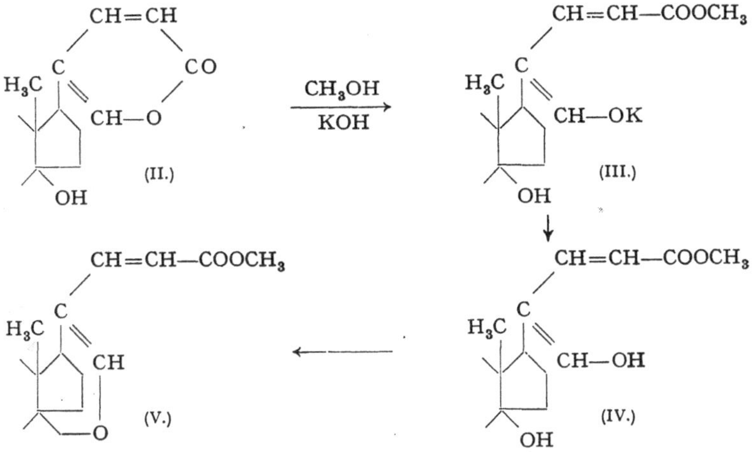

In the case of bufotalinin the methyl ester was obtained in crystalline form (69); it reduced TOLLENS' reagent in pyridine showing the presence of an aldehyde group.

In the molecules of cinobufagin or marinobufagin the lactone ring opens under the action of methanolic barium hydroxide (65). The new product gave a strong reaction for aldehydes with SCHIFF's reagent and a positive reaction with ferric chloride, showing the presence of an enol group.

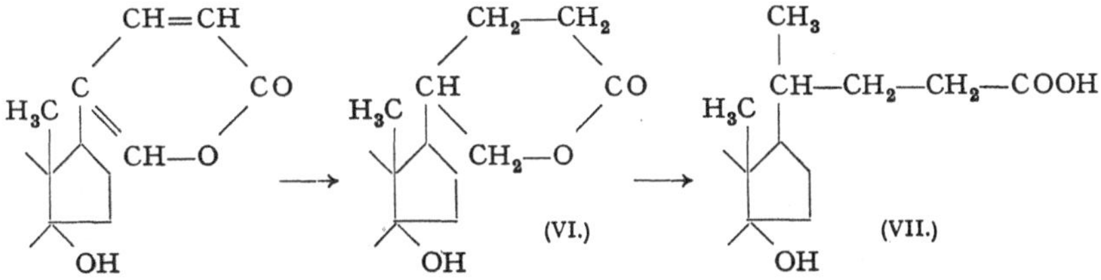

The two double bonds of the lactone ring are reduced with great facility to saturated structures (VI). On stronger reduction the ring opens and an acid (VII) is obtained possessing a side chain similar to that in cholanic acids. Examples of such reactions will be given later.

A common feature of the bufagins is a tertiary hydroxyl at $C_{(14)}$. This follows from the isolation, after the opening of the lactone ring, of products which have lost one molecule of water and give neither aldehyde nor enol reactions (V). This tertiary hydroxyl group is easily split from the bufagins by treating them with hydrochloric acid after which a double bond appears in the D ring:

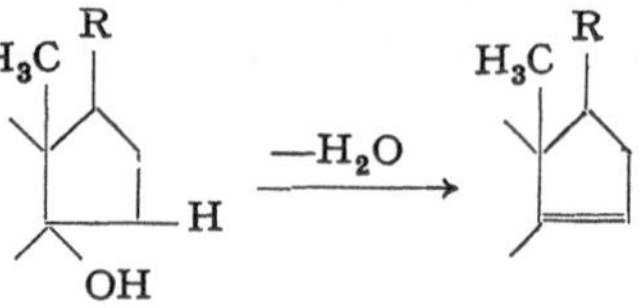

Main representatives of the bufotalin class.

Bufotalin. This major bufagin constituent of the poison of the European toad, *B. vulgaris*, was obtained as crystals by Wieland and Weil (74) who kept the name proposed earlier by Faust (22, 23) for a pharmacologically active, N-free, amorphous substance which he had isolated from the same species. Kotake (48) isolated bufotalin from *B. vulgaris formosus*.

Bufotalin, for which Wieland and Behringer (67) proposed the structure (IX), can be considered as the best studied of all bufagins.

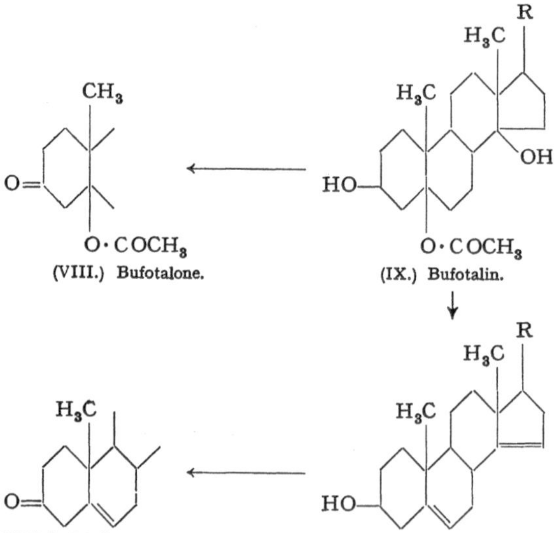

(VIII.) Bufotalone.

(IX.) Bufotalin.

(XI.) Bufotalienone.

(X.) Bufotaliene.

Bufotalin possesses a secondary alcohol group, since when oxidized with chromic anhydride it yields a ketone, bufotalone (VIII). When treated with hydrochloric acid, one molecule of water and one molecule of acetic acid are removed and a new compound, bufotaliene (X) is formed, which contains two double bonds.

The structure of bufotaliene was derived by WIELAND and BEHRINGER from the position of the tertiary hydroxyls in bufotalin. One hydroxyl is at $C_{(14)}$ because when bufotalone (VIII) is treated with alkali (68), the lactone ring opens and, on acidification, anhydro-bufotalonic acid (V) is produced which has the same composition as the original ketone and does not give any enol or aldehyde reaction. That double bond which appears on the elimination of this hydroxyl must be assigned to $C_{(14:15)}$ because it is easily hydrogenated, which would not be the case with a double bond at $C_{(8:14)}$.

Of the remaining tertiary carbon atoms WIELAND and BEHRINGER selected $C_{(5)}$ for the position of the acetoxy group since by its elimination a double bond is formed which can be easily hydrogenated. This would be the case for a double bond in $C_{(5:6)}$, a well known fact from sterol chemistry. If the acetoxy groups were located at $C_{(8)}$ or $C_{(9)}$, double bonds $C_{(8:9)}$ or $C_{(7:8)}$ would be formed which are relatively resistant to hydrogenation.

Some details remain to be clarified in the case of bufotaliene. WIELAND and BEHRINGER (67) noticed that bufotaliene is oxidized to a ketone, bufotalienone (XI), during which process the usual migration of the double bond from $C_{(5:6)}$ to the $C_{(4:5)}$ position would be expected. However, bufotalienone gives a spectrum with a maximum at 300 mμ. which corresponds to the unsaturated lactone ring; the usual maximum at 235 mμ. is missing, which appears in the cholestenone spectrum and is considered as being characteristic for a double bond at $C_{(4:5)}$ in conjugation with a keto group.

The secondary hydroxyl had been assigned the $C_{(3)}$ position by analogy with the sterols. WIELAND and BEHRINGER have recently proved its location in ring A as follows.

When bufotaliene (XII) is hydrogenated with palladium, it binds more than five moles of hydrogen, giving two neutral, stereoisomeric

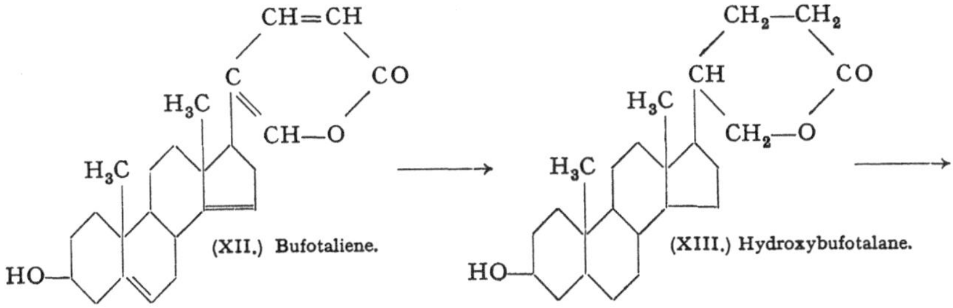

(XII.) Bufotaliene.   (XIII.) Hydroxybufotalane.

compounds, α- and β-hydroxybufotalane (XIII) in 80% yield. The
remaining 20% appears in the form of acidic products.

(XIV.)   (XV.)   (XVI.)   (XVII.)

The α- and β-hydroxybufotalanes lose water by treatment with
boron trioxide to give two bufotalenes (XIV). β-Bufotalene yielded on
oxidation with osmium tetroxide a glycol (XV) which was then oxidized
to a dicarboxylic acid (XVI). The latter produced a ketone by thermic
decomposition. Considering our knowledge of sterol chemistry, only a
dicarboxylic acid formed by opening the ring A can react in the indicated
manner. Although such an acid could also be obtained with a hydroxyl
group in the $C_{(2)}$ or $C_{(4)}$ position, the authors prefer $C_{(3)}$ because this
position is the one which is hydroxylated in natural cholane derivatives.
From the 20% acidic substances obtained by the hydrogenation of
bufotaliene, they isolated a hydroxy-isocholanic acid (XVIII) which
is identical with that obtained by Wieland, Hesse and Meyer (70)
by hydrogenation of acetyl-bufotaliene.

(XVIII.)   (XIX.)

(XX.) Iso-bufocholanic acid.

Evidently, this acid is produced by hydrogenolysis of the lactone ring. When distilled in vacuo, it loses water and is converted into an unsaturated acid (XIX) which, by hydrogenation yields a cholanic acid (XX). This acid was termed isobufocholanic acid and is different from all known cholanic acids (70). Since during the dehydrations and hydrogenations asymmetric centers are involved, this acid cannot represent the original stereochemical configuration of bufotalin.

Marinobufagin. This was the first bufagin isolated in crystalline form. It was obtained by ABEL and MACHT (*1*) from the dried secretion of the tropical toad, *B. marinus*, which in their paper was named *B. agua.* The isolation was repeated by JENSEN and CHEN (*37*) and a formula $C_{24}H_{32}O_5$ was finally accepted, in part, as a result of an X-ray crystallographic study made by CROWFOOT (*13*). The same compound has also been isolated from an Argentine species of toad, *B. paracnemis* by DEULOFEU and MENDIVE (*20*).

Marinobufagin adds three moles of hydrogen (*65, 34*); it contains two double bonds in the lactone ring and one elsewhere in the molecule. By the action of alkali the lactone ring was opened (*65*) but the acid produced has not been identified. Under the usual conditions marinobufagin is converted into a diacetate (*19*) and not a monoacetate as was believed earlier. From our knowledge of the reactions of other bufagins it seems that one of the hydroxyls is tertiary and is located at $C_{(14)}$ while the other two form secondary alcohol groups.

Gamabufagin (gamabufotalin, gamabufogenin). This bufagin was first isolated by KOTAKE (*47*) from a Japanese toad species which he took first for *B. bufo japonicus* and afterwards identified as *B. vulgaris formosus* (*48*). It seems to be identical with the gamabufagin found in the poison of the same toad by CHEN, JENSEN, and CHEN (*12*), and also with the gamabufogenin of WIELAND and VOCKE (*73*). KOTAKE and KUWADA (*52*) isolated it from *Ch'an Su*. KOTAKE and KUBOTA (*50*), who accepted the formula $C_{24}H_{34}O_5$ as proposed by WIELAND and VOCKE (*73*), suggested structure (XXI) for gamabufagin.

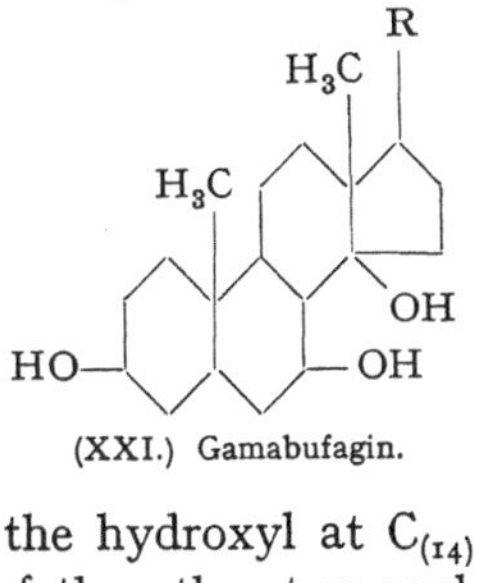

(XXI.) Gamabufagin.

While the position of the hydroxyl at $C_{(14)}$ has been established by the usual reactions, that of the other two such groups is based only on analogies. They must be of secondary nature since a diacetate was obtained (73), and on oxidation a diketone appeared (50). $C_{(3)}$ has been selected for an OH-group because this place is always hydroxylated in the steroids; furthermore, $C_{(7)}$ is preferred to $C_{(11)}$, because of the easy acetylation and negative Hammarsten reaction (as modified by Yamasaki). Kondo and Ohno (52), however, prefer to place the secondary hydroxyl group at $C_{(6)}$ instead of $C_{(7)}$.

Cinobufagin. In 1928 Kotake (46) isolated from *Senso* a crystalline, neutral constituent which he termed "bufagin". Similar substances were formerly isolated from the same source by Shimizu (57) and Kodama (41). In 1930 Jensen and Chen (36) also obtained a crystalline compound from *Senso* and named it "cinobufagin". This was later identified with Kotake's bufagin. This bufagin was also isolated from a Chinese toad, *B. bufo gargarizans,* by Chen and Chen (6) and this result was regarded as a proof that *Senso* is the dried poison of that toad or that at least *B. gargarizans* is one of its principal sources.

Cinobufagin, as described by many authors, melted at 221–223° and was considered a homogeneous substance. However, in 1937 Kotake and Kuwada (51) resolved this bufagin by chromatography on aluminum oxide into two compounds. The major constituent, (m. p. 212–213°) was termed "new" cinobufagin, and the minor component (m. p. 248–249,6°) was named "cinobufotalin". The name cinobufagin or "old" cinobufagin is used here for the mixture of the two bufagins with which all work prior to 1937 was done. Derivatives of the mixture and of the new cinobufagin have practically the same melting points (51).

For the "new" cinobufagin Kotake and Kuwada (52) proposed the formula $C_{26}H_{34}O_6$ which is identical with that of the "old" one as, established by Tschesche and Offe (65) as well as by Crowfoot and Jensen (14). Cinobufagin is an acetylated bufagin, with three double bonds. A structural formula leaving open the position of one double bond and two hydroxyls was proposed by Tschesche and Offe while Kotake and Kuwada (52, 54) suggest structure (XXII) for the "new" cinobufagin.

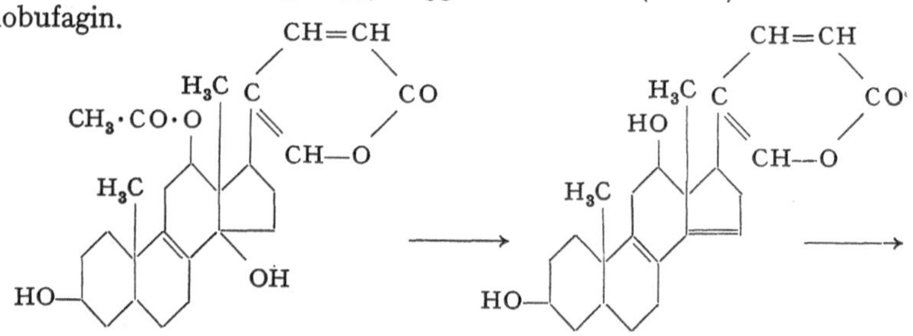

(XXII.) "New" cinobufagin.   (XXIII.) Deacetyl-anhydrocinobufagin.

On acetylation cinobufagin gives an acetyl-derivative and on oxidation a monoketone, cinobufagone (40). One of the hydroxyls is thus secondary and has been placed at $C_{(3)}$ by analogy. The location of the hydroxyl at $C_{(14)}$ was deduced by TSCHESCHE and OFFE (65) because both α- and

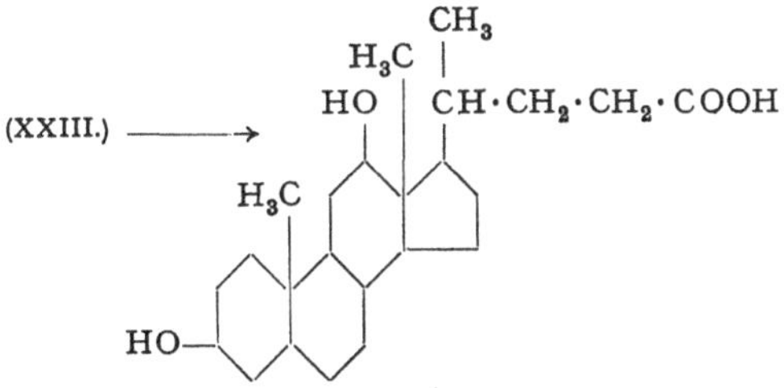

(XXIII.) $\longrightarrow$

CH$_3$
H$_3$C |
HO | CH·CH$_2$·CH$_2$·COOH
H$_3$C

HO—

(XXIV.) Cinobufagin dihydroxycholanic acid.

β-hexahydro-cinobufagin acetates (XXV) gave two isomeric desacetyl-hexahydrocinobufaginic acids (XXVI) when treated with alkali.

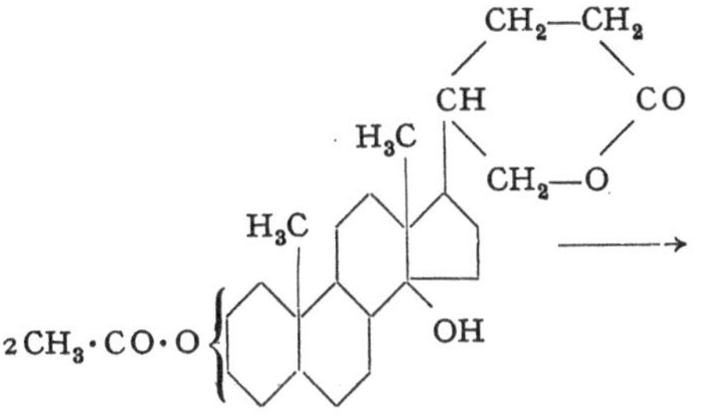

CH$_2$—CH$_2$
CH   CO
H$_3$C
CH$_2$—O

H$_3$C

2 CH$_3$·CO·O

OH

$\longrightarrow$

(XXV.) Hexahydro-cinobufagin acetates.

CH$_2$—CH$_2$·COOH
CH
H$_3$C
CH$_2$OH

OH

$\longrightarrow$

CH$_2$—CH$_2$·COOH
CH
H$_3$C
CH$_2$

—O

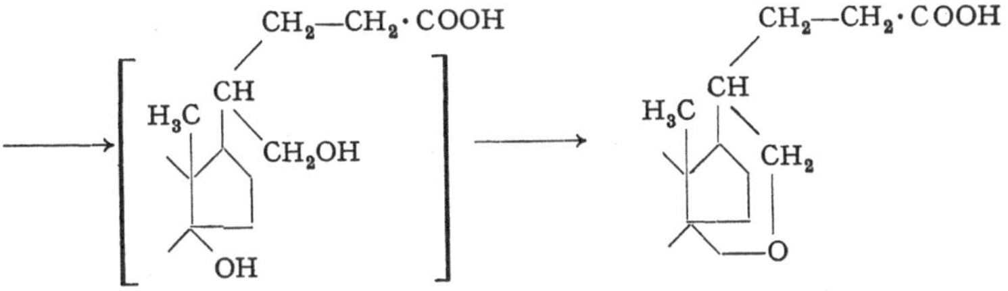

(XXVI.) Deacetyl-hexahydro-cinobufaginic acids.

A similar series of reactions was carried out with "new" cino-bufagone (54).

The location of the acetylated hydroxyl and the double bond, by KOTAKE and KUWADA (52) was based on color reactions only, and thus both positions must be considered doubtful. A double bond at $C_{(8:9)}$ should be resistant to hydrogenation; however, this is not the case.

A proof that the third hydroxyl group is secondary was derived from the observation that, when the new cinobufagin is treated with hydrochloric acid, a deacetyl-anhydrocinobufagin (XXIII) is obtained, which on hydrogenation gives a cinobufagin dihydroxycholanic acid (XXIV); the latter can be oxidized to a diketo-acid.

Cinobufotalin. KOTAKE and KUWADA (*52*) proposed structure (XXVI a) for cinobufotalin. The acetyl appears at the same carbon atom as in cinobufagin because of the ease with which both compounds give mixed crystals. In locating the other hydroxyls the positive HAMMARSTEN reaction has been considered.

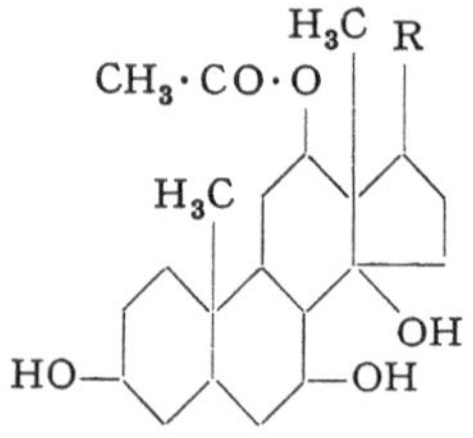

$$H_3C \quad R$$
$$CH_3 \cdot CO \cdot O$$
$$H_3C$$
$$OH$$
$$HO— \quad —OH$$

(XXVI a.) Cinobufotalin.

Bufalin. This bufagin has been isolated by KOTAKE and KUWADA from *Senso* (*53*) and they proposed the symbol (XXVIII) for it. Bufalin contains double bonds only in the lactone ring; and by hydrogenation of the acetyl derivative, tetrahydro-acetylbufalin is obtained. Oxidation yields a ketone, termed bufalone (XXVII). A treatment with hydrochloric acid eliminates water giving anhydro-bufalin (XXIX). The position of the hydroxyls at $C_{(3)}$ and $C_{(14)}$ is based on analogies with other bufagins and sterols.

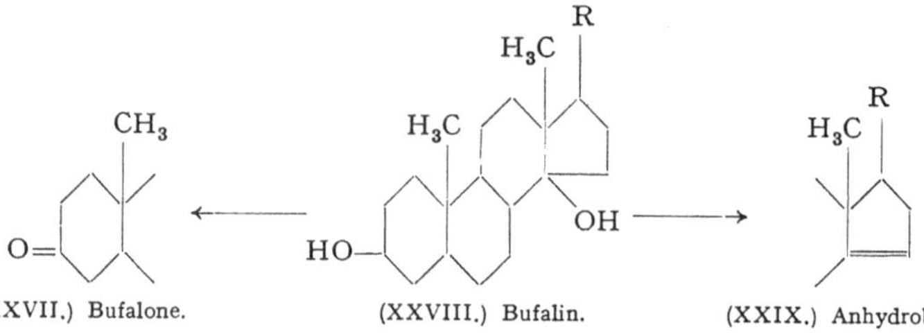

$$CH_3 \qquad\qquad H_3C \quad R \qquad\qquad R$$
$$\qquad\qquad H_3C \qquad\qquad H_3C$$
$$O= \qquad\qquad HO— \quad OH \qquad\qquad$$

(XXVII.) Bufalone.   (XXVIII.) Bufalin.   (XXIX.) Anhydrobufalin.

Pseudo-desacetyl-bufotalin has been isolated from *Senso* by KONDO (*43*) and is referred to in his early papers as pseudo-bufotalin. It is known only as an amorphous powder. Its acetate is likewise amorphous, the simplest crystalline derivatives being the *p*-nitrobenzoate (m. p. 227–229°) and the 3,5-dinitrobenzoate (m. p. 238–240°). Much work has been devoted by KONDO and collaborators to establish the structure of pseudo-desacetylbufotalin which displays a high pharmacological activity.

OHNO (*55*) gives (XXX) as the most probable structure for acetyl-pseudo-desacetylbufotalin.

By permanganate oxidation, according to STEIGER and REICHSTEIN (*60*), an etiocholanic acid (XXXI) was obtained which yielded easily the lactone (XXXII). This is interpreted as a proof of the location of a hydroxyl group at $C_{(14)}$. The acid (XXXI) was converted into its methyl ester and the latter degraded by current methods to yield the corresponding ketone (XXXIII). On oxidation, this ketone gave a mixture of a di-

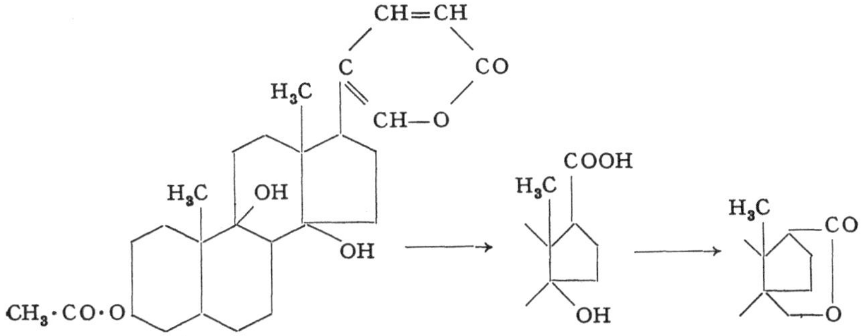

(XXX.) Acetyl-pseudo-desacetylbufotalin. (XXXI.) (XXXII.)

carboxylic acid (XXXIV) and a monoacid (XXXV). They were difficult to purify but when distilled they were converted into the homogeneous lactone (XXXVI). By studying spatial models it was found that only a hydroxyl in $C_{(10)}$ could form a lactone in an acid of type (XXXV). Although it is difficult to understand the resistance of the $C_{(14)}$ hydroxyl group to vacuum distillation, the authors demonstrated in a series of reactions that it remained unaltered in the lactone molecule.

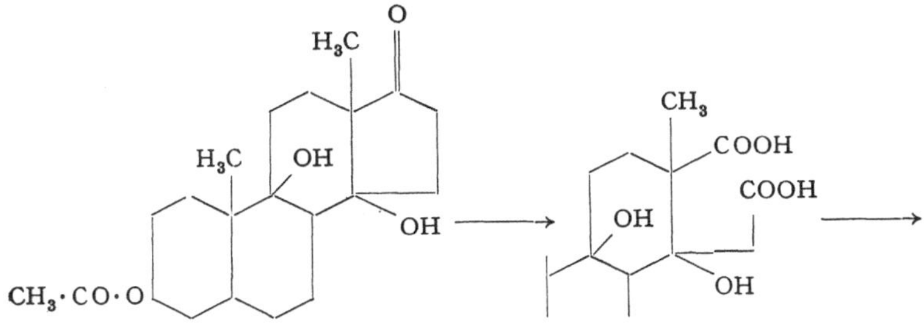

(XXXIII.) (XXXIV.)

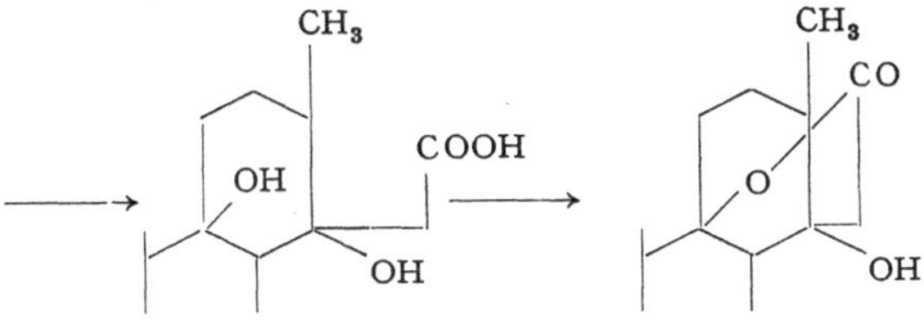

(XXXV.) (XXXVI.)

Pharmacology of the bufagins. VULPIAN must be credited with having noted first the cardiac activity of the toad poisons (1850). A large volume of work was then done in this field using the crude or only slightly purified secretion, until FAUST (*22, 23*) prepared in 1902 from *B. vulgaris* an

amorphous but rather pure bufagin which he named bufotalin. He
found that bufotalin has a digitalis-like effect on frog's heart. It slows
down the heart rate, intensifies the systole and finally, produces systolic
standstill.

Later, many investigators have assayed the pharmacological activity
of different crystalline bufagin preparations and found effects similar
to those described by Faust. The paper of Abel and Macht (1) on the
pharmacology of marinobufagin and those of Chen, Jensen, and
Chen (2–12) should be consulted on the individual bufagins. These
investigators have found that the usual symptoms of digitalis poisoning
are produced in cats, viz. arrhythmia, secondary tachycardia and
ventricular fibrillation.

Bufagins taste bitter; they cause numbness of the tongue and lips
and also vomiting.

Chen and Chen (8) have published a comparative study of the
different bufagins and by employing Hatcher's method they found
that the "cat unit" varies from 0,09 mg. per kg. in the case of arenobufagin to 0,77 mg. for marinobufagin.

The elimination of hydroxyl groups from the bufagin molecule or
acetylation or oxidation to ketones reduces the original potency although
the specific cardiac action is retained. Hydrogenation or opening of
the lactone ring causes an almost total loss of activity [Chen and
Chen (7)]. The same authors attribute the lack of persistence of the
bufagin action as compared to that of the cardiac glycosides obtained
from plants, to the absence of sugar.

It is interesting that the toad is more resistant than the common
frog to its own poison, but is not immune to it. When marinobufagin
is injected into B. marinus, although an immediate systolic arrest does
not take place the rhythm of the heart is altered, and, finally, a perfectly
developed systolic contracture appears (1, 25, 26).

Geszner (26) stresses that toad's blood contains a certain amount
of substances with an activity similar to those isolated from the venom,
and that they are important for the behavior of the isolated heart.

II. Bufotoxins.

The bufotoxins are those components of the toad poison which upon
hydrolysis yield suberic acid, arginine and a substance of the bufagin
type; for the latter the general term "genin" will be used as proposed
by Wieland.

While we must assume in some instances that the genin component
of a bufotoxin is identical with the free bufagin which occurs in the
same venom, the acid sensitivity of the bufagins prevents their isolation

from the hydrolysate. Thus, in the cases so far investigated, a conversion product of the bufagin was isolated instead of the native genin.

WIELAND and BEHRINGER (67) found 7 mg. of crude bufotoxin per gram of dried secretion *ex B. vulgaris* male animals, corresponding to 1,1 mg. per toad. In females the amounts were, 4 mg., and also 1,1 mg.

The bufotoxins are scarcely soluble in the usual organic solvents with the exception of methanol, ethanol, and pyridine. The solubility is greater in aqueous than in anhydrous alcohols.

The term "bufotoxin" includes this whole class of compounds and, in particular, this is the name of the toxin isolated by WIELAND from *B. vulgaris*. In other instances it is usual to prefix the name with that of the toad species.

A list of the bufotoxins so far isolated is given in Table 2.

Table 2. Bufotoxins.

Name	Toad Species	Formula	M. p.	References
Bufotoxin.........	*B. vulgaris*	$C_{26}H_{35}O_6 \cdot C_{14}H_{25}O_4N_4$ (a)	204–205°	(66)
Marinobufotoxin ..	*B. marinus*	$C_{24}H_{31}O_5 \cdot C_{14}H_{25}O_4N_4$	204–205°	(40)
Gamabufotoxin	*B. vulgaris formosus*	$C_{24}H_{33}O_5 \cdot C_{14}H_{25}O_4N_4$	210°	(18, 73)
Arenobufotoxin ...	*B. arenarum*	$C_{24}H_{30}O_5 \cdot C_{14}H_{25}O_4N_4$	214°	(67)
Cinobufotoxin	(Senso)	$C_{24}H_{39}O_8 \cdot C_{14}H_{25}O_4N_4$ (b)	200°	(6, 36)
Viridobufotoxin....	*B. viridis viridis*	$C_{23}H_{35}O_6 \cdot C_{14}H_{25}O_4N_4$ (b)	198–199°	(10)
Regularobufotoxin .	*B. regularis*	$C_{25}H_{35}O_7 \cdot C_{14}H_{25}O_4N_4$ (b)	205°	(32)

(a) Contains one acetyl group.

(b) Bufotoxins with genins which seem to contain a number of carbon atoms different from 24 or 26, are of doubtful purity.

Main representatives of the bufotoxin class.

Bufotoxin obtained by WIELAND and ALLES (66) from *B. vulgaris* was the first toxin isolated and is the best studied representative of this class. In their early paper WIELAND and ALLES found that on hydrolysis bufotoxin yielded arginine, suberic acid and bufotaliene (X); this unsatured product was also obtained when bufotalin was treated with hydrochloric acid.

For many years it was thought that the genin occurring in bufotoxin had one double bond less than bufotalin; however, hydrogenation showed that it binds two moles of hydrogen in the lactone ring. The presence of the same lactone ring as in bufagins can also be inferred from the spectrum which shows a maximum at 300 mμ. (69).

WIELAND and BEHRINGER (67) proposed for bufotoxin the structure (XXXVII), which contains bufotalin as the genin; the latter is converted into bufotaliene during hydrolysis. This formula is based, in part, on the fact that bufotoxin forms a salt with two equivalents of barium when treated with methanolic baryta, and that this salt contains one methoxyl group. One of the acidic hydrogens originates from the enol produced by opening the lactone ring, and the second is present in the suberic acid or arginine part of the molecule. That the second carboxyl group is located in the arginine part follows from the titration of bufotoxin which requires only 0,22 equivalents of alkali. This is explained by the weak acidity of the arginine section of the molecule.

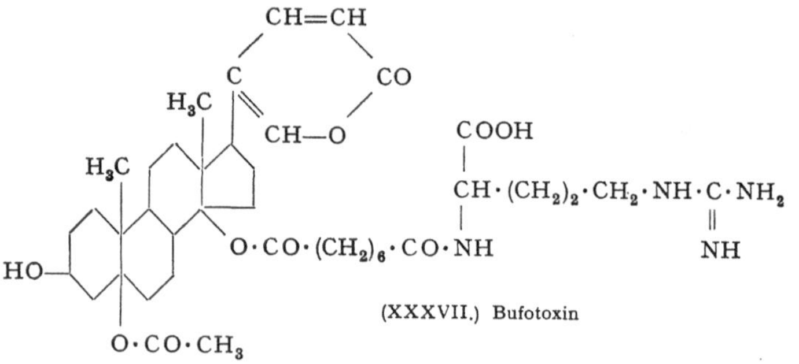

(XXXVII.) Bufotoxin

Arenobufotoxin is the toxin isolated from *B. arenarum* (*11*). By its hydrolysis Wieland and Behringer (*67*) obtained suberic acid and a genin, $C_{24}H_{26}O_3$, (m. p. 195°). No arginine was isolated because of the small amounts of starting material. The empirical formula of areno-bufotoxin shows that its genin must be different from all of the described arenobufagins.

Gamabufotoxin was isolated by Wieland and Vocke (*73*) from *B. vulgaris formosus*. Its hydrolysis yielded suberic acid, arginine, and the same anhydro-gamabufagin which can be prepared by a hydrochloric acid treatment of gamabufagin. Although no further proof is available, it can be assumed that the genin in gamabufotoxin is gamabufagin.

Marinobufotoxin. This toxin was isolated from *B. marinus* by Jensen and Chen (*37*). By hydrolysis Jensen and Evans (*40*) obtained suberic acid, arginine, and a genin, $C_{24}H_{28}O_3$ (dianhydro-marinobufagin) which can also be obtained by treating marinobufagin with sulfuric acid.

Pharmacology of the bufotoxins. The study of different bufotoxins as carried out by Chen and his collaborators has shown that the pharmacological effects of these toxins are similar to those of the bufagins. They slow down the rate of the frog's heart, and bring about ventricular systolic standstill. In cats, bradycardia, arrhythmia, secondary tachycardia and circulatory collapse due to ventricular fibrillation were observed; a vomiting effect was also noted.

The effective doses of the bufotoxins in the cat, range, when HATCHER's method is used, from 0,27 mg. per kg. in the case of viridobufotoxin to 0,80 mg. per kg. with fowlerobufotoxin [CHEN and CHEN (8)].

III. Basic Constituents of toad venoms.

The toad poisons contain a number of basic constituents some of which are very active pharmacologically.

The presence in the venom of substances giving the reactions of alkaloids was noted by PHISALIX and BERTRAND (56) in 1893 but crystalline bases were not isolated until 1912 when ABEL and MACHT (1) made the surprising discovery that the secretion of B. marinus contains large amounts of adrenaline. In 1920 HANDOVSKY (28) separated from the secretion of the common European toad, B. vulgaris, a base that gave a crystalline oxalate and picrate. For this base the name "bufotenine" was retained following the earlier suggestion of PHISALIX and BERTRAND.

A detailed investigation of the basic constituents of this venom revealed the presence of other bases which are all related to bufotenine. They belong to the indole class and are, thus, also related to tryptamine as had been assumed by JENSEN and CHEN (38).

A study of the secretions of many toad species has shown that in most cases they contain more than one base; and no species has ever yielded all of the known basic constituents. Only further work may decide whether the missing bases were originally present but have been destroyed during the isolation process. The possibility of an enzymatic destruction while the skins are dried or by chemical agents later, follows from an investigation by DEULOFEU and DUPRAT (18) who worked out a scheme for the systematic isolation of these bases both from the skin and secretion. It was found that adrenaline could be isolated from certain species, starting from the dried secretion; however, it could not be obtained from the skins. In contrast, bufothionine, a sulfo-conjugated base (see below), can be isolated easily from the dried skins but with difficulty from the poison.

Bufotenine was the base isolated by HANDOVSKY (28) from the secretion of B. vulgaris in form of a crystalline oxalate and picrate. The free base was prepared in crystalline form by WIELAND, KONZ, and MITTASCH (72) after purification by distillation. It has the formula, $C_{12}H_{16}ON_2$, and melts at 146–147°. It is easily soluble in alcohol, less in acetone, very slightly in ether; it is almost insoluble in water. Bufotenine gives a blue color reaction with ferric chloride which seems to be an oxidation since the color only appears after several minutes. FOLIN's adrenaline reagent also produces blue color.

Bufotenine gives a red monopicrate which at 140° changes into a yellow modification, m. p. 180–181° (75); a dipicrate, m. p. 177–178° (29, 75);

a picrolonate, m. p. 120–121° (*17*), and a flavianate, m. p. 130–131° (*38*) which all are useful in the identification of this base.

Bufotenine is converted by methyl iodide into a methiodide which was found to be identical with bufotenidine iodide, $C_{13}H_{19}ON_2I$, the derivative of another toad venom base.

Bufotenine contains a phenolic hydroxyl group. Its structure and that of bufotenidine was clarified by the synthesis of bufotenidine iodide methyl ether. Following the working hypothesis that both bases contained an indole ring, WIELAND, KONZ, and MITTASCH (*72*) proposed for bufotenine the formula (XXXVIII) in which the phenolic group was located by synthesis. Bufotenidine is then the corresponding quaternary base.

WIELAND and his colleagues first synthesized the compound (XXXIX), with the phenolic hydroxyl at $C_{(6)}$ as in the harmane alkaloids. This compound was different from the methiodide obtained by exhaustive methylation of bufotenine or bufotenidine; however, the synthetic compound (XL), with the phenol group at $C_{(5)}$ as in physostigmine, was found to be identical with the methiodide prepared from the natural product. Consequently, bufotenine has the structure (XLI) and bufotenidine, the structure (XLII); the latter has been formulated as a betaine, in accordance with the analytical data.

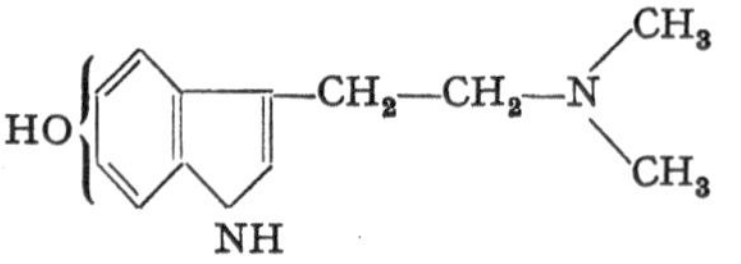

(XXXVIII.) 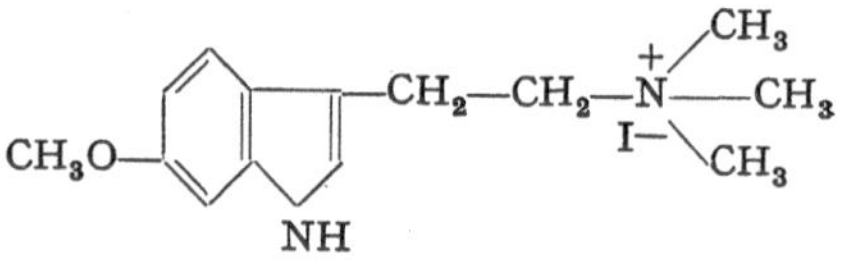   (XXXIX.)

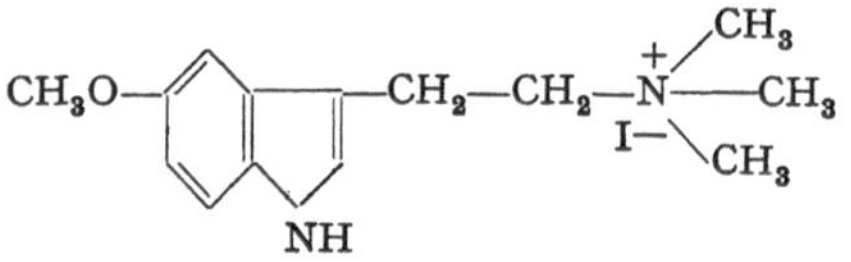

(XL.) 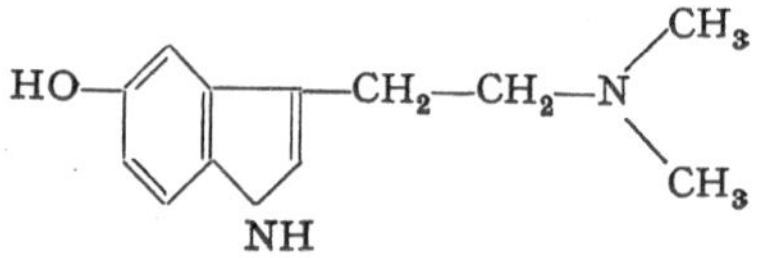   (XLI.) Bufotenine.

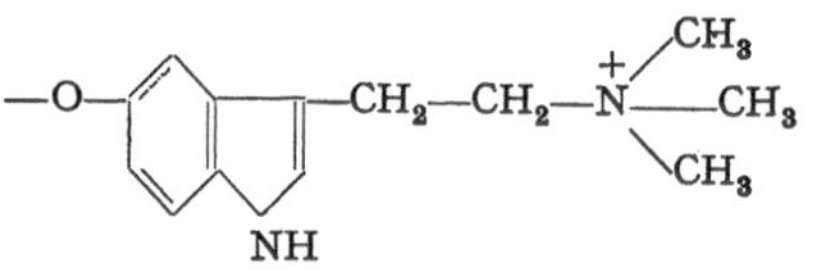

(XLII.) Bufotenidine.

The total synthesis of bufotenine was carried out by HOSHINO and SHIMODAINA (*29*) who prepared 5-ethoxytryptophol (XLIII), which yielded on treatment with phosphorous tribromide the corresponding bromide (XLIV). When this bromide was treated with dimethylamine, bufotenine ethyl ether was obtained (XLV) which on de-ethylation with aluminum chloride yielded bufotenine (XLVI). The sample was found to be identical with the natural base.

C$_2$H$_5$O—⟨ ⟩—CH$_2$—CH$_2$OH $\xrightarrow{\text{PBr}_3}$ C$_2$H$_5$O—⟨ ⟩—CH$_2$—CH$_2$Br $\xrightarrow{\text{NH(CH}_3)_2}$

NH   NH

(XLIII.) 5-Ethoxytryptophol.   (XLIV.)

C$_2$H$_5$O—⟨ ⟩—CH$_2$—CH$_2$—N(CH$_3$)$_2$ $\xrightarrow{\text{AlCl}_3}$ HO—⟨ ⟩—CH$_2$—CH$_2$—N$\Big\langle{}^{CH_3}_{CH_3}$

NH   NH

(XLV.) Bufotenine ethyl ether.   (XLVI.) Bufotenine.

Bufotenine has also been found in: *B. vulgaris* (*28*, *39*), *B. arenarum* (*38*, *72*), *B. viridis viridis* (*10*, *38*), *B. chilensis*, *B. crucifer*, *B. paracnemis* (*18*), *B. marinus* (*4*), and in *Ch'an Su* (*36*).

In the course of his extensive work on bufotenine *ex B. vulgaris*, WIELAND (*72*) obtained in one instance 40 mg. of the base per gram of dried secretion; furthermore, in collaboration with BEHRINGER (*67*), about 3 mg. per gram dried venom from the male toad, corresponding to 0,05 mg. of bufotenine per animal; females gave almost identical figures. From *B. marinus* DEULOFEU and MENDIVE (*20*) isolated 13 mg. of bufotenine per gram of dry venom.

Pharmacology of Bufotenine. Using crystalline bufotenine salts, HANDOVSKY (*28*) observed that they produce in cats and rabbits an initial lowering of the blood pressure, with a subsequent rising. This action is peripheral and can be also obtained in the decapitated animal. Pupillar contraction was also observed. Bufotenine produces on the isolated intestine or uterus of the guinea pig contractions which soon disappear.

The frog is rather immune to the action of bufotenine. When injected, an increase in the tone of the heart contractions is observed with a decrease of the rate and ventricular standstill.

The effect of bufotenine in raising the blood pressure is less marked than that of bufotenidine.

CHEN and his collaborators have assayed the flavianate of a base melting at 130–131° that we now know to be identical with bufotenine. While in the paper published with JENSEN (*38*) they did not find pressor activity, in subsequent articles (*10*, *11*) a weak effect was described.

Bufotenidine. This base was isolated by WIELAND, HESSE, and MITTASCH (*71*) from *Ch'an Su* and from the dried secretion of the European toad, *B. vulgaris*.

The free base is not known in crystalline form. It can be separated from bufotenine when present in a slightly alkaline solution, since the

former cannot be extracted with ether, while bufotenine is somewhat
ether soluble. It gave a monopicrate, m. p. 198° (71); a picrolonate,
m. p. 255° (17); a flavianate, m. p. 200° (71); and an iodide, m. p.
217,5° (75).

The constitution of bufotenidine follows from its relationship with
bufotenine.

Bufotenidine has been observed in: B. vulgaris (71), B. bufo
gargarizans, B. fowleri, B. americanus, B. quercicus (5), B. formosus (39),
Xenopus laevis (32), and in Ch'an Su (36, 71).

WIELAND and BEHRINGER (67) found about 0,07 mg. of bufotenidine
per toad (B. vulgaris), irrespective of sex.

Pharmacology of Bufotenidine. The pharmacological activity of
this base has been studied by CHEN, JENSEN, and CHEN (2–6, 9–12, 38)
who used the flavianate mentioned. Bufotenidine exerts a notable pressor
action in cats, rabbits, and dogs. This effect is much like that of adrenaline;
and, calculated on a weight basis, bufotenidine is about one tenth as
potent as adrenaline hydrochloride. The pressor action is due in part
to vasoconstriction and in part to cardiac stimulation. Bufotenidine
produces, on perfusion of an isolated frog heart, an increase in tone,
decrease in the heart rate, and, finally standstill.

It contracts the isolated small intestine of the rabbit or the isolated
uterus of the rabbit or guinea pig.

Bufothionine was discovered by WIELAND and VOCKE (73) while
working up the skins of the Japanese toad, B. vulgaris formosus.

It forms long prisms which melt at 250°. It shows very poor solubility
in boiling water, and is practically insoluble in common organic solvents.
Bufothionine has the composition, $C_{12}H_{14}O_4N_2S$, and is a very weak
base. It is resistant to boiling with 0,1 N sodium hydroxide. On boiling
with hydrochloric acid, bufothionine suffers hydrolysis to give sulfuric
acid and the hydrochloride of a base, $C_{12}H_{14}ON_2$. The latter differs from
bufotenine by having two hydrogen atoms less. It was designated as
dehydro-bufotenine by WIELAND and WIELAND (75). The same base was
later found in the secretion of many other species.

(XLVII.) Dehydro-bufotenine.   (XLVIII.) Bufothionine.

Assuming for dehydro-bufotenine the tryptamine structure as present
in bufotenine, its only possible symbol is (XLVII). The correctness
of this symbol was proved by WIELAND and WIELAND (75) by oxidation
of bufothionine with permanganate (when dimethylamine and formic

acid were isolated), and by reduction of (XLVII) to bufotenine. Bufo-thionine is then the sulfo-conjugate of the base (XLVIII). The dipolar nature explains its insolubility.

While many examples of sulfo-conjugation with phenolic substances are known in higher animals, bufothionine is one of the few such compounds which occur in lower animals.

Bufothionine has also been found in: B. vulgaris formosus (73), B. arenarum (32), B. chilensis, B. crucifer, B. spinulosus, and B. para-cnemis (18).

Dehydro-bufotenine is produced by the acid hydrolysis of bufothionine. It has been found in the free state in the dried secretion and skins of many toads species. When crystallized from water it gives plates melting at 199° and at 218° from ethanol. It gives a picrate, m. p. 187° (75); a picrolonate, m. p. above 300° (17); and a flavianate, m. p. 271–272° (38). Its hydrochloride melts at 242° and the sulfate at 209° (73).

Dehydro-bufotenine can be separated from bufotenine by its insolubi-lity in ether.

Dehydro-bufotenine has been observed in B. arenarum (11, 75), B. valiceps, B. marinus (38, 58), B. regularis (32), B. chilensis, B. crucifer, B. spinulosus, B. paracnemis (18), and in Ch'an Su (39).

It is still doubtful whether dehydro-bufotenine is present as such in the secretions and skins or if it is produced during the hydrolysis of bufothionine.

Pharmacology of Dehydro-bufotenine. CHEN and his collaborators have studied the pharmacological activity of some flavianate samples which, as we now know, are identical with dehydro-bufotenine flavianate. With the purest preparations (m. p. 271–272°) no increase in blood pressure was observed. Injected into frogs, a decrease in the heart rate with stoppage of cardiac contraction was produced. No effect could be observed on the isolated rabbit intestine but contraction of the guinea pig virgin uterus took place.

L-Adrenaline, identical with the hormone produced by the medullar portion of the suprarenals was first isolated by ABEL and MACHT (1) from the dried secretion of B. marinus and later it was also found in the secretions of B. arenarum (15, 32), B. regularis (32), B. paracnemis (20), and in Ch'an Su (35).

The isolated adrenaline from B. marinus represented 4,48% of the dried venom, but by pharmacological methods ABEL and MACHT (1) determined the presence of 5–7% adrenaline in the dried secretion.

While CHEN and CHEN (4) obtained similar figures pharmacologically, and calculated a content of 23,8–58,8 mg. of adrenaline in the glands of the toad, SLOTTA, VALLE, and NEISSER (58) who studied B. marinus collected in Brazil, found only 2% in the dried secretion and isolated

1,35% adrenaline from it. In *B. regularis*, Chen and Chen (*3*) determined pharmacologically a content of 10,7 mg. of adrenaline per animal or 4,3–5,0% of the dried secretion.

IV. Sterols.

From all the toad skins or dried secretions investigated a sterol fraction has been isolated in every case. For many years it was considered to be composed almost exclusively of cholesterol, until Hüttel and Behringer (*30*) reported that the major constituent of the sterol fraction obtained from the dried poison of *B. vulgaris* was γ-sitosterol, identical with one of the most widespread vegetable sterols.

The sterol fraction of skin extracts *(B. vulgaris, B. vulgaris formosus, B. arenarum)* is a mixture of cholesterol with 5–20% γ-sitosterol. An accumulation of this phytosterol takes place in the poisonous gland, while elsewhere in the skin the typical animal sterol is found.

The toad seems to be the only vertebrate animal in which a plant sterol occurs in a relatively high concentration. Ergosterol has been found in egg yolk cholesterol samples by Windaus and Stangel (*76*) but in much smaller amounts.

A spectrographic study of sterol samples originating from different toad secretions by Chen, Jensen et al. showed the existence of absorption bands which the authors attributed to ergosterol. However, the presence of 7-dehydrocholesterol cannot be excluded since it is spectroscopically almost identical with ergosterol.

V. Suberic Acid.

From many species of toad skins and dried poisons, suberic acid has been isolated. The presence of the free acid is evidently connected with its occurrence as a component of the bufotoxin molecules in combination with arginine and a bufagin-like substance. No free arginine has been reported to occur in toad secretions so far.

References.

1. Abel, J. J. and D. I. Macht: Two Crystalline Pharmacological Agents Isolated from *B. agua.* J. Pharmacol. exp. Therapeut. 3, 319 (1911–12).
2. Chen, K. K. and A. L. Chen: Notes on the Poisonous Secretions of Twelve Species of Toads. J. Pharmacol. exp. Therapeut. 47, 281 (1933).
3. — — The Physiological Action of the Principles Isolated from the Secretion of the South African Toad *(B. regularis).* J. Pharmacol. exp. Therapeut. 49, 503 (1933).
4. — — The Physiological Action of the Principles Isolated from the Secretion of the Jamaican Toad *(B. marinus).* J. Pharmacol. exp. Therapeut. 49, 514 (1933).
5. — — A Study of the Poisonous Secretion of Five Northamerican Toads. J. Pharmacol. exp. Therapeut. 49, 526 (1933).

6. CHEN, K. K. and A. L. CHEN: The Parotid Secretion of *B. bufo gargarizans* as a Source of *Ch'an su*. J. Pharmacol. exp. Therapeut. 49, 543 (1933).

7. — — The Active Groupings in the Molecules of Cino- and Marinobufagins and Cino- and Vulgarobufotoxins. J. Pharmacol. exp. Therapeut. 49, 548 (1933).

8. — — Similarity and Dis-similarity of Bufagins, Bufotoxins and Digitaloid Glucosides. J. Pharmcol. exp. Therapeut. 49, 561 (1933).

9. CHEN, K. K., H. JENSEN and A. L. CHEN: The Pharmacological Action of the Principles Isolated from *Ch'an su*, the Dried Venom of the Chinese Toad. J. Pharmacol. exp. Therapeut. 43, 13 (1931).

10. — — — The Physiological Action of the Principles Isolated from the Secretion of the European Green Toad *(B. viridis viridis)*. J. Pharmacol. exp. Therapeut. 49, 14 (1933).

11. — — — The Physiological Action of the Principles Isolated from the Secretion of *B. arenarum*. J. Pharmacol. exp. Therapeut. 49, 1 (1933).

12. — — — The Physiological Action of the Principles Isolated from the Secretion of the Japanese Toad *(B. formosus)*. J. Pharmacol. exp. Therapeut. 49, 26 (1933).

13. CROWFOOT, D.: A Note on the X-ray Crystallography of the Toad Poisons, Bufagin and Cinobufagin and Strophantidin. Chem. and Ind. 54, 508 (1935).

14. CROWFOOT, D. and H. JENSEN: The Molecular Weight of Cinobufagin. J. Amer. chem. Soc. 58, 2018 (1936).

15. DEULOFEU, V.: Adrenalin in Gift von *Bufo arenarum*. Hoppe-Seyler's Z. physiol. Chem. 237, 171 (1935).

16. — Quimica de los venenos de sapo, en especial de las bufaginas. Bol. Soc. quím. Perú 6, IV, 27 (1940).

17. DEULOFEU, V. and B. BERINZAGHI: Picrolonates of Bufotenine, Bufotenidine and Dehydrobufotenine. J. Amer. chem. Soc. 68, 1665 (1946).

18. DEULOFEU, V. and E. DUPRAT: The Basic Constituents of the Venom of Some South American Toads. J. biol. Chemistry 153, 459 (1944); Los constituyentes básicos de los venenos de algunos sapos sudamericanos. An. Asoc. quim. argent. 32, 75 (1944).

19. DEULOFEU, V., R. LABRIOLA and E. DUPRAT: Acetyl Content of Marinobufagin, Arenobufagin and Acetyl-marinobufagin. Nature [London] 145, 671 (1940).

20. DEULOFEU, V. und J. R. MENDIVE: Über das Gift einer Argentinischen Kröte. Liebigs Ann. Chem. 534, 288 (1938).

21. DUPRAT, E.: Estudios químicos de venenos de sapos sudamericanos. Tesis. Facultad de Ciencias Exactas, F. y N. Buenos Aires, 1941.

22. FAUST, E. S.: Über Bufonin und Bufotalin, die wirksamen Bestandteile des Krötenhautdrüsensekretes. Naunyn-Schmiedebergs Arch. exp. Pathol. Pharmakol. 47, 278 (1902).

23. — Weitere Beiträge zur Kenntnis der wirksamen Bestandteile des Krötenhautdrüsensekretes. Naunyn-Schmiedebergs Arch. exp. Pathol. Pharmakol. 49, 1 (1902).

24. GERZENSTEIN, D.: Determinación del índice de acetilo por el método de KUHN y ROTH. Tesis, Facultad de Ciencias Exactas, F. y N. Buenos Aires, 1943.

25. GESZNER, O.: Über die Wirkung des Krötengiftes auf das isolierte Kaltblüterherz. I. Die Wirkung des Krötengiftes auf das isolierte Froschherz. Naunyn-Schmiedebergs Arch. exp. Pathol. Pharmakol. 114, 218 (1926).

26. — Über die Wirkung des Krötengiftes auf das isolierte Kaltblüterherz. II. Die Wirkung des Krötengiftes (und der Digitalissubstanzen) auf das isolierte Krötenherz. Naunyn-Schmiedebergs Arch. exp. Pathol. Pharmakol. 118, 325 (1926).

27. Gils, G. E. von: Über die chemische Zusammensetzung des Parotissekretes von *B. melanostictus*. Acta brevia neerl. Physiol., Pharmacol., Microbiol. E. A. 8, 84 (1938).

28. Handovsky, H.: Ein Alkaloid im Gifte von *Bufo vulgaris*. Naunyn-Schmiedebergs Arch. exp. Pathol. Pharmakol. 86, 138 (1920).

29. Hoshino, T. and K. Shimodaira: Synthese des Bufotenins und über 3-Methyl-3-β-oxyäthyl-indolenin. Liebigs Ann. Chem. 520, 19 (1938).

30. Hüttel, R. u. H. Behringer: Über das Vorkommen von Pflanzensterinen in Kröten. Hoppe-Seyler's Z. physiol. Chem. 245, 175 (1937).

31. Ikawa, S.: Untersuchung über die Bestandteile des „Senso". V. Über die Konstitution des Pseudobufotalins. J. pharmac. Soc. Japan 55, 748 (1935); Chem. Zbl. 1935 II, 2963.

32. Jensen, H.: Chemical Studies on Toad Poisons. VII. *Bufo arenarum, Bufo regularis* and *Xenopus laevis*. J. Amer. chem. Soc. 57, 1765 (1935).

33. — Chemical Studies on Toad Poisons. VIII. The Dehydrogenation of Cinobufagin. J. Amer. chem. Soc. 57, 2733 (1935).

34. — Chemical Studies on Toad Poisons. Further Contributions to the Chemical Constitution of Marinobufagin, Cinobufagin and Gamabufagin. J. Amer. chem. Soc. 59, 767 (1937).

35. Jensen, H. and K. K. Chen: A Chemical Study of *Ch'an su*, the Dried Venom of the Chinese Toad, with Special Reference to the Isolation of Epinephrine. J. biol. Chemistry 82, 397 (1929).

36. — — Chemical Studies on Toad Poisons. II. *Ch'an su*, the Dried Venom of the Chinese Toad. J. biol. Chemistry 87, 741 (1930).

37. — — Chemical Studies on Toad Poisons. III. The Secretion of the Tropical Toad, *B. marinus*. J. biol. Chemistry 87, 755 (1930).

38. — — Chemische Studien über Krötengifte. V. Die basischen Bestandteile des Krötensekrets. Ber. dtsch. chem. Ges. 65, 1310 (1932).

39. — — The Chemical Identity of Certain Basic Constituents Present in the Secretions of Various Species of Toads. J. biol. Chemistry 116, 87 (1936).

40. Jensen, H. and E. A. Evans: Chemical Studies on Toad Poisons. VI. *Ch'an Su*, the Dried Venom of the Chinese Toad, and the Secretion of the Tropical Toad, *Bufo marinus*. J. biol. Chemistry 104, 307 (1934).

41. Kodama, K.: Beiträge zur Pharmakologie von *Senso*. II. Über Bufagin. Acta Scholae med. Univ. imp. Kioto 4, II, 201 (1921).

42. Kondo, H. u. S. Ohno: Die Untersuchung über die Bestandteile des „Senso". X. Zur Kenntnis der Isomeren Anhydrogamabufotalins. J. pharmac. Soc. Japan 56, 186 (1939).

43. — — Die Untersuchung über die Bestandteile des „Senso". VI. J. pharmac. Soc. Japan 58, 15 (1938).

44. — — Die Untersuchung über die Bestandteile des „Senso". VII. Ein neuer Bestandteil des einheimischen Krötengiftes. „F₃-Bufotalin." J. pharmac. Soc. Japan 58, 102 (1938).

45. — — Die Untersuchung über die Bestandteile des „Senso". IX. Über eine neue Begleitsubstanz des Cinobufagins, „Cinobufotalidin". J. pharmac. Soc. Japan 58, 235 (1939).

46. Kotake, M.: Über das Krötengift. I. Die Zusammensetzung des chemischen Arzneimittels „Senso". Liebigs Ann. Chem. 465, 1 (1928).

47. — Über das Krötengift. II. Die giftigen Bestandteile des Sekretes der japanischen Kröte *(B. bufo japonicus)*. Liebigs Ann. Chem. 465, 1 (1928).

48. KOTAKE, M.: Über das Krötengift. III. Nachtrag der Mitt. ,,Über giftige Bestandteile des Sekrets der Japanischen Kröte. Sci. Pap. Inst. physic. chem. Res. **9,** 233 (1928).

49. — Chemische Versuche über Krötengift. V. Zur Kenntnis des Bufagins und Bufalins. Sci. Pap. Inst. physic. chem. Res. **24,** 39 (1934).

50. KOTAKE, M. u. T. KUBOTA: Chemische Versuche über Krötengift. VIII. Über die Konstitution des Gamabufotalins. Sci. Pap. Inst. physic. chem. Res. **34,** 824 (1938).

51. KOTAKE, M. u. K. KUWADA: Chemische Versuche über Krötengift. VI. Zur Kenntnis der Bestandteile des *Ch'an su (Senso).* Sci. Pap. Inst. physic. chem. Res. **32,** 1 (1937).

52. — — Chemische Versuche über Krötengift. VII. Zur Kenntnis der Bestandteile des *Ch'an su* sowie der Konstitution des Cinobufagins und Cinobufotalins. Sci. Pap. Inst. physic. chem. Res. **32,** 79 (1937).

53. — — Chemische Versuche über Krötengift. X. Chemische Konstitution des Bufalins. Sci. Pap. Inst. physic. chem. Res. **36,** 106 (1939).

54. KUWADA, K. u. K. KOTAKE: Chemische Versuche über Krötengift. IX. Die Versuche der Konstitution des Cinobufagins. Sci. Pap. Inst. physic. chem. Res. **35,** 419 (1939).

55. OHNO, S.: Die Untersuchung über die Bestandteile des ,,*Senso".* XI. Über die Konstitution des Acetyl-ψ-desacetylbufotalins. J. pharmac. Soc. Japan **60,** 226 (1940).

56. PHISALIX, C. et G. BERTRAND: Sur les principes actifs du venin de crapaud commun *(Bufo vulgaris).* C. R. Séances Soc. Biol. Filiales Associées **54,** 932 (1902).

57. SHIMIZU, S.: Pharmacological and Chemical Studies on *Senso,* the Dried Venom of the Chinese Toad. J. Pharmacol. exp. Therapeut. **8,** 347 (1916).

58. SLOTTA, C. H. and C. NEISSER: Estudos sobre os venenos de sapos brasileiros. 1. Composicao do veneno de *B. marinus.* Mem. Inst. Butantan **11,** 89 (1937).

59. SLOTTA, C. H., J. R. VALLE and C. NEISSER: Estudos sobre os venenos de sapos brasileiros. 2. Sobre a adrenalina no veneno de *B. marinus.* Mem. Inst. Butantan **11,** 101 (1937).

60. STEIGER, M. u. T. REICHSTEIN: Ein neuer Abbau des Digoxigenins. Helv. chim. Acta **21,** 828 (1938).

61. STOLL, A. u. A. HOFMANN: Umsetzungsprodukte von Scillaren A. Helv. chim. Acta **18,** 82 (1935).

62. STOLL, A., A. HOFMANN u. A. HELFENSTEIN: Die Natur der Sauerstoffatome im Scillaridin A. Helv. chim. Acta **17,** 641 (1934).

63. — — — Die Identität der α-Scillansäure mit Allocholansäure. Helv. chim. Acta **18,** 644 (1935).

64. TSCHESCHE, R. u. H. A. OFFE: Über Krötengifte. I. Die Selen-Dehydrierung des Cinobufagins. Ber. dtsch. chem. Ges. **68,** 1998 (1935).

65. — — Über Krötengifte. II. Zur Kenntnis des Cino- und Marinobufagins. Ber. dtsch. chem. Ges. **69,** 2361 (1936).

66. WIELAND, H. u. R. ALLES: Über den Giftstoff der Kröte. Ber. dtsch. chem. Ges. **55,** 1789 (1922).

67. WIELAND, H. u. H. BEHRINGER: Zur Konstitution des Bufotalins. XI. Über Krötengiftstoffe. Liebigs Ann. Chem. **549,** 209 (1941).

68. WIELAND, H. u. G. HESSE: Zur Konstitution der Giftstoffe der einheimischen Kröte. VIII. Liebigs Ann. Chem. **517,** 22 (1935).

69. WIELAND, H., G. HESSE u. R. HÜTTEL: Zur Kenntnis der Krötengiftstoffe. IX. Weiteres zur Konstitutionsfrage. Liebigs Ann. Chem. **524,** 203 (1936).

70. Wieland, H., G. Hesse u. H. Meyer: Der Abbau des Bufotalins zu einer Cholan-säure. Über Krötengiftstoffe. VI. Liebigs Ann. Chem. **493**, 272 (1932).
71. Wieland, H., G. Hesse u. H. Mittasch: Über basische Inhaltsstoffe des Haut-sekrets der Kröte. V. Ber. dtsch. chem. Ges. **64**, 2099 (1931).
72. Wieland, H., W. Konz u. H. Mittasch: Die Konstitution von Bufotenin und Bufotenidin. Über Krötengiftstoffe. VII. Liebigs Ann. Chem. **513**, 1 (1934).
73. Wieland, H. u. F. Vocke: Über die Giftstoffe der japanischen Kröte. IV. Über Krötengiftstoffe. Liebigs Ann. Chem. **481**, 213 (1930).
74. Wieland, H u. F. J. Weil: Über das Krötengift. Ber. dtsch. chem. Ges. **46**, 3315 (1913).
75. Wieland, H. u. Th. Wieland: Zur Kenntnis der Krötengiftstoffe. X. Die Konstitution des Bufothionins. Liebigs Ann. Chem. **528**, 234 (1937).
76. Windaus, A. u. O. Stangel: Über das Provitamin des Eiersterins. Hoppe-Seyler's Z. physiol. Chem. **244**, 218 (1936).

Addendum.

Recently, it was shown by K. Meyer [Zur chemischen Konstitution der herz-aktiven Krötengifte; Experientia **4**, 385 (1948)], by the investigation of some degradation products, that bufalin and digitoxigenin differ only in the lactone ring. The positions of the two hydroxyl groups and the spatial configurations were found to be identical in both compounds.

(Received, May 12, 1948.)

Biochemistry of Fish Proteins.

By E. GEIGER, Los Angeles, California.

With 1 Figure.

Introduction.

At the end of the last century, MIESCHER (66) made a careful study of fish protein in his series of classic investigations on protein metabolism in the Rhine salmon during milting. These experiments, their extension by KOSSEL (55), and the valuable contribution of GREENE (39), GREENE (40) and WEISS (106) constitute one of the most stimulating chapters in "dynamic" biochemistry. Except for this unique achievement, the literature related to the biochemistry of fish proteins unfortunately has very few highlights. For the most part, the work done is a recapitulation of mammalian biochemistry; some fish tissues have been analyzed for the presence or absence of compounds previously demonstrated in mammalian or avian tissues. Special investigation of the peculiarities of fish protein, studies of the physical and chemical properties which differentiate these particular substances as "fish proteins" have not yet been attempted.

It is surprising that so little effort has been made to correlate the characteristics of fish protein with the internal and external milieu. Although the profound influence of ionic strength and urea content of the medium on the solubility and other properties of protein are well known, practically no study has been published comparing the physico-chemical properties of protein present in the "euryhaline" or "steno-haline" fishes [BALDWIN (7)] or in species with a physiological uremia.

The environmental extremes in which fish are found are such as to expose the tissues of the poikilotherm fishes, without heat regulation, in the Arctic waters and in equatorial streams to vastly different temperatures; species which live in shallow mud are subject to very different pressures from those which live at the bottom of the deep sea; and, finally, representative fishes may be found in surroundings with widely varying osmotic characteristics. Careful investigation of the protein of so diversely adapted organisms may divulge much information on the biological significance and the adaptative changes of the proteinaceous substances.

The term "fish" comprises nearly 40 000 different kinds, so that in presenting the following review on fish proteins, we wish to emphasize with McCay (62) that, "the carp or goldfish is as different from the trout ... as the cat is from the cow; this restricts generalization".

The papers reviewed below have been grouped relative to the source of protein investigated; thus, we will first consider the proteins derived from fish muscle, then the blood proteins, proteins with specialized functions such as collagens and enzymes, and finally, the proteins of the generative organs.

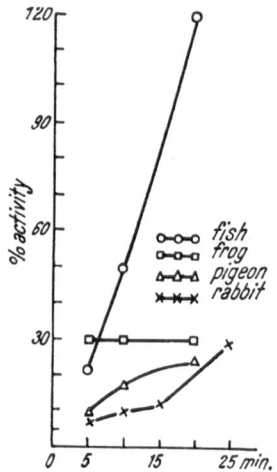

Fig. 1. Relation between Activity of Myosin and Time of Extraction. Minced muscle suspended in 0,6 M KCl at 0° C. The muscle extracts from *Amiurus nebulosus* and *Esox lucius* show, in contrast to other muscle extracts, 120% activity after as short a time as 20 min. (GUBA).

Composition of Fish Muscle Protein.

The protein content of fish muscle tissue seems to vary widely depending on the species investigated, as demonstrated in Table 1. A major difficulty in the evaluation of the available data is that some authors base their calculations on the total nitrogen content of the fresh muscle without taking into consideration the non-proteinic nitrogen constituents such as, peptides, amino acids, nucleic acids, and other extractives. Other authors determine the nitrogen content of the acetone-dehydrated and extracted material. Finally, some investigators extract the dehydrated, solvent extracted material with hot water, a treatment which removes a considerable amount of gelatin like proteinaceous substances from some fish. This also happens when the muscle protein is heat-coagulated from the tissue in slightly acid water (pH 4–6).

The magnitude of the error thus introduced may be judged from the results of ATWATER (4), who found that the hot water soluble, gelatinaceous material which on cooling sets to a firm jelly, was 9,46% of the protein in the case of herring, 16,36% in haddock, and 5,27% in salmon. Another requirement which has not been considered seriously enough is the absolute freshness of the investigated material. It is well known that chemical changes occur very rapidly in fish muscle tissue [GEIGER *et al.* (36); SIGURDSSON (89)].

REAY and KUCKEL (77) found in haddock, and BAILEY (5) in torpedo fish that the amount of protein which is soluble in LiCl or in $N/100$ HCl decreases considerably during storage. SCHMIDT-NIELSEN and STENE (86) found 5,5% of the cod and plaice muscle protein to be soluble in 1% NaCl; on keeping the fish, this fraction decreased.

Table 1. Protein Content of Fish Muscle.

Fish	Protein in Fish Muscle % N × 6,25	Water Content (%)	Protein in Dry Muscle (%)	Reference
Cod steak................	18,7	79,7	91,1	(25)
Cod (edible portion)	16,7	82,5	93,1	(25)
Cod (raw)................	16,5	82,6	94,8	(18)
Herring (edible portion)	19,5	72,5	69,0	(25)
Herring..................			87,2—87,6	(54)
Haddock.................			86,6—89,3	(54)
Halibut..................	18,6	75,4	75,5	(60)
Salmon	17,4	63,4	48,0 (?)	(60)
Cockle...................	12,31—15,62	80,0—84,0	78,0	(45)
Bengal fish	14,09	70,87	48,0	(84)
Swordfish................	18,7	76,6	80,0	(61)
Swordfish................	18,8	74,9	75,0	(18)
Carp sucker..............	19,2	76,2	80,6	(18)
Carp or German carp......	18,2	77,9	82,3	(18)
Bass (Atlantic Black Sea) ..	19,2	79,3	92,75	(18)
Bass (Black, Large and Small Mouthed)	20,6	76,7	88,4	(18)
Bass (California)..........				
Bass (White Sea).........	21,4	76,3	90,3	(18)
Bass (Striped)............	18,9	77,7	84,7	(18)
Flounders (Summer and Winter)................	14,9	82,7	86,1	(18)
Tuna Yellowfin............	24,7	71,5	87,0	(60)
Tuna Yellowfin............	23,28	74,7	92,0	(44)
Albacore.................	25,7	72,99	94,5	(44)
Pilchard (canned).........			64,4	(44)
Spanish Mackerel (canned) .			89,1	(44)
Bonito (including California, Atlantic and striped).....	24,0	67,6	74,07	(18)
Butterfish and Dollarfish ...	18,1	71,4	63,2	(18)

The rapidly occurring changes in the composition of fish after death may also be responsible for the observation of BAILEY (6) that different samples of fish muscle, in contradistinction to mammal tissues, show large differences in the myosine composition *inter se*.

More consistent data can be expected only if absolutely fresh muscle tissue will be used in all cases, and if the calculations will uniformly be based on the nitrogen content of well defined fractions which were separated from the total nitrogenic constituents present in the whole muscle, on the basis of solubilities or other physicochemical properties. Such attempts were made by FÜRTH (31) who, as early as 1895, isolated different protein fractions from the fish muscle using methods based on the experiments of KÜHNE and HALLIBURTON. In his classical paper, "Über die Eiweiß-Körper des Muskel-plasmas", FÜRTH described the separation from carp muscle, as well as from other "commercial" salt water fishes, of a substance termed "Myoproteid" which he did not find in mammalian muscle. "Myoproteid" does not precipitate even if heated to 100° C. at neutral reaction, but only in a strongly acetic acid medium. According to recent investigations it appears to be a denaturation and dissociation product of myogen [ROTH (82)]. The properties of "Myoproteid", as well as some other substances isolated by earlier authors from fish muscle (for example, the soluble "myogen-fibrin" of PRZIBRAM) are discussed in detail by SAMUELY (85) in ABDERHALDEN's Biochemisches Handlexikon.

Recent developments in the bio- and physicochemistry of the muscle have made the value of these earlier attempts obsolete.

In 1930, LOGAN (59) investigated the soluble muscle proteins of haddock and extracted brine frozen fish with solutions of different ionic strength.

5 g. of ground tissue was added to 50 ml. of water or salt solution in tubes and kept at 10° C. for several days. The nitrogen content of the filtered samples was determined after making due allowance for the non-protein-N (by precipitation with trichloroacetic acid), and it was shown that a considerable protein fraction dissolves in water alone. With increase of the salt concentration of the extracting fluid, there was first an increase in dissolved protein, then the solubility curve flattened. A maximum solubility was obtained using 0,3–0,6 molar neutral potassium phosphate. The author found that the water soluble proteins which coagulate at 70° C. at p_H 6 (another slight flocculation occurs at p_H 5), were different from the protein which was extracted with molar NaCl solution and heat-coagulated at p_H 5 (but not at 6). Unfortunately, LOGAN did not consider the possibility of changes by proteolytic enzymes which might have even at 10° C., taken place during the incubation period of several days.

Recent research in this field seems to be characterized by the tendency to compare the fish muscle fractions with the better investigated mammalian muscle, and most authors seem to be satisfied with the demonstration of identities, similarities, or dissimilarities between the

Table 2. Distribution of Nitrogen between the Proteins of Skeletal Muscle [BAILEY (5)].
(g. N/100 g. of total coagulable nitrogen.)

Muscle	Composition of Intracellular Fraction (N as percentage of total intracellular nitrogen)					
	Stroma Protein	Intracellular Protein	Myosin	Myogen	Globulin X	Myo-albumin
Rabbit	15	85	67,5	9	22,5	1
Rabbit	17	83	47	26,5	26,5	—
Haddock........	3	97	69	18,6*	Residual Intra-cellular 12,4	

* This figure includes globulin X if present in fish muscle.

fractions investigated. Table 2, taken from a paper by BAILEY (5), shows the results of such comparison. BAILEY states that fish myosin is more difficult to prepare than mammalian myosin [see below, ROTH (82)] since the KCl solution often becomes viscous and difficult to filter.

Analysis has shown that while the values of the tyrosin content of fish myosin (3,90–4,40%) are somewhat higher than those of other animals (3,22–3,40%), the figures obtained with different fish samples did not agree well *inter se*. The total sulfur content in mammal or chicken myosins varies between 1,06 and 1,12%, but in fish it rises to 1,21%. We may conclude that not only is the relative ratio of the proteins in the fish muscle somewhat different from that in the rabbit, but the chemical composition of the fish myosin itself is also different from that in other animals.

GALVYALO and GORYUKHINA (32) compared the muscle proteins of the rabbit with those of the mirror carp and found by electrodialysis that the aqueous extract of the fish muscle yielded only 50% as much albumin-nitrogen as did the rabbit muscle.

The fractions obtained from mirror carp by successive extractions with (a) water, 0,25% acetic acid and 0,25% NaOH, (b) water, borate buffer of pH 9, and NaOH, and, (c) NaOH alone, contained the same percentage of nitrogen as corresponding fractions of rabbit muscle.

According to a private communication by Dr. A. SZENT-GYÖRGYI, GUBA (42), found that the "actin" of the fish muscle goes into solution more easily than that of the mammal muscle. Some of GUBA's results are demonstrated in Fig. 1, p. 268.

More recently, ROTH (82) investigated the muscle proteins of some teleost fish such as carp and tench.

She found that at pH 8, three fractions of the fish protein are soluble in aqueous 0,6 M-KCl solution. After only two hours this extract was as viscous as mammalian muscle extracts after twenty-four hours [cf. BAILEY (5)]. When the extract is dialyzed against double-distilled water at pH 7, then 35 to 40 per cent of the extracted protein precipitates at 0,24 M-KCl (fraction 1). The isoelectric point of this protein is pH 5,4 and when re-dissolved in 0,36 to 0,6 M-KCl, it is characterized by high viscosity, flow birefringency, and filament formation. Because of these properties, and on the basis of its sedimentation constant, this fraction was assumed to be identical with SZENT-GYÖRGYI's (93) "heavy" myosin (acto-myosin); "light" myosin was not present. This seems to indicate, however, a difference in the adenosin-triphosphate metabolism rather than in the protein composition itself.

On further dialysis, or on extraction at a lower ionic strength, another protein (fraction 2) has been isolated which seemed to be identical with the "globulin X" fraction of WEBER. After precipitation of this fraction, only albumin remained in solution (fraction 3). This albumin is characterized by a rapid, spontaneous denaturation, by its specific rotation measured at pH 6 and pH 2,5, and its lability in acid media. Fraction 3 seems to be identical with frog muscle myogen but different from mammalian myogen.

Summarizing these investigations, they indicate that both the relative quantity of the stroma protein (BAILEY) and the quality of the albumin fraction of the myogen, seem to be different from corresponding fractions obtained from mammalian muscle. In addition, BAILEY's amino acid analyses suggest that, in spite of the similarity in the physicochemical properties, the fish myosin may also be different from that of the other muscles.

Amino Acid Content of Fish Muscle Protein.

Considerably more work has been done in the field of the amino acid composition of muscle proteins. This work was governed mainly by practical considerations, and was carried out in order to decide whether the protein fractions of the edible commercial fish are adequate to supply our organism with all necessary building stones for protein synthesis. This nutritional viewpoint seems to be responsible for the fact that most

Table 3. Amino Acid Composition

	Mackerel (36)	Swordfish (61)	Fish (?) Muscle (11)	Cod (1)	Cod (8)	Cod (2)	Sprat (2)	Roach (2)	Salmon (8)	Halibut (19)	Beef (8)
1. Arginine	7,1	9,1	5,6	6,8	6,26	9,0	7,20	7,30	6,39	6,34	6,91
2. Histidine	3,8	3,3	1,9	4,8	2,04	1,6	1,70	1,45	2,30	2,55	2,25
3. Lysine	11,4	7,2	6,8	8,0	8,40	4,8	6,0	4,08	9,00	7,45	8,11
4. Tyrosine	1,55		4,0	2,0	4,54	4,75	3,60	4,60	4,38	2,39	4,30
5. Tryptophane	1,7	1,3	1,3	2,1	1,28	0,67	0,85	0,68	1,38	+	1,35
6. Phenylalanine	4,74	4,1	4—5	1,1	4,28	14,4	14,0	14,8	4,52	3,04	4,92
7. Cystine and Cysteine	2,6		1,2	0,6	1,04 1,19	1,8	1,6	1,6	0,74 1,22		0,94 1,29
8. Methionine	4,4	3,8	3,4	0,3	3,16 3,78	2,2	2,3	2,2	3,17 3,16		3,11 3,17
9. Serine			4,5	1,8	4,94	3,50	2,45	4,45	3,96		5,43
10. Threonine	5,2	2,6	4,4		4,52	0,57	0,58	0,58	4,19		4,57
11. Leucine	6,9	7,8		7,5		16,3	15,5	18,0		10,33	
12. Isoleucine	5,9	7,5		1,5							
13. Valine	6,2	4,7		3,7		0,66	0,55	0,64		0,79	
14. Glutamic acid	9,6		14,0	7,5		22,3	24,4	20,4		10,13	
15. Aspartic acid	9,5			0,6!						2,73	
16. Glycine	5,2					1,6	1,80	2,70		0,0	
17. Alanine			7,0	5,7						?	
18. Proline			3,0	2,8		5,1	5,0	7,1		3,17	
19. Hydroxyproline				0,9			5,0				
20. Norvaline				0,4							
21. Norleucine				0,4							
22. Hydroxylysine						1,15	1,40	1,32			
23. Ammonia						1,50	1,95	1,80			
Total						91,2	90,9	93,7			

investigators restricted their determinations to the "essential" amino acids. Some results of such analyses are condensed in Table 3. They have only restricted value; it should also be noted that ÅGREN's figures differ sufficiently from other data to require a re-investigation; it seems doubtful whether the composition of the Swedish fish is responsible for the results, or rather the methods employed.

The papers of ROSEDALE (81) and of KERNOT, KNAGGS and SPEER (53) on the diamino-acid content of fish muscle are mainly of methodical interest.

The reasons for the restricted value of the data in Table 3 follow: (a) In most cases, the proteins have not been isolated and purified, and hence, the figures also include the amino acids present in the non-protein fraction. (b) It has not yet been satisfactorily established to what extent the quantity of the individual amino acids in the hydrolysate corresponds to the amounts which were originally present in the protein. We do not

of Fish Muscle Protein.

Cod (73)	Croaker (73)	Haddock (73)	Halibut (73)	Sea Herring (73)	Lake Trout (73)	Boston Mackerel (73)	Spanish Mackerel (73)	Mullet (73)	Pilchard (73)	Red Snapper (73)	Chum Salmon (73)	King Salmon (73)	Silver Salmon (73)	Shad (73)	Squeteaque or Sea Trout (73)
5,58	5,81	5,70	6,00	5,09	5,73	5,78	5,27	5,78	5,60	6,18	5,55	5,02	5,68	4,54	5,90
1,72	1,37	1,17	1,66	1,56	1,40	1,93	1,48	1,61	1,23	1,57	1,30	1,41	1,87	1,09	1,42
6,83	6,10	6,41	6,16	7,03	7,15	7,13	6,53	6,74	6,78	6,72	5,69	6,27	6,57	6,45	6,78
1,06	1,24	0,85	1,64	1,23	1,17	1,36	1,37	1,36	1,30	1,22	1,33	1,20	1,44	1,22	1,01
1,41	1,15	1,16	1,45			1,18	1,25	1,29		1,29		1,27	1,39	1,17	

know how much of the amino acids is destroyed by hydrolysis, since this destruction depends not only on the method employed but also on the presence or absence of non-proteinic compounds such as carbohydrates, fatty acids, etc. (c) The methods used for the estimation of the amino acids are not very accurate. A marked advance has been made in the development of microbiological methods, but th se are not yet applicable for the determination of all amino acids. It seems reasonable, however, to assume that the combination of chromatographic, microbiological, and enzymatic procedures (decarboxylase) may soon give a reliable picture of the amino acids present in fish proteins.

For the time being at least, comparable results can be expected only from investigations in which the same author determines the amino acid content of fish muscle and other proteins simultaneously, using identical methods not only for the estimation of the amino acids, but also for the isolation and hydrolysis. To this class of contributions belong the studies of BLOCK and BOLLING (*11*), POTTINGER and BALDWIN (*73*), and BEVERIDGE (*9*). The latter author determined the sulfur distribution in fat-free, vacuum dried samples according to BAERNSTEIN's method as modified by KASSELL and BRAND; these analyses were carried out in triplicate and quadruplicate with the results listed in Table 4.

Table 4. Distribution of Sulfur in Some Fish Proteins [BEVERIDGE (*9*)].

Source of Protein	Nitrogen (%)	Sulfur (%)	Methionine (%)	Cystin (%)
Lingcod................	16,51	1,21	3,82	1,46
Halibut................	16,50	1,18	3,66	1,47
Lemon sole	16,62	1,18	3,72	1,40
White Spring Salmon ...	16,60	1,17	3,64	1,36
Casein			3,15	0,53
Beef muscle............			3,19	0,97
Egg albumin			5,23	1,78

Table 4 shows that egg albumin is the richest in both sulfur-containing amino acids; in fact, BLOCK and BOLLING (*11*) list egg albumin as the protein with the highest methionine content. Fish flesh seems to be the next best source. According to SHARPENAK et al. (*87*), *Luciopera sandra* contains more methionine and cystin than beef.

Valuable are the investigations of BEACH, MUNKS and ROBINSON (*8*).

These authors cooked flesh samples of different animals for a short time in order to prevent enzymatic decay. The flesh was then ground, frozen in dry ice, desiccated by the "chryochem" procedure, and alternately extracted with alcohol and alcohol-ether in a SOXHLET apparatus. To remove the extractives, the minerals, and the carbohydrates, the samples were then given three 15-minute treatments with hot water. By this technique, 8 to 14 per cent of the total nitrogen was lost. The end products contained 14 to 17 per cent nitrogen (on a moisture and ash-free basis).

Comparing the respective data for cod, salmon, beef, veal, lamb, pork, chicken, turtle, frog and shrimp, the authors concluded: "In general, it can be seen that muscle tissues of these different classes of animals do not differ widely in their amino acid patterns, which implies that the same amino acid composition of muscle proteins is repeated throughout the animal kingdom and indicates that, as far as these ten amino acids are concerned, the protein of one muscle is as good as that of another in supplying amino acids in the diet."

Assuming, with due reservations, that the nutritive value of a protein is a function of only the ten "essential" amino acids estimated by the authors mentioned, a final conclusion could be made only if after determination of all the amino acids present, the sum of the determined amino acids would agree with the total-nitrogen and amino-nitrogen content of the initial protein. But even then, we would have achieved only the important inventory of the protein building stones. The pattern according to which these building stones are arranged in the individual fish proteins and which is responsible for the specificity of the respective proteins, would still be unknown. Neither X-ray investigations nor studies of the stepwise enzymatic degradation of fish proteins are available. Immunobiological studies referring to the antigenic specificity of the different fish proteins are also lacking.

Nutritive Value of Fish Protein.

Since prehistoric times, fish has been the standard food for many human races. For the nations of the Western civilization, however, fish has meant either an easily accessible, cheap substitute of meat or a food prescribed by religious rites or, at best, a delicacy for connoisseurs; it has not been regarded as equivalent to butcher's meat. The critical meat shortage during the first World War and, furthermore, the rapid development of biochemical knowledge has directed the attention of nutritionists, economists, and dietitians to fish as an important protein source. Stimulated by this practical interest, a long series of important studies on the nutritive value of fish proteins was initiated.

Feeding Experiments.

The first extended investigations were carried out in 1918 by DRUMMOND (25) in England. Based on the classical experiments on the biological value of foodstuffs of McCOLLUM and co-workers, DRUMMOND investigated the true food value of fish which he said, "will figure most prominently as a substitute ... in the diet of the nation in the near future". For this study, DRUMMOND selected representatives of three types of fish, viz. the herring *(Clupea harengus)* from the group of the so-called "fat" fishes; the cod *(Gadus morrhua)* as a representative of the "lean" fishes; and as a preserved fish, the canned salmon *(Oncorrhynchus chouicha)*. The protein of the meat was separated by heat coagulation in the presence of some acid and the coagulum was extracted with water, twice with hot alcohol, and finally with ether. The coagulated fish proteins showed growth promoting power when fed on a 6% basis, i.e. 6% of the total diet. This effect was as high as that of beef proteins and even superior to that of caseinogen.

DRUMMOND concluded that the high nutritive value of the fish was proved, and if taken advantage of, it could be the means of affecting considerable economies in the national protein requirements.

In the same year HOLMES (46) reported experiments from the United States in which the food value of the butterfish, the Boston mackerel, the grayfish, and of the salmon was investigated. He determined the nutritive value of the whole fish and not that of the isolated protein; therefore, his experiments should not be discussed further in the present review; however, it might be pointed out that according to HOLMES, the fish varieties investigated by him on humans represented a food of high digestibility and nutritive value.

SUZUKI's assays (92) in which the growth promoting value of the muscle protein of marine animals was investigated were not available to the writer, but according to the Chemical Abstracts, 5% of sardine and cuttlefish protein in the diet was sufficient to maintain the normal weight of rats, and 7% of the bonito protein was sufficient to promote growth. These results indicate that SUZUKI's fish proteins had a high biological value.

An important paper was published in 1928 by KIK and McCOLLUM (54) who investigated (a) the "food value" of protein present in fish meat and (b) the "supplementary value" of the fish protein when fed in combination with other proteins. These authors prepared the protein samples by cooking the fresh boned fish; the coagulated proteins, together with the cooking juices, were then dried and powdered.

KIK and McCOLLUM found that haddock and herring protein mixed with otherwise adequate basal diets at 9% and 15% levels promoted growth in rats over an extended period of time. Herring proteins seemed to be somewhat superior to those of haddock. In a second group of experiments, the critical protein level of 9% was used with an otherwise complete basal diet. In these assays, haddock or herring protein supplied 3% of the total protein intake and proteins derived from cereals and legumes supplied the other 6% (Table 5).

Table 5. Growth Promoting Effect of Some Fish Proteins.

6% Vegetable Protein in the Diet	Growth Promotion When Fed With 3% of	
	Haddock Protein	Herring Protein
Peas	Some	Less than haddock
Navy Beans	None	None
Wheat	Good	Better than haddock
Oats	Fair	Fair

We see from Table 5 that the fish protein mixed with navy bean protein had no growth promoting power. Mixed with pea protein, the herring protein showed less growth promotion than the haddock. In a mixture containing oat protein, the growth of the rats when fed on haddock and herring was fair, and in combination with wheat protein, the haddock showed a good growth promoting power. The herring protein proved to be even superior and compared favorably with equal amounts of protein derived from liver, steak or kidney. The most surprising result of the authors mentioned is that (excepting the satisfactory supplementary value of herring protein when fed in combination with wheat protein) the investigated fish protein proved to be inferior to beef-liver, steak or kidney proteins in all cases, and could not supplement the protein of legumes to any appreciable extent. Some of these results are in variance with more recent experiments and require re-investigation.

LANHAM and LEMON (56) obtained proteins from fish and from beef (ground round steak) by repeated acetone extraction. The protein content of their preparations was very high and the ether extractable material low. They proved first, in experiments on rats, that the acetone treatment did not destroy the nutritive value of their preparations, since the ratio, gain in weight/gram of consumed protein, for a group fed acetone-dehydrated haddock, was not significantly different from the value observed in another group which was fed haddock dried in a steam bath. The basal diets used proved to be complete so that the protein level was the only growth limiting factor. Male albino rats of a starting weight of 49–54 grams were fed the diets containing 9% protein *ad libitum* for a ten week period. The gains in weight were adjusted for the differences in protein intake by employing FISHER's method of covariant analysis according to TITUS and HARSHAW (97). Table 6 shows that these fish proteins were of excellent nutritive qualities. The nutritive values in rats of the respective proteins studied can be expressed relative to that of the protein derived from the oyster:

100%	90%	80%	63%
Oyster	Pilchard	Shad	Beef
	Red-snapper	Cod	
	Boston mackerel	Croaker	
	Shrimp	Silver salmon	

Table 6. Ratio of Gain in Weight per gram of Protein and Adjusted Gains in Weight for Different Protein Intakes During a Ten-week Period [LANHAM and LEMON (56)].

Protein Source	Number of Rats	Average Gain in Weight (g.)	Average Protein Intake (g.)	Gain in Wt. per g. Protein	Standard Error	Adjusted Gains in Weight
Shad	16	117,8	59,43	1,96	0,059	100,4
Cod	16	105,9	52,10	1,96	0,072	102,4
Croaker........	16	121,8	59,52	2,03	0,052	104,1
Boston Mackerel.	12	122,7	54,62	2,23	0,064	119,0
Silver Salmon ...	11	104,4	49,40	2,14	0,034	107,2
Pilchard	11	96,9	46,75	2,03	0,054	115,6
Red-snapper.....	10	88,4	46,72	1,88	0,027	115,6
Beef...........	12	95,2	58,00	1,64	0,048	81,9

Some feeding experiments with purified cod protein were reported briefly by ABDERHALDEN (*1*) in 1936 who did not give detailed information on the level at which this protein or the control casein was fed. Some other data [published in the Nova Acta Leopoldina, N. F. 3, 612 (1936)] were not available to the present writer.

ABDERHALDEN states in the note mentioned that during a 59-day period, the 22 rats fed fish protein increased their weight by 121%, and 23 rats kept on a casein diet, by 103%; this proves the high nutritive value of the cod protein which seems to be identical with the "Viking Eiweiß" to be discussed below.

A detailed study on the nutritive value of fish proteins was carried out by NILSON, MARTINEK and JACOBS (*69*). The protein samples used were prepared by thorough acetone extraction of fish meat, and were of the same origin and quality as those used by POTTINGER and BALDWIN (*73*) in their studies on the amino acid composition of fish proteins (Table 4).

The protein content of the samples was between 86 and 97%. The diet contained 9% protein derived from fish flesh and 1,2% from supplementary sources such as liver extract, dried yeast, wheat germ, and corn starch. Rats of an initial weight of 49 to 57 grams were used in groups containing 10 to 22 animals of both sexes. The animals were kept in individual cages and fed *ad libitum*.

Besides the nutritive value, the "apparent" digestibility of the total protein of the diet was also investigated. This was determined by collecting the feces during one whole week of the assay period. All of the proteins were well digested (Table 7) and very little deviation from the mean "apparent" digestibility values was observed (highest value, 7% over, and lowest, 2,4% under the mean value). The growth data as presented in Table 7 indicate that all the rats fed fish protein gained more in weight than those fed beef protein. When, however, the growth data were re-calculated on the basis of equal food intake, then there was no difference between the growth response to the beef and the marine proteins.

Such re-calculated results fail to agree with those of LANHAM and LEMON (*56*) who found growth on fish protein superior to that of beef. NILSON, MARTINEK and JACOBS (*69*) discuss the reasons which may be responsible for this contradiction. They point out that the rats in the experiments of LANHAM and LEMON did not grow as well as in their own assays, suggesting that some factors other than protein may have been the limiting growth factors in the earlier experiments, for example, insufficient vitamin supplementation or unsatisfactory control of the surrounding temperature. As a further possibility, NILSON *et al.* point out that their diet contained, in addition to the 9% fish protein, 1,2% protein derived from liver extract, dried yeast, wheat germ, and corn starch. Therefore, their study supplied information for the "balancing" rather than for the "true" nutritive value of fish proteins. The proteins

Table 7. Mean and Estimated Gain in Weight of Rats Fed for a 10-Week Period with Diets Containing 9 per cent. of Protein from Fishery Products [NILSON, MARTINEK and JACOBS (69)].

Source of Protein	Number		Mean Gain in Weight			Estimated Mean Gain for Group on Basis of Equal Food Intake (g.)	Apparent Digestibility (Mean)
	Males	Females	Males (g.)	Females (g.)	Group (g.)		
Beef round	12	10	136,3	121,8	129,7	146,2	88,5
Bonito	4	6	166,0	141,7	151,4	141,5	89,7
Catfish	4	6	190,8	117,3	146,7	150,7	90,6
Halibut	4	6	144,2	136,2	139,4	147,1	88,6
Herring:							
Lake........	7	4	152,7	113,5	138,5	152,7	96,2
Sea	6	4	183,7	130,3	162,3	163,2	89,6
Mullet	4	6	183,5	123,5	147,5	151,3	87,7
Salmon:							
Chinook	6	9	147,5	110,3	125,2	139,3	91,3
Chum	4	6	180,3	142,7	157,7	146,0	88,7
Coho	5	7	135,8	123,6	128,7	140,3	90,2
Pink	4	6	174,5	127,0	146,0	142,1	88,4
Sockeye	4	6	170,5	130,7	146,6	150,6	88,7
Squeteague	4	6	133,3	113,7	121,5	139,9	91,6
Trout, lake	4	6	177,0	126,7	146,8	133,5	89,6
Tuna:							
Albacore	3	7	150,0	154,0	152,8	138,2	90,0
Bluefin......	4	6	186,0	150,5	164,7	142,2	89,9
Skipjack	4	6	154,3	149,3	151,3	134,8	88,4
Yellowfin	4	6	164,8	129,5	143,6	140,6	90,7
Mean			162,8	130,1	144,5	144,5	89,9

of different origin (from yeast, wheat, etc.) may have supplemented the beef diet with such balancing factors which have masked the higher biological value of fish protein.

The observed data show a superior growth effect of fish protein, and the question arises, whether a re-calculation of the growth data by NILSON et al. is really justified. The theoretical background of this calculation will not be discussed but it should be emphasized that as long as we do not know which factors are responsible for the higher food intake or for the better appetite, we have to assume that proteins which have appetite stimulating properties are, at least for the growing rat, ipso facto of higher biological value.

The superior value of fish proteins was also demonstrated in careful, recent experiments by BEVERIDGE (9, 9a). Commercially important fishes such as lingcod (Ophiodon elongatus), halibut (Hypoglossus steno-lepsis), lemon sole (Parophris vetulus), white spring salmon (Oncorhynchus tshawytsha), herring (Clupea palasii), and red-snapper (Sebastodes ruberrimus) were investigated.

The beef sample was composed of cuts roughly representative of the whole steer. The meat samples were autoclaved for one hour at 99° to 100° C., pressed, minced, and partially dehydrated at 60° C. The cooking liquors were separated from oil or fat, evaporated to a thick syrup; after combining with the partially dehydrated meat, re-dried at 60° to a moisture content of 2 to 4 per cent.; and finally, ground to a fine meal.

Wistar strain rats were fed a diet containing a protein level of 8% and thus moderate yet sub-optimal growth was obtained. Paired feeding technique was applied and, besides the weight gain, the nitrogen retention was determined. A statistical analysis of the percentage nitrogen composition of the carcasses revealed that the gain in weight afforded

Table 8. Biological Values of Proteins of Fish and Beef Flesh and Egg Albumin. (Average initial weight of rats: 67,6 g. Range: 40–97 g. Days on test diets: 28. Number of rats per group: 10. The rats were paired-fed on the corresponding animals eating the lingcod diet which was given *ad libitum*.) [Beveridge (9a).]

Source of Dietary Protein	Daily Food Intake (g.)	Average Gain in Weight (g.)	Gain in Weight per g. Protein Fed (g.)
Lingcod	9,6	56,3	2,62
Halibut	9,0	50,3	2,47
Lemon Sole	9,0	50,1	2,50
Salmon	9,1	52,0	2,56
Beef	9,0	46,6	2,31
Egg albumin	8,7	45,6	2,34
Control diet E............	9,0	76,6	—

a true indication · of protein anabolism (Table 8). In Beveridge's experiments, the gains on fish flesh protein (corrected for varying food intake) were significantly greater than those achieved by beef flesh

Table 9. Biological Values of Fish and Beef Flesh Proteins. (Initial weight of rats: 50 ± 2,5 g. Days on test diets: 28.) [Beveridge (9a).]

Source of Dietary Protein	Sex	Number of Rats	Daily Food Intake (g.)	Average Gain in Weight (g.)	Gain Corr. for Food Intake (g.)	Gain in Weight per g. Protein Fed (g.)
Halibut............	M	20	10,3	75,5	74,9	3,27
Halibut............	F	20	10,5	70,1	68,1	2,98
Beef...............	M	20	10,2	65,3	66,0	2,86
Beef...............	F	20	10,1	59,7	61,6	2,63
Lingcod............	M	15	8,9	71,7		3,60
Lemon sole	M	15	8,9	72,8		3,66
White spring salmon.	M	15	9,3	77,2		3,68
Red-snapper........	M	15	9,0	78,8		3,86
Herring............	F	15	8,8	61,4		3,18

protein or egg albumin. These findings were corroborated in another group of experiments in which the growth of male and female rats was separately investigated (Table 9). Statistical analysis proved that the observed differences at a 9% level were highly significant for both sexes. At a level of 12% protein, only the difference in males was significant. It seems that at this level, the protein content was no longer the only limiting growth factor. The superiority of fish and beef protein over casein was evident, however, even at this level.

SAHA (84) found the proteins of Bengal fishes superior to casein in growth promotion experiments.

The biological value of the protein of the commercially important swordfish was recently investigated by LOPEZ-MATAS (60). Rats kept on paired feeding for 6 or 7 weeks received swordfish, beef, and chicken protein on an 8% basis. The condensed Table 10 includes the average

Table 10. Nutritional Value of Swordfish Protein. [LOPEZ-MATAS (60).]

	Total Diet Intake (g.)	Total Protein Intake (g.)	Total Gain in Weight (g.)	Gain in Weight per gram of Protein Intake
Swordfish Diet	417,7	33,4	107,0	3,22
Beef Diet	418,2	33,4	100,1	3,01
Chicken Diet	418,4	33,4	109,5	3,30

results obtained by the three groups, each comprising 21 animals. The *in extenso* published data show that swordfish protein in all cases is at least equivalent in growth promotion to chicken protein, and probably also to beef protein.

Evaluation of Results.

A critical evaluation of the experiments discussed in the present review seems to indicate that *fish protein has a high biological value* which is probably somewhat above that of beef protein. When investigating the causes, we have first to consider the amino acid composition. Comparison of the available analytical data gives no indication (Table 3), however, for any significant difference which could be made responsible for the superior growth of rats fed on fish protein, nor is it likely that the somewhat higher concentration in sulfur containing amino acids could be the cause, especially since fish protein was found by BEVERIDGE to be superior even to egg albumin which contained the highest amount of such amino acids. Before dismissing the difference in the amino acid composition, one should also consider that further improvement in the analytical methods and more comparative data comprising also the "non-essential" amino acids are needed; only then will we be able to decide which amino acids and in which relative doses are required in order to secure optimum growth in rats.

A second factor which may be responsible for the superior growth on fish protein is the high *digestibility* and the speed with which the individual amino acids are liberated from this protein. It was recently pointed out [MELNICK, OSER and WEISS (64); GEIGER (33)] that the rates at which amino acids are enzymatically released from the protein and absorbed from the intestines determine to a marked degree their growth promoting effect.

Unfortunately, there are no reliable quantitative data available to establish the generally assumed easier digestibility of fish proteins in mammals. The relatively low content of fish muscle on "stroma-protein" (3% in fish muscle against 15 to 17% in rabbit muscle; Table 2) may be at least partially responsible for the better digestibility.

Some physicochemical investigations by GUBA (42) suggest a further reason for such a property. In unpublished experiments in the present writer's laboratory, it was found that tryptic enzymes liberate tryptophan faster from fish protein than, for instance, from casein; and also that the bacterial histamine formation from fish protein proceeds at a higher rate than that from beef protein. These observations suggest that the grouping of some amino acids in the fish proteins is such that they are more rapidly released by enzymatic cleavage than from some other proteins. This question needs further investigation.

Finally, some specific *growth promoting factors* may be present in fish protein. Such a factor was recently isolated from the "fish solubles" by ROBLEE et al. (79). It is water soluble, acetone and ether insoluble, dialyzable, and heat resistant over a pH range from 3,0 to 9,0. A similar factor may be present in all fish protein samples which were prepared by acetone dehydration and not extracted with water. This point was investigated by the present writer by feeding rats acetone-dehydrated yellowfin and beef round steak protein respectively.

The proteins were incorporated in a complete basal diet at a level of 12%. Paired feedings were used. The growth curve showed in agreement with earlier experiments that the growth on fish protein was superior to that on beef. The finely ground beef and fish protein samples were then suspended in water and dialyzed through a cellophane casing against frequently changed distilled water (covered with toluene). After one week's dialysis, the protein was filtered, dehydrated by acetone and dried in vacuo. The feeding experiments were then repeated using the dialyzed samples.

The experiments show that the beef protein had the same growth-promoting power before and after dialysis; however, the growth on the dialyzed fish protein was somewhat lower than before but still superior to that on the beef protein diet. Evidently, some growth factor was lost by dialysis from fish but at least one part of the growth factors present in fish protein was not dialyzable and is probably not identical with the factor described by ROBLEE et al. (79).

This survey shows that the high biological value of fish protein in feeding animals has been definitely established; however, further experiments seem to be necessary on humans. The results mentioned above confirm the experiences collected in countries where fish meat represents the main source of protein, and also the following conclusions of HOLMES (46): "considering the experiments as a whole, the very complete utilization of the protein . . . supplied by the fishes studied, offer additional experimental evidence that fish is a very valuable food and that its extensive use in the dietary is especially desirable."

A more general use of fish proteins in our normal diet would relieve the world-wide protein hunger and replenish the steadily dwindling supply of meat. This was inhibited for centuries by geographical factors and by the serious problem of transportation because of the ease with which fish spoils. The modern canning and freezing techniques, however, eliminate this difficulty and thus, fish can be transported and stored without loss of its flavor or biological value. Unfortunately only a relatively small proportion of the fish is ·used for human consumption in the form of fresh, dried, frozen or canned products [see also WILDER and KEYS (107)]. The major portion is transformed into fish meal or fertilizers and is lost for *direct* human consumption. In some parts of the world, for instance in some canneries of Alaska, the not canned fish meat is often entirely wasted. With the increasing protein shortage, the problem arose whether it would be possible to save the proteins from the waste products. In order to produce a suitable edible material, the proteins have to be separated from the easily perishable components and from the factors which make such by-products unpalatable. Several years ago such proteins were prepared in the present writer's laboratory from cannery waste, using mackerel, sardine, or tuna, by isoelectric precipitation.

The biological value of such proteins was investigated in collaboration with DEUEL et al. (22) in rat growth assays and in experiments in which the recovery from hypoproteinemia was studied using CANNON's method. The growth of rats on these fish protein samples was superior to the growth on casein.

In the rats receiving 9% of the diet as tuna waste protein, the average of 157,6 grams of food was consumed over a 24-day period, and the average weight increase was 30,5 grams. In contrast, an average of 114,9 grams of food was consumed in the control casein tests, and the average gain in weight over the whole period amounted to only 8 grams. Even at a 21% level, at which casein shows optimum growth effect, the mackerel protein produced a 35% higher growth.

The results of the experiments with hypoproteinanemic rats were as follows: The weight increase on the casein diet was 20,2 grams higher, and on mackerel,

28,8 grams higher than the value which would have been expected, had the rats continued on the basal diet. The hemoglobin showed a rise from the sub-normal value, 8,37 grams, in the control rats to 9,83 grams in the casein group, and to 10,42 grams in the mackerel protein group.

Mackerel protein proved to be superior in growth promotion of protein depleted rats and in hemoglobin formation, but had a potency equal to casein in the regeneration of plasma proteins. The experiments were extended to vitamin A depleted rats using mackerel protein instead of casein in the diet prescribed by the U. S. P. method for bioassay of vitamin A (23). When uniform doses of vitamin A were administered daily, greater growth was obtained in rats receiving the fish protein diet than in the litter mates fed a corresponding casein diet. This indicates also that greater growth results with limited amounts of vitamin A when a protein of higher nutritive value than casein is employed.

The isolation of protein from fish on a commercial scale has been performed for several years in Germany. The "Viking-Eiweiß" or the so-called "synthetic egg white" was produced from cod by removing the relatively soluble material from diminuted fish with dilute acetic acid, extracting the lipids with a solvent, and treating the extracted tissue with warm dilute alkali (83). After neutralization with acetic acid and spray drying, a white powder was obtained which contained 93,9% protein and 4,65% moisture plus ash. The amino acid composition was the same as published by ABDERHALDEN (1) who apparently used this material for his analytical work. This fish protein was widely used mainly as a substitute of egg white.

Fish Blood Proteins.

The nature of the proteins present in fish blood seems to have been investigated first by NOLF (70) who established the presence of both globulin and albumin-like fractions. The problem was re-investigated with HOWE's more modern methods by LEPKOVSKY (57).

He isolated some globulin fractions from lithium oxalate plasma at 14%, 16% and 22% Na_2SO_4 concentration. Albumin was isolated at a 26% Na_2SO_4 concentration. Fibrinogen was determined separately by the addition of $CaCl_2$ and subtracted from the euglobulin value. The values for the total-N, albumin-N and globulin-N expressed in mg. per cc. were, 22,8, 1,28, and 0,86 for menhaden; 12,8, 0,62, and 2,14 for the goosefish; and 27,2, 1,00, and 2,8 for the dogfish.

These results are remarkable in the great variation which they indicate for different fish species. It should be pointed out that the variations in plasma proteins among teleosts themselves are greater than between some of the teleosts and the dogfish.

ZUNZ (109), using GRAM's method, investigated the fibrinogen content of different fish plasmas and according to his data, the values for teleosts were on the average somewhat higher than those for the selachians. The figures expressed as mg. fibrinogen per hundred cc. of plasma were, 273 to 355 for teleosts, and 109 to 263 for selachians. The corresponding values were given by the author as 224 for man, 293 for rabbit, and 352 for dog. Some data collected by FONTAINE (29), BOUCHET-FIRLY (13), and DEMENIER (21) are listed in Table 11 which demonstrates a considerable variation in the albumin: globulin ratio, as well as in the individual plasma protein concentrations. The authors state that differences in age, sex, nutritional status and seasonal variations may influence the concentration of plasma proteins. ROCHE, DERRIEN and CHOUAIECH (80) have recently demonstrated (what was already indicated by the results of Table 11), that it is possible to isolate the typical protein fractions which are present in mammalian plasma also from fish blood. These authors investigated the solubility of the serum protein fractions of teleost and selachian fishes at pH 6,5, in ammonium sulfate solutions of varying concentrations. Using the COHN equation, they were able

Table 11. Protein Content of Blood Serum in Fish.

Fish		Serum Proteins in g. per liter				Reference
		Total	Albumin	Globulin	A/G	
Anguilla vulgaris (from salt water)	T^1	56,25	17,81	38,44	0,46	(21)
Anguilla vulgaris (from fresh water)	T	58,57	15,62	42,95	0,36	(21)
Anguilla vulgaris (from salt water)	T	55,80				(29)
Anguilla vulgaris (from fresh water)	T	70,0				(29)
Labrus begryla..................	T	25,85				(29)
Gadus luscus	T	25,25				(29)
Conger vulgaris	T	44,50				(29)
Silliorhinus Cami	S	28,0				(29)
Scyllium catulus	S	28,0				(29)
Scyllium catulus	S	23,95	13,18	10,77	1,22	(21)
Raja clavata....................	S	24,78	3,37	21,41	0,15	(21)
Raja	S	34,0				(19)
Trygon pasticana	S	20,0	5,81	14,19	0,3	(21)
Torpedo marmorata	S	24,55	3,56	20,99	0,16	(21)
Petromyzon marinus.............	↑♂	39,34	3,87	35,47	0,11	(47)
Petromyzon marinus.............	♀	39,91	12,50	27,41	0,45	(47)
Alosa vulgaris	↑ T ♂	33,72	4,44	29,28	0,15	(47)
Alosa vulgaris	♀	37,74	3,37	34,37	0,10	(47)
Cyprinus carpio	↑ T	31,7				(29)
Esox lucius	↑ T	45,2				(29)
Tinca vulgaris..................	↑ T	29,0				(29)
Tinca vulgaris.................	T	29,85	10,50	19,35	0,50	(21)
Values for man................		65–80	40–60	15–25	2–2,3	

[1] T = Teleost. ↑ = freshwater fish. S = Selachian.

to demonstrate the identity of the precipitation constant (K') for euglobulin, pseudoglobulin, and the albumin fractions of different fish sera. They stress that the proteins of fish sera do not differ essentially from those of mammalian origin since the characteristic conditions for precipitation of each fraction are independent of the species from which the plasma was obtained. By means of the same technique, a close relationship between fish chromoprotein and horse hemoglobin was demonstrated.

Gon (38) found that the protein content of the blood of fresh water fishes showed seasonal variations. The lowest protein content was noted during the summer, while the minimum values for non-protein-N were found during the winter season. The seasonal variations in the blood protein level for the carp (Cyprinus carpio), the eel (Anguilla japonica), and the kamulchi (Ophiocephalus tradianus) are summarized in Table 12.

Table 12. Seasonal Variations in the Protein Level in Fish Blood [Gon (38)].

Season	Blood Protein (g. in 100 ml.) of		
	Carp	Eel	Kalmuchi
Winter.....	13,47	14,95	14,39
Spring	11,10	12,23	13,18
Summer....	10,77	9,55	10,12
Autumn....	11,78	12,82	13,92

Unfortunately, there are no indications whether the seasonal differences observed reflected changes in the environmental temperature or changes in the nutritional status of the fish, or possibly, some endocrine factors related to spawning. It would also be of interest to investigate possible seasonal variations in the albumin : globulin ratio. It is imperative that all data relative to the proteins of fish blood be reconsidered in the light of these seasonal fluctuations.

Bouchet-Firly and Fontaine (13) have presented evidence for fluctuations of another type in the blood protein concentration. They observed that the blood of the "euryhaline"[1] eel (Anguilla vulgaris) contained 5 to 7% protein when the fish had been taken from fresh water, but only 4 to 5,6% when taken from sea water. In contrast, the closely related but "stenohaline" conger, has a blood protein level which is independent of the salinity of the water. The authors mentioned conclude that the

Table 13. Blood Serum Concentration (g. per liter) of Anguilla vulgaris After Transfer from Fresh Water to Salt Water (47).

Days After Transfer	Total Protein	Albumin	Globulin	A/G Ratio
5	55,86	18,19	37,67	0,48
11	49,33	20,31	29,02	0,70
13	43,79	16,12	27,67	0,58
65	31,83	13,00	18,83	0,69

[1] "Euryhaline": to be capable of withstanding wide variations of external salinity, as opposed to "stenohaline" [Baldwin (7)].

differences observed may indicate an adaptation of the eel to changing salinity. IBANEZ and FONTAINE (47) tested this possibility and were able to demonstrate a progressive decline in the serum protein concentration after some "euryhaline" animals were transferred from fresh to salt water (Table 13).

The contention of FONTAINE and BOUCHET-FIRLY (29), "La concentration in protéines semble bien jouer un rôle important dans l'équilibre minéral du sérum des poissons.", however, remains somewhat ambiguous. The more marked increase in the concentration of the globulins, with a lesser change in the concentration of the osmotically more significant albumins would seem to throw some doubt on the validity of their claim. Furthermore, the data published by DEMENIER (21) on *Anguilla* do not conform to the theory outlined above.

The general problem of how the blood proteins have been adjusted to environments of varying salinity, and the problems raised by the highly variable blood urea levels in some fish, seem to offer a promising field for further biochemical investigations.

Fish Collagens.

The material present in the outer cover, in the scales, and in the swim bladders of several fishes is representative of the albuminoids known as collagens. Of these, the most thoroughly investigated is the commercially important *"Isinglass"* or "Ichthyocolla" which is prepared from sounds or swim bladders. The highest quality Isinglass originates from the swim bladders of the sturgeon, while the material extracted from cod, ling, hake or weakfish *(Otolithus regalis)* is less valuable because of its inferior appearance, fishy taste and strong odor. Isinglass is manufactured from the cleaned, salted and dried sounds by hot water extraction and dehydration under carefully controlled conditions. The methods of preparation have been discussed recently by PROCTER (73a).

The amino acid composition of Isinglass from hake swim bladder has been determined by BEVERIDGE and LUCAS (10). According to Table 14, it seems to be characterized by the presence of phenylalanine traces only, and by the absence of tyrosine and tryptophane.

Isinglass swells in cold water; it dissolves easily in hot water, dilute acids or alkalis, but is insoluble in alcohol. Prolonged boiling hydrolyzes this collagen to gelatin. Solutions of Isinglass in tartaric or sulfurous acid are known as "finnings", and are used in the clarification of wine and beer. This clarifying action probably results from adsorption of suspended particles, the formation of insoluble tannates, and the precipitation of colloids [cf. FAURE-FROMIET and COUGNEY (25a); CARPENTER (14)].

Table 14. Amino Acid Composition of "Isinglass" [Beveridge and Lucas (10)].

	% of total Nitrogen
Arginine..........	16,5
Histidine	1,62
Lysine	4,58
Hydroxylysine	0,34
Tyrosine	0
Tryptophane.......	0
Phenylalanine......	0,85
Cystine...........	0,06
Methionine	1,43
Serine............	3,03
Threonine	2,08
Leucine	2,21
Isoleucine	1,76
Valine	1,50
Glutamic acid	4,27
Aspartic acid	2,73
Glycine	9,79
Alanine	12,7
Proline...........	11,88
Hydroxyproline	2,75
Ammonia..........	3,09
Total...	83,17

In the search for a suitable blood substitute during the Second World War, Taylor (94, 95), Waters (105), Pugsley and Farquharson (74) considered the use of intravenously injected Isinglass solution. First they established the fact that Isinglass was non-antigenic, i. e. neither animals nor humans could be sensitized by injection of the material. Further experiments showed that 4 to 7% Isinglass solutions in isotonic saline were capable of restoring the circulation of animals in experimental traumatic or hemorrhagic shock. The material proved to be also effective when injected into patients during shock or acute hemorrhage. It was also established that Isinglass injections do not interfere with the regeneration of plasma proteins.

The assumption, made by Taylor and Moorhouse (94), that injected Isinglass may be utilized for the building of body tissues seems, however, to be open to criticism. First of all, Isinglass is an incomplete protein lacking some of the essential amino acids [Beveridge (10)]; secondly, according to Waters (105), the renal excretion of this material is so rapid that the urine may contain up to 11,5% Isinglass and have a tendency for gelatination on cooling.

Practically nothing is known about the properties and composition of the proteinaceous material originally present in the skin, scales or bones of living fish. Post mortem studies are complicated by the ease with which it is enzymatically hydrolyzed, decomposed by bacteria or degra ed by careless extraction methods. The degradation products of collagen, viz. "fish gelatins" or "fish glues", are manufactured from fish refuse including skin, head and bones. The possibility of producing high-grade gelatine from fish has been investigated, especially by some Japanese authors.

According to Smith (90), Kikuti prepared an excellent gelatin from dried sharkskin. This material contained 16,9% N; its 10% gel melted at 29,5° and solidified at 23,3° C. The X-ray diffraction diagram was identical with that obtained from high-grade gelatin used for photographic purposes.

Fish gelatins and glues are of technological importance but of restricted interest to the biochemist. Re-investigation by modern methods will be necessary in order to determine how far the "Ichthylepidin" isolated from fish scales or the "Elastoidin" prepared from the fins of the shark is a representative of a special class of compounds. For details see SAMUELY (*85*).

Fish Proteins with Enzymatic Activity.

NORRIS and his collaborators (*71*, *72*) have isolated crystalline proteins with strong proteinase-activity from the stomach lining of the king salmon and the yellowfin tuna *(Neothunnus macropterus)*. The methods used were modifications of the procedures formerly used on pepsin. The isoelectric point of the salmon pepsin as determined by electrophoresis, was found to be p_H 3,0, while the minimum solubility method gave p_H 2,9; the isoelectric point of the tuna pepsin determined by the latter method was p_H 3,0. The non-identity of the fish enzyme samples and those isolated by NORTHRUP from mammals were suggested by the needle-type crystals of the former. Further analysis of the amino acid composition showed that the salmon enzyme contained about twice the cysteine, half the tryptophane and about 0,7 as much tyrosine as the swine pepsin.

The elementary analysis of salmon pepsin gave the following results: C: 51,9%; H: 61,48%; N (DUMAS): 15,62%; N (KJELDAHL): 15,2%; S: 1,58%; P: 0,031%; and Ash: 0,08%.

By the method of ANSON and MIRSKY, tuna pepsin was shown to have an activity approximately twice that of swine pepsin with a p_H optimum at 2,5 to 2,6. The gastric enzymes of halibut, bluefin tuna and albacore were also crystallized, according to a private communication with Dr. E. R. NORRIS.

Proteins Isolated from Fish Sperm.

Although protamines were first obtained from fish sperm by MIESCHER (*66*) as early as 1868, their proteinaceous nature was not recognized until 1884. Then KOSSEL (*55*) showed by analysis of protamin hydrolysates that the protamines like typical proteins are built of amino acids in peptide linkage. It was further demonstrated that protamines are biologically derived from ordinary proteins which, in the course of spermatogenesis, are converted first to histones and, finally, to protamines. These changes are characterized by a loss of some mono-amino acid units and a relative increase of the diamino acid content which results in a strongly basic character.

In contrast to genuine proteins, histones and protamines are acid soluble and are not coagulated by heat under the usual conditions. The contention that proteins are converted to histones and protamines was supported by GILL (37), who showed that during mating, the arginine content of the proteins of the male gonad in the herring increases by 30 to 50%; no corresponding changes were observed in the ovaries. KOSSEL (55) working on *Gadus*, and HAGERTY and GEIGER (44) working on yellowfin tuna, have isolated histones which were probably intermediary products in the degradation of protein to protamines. These histones are characterized by a diamino acid content of 28 to 30% compared with that of 70 to 90% in protamines.

The composition, structure, and biological significance of the protamines and histones has been presented by KOSSEL in his important monograph (1928). KOSSEL's observations on the composition and structure of protamines have been generally confirmed and extended by WALDSCHMIDT-LEITZ (101, 103), RASMUSSEN (75), LINDERSTRÖM-LANG (76), FELIX (26, 27) and other authors. Interest in these compounds has been revived by the investigations of HAGEDORN (43) on protamin zinc insulin, and by the discovery of their effect on blood clotting.

WALDSCHMIDT-LEITZ et al. (103) isolated the enzyme "protaminase" from the mammalian pancreas which specifically splits only those proteins of small molecular weight which contain a terminal free carboxyl group belonging to an arginine unit. Fractional degradations with this enzyme have established the fact that the terminal carboxyl group of the protamines investigated is indeed provided by an arginine group. On the other end of the peptide chain, a proline residue contains a free imino group [RASMUSSEN and LINDERSTRÖM-LANG (76)].

FRAENKEL-CONRAT and OLCOTT (30) have recently re-investigated the structure of salmine (protamine from salmon sperm) and claimed, on the basis of esterifications and titrations, that it contains no free carboxyl. Experiments with formol titration, with ninhydrin and with manometric techniques further failed to demonstrate the presence of free amino or imino groups. They claim that "salmine may have a cyclic structure similar to those suggested for gramicidine and tyrocidine". On the other hand, TRISTRAM (98) repudiates this conclusion of FRAENKEL-CONRAT and OLCOTT, because their methods are not applicable to a protein with such "singular properties" as salmine. According to TRISTRAM, the earlier concept that salmine consists of a normal peptide chain containing 58 amino acid residues, with proline occupying the terminal imino position, is still valid.

WALDSCHMIDT-LEITZ et al. (103) have calculated the molecular weight of clupein to be 2021, and for salmin 2855 which values agree well with those determined by sedimentation. The molecular weights suggested by RASMUSSEN and LINDERSTRÖM-LANG (75, 76), by FELIX and MAGER (26), as well as by BLOCK and BOLLING (12) are twice to four times as high.

The sequence of the amino acids within the protamine molecules has been discussed by WALDSCHMIDT-LEITZ et al. (103), FELIX and MAGER (26), BLOCK and BOLLING (12), and GREENSTEIN (41).

According to MUJAKE (65), the isoelectric point for the various protamines lies between pH 9,7 and 12,4. Cyprinin (protamine from carp) which contains the least arginine also has the lowest isoelectric point.

Recent data on the amino acid composition of some protamines are presented in Table 15. Of particular interest are the observations of BLOCK and BOLLING (12) who demonstrated for the first time the presence of alanine and isoleucine in salmine.

Table 15. Amino Acid Composition of Protamines.

| Protamin | % of Total N | | | | | | | | | | Reference |
	Arginin	Histi-din	Lysin	Imino-N	Alanin	Serine	Valine	Pro-line	Gly-cine	Isoleu-cine	
Clupein	89,39			3,76	1,84	1,73	3,67				(26)
Iridin[1].........	86,95			6,11	2,08	2,02	3,50				(102)
Acipenserin[2] ...	70,8 to 78,6	9,4 to 11,2	5,3 to 7,2								(58)
Salmin	90,2				1,8	3,0	1,6	3,0		0,5	(12)
Salmin	89,0				0,56	3,94	1,20	2,29	1,78	0,56	(98)

Pharmacological Effects of some Histones and Protamins.

It is somewhat surprising that histones and protamines which are constituents of normal tissues, show a relatively high toxicity if administered parenterally.

Working in KOSSEL's laboratory, THOMPSON (96) demonstrated that as little as 0,2 grams of clupein carbonate injected into an anaesthetized 10 kg. dog, proved to be fatal. The main symptoms were, depression of blood pressure, leucopenia and retardation of blood coagulation. Salmine, sturine or histones had the same effect. This observation was confirmed by JAPPELLI (48) for the unanaesthetized dog. JAQUES, CHARLES and BEST (49) described the toxic effect of salmine (5–20 mg./kg.) in dogs as a decrease in blood pressure, dyspnea, retching, vomiting and diarrhea. VARTIAINEN and MARBLE (99) found that subcutaneous injection of salmin to mice, in doses of 250 mg./kg., leads to respiratory distress, muscular rigidity, opistothonus, polyuria, hypothermia, and hyperglycemia. A histological examination revealed vascular damage with hemorrhages into the thymus, lungs and kidney; furthermore, toxic tubular changes in the kidney and focal necrosis of the liver. According

---

[1] From trout sperm.
[2] From *Acipenser huso, guldenstadtii* and *stellatus*.

to Reiner, de Beer and Green (78), the LD_{50}, i. e. the medium lethal dose for mice was 94 mg./kg. when administered intraperitoneally. It was also observed that the oxygen consumption of minced rat liver when suspended in glucose-rat serum could be inhibited by the addition of protamine. Protamine has been found to be very toxic also to Trypanosoma equiperdum.

As early as 1931, McClean (63) stated that clupeine sulfate "inhibits the growth of B. typhosus in dilutions considerably higher than those in which phenol shows any effect".

Annau (3) suggests that these pharmacologic effects result from the imidazol and the arginine groups of the histidine and arginine building stones. This possibility was considered earlier by Kossel but it was discarded since globin, with basic character due to an imidazol group, is not toxic, and because the toxicity of protones which are produced by protamine hydrolysis is negligible, in spite of the presence of guanidin groups.

Another possibility was suggested by Shelley, Hodkins and Vischer (88) who showed that addition of salmin to blood caused precipitation and hemagglutination. When isolated organs were perfused with blood, the flow rates were markedly decreased after the addition of salmin. "These experiments suggest the possibility that protamines exert toxic effects through embolic vascular phenomena." This may in part explain the findings of Vartiainen and Marble (99); nevertheless, the effect of protamine on isolated tissues and lower organisms requires further investigation.

The delayed blood coagulation after injection of protamines, first noted by Thompson (96), was also studied by Waldschmidt-Leitz, Stadler and Steigenwaldt (100) who demonstrated an increased clotting time in vitro, after the addition of clupein, scombrin or salmin sulfate. Two important contributions refer to the probable mechanism of this action. Chargaff and Olson (17) found that stable emulsions of highly purified cephalin are precipitated immediately by the addition of salmin, over the wide range pH 2 to 11. This precipitate is not an adsorbate but a protamine salt, formed with the monobasic acid, cephalin. A similar compound is produced by lecithin, but only at pH 10 to 11. Since cephalin is one of the requisite physiologic agents in blood coagulation (Howell), Chargaff and Olson assume that the precipitation of cephalin may be responsible for the anticoagulant effect of protamines.

Mylon, Winternitz and Sütő-Nagy (67) further demonstrate the formation of a precipitate involving fibrinogen upon the addition of protamin to plasma. This precipitate may be redissolved by 3% NaCl from which it is coagulated at 54–56° C. The solution of the protamine-fibrinogen complex forms a gel only several hours after the addition of fresh serum or thrombin. Evidently, the precipitation of fibrinogen could also well explain the anticoagulant action of protamines. Either the cephalin or the fibrinogen precipitation may also be responsible for the formation of the emboli observed by SHELLEY et al. (88).

Recent investigations of CHARGAFF (15, 16) concerning the effect of protamin on blood coagulation disclosed the important fact that a small dose of salmin (11 mg./kg.) when injected into dogs, although it has no direct effect on blood coagulation, nevertheless interferes with the anticoagulant effect of heparin. In vitro experiments with chicken plasma confirmed the inactivating effect of protamine on some anti-coagulants such as heparin, polyvinyl and cellulose sulfuric acids, and cerebrons; however, it shows a weaker effect on hirudin and no influence et all on the anticoagulant properties of oxalate and citrate.

Using this antagonism as a tool, CHARGAFF (16) studied the delayed blood coagulation in patients with obstructive jaundice. Since the changed coagulation time remained abnormal after addition of protamine, the conclusion that heparin was not the responsible factor seemed to be justified. By a similar technique, WATERS, MARKOVITZ and JAQUES (105a) were able to demonstrate an increased heparin content of the blood in the hemorrhagic sequelae to anaphylactic or peptone shock.

CHARGAFF, as well as JORPES, PEHR and THANING (51) suggested the parenteral administration of protamin for the complete inactivation of administered heparin in the event of dangerous bleedings. They found a quantitative relation, with 1 mg. of protamin exactly "neutralizing" 0,3 mg. of heparin. Such small doses of protamin can be injected with no toxic effects, although larger quantities should be avoided since a slight excess of protamin may in itself produce anticoagulant action [THOMPSON (96); WALDSCHMIDT-LEITZ et al. (100)].

KERN and LANGNER (52) found that purified salmin shows no antigenic properties when injected into guinea pigs which is consistent with extended clinical observations using protamine insulins. Some conflicting data of WALTHER and AMMON (104) require re-investigation.

STEDMAN and STEDMAN (91) recently isolated a protein from cod sperm which, in contrast to the protamines and histones, is soluble in alkalis and precipitates with acids. It represents 60% of the total sperm; nucleic acid accounts for 28%, while the remaining 12% probably consists of protamines and histones.

The isoelectric point of this new protein lies between p_H 3,0 and 5,0. Preliminary analyses show that it contains about 25% basic amino acids, viz. arginine, 9,5%; histidine, 5%; and lysine, 11%. The presence of tryptophane and cystine as well as considerable amounts of glutamic and aspartic acids has also been established.

The authors assume that their "chromosomin" is identical with chromatin of the cell nucleus.

Proteins Isolated from Female Reproductive Organs of Fish.

Studies of these proteins have yielded relatively little information of biochemical interest. This field has been reviewed by Samuely (*85*), and by Needham (*68*) who discuss the comparative biochemical and morphogenetic aspects. More recently, Masuda (*61 a*) investigated fish roe proteins by complement and precipitin reactions and found a marked racial and organ specificity.

A large number of more or less clearly defined protein-like substances have been prepared from the roe-sack and from various fish eggs. The proteins derived from the egg membrane are probably pseudokeratins of mesodermal origin. Young and Inmann (*108*) have recently isolated from the egg membrane of the Atlantic salmon *(Salmo salar)* a compound belonging to this group which contains 15,3% total-N, 1,84% cystine, 1,42% tryptophane, 5,12% histidine, 3,51% lysine, 5,79% arginine, and 1,04% glucosamine.

The ichthulins which have been obtained from fish egg yolks appear to be chemically related to the vitellins found in avian yolks. The ichthulins and percaglobulin from the roe-sack of the perch, the clupeovin from the roe of *Clupea harengus* and the ovomucoids from the different fishes are chemically not well defined. Kossel's (*55*) following statement relative to protamines seems to be valid also with reference to these substances, "Many of the data go back to the years when there was no clear standpoint for the criticism of results and particularly, the analytical methods were not developed. Therefore, re-examination is desirable."

References.

1. Abderhalden, E., E. Baertich u. W. Ziesecke: Über den Gehalt des aus Kabeljau gewonnenen Muskeleiweiß an Aminosäuren. Hoppe-Seyler's Z. physiol. Chem. **240**, 152 (1936).

2. Ågren, G.: Amino Acid Composition of Some Species of Swedish Fish. Acta physiol. scand. **7**, 134 (1944).

3. Annau, E.: Das Verhalten der Histone im biologischen Versuche. Hoppe-Seyler's Z. physiol. Chem. **205**, 154 (1932).

4. Atwater, W. O.: The Chemical Composition of Food Fishes. Washington. 1891.

5. Bailey, K.: Composition of the Myosins and Myogens of Sceletal Muscle. Biochemic. J. **31**, 1406 (1937).

6. — The Proteins of Skeletal Muscle. Adv. in Protein Chemistry **1**, 289 (1944).

7. Baldwin, E.: Introduction to Comparative Biochemistry. Cambridge Univ. Press. 1937.

8. Beach, E. F., B. Munks and A. Robinson: The Amino Acid Composition of Animal Tissue Protein. J. biol. Chemistry **148**, 431 (1943).

9. Beveridge, J. M. R.: Sulphur Distribution on Fish Flesh Proteins. J. Fisheries Res. Board Canada **7** (2), 51 (1947).

9a. BEVERIDGE, J. M. R.: The Nutritive Value of Marine Products. J. Fisheries Res. Board Canada 7 (1), 35 (1947).

10. BEVERIDGE, J. M. R. and C. C. LUCAS: Amino Acids of Isinglass. J. biol. Chemistry 155, 547 (1944).

11. BLOCK, R. J. and D. BOLLING: The Amino Acid Composition of Proteins and Foods. Springfield: Ch. C. Thomas. 1945.

12. — — The Constitution of Salmine. Arch. Biochemistry 6, 419 (1945).

13. BOUCHER-FIRLY, S.: Sur la teneur en protéines du sérum d'anguille et de congré et ses variations au cours des changements de salinite. C. R. Séances Soc. Biol. Filiales Associées 115, 952 (1932).

14. CARPENTER, C. F.: Isinglass and Finnings. J. Incorporated Brewers Guild 23, 110 (1937).

15. CHARGAFF, E.: Protamine Salts of Phosphatides. J. biol. Chemistry 125, 661 (1938).

16. — Protamines and Blood Clotting. J. biol. Chemistry 125, 671 (1938).

17. CHARGAFF, E. and K. B. OLSON: Chemistry of Blood Coagulation. J. biol. Chemistry 122, 153 (1938).

18. CHATFIELD, CH. and A. GEORGIAN: Proximate Composition of American Food Materials. Circular No. 549, U. S. Dept. of Agriculture (1940).

19. CLARK, E. D. and R. W. CLOUGH: Nutritive Value of Fish and Shellfish. Report of the Bureau of Fisheries No. 1000, U. S. Dept. Commerce (1926).

20. DEAS, C. P. and H. L. A. TARR: The Value of Fish and Fish Proteins in Foods. Progress Rep. Fisheries Res. Board Canada 69, 66 (1946).

21. DEMENIER, G.: Sur la teneur en sérine et en globuline du sérum de quelque Poissons. C. R. Séances Soc. Biol. Filiales Associées 115, 555 (1934).

22. DEUEL, H. J., Jr., M. C. HRUBETZ, C. H. JOHNSTON, R. J. WINZLER, E. GEIGER and G. SCHNAKENBERG: Studies on the Nutritive Value of Fish Proteins. I. J. Nutrit. 31, 175 (1946).

23. DEUEL, H. J., Jr., M. C. HRUBETZ, C. H. JOHNSTON, H. S. ROLLMAN and E. GEIGER: The Use of Mackerel Protein in the Bioassay Test for Vitamin A. J. Nutrit. 31, 187 (1946).

24. DIRR, K. u. K. FELIX: Über Clupein. Hoppe-Seyler's Z. physiol. Chem. 205, 83 (1932).

25. DRUMMOND, J. C.: The Nutritive Value of Certain Fish. J. Physiology 52, 94 (1918).

25a. FAURE-FRÉMIET, E. and A. COUGNEY: Properties of Ichthycolla. Bull. Museum natl. hist. nat. (Paris) 9, 188 (1937).

26. FELIX, K. u. A. MAGER: Über Clupein. VIII. Hoppe-Seyler's Z. physiol. Chem. 249, 111 (1937).

27. FELIX, K., L. BAUMER u. E. SCHÖRNER: Über das Schicksal des Protamins in der befruchteten Eizelle. Hoppe-Seyler's Z. physiol. Chem. 243, 44 (1936).

28. FIRLY, S. et M. FONTAINE: Sur la relation existant dans le sérum d'Anguille entre le teneur en protéines et le rapport de la pression osmotique due au NaCl à la pression osmotique totale. C. R. Séances Soc. Biol. Filiales Associées 110, 471 (1932).

29. FONTAINE, M. et S. BOUCHET-FIRLY: Sur la teneur en protéine du sérum des POISSONS. Bull. Inst. Oceanography (Monaco) No. 610 (1932).

30. FRAENKEL-CONRAT, H. and H. S. OLCOTT: Possible Cyclic Structure of Salmine. Federation Proc. (Amer. Soc. exp. Biol.) 6, 253 (1947).

31. FÜRTH, O. VON: Über die Eiweißkörper des Muskelplasmas. Naunyn-Schmiedebergs Arch. exp. Pathol. Pharmakol. 36, 231 (1895).

32. Galvyalo, M. Y. and T. Goryukhina: The Properties of Muscular Proteins in Fish. Chem. Abstr. **37**, 1492 (1943).

33. Geiger, E.: Experiments with the Delayed Supplementation of Incomplete Amino Acid Mixtures. J. Nutrit. **34**, 97 (1947).

34. — Histamine Content of Fish. A Tentative Method for Quantitative Determination of Spoilage. Food Research **9**, 293, (1944).

35. — Physiology of the Utilization of Dietary Amino Acid Mixtures. Thesis, Univ. Southern California, Los Angeles (1946).

36. Geiger, E., G. Courtney and G. Schnakenberg: The Content and Formation of Histamine in Fish Muscle. Arch. Biochemistry **3**, 311 (1944).

37. Gill, R.: Variations of Amino Acid Composition of Herring. Dove Marine Lab. Report, N. S. **15–16**, 33 (1926).

38. Gon, S. K.: Blood Chemistry of Fish. III. Nonprotein Nitrogen, Chloride and Total Protein Contents of the Blood of Fresh-Water Fish in Winter; IV. In Summer; V. in Spring; VI. in Autumn. Chem. Abstr. **35**, 2616 (1941).

39. Greene, C. W.: Biochemical Changes in the Muscle Tissue of King Salmon During the Fast of Spawning Migration. J. biol. Chemistry **39**, 435 (1919).

40. Greene, K. H.: Changes in the Nitrogenous Extractives in the Muscular Tissue of the Fast of Spawning Migration. J. biol. Chemistry **39**, 457 (1919).

41. Greenstein, J. P.: Nucleoproteins. Adv. in Protein Chemistry **1**, 238 (1944).

42. Guba, T.: Observations of Myosin and Actymosin. Studies Inst. Med. Chem. Univ. Szeged (Hungary) **3**, 40 (1943).

43. Hagedorn, H. D., B. N. Jensen, N. B. Krarup and I. Wodstrup: Protamine Insulinate. J. Amer. med. Assoc. **106**, 177 (1936).

44. Hagerty, E. B. and E. Geiger (unpublished).

45. Heras, de la, A. R. and R. L. Costa: Spanish Fish. Chemical Composition of the Cockle (*Cardium edule* L.). Chem. Abstr. **41**, 5229 (1947).

46. Holmes, A. D.: Experiments on the Digestibility of Fish. U. S. Dept. agric. No. 649 (1918).

47. Ibanez, O. G. et M. Fontaine: Recherches sur les protéines du sérum sanguin de quelques Poissons migrateurs. Bull Inst. Oceanographique (Monaco) No. 679 (1935).

48. Japelli, A.: Sull'Azione Fisiologica di un Solfato di Protamino. Bull. Soc. Ital. Biol. Sper. **8**, 778 (1933).

49. Jaques, L. B., A. F. Charles and C. H. Best: The Administration of Heparin. Acta med. scand., Suppl. **90**, 190 (1938).

50. Jaques, L. B. and E. T. Waters: The Identity and Origin of the Anticoagulant of Anaphylactic Shock in Dog. J. Physiology **99**, 454 (1941).

51. Jorpes, E., E. Pehr and T. Thaning: Neutralization of Action of Heparin by Protamine. Lancet **237**, 975 (1939).

52. Kern, A. R. and P. H. Langer: Protamine and Allergy. J. Amer. med. Assoc. **113**, 199 (1939).

53. Kernot, J. C., J. Knaggs and N. E. Speer: Diamino-N of Cod Muscle. Biochemic. J. **24**, 378 (1930).

54. Kik, M. C. and E. V. McCollum: The Nutritive Value of Haddock and Herring. Amer. J. Hyg. **8**, 671 (1928).

55. Kossel, A.: The Protamines and Histones. London: Longmans Green and Co. 1928.

56. Lanham, W. B., Jr. and J. M. Lemon: Nutritive Value for Growth of Some Proteins of Fishery Products. Food Res. **8**, 549 (1938).

57. Lepkovsky, S.: Serum and Plasma Proteins in Fish. J. biol. Chemistry **85**, 667 (1930).

58. LISSITZIN, M. A. u. N. S. ALEXANDROWSKAJA: Chemische Zusammensetzung der Stöhrprotamine. Hoppe-Seyler's Z. physiol. Chem. 238, 54 (1936).
59. LOGAN, J. F.: The Soluble Proteins of the Muscle Tissue of the Haddock. Contr. Canad. Biol. and Fisheries (New Series) 6, 1 (1930).
60. LOPEZ-MATAS, A.: The Utilization and Nutritive Value of Swordfish. Thesis, Univ. of Massachusetts, Amherst (1947).
61. LOPEZ-MATAS, A. and C. R. FELLERS: Swordfish in Human Nutrition. J. Amer. Dietetic Assoc. 23, 1055 (1947).
61a. MASUDA, Y.: Immunological Studies of Fish-roe Protein. Jap. Z. Mikrobiol. Pathol. 27, 366 (1933); Chem. Abstr. 27, 4297 (1933).
62. McCAY, C. M.: The Biochemistry of Fish. Annu. Rev. Biochem. 4, 445 (1937).
63. McCLEAN, D.: Further Observation on Testicular Extract and its Effect upon Tissue Permeability. J. Pathol. Bacteriology 34, 459 (1931).
64. MELNICK, D., L. B. OSER and S. WEISS: Rate of Enzymatic Digestion of Proteins as a Factor in Nutrition. Science [New York] 103, 326 (1946).
65. MUJAKE, S.: Isoelektrische Punkte der Protamine. Hoppe-Seyler's Z. physiol. Chem. 172, 225 (1927).
66. MIESCHER, F.: Statistische und biologische Beiträge zur Kenntnis vom Leben des Rheinlachses im Süßwasser. Leipzig. 1897.
67. MYLON, E., M. C. WINTERNITZ and G. J. SÜTÖ-NAGY: Determination of Fibrinogen with Protamine. J. biol. Chemistry 143, 21 (1942).
68. NEEDHAM, J.: Chemical Embryology. Cambridge Univ. Press. 1931.
    — Biochemistry and Morphogenesis. Cambridge Univ. Press. 1942.
69. NILSON, H. W., W. A. MARTINEK and B. JACOBS: Nutritive Value for Growth of Some Fish Proteins. U. S. Dept. of Interior, Fish and Wildlife Service, No. 178 (1947).
70. NOLF, P.: De la richesse du sang de quelques animaux marines en globuline et albumine. Arch. int. Physiol. 4, 98 (1906).
71. NORRIS, E. R. and W. D. ELAM: Preparation and Properties of Crystalline Salmon Pepsin. J. biol. Chemistry 134, 443 (1940).
72. NORRIS, E. R. and J. C. MATHIES: The Gastric Properties of Yellowfin Tuna. Federation Proc. (Amer. Soc. exp. Biol.) 6, 281 (1947).
73. POTTINGER, S. R. and W. H. BALDWIN: The Content of Certain Amino Acids in the Edible Portions of Fishery Products. Proc. Sixth Pacific Science Congress 3, 453 (1939).
73a. PROCTER, F. G.: "Isinglass". J. Inst. Brewing 50, 117 (1944).
74. PUGSLEY, H. E. and R. F. FARQUHARSON: The Clinical Use of Isinglass. Canad. med. Assoc. J. 49, 262 (1943).
75. RASMUSSEN, K. E.: Clupeinuntersuchungen. Hoppe-Seyler's Z. physiol. Chem. 224, 97 (1933).
76. RASMUSSEN, K. E. u. K. LINDERSTRÖM-LANG: Clupeinuntersuchungen II. Hoppe-Seyler's Z. physiol. Chem. 227, 181 (1934).
77. REAY and KUCKEL: cf. BAILEY (5, 6).
78. REINER, L., E. J. DE BEER and M. GREEN: Toxic Effects of Some Basic Proteins. Proc. Soc. exp. Biol. Med. 50, 71 (1942).
79. ROBLEE, A. R., C. A. NICHOL, W. W. CRAVENS, C. A. ELVEHJEM and J. G. NALPIN: Some Properties of an Unidentified Chick Growth Factor Found in Condensed Fish Solubles. J. biol. Chemistry 173, 117 (1947).
80. ROCHE, J., Y. DERRIEN et M. S. CHOUAIECH: La précipitation des protéines sériques par le sulfate d'ammonium. La sérum des vertèbres marins. C. R. Séances Soc. Biol. Filiales Associées 130, 1301 (1939); Ann. Inst. océanograph. 79, 20 (1940).

81. Rosedale, J. L.: The Diamino Acid Content of Fish. Biochemic. J. 23, 160 (1929).
82. Roth, E.: Die Muskeleiweißkörper des Fisches. Biochem. Z. 318, 74 (1947).
83. Rudolph, W.: Eiweiß aus dem Meer. Med. Mschr. 1, 313 (1947).
84. Saha, K. C.: Biological Value of Proteins of Bengal Fish. J. Indian. chem. Soc. 17, 223 (1940); Chem. Abstr. 34, 5902 (1940).
85. Samuely, F.: Proteine der Tierwelt. Abderhaldens Biochem. Handlexikon, Bd. IV, S. 51. 1911.
86. Schmidt-Nielsen, S. and J. Stene: Biochemical Researches on Fish Muscle. Kong. norske Vidensk. Selsk., Forh. 20, 4 (1931); Chem. Abstr. 26, 2248 (1932).
87. Sharpenak, A. E., O. N. Balashova and I. P. Gureva: The Amino Acids of Fish Proteins. Chem. Abstr. 32, 5092 (1938).
88. Shelley, W. B., M. P. Hodgkins and M. B. Vischer: Studies on the Toxicity of Protamine. Proc. Soc. exp. Biol. Med. 50, 300 (1942).
89. Sigurdsson, G. J.: Chemical Tests of the Quality of Fish. Ind. Engng. Chem., analyt. edit. 19, 893 (1947).
90. Smith, P. J.: Glue and Gelatine. Brooklyn, Chem. Publ. Co. (1943).
91. Stedman, E. and E. Stedman: Chromosomine, a Protein Constituent of Chromosomes. Nature [London] 152, 267 (1943).
92. Suzuki, M., T. Jokuda, T. Okimoto and T. Nagasawa: Nutritive Value of Muscle Proteins of Marine Animals. J. Tokyo chem. Soc. 40, 385 (1919); Chem. Abstr. 14, 76 (1920).
93. Szent-Györgyi, A.: Chemistry of Muscular Contraction. New York: Acad. Press. 1947.
94. Taylor, N. B. and M. S. Moorhouse: The Use of Isinglass as a Blood Substitute in Haemorrhage and Shock. Canad. med. Assoc. J. 49, 251 (1943).
95. Taylor, N. B. and E. T. Waters: Isinglass as a Transfusion Fluid in Haemorrhage. Canad. med. Assoc. J. 44, 547 (1941).
96. Thompson, W. H.: Die physiologische Wirkung der Protamine und ihrer Spaltungsprodukte. Hoppe-Seyler's Z. physiol. Chem. 29, 1 (1900).
97. Titus, H. W. and H. M. Harshaw: A Method of Analyzing the Data of Nutrition Experiments. Poultry Sci. 14, 3 (1935).
98. Tristram, G. R.: Constitution of Salmine. Nature [London] 160, 637 (1947).
99. Vartiainen, I. and A. Marble: The Effect of Subcutaneous Administration of Protamine (Salmine) to Rabbits and Mice. J. Lab. clin. Med. 26, 1416 (1940/41).
100. Waldschmidt-Leitz, E., P. Stadler u. F. Steigerwaldt: Über Blutgerinnung. Hoppe-Seyler's Z. physiol. Chem. 183, 39 (1929).
101. Waldschmidt-Leitz, E., F. Ziegler, A. Schaffner u. L. Weill: Über die Struktur der Protamine. Hoppe-Seyler's Z. physiol. Chem. 197, 219 (1931).
102. Waldschmidt-Leitz, E. u. E. Kofranyi: Über die Struktur der Protamine. Hoppe-Seyler's Z. physiol. Chem. 236, 181 (1935).
103. Waldschmidt-Leitz, E. u. T. Kollmann: Über die enzymatische Spaltung der Protamine. Hoppe-Seyler's Z. physiol. Chem. 166, 262 (1927).
104. Walther, G. u. R. Ammon: Anaphylaktischer Schock nach Protamin. Klin. Wschr. 18, 288 (1933).
105. Waters, E. T.: Isinglass and Gelatine as Blood Substitutes. Canad. med. Assoc J. 44, 395 (1941).
105a. Waters, E. T. and L. B. Jaques: The Identity and Origin of the Anticoagulant of Anaphylactic Shock in Dog. J. Physiology 99, 454 (1941).
106. Weiss, F.: Untersuchungen über die Bildung des Lachsprotamins. Hoppe-Seyler's Z. physiol. Chem. 52, 107 (1907).

107. WILDER, R. U. and TH. A. KEYS: Foods for Emergencies. J. Amer. med. Assoc. **136**, 322 (1948).

108. YOUNG, E. G. and W. R. INMAN: Protein of Casing of Salmon Eggs. J. biol. Chemistry **124**, 189 (1938).

109. ZUNZ, E.: De la teneur fibrinogène du plasma chez les Poissons. Arch. int. Physiol. **37**, 274 (1933).

(Received, June 1, 1948.)

Some Recent Developments in Chemical Genetics.

By G. W. BEADLE, Pasadena, California.

With 9 Figures.

The purpose of this survey is to give non-biologists, particularly chemists, a brief view of the field of chemical genetics—its basic aims, its methods, and its accomplishments. It is not to be an exhaustive review but is written from what is admittedly a limited point of view. Since it is planned primarily for readers with a background in chemistry, an attempt is made to make it autonomous biologically—that is, to assume little specific biological knowledge.

Classical Genetics.

After MENDEL's famous paper of 1865 was first fully appreciated in 1900, the science of heredity entered a period of rapid development (64). Genes, the units of inheritance postulated by MENDEL, were shown to be carried in the chromosomes of the cell nucleus, and the manner in which they are transmitted from parent to offspring was investigated in great detail. It is convenient to refer to this phase of the investigations as "classical genetics", dealing as it does mainly with the mechanisms by which genes are passed on from generation to generation with relatively little regard for their nature or their exact role in development and function.

Albinism, lack of *melanin pigment*, may be used as an example of an inherited trait that illustrates several properties of genes as seen in terms of classical genetics. Melanins are the pigments derived from tyrosine which in man and many other animals are characteristically found in the hair, skin and eyes. Their chemical nature is incompletely understood (27, 58). In many species of animals mutant individuals or varieties are known in which melanin is partly or completely absent. The pink-eyed white rabbit is such a type. Analogous traits occur in man.

One form of human albinism is inherited as a typical mendelian recessive trait. This means that its inheritance is accounted for by assuming two forms of a given gene, one normal and the other mutant or defective. Since in the fertilized egg and in all cells descended from it in an individual there are two representatives of every gene, one derived from each parent, there are three possible types of individuals with respect to the gene for albinism. Representing the normal form of this gene with *A* and the mutant form with *a*, these three genetic types would be *AA*, *Aa* and *aa*. The second type is indistinguishable from the first—a situation covered by saying that the *a* form of the gene is *recessive* or, conversely, that the *A* form is *dominant*.

Anticipating a chemical interpretation, we may say that the *A* form or *allele* of the gene promotes the formation of melanin while the mutant *a* form is inactive. In an individual carrying both forms—a *heterozygote*— the active form of the gene does its work and melanin is formed.

There are many traits which show this type of inheritance. When the genetic behavior of two or more of these is investigated simultaneously in the same group of individuals, the results show that often different genes are at work. In the vinegar fly *Drosophila*, the genetics of which has been studied in great detail, there are some 500 different mutant genes known. These have been shown to be located at specific places in the four kinds of chromosomes which this organism has in its cells. Exactly how many different genes a fly possesses is unknown, but there is evidence for believing that the number is of the order of 10 000 (*64*). The number is probably not far different for man.

In man there are two sets of chromosomes, one set of 24 from each parent. Since in each set there are several thousand genes, there are obviously many genes per chromosome. Those carried in a single chromosome exhibit the phenomenon known to geneticists as *linkage*, *i. e.*, they tend to be inherited together. This tendency is not complete, because exchanges of corresponding segments of the two chromosomes of a given pair can occur during certain cell divisions. These are the divisions in which chromosomes are reduced to a single set per cell—the time of formation of eggs and sperms in animals. Such exchange, known as *crossing over*, recombines linked genes with a frequency which varies from 0 to 50 new combinations per hundred cells, depending on the distance apart of the genes concerned. Making use of this relation, the relative positions of genes in chromosomes have been mapped. This is relatively simple to do in the vinegar fly because of its few chromosomes and its short life cycle. In man, an organism notoriously unfavorable for genetic study, it is much more difficult.

The linear arrangement of the genes in chromosomes, while relatively permanent, is subject to accidental change. Segments of genes may be dropped out, duplicated, inverted, or exchanged with non-corresponding segments from different chromosomes. This latter phenomenon—*trans-*

location—may be regarded as a kind of illegitimate crossing over. All of these types of rearrangements occur spontaneously with a very low frequency—once in many thousands of generations. Their frequencies are greatly increased by ionizing radiations like X-rays or neutrons. These chromosomal changes have predictable and detectable genetic results, and in material favorable for direct cytological observation, such as maize and *Drosophila*, can be seen under the microscope. In those instances in which they can be studied by both techniques, perfect correlation is found between observed chromosome aberrations and deviations from the normal genetic behavior. These correlations are the classical proof of the theory that genes are carried in chromosomes.

A gene must exist in at least two forms to be studied by the method of classical genetics. This implies *mutability*—a property that is known for many genes. Presumably, mutability is a fundamental property possessed by all genes. But the frequency with which mutational changes occur varies from gene to gene within a single environment, and from environment to environment for a single gene. High energy radiations, and certain chemicals are known to increase mutation frequencies greatly. Spontaneous gene mutation frequencies of 10^{-5} to 10^{-8} per generation are not uncommon, but these limits are transgressed in both directions for certain genes.

Chemical Genetics.

In contrast with classical genetics, chemical genetics is concerned with what genes are and *how they function at a molecular level*, rather than with the mechanisms of their transmission. Experience shows that not all organisms are equally favorable for the study of gene nature and function. Neither do all genes of a single organism yield to the same experimental approach. To consider the last point first: heritable traits in a single species can be arranged in an order which appears to bear some relation to the directness with which they are related to the controlling genes. As an example, complex patterns of instinctive behavior may be considered as representing one extreme. WRIGHT (*82*) has illustrated this by considering the form of the web spun by spiders. This is with little doubt subject to gene control, but the chain of events intervening between initial gene action and final web pattern is not one that a realistic chemist would expect to unravel in a year's work—or even in a lifetime. At the other extreme are simple chemical differences in which a specific chemical reaction is carried out only in the presence of one form of a gene. Several examples of traits that are apparently simple chemically will be discussed in this survey. Between the two extremes lie many situations of intermediate complexity. Characters involving the morphology of multicellular organisms seem more complex than those dependent on differences at a molecular level and less so than those involving psychological differences. Some knowledge and a small amount of deliberation, however, leads one quickly to the conclusion that, while there is no doubt some basis for a classification of the kind just given, it must not be taken too seriously. To illustrate: a specific form of heritable feeble mindedness in man—a trait that would appear to lie at the complex end of the above series—has recently been shown to be completely correlated with inability to oxidize phenylpyruvic acid, $C_6H_5 \cdot CH_2 \cdot CO \cdot COOH$, in the *para* position. It is true that the connecting links between the specific gene-controlled reaction and the mental disorder are unknown—and there may be many of them; nevertheless, this interesting situation, which will be discussed in more detail later, suggests what may be regarded as a general hope of chemical genetics—that such apparently complex traits can eventually be reduced to the molecular level and chemical structures.

It would seem evident in the light of our present knowledge that the simplest living system in which genes are an essential part would yield most readily to experimental study. While this may now appear as a fairly obvious conclusion, it has not always been self-evident. It must be remembered that a study of genes by the methods of classical genetics is dependent on sexual reproduction as a basis of the phenomena of segregation and recombination. Only higher forms of plants and animals possess readily discernible sexual mechanisms. This fact eliminated from consideration by early geneticists all asexually reproducing forms such as blue-green algae and bacteria. Living systems below the cellular level of organization—the viruses—which now appear to offer many advantages for chemical genetics, were discovered little more than a quarter of a century ago and have had their chemical and biological properties revealed clearly only within the last decade. Only within the last few years have they been made the object of genetic study.

Two recent developments have served to focus the attention of geneticists on the bacteria and viruses. The first is an increasing feeling of confidence that genes are subject to study in terms of function as well as in regard to the mechanism of their inheritance. The second is the discovery that at least some bacteria and viruses possess a form of reproduction involving mechanisms by which the genes of different individuals can be recombined. Because they probably represent living systems reduced to their simplest terms—one or relatively few molecules capable of duplicating themselves and of mutating—the smaller viruses would seem to offer great hope for future advance in our knowledge of the nature of genes and their manner of multiplication.

Phenylalanine Metabolism.

In man there occur several metabolic defects involving derivatives of phenylalanine, $C_6H_5 \cdot CH_2 \cdot CH(NH_2) \cdot COOH$, an amino acid found in the proteins of all protoplasms. The longest known of these defects, and one included in GARROD's now classical list of "inborn errors of metabolism" (31), is *alcaptonuria*. In this rare hereditary disease, alcapton, or 2,5-dihydroxyphenylacetic acid (II), is not oxidized along the normal metabolic paths, but instead is excreted in the urine. On exposure to air, urine containing this compound turns dark—the characteristic symptom of alcaptonuria. The chemical basis of the disease was established by BÖDEKER in 1858. That it is a recessive mendelian character was suggested by BATESON in 1902, soon after the "rediscovery" of MENDEL's paper. In 1914 GROSS (33) presented evidence indicating that alcaptonurics lack a specific enzyme, possessed by normal individuals, which catalyzes the oxidation of alcapton. GARROD (31) has summarized evidence that alcapton is a product of phenylalanine and tyrosine metabolism. The situation may be summarized by scheme (I).

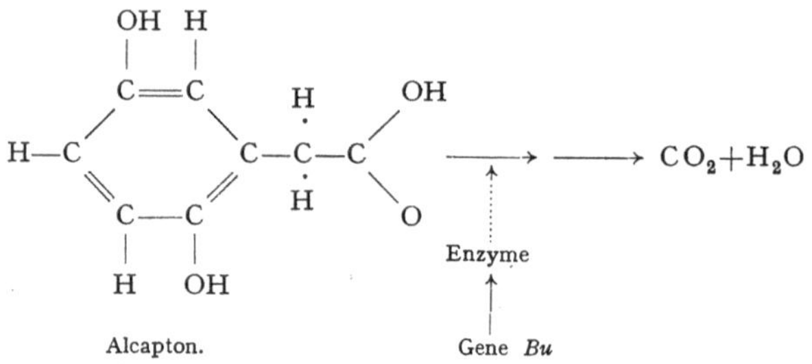

Alcapton.

$\longrightarrow \longrightarrow CO_2 + H_2O$

Enzyme

Gene *Bu*

(I.)

Apparently, the normal form of the *bu* (for *b*lack *u*rine) gene is somehow concerned with the elaboration of the enzyme "alcaptonase" which catalyzes the breakdown of alcapton. In the absence of the dominant form of the gene, the reaction by which alcapton is oxidized is blocked, and alcapton piles up—figuratively like water accumulates behind a dam.

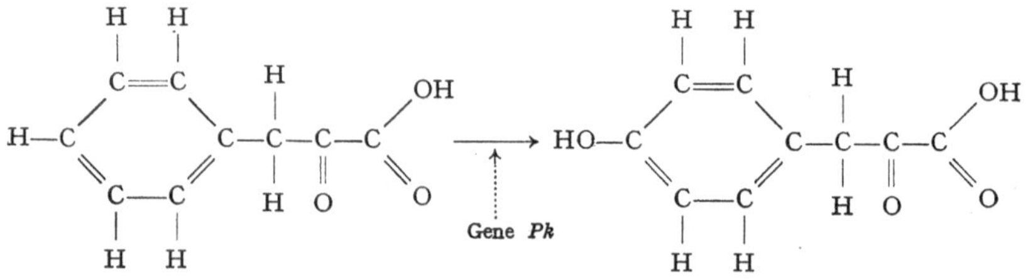

Phenylpyruvic Acid.

Gene *Pk*

p-Hydroxyphenylpyruvic Acid.

(II.)

A second trait in man, investigated by FÖLLING (29), PENROSE (66), and others (30), is characterized by inability to oxidize phenylpyruvic acid to its *para*-hydroxy derivative. This has the consequence that phenylpyruvic acid is excreted in the urine. It may be detected and estimated colorimetrically on addition of ferric chloride. Like alcaptonuria, "phenylketonuria" is inherited as a single-gene recessive character. Chemically it is represented as follows (II).

As already mentioned, phenylketonurics are invariably feebleminded—they range from imbeciles to low grade morons. The relation to the specific chemical reaction that is blocked in this hereditary disease to the associated mental defect is not understood.

Biochemical studies on phenylketonurics indicate that phenylpyruvic acid is biologically derived from phenylalanine. Its relation to alcaptonuria is indicated in the scheme of phenylalanine metabolism given in Figure 1.

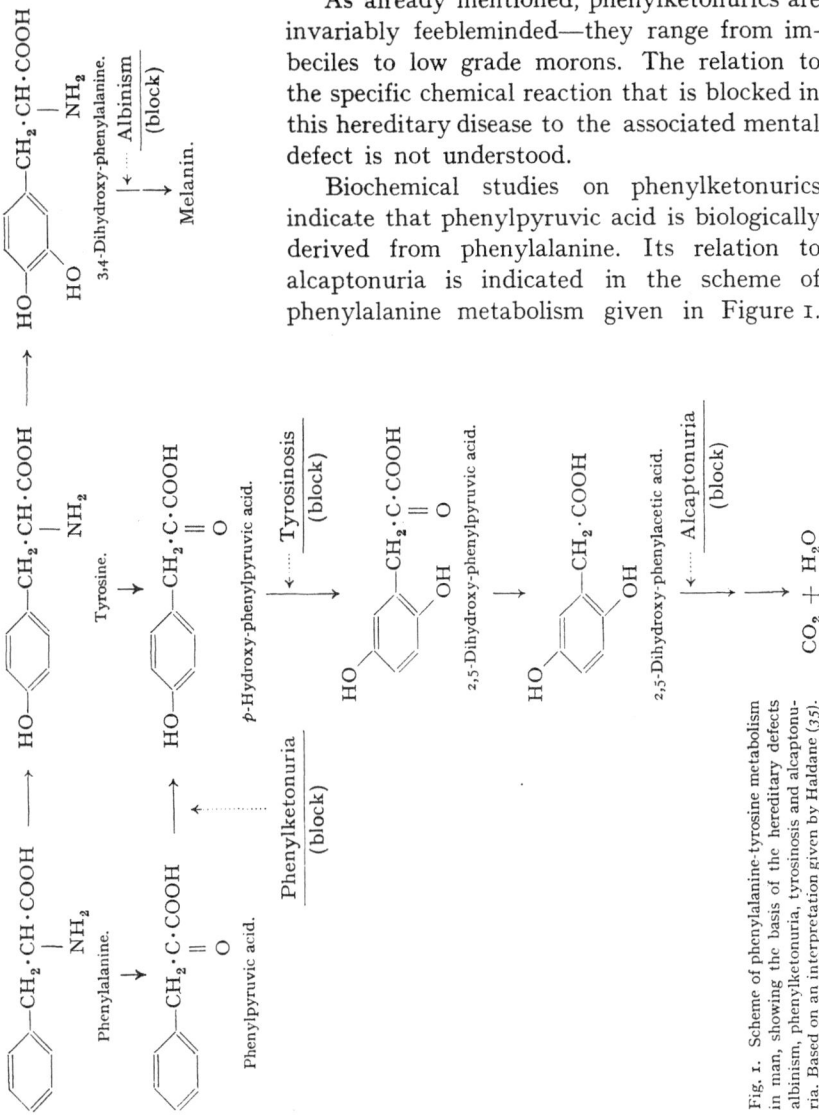

Fig. 1. Scheme of phenylalanine-tyrosine metabolism in man, showing the basis of the hereditary defects albinism, phenylketonuria, tyrosinosis and alcaptonuria. Based on an interpretation given by Haldane (35).

Not all of the relations shown in this diagram are firmly established—but they do represent an interpretation for which there is at least some evidence.

"Tyrosinosis", inability to oxidize p-hydroxyphenylpyruvic acid, has been reported in only a single instance (59). Under these circumstances it is not possible to say anything about its inheritance.

The specific chemical reaction blocked in human albinism is not known. In the corresponding deficiency in certain other animals, however, there is evidence that dopa-oxidase, the enzyme that catalyzes the oxidation of 3,4-dihydroxyphenylalanine, $(HO)_2 \cdot C_6H_3 \cdot CH_2 \cdot CH(NH_2) \cdot COOH$, is lacking (71).

The most complete genetic study of melanin formation that has so far been made is that of WRIGHT and his collaborators (83) on guinea pigs. Since it has not yet been possible to interpret this work in terms of known chemical reactions, it will not be reviewed in detail here.

Antigen Specificity.

In many instances the elaboration of specific antigens is known to bear a relatively simple relation to the presence of particular genes.

In the *A-B blood groups* in man, for example, there exists a series of alleles (p. 301) of a specific gene showing a simple relation to the antigens carried by red blood cells. Disregarding minor complications (78), there are six genetic types of individuals possible and these have the antigen and antibody characteristics as listed in Table 1.

Table 1. Genetic types in the *A-B* blood groups in men.

Genetic type	Antigens present in red blood cells	Antibodies normally present in blood serum
OO	none	α, β
AO	A	β
AA	A	β
BO	B	α
BB	B	α
AB	AB	none

When cells and sera are mixed, agglutination of the cells occurs if corresponding antigens and antibodies are present, say when B-carrying cells are mixed with serum containing β antibody.

Relations similar in principle hold for the *M-N* blood types in man where the gene-antigen relations are as listed in Table 2. Here antibodies are not normally present in blood serum but may be induced to form if cells bearing antigens of another type are introduced—*e. g.*, antibodies directed against M cells are produced in an individual of N type when M cells are introduced (78).

Table 2.

Genetic constitution	Antigens in red blood cells
MM	M
MN	N
MN	M, N

The *Rh* series of blood type in man has achieved considerable attention from both geneticists and medical men because of its relation to the disease *erythroblastosis fetalis*. At times when a mother without Rh antigens carries a fetus producing them, antigens may leak through the placenta, inducing antibody formation in the mother. In a subsequent pregnancy, these may leak back to an Rh-positive fetus and cause red blood cell destruction. While the genetic situation here is complex in detail; it is again simple in principle, in that the one-gene-one-antigen relation appears to hold (*28*).

Similar gene-antigen relations have been found to be valid in many other animals (*45*). IRWIN and his associates (*45*) have made extensive studies in pigeons, doves and their hybrids and have found that in general for every antigenic difference between two individuals there is a corresponding genic difference. In species hybrids, there sometimes appear so-called "hybrid" antigens that are not found in either parent. The significance of these is not entirely clear. Cattle vary in cellular antigens to such an extent that the probability of any two individuals arising from separate fertilized eggs being identical is very small. There are known in this species at least 40 immunologically distinct antigens (*45*).

In the protozoan, *Paramecium*, antigenic differences are known. In some instances such differences are correlated with nuclear gene differences but in others they are inherited cytoplasmically (*75*).

Hemophilia.

A specific disease of man, known as hemophilia and characterized by slow clotting of the blood, is inherited in a simple fashion. The gene concerned, unlike those concerned with the human diseases mentioned above, is carried in the so-called "X" chromosome. This is one of a pair of chromosomes associated with sex determination, XX individuals being female while XY are male. Males produce equal numbers of X- and Y-bearing sperms which, when they fertilize eggs (all X-bearing) give daughters and sons respectively. The Y-chromosome is at least partly empty genetically; that is, at least a large section of it carries no counterparts of genes present in the X chromosome. The gene for hemophilia has no allele in the Y chromosome. Its mutant form is recessive. When a male receives an X chromosome carrying the mutant gene, he shows the disease hemophilia and has a relatively small chance of surviving to reproduce.

Queen Victoria of England was a carrier of hemophilia, *i. e.*, she had one X carrying a normal form of the gene and one carrying the defective form. She transmitted the defective allele through at least two daughters and a son to later generations (*35*).

There is evidence in the work of LAWRENCE and CRADDOCK (*50*) that hemophiliacs lack a specific γ-globulin blood protein that is essential for normal blood clotting. This suggests that the normal form of the hemophilia gene is somehow concerned with the elaboration of this particular protein.

Hemophilia illustrates both the phenomenon of mutation and natural selection. HALDANE (34) has estimated that about one male in 10000 carries a defective allele of the hemophilia gene. Since males carry a random sample of all X chromosomes, 1 in 10000 represents the proportion of all X chromosomes that carry the defective allele of this gene. Of the defective X chromosomes carried by heterozygous females—who do not of course show any direct signs of hemophilia—half will appear in sons in the next generation. Since these seldom live to reproduce, the frequency of defective forms of the gene carried by females will be somewhere near halved each generation. This would mean that the frequency of hemophilia should decrease rapidly. As HALDANE (34) has pointed out howeuer, it does not decrease because defective genes arise from the normal form at a rate that just balances their rate of elimination through natural selection. Since the rate of elimination can be estimated, the mutation rate, normal to defective, can be determined. It is of the order of one mutation per 50000 X chromosomes per generation (34).

Gene-directed Specific Reactions in Other Organisms.

Many instances are now known in which something can be said about the relation of genes to known chemical reactions. Among the earliest of these to be studied by both geneticists and biochemists were those involving the synthesis and interconversion of the anthocyanin-like pigments in higher plants (51). The three principal types of these are the anthocyanins, flavonols and chalkones of which the following specific compounds are examples (III–V).

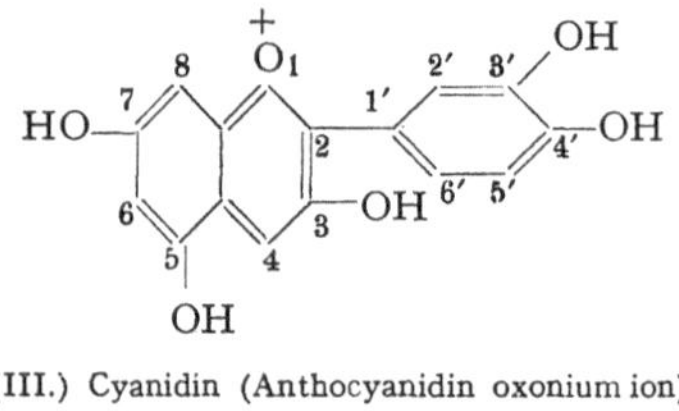

(III.) Cyanidin (Anthocyanidin oxonium ion).

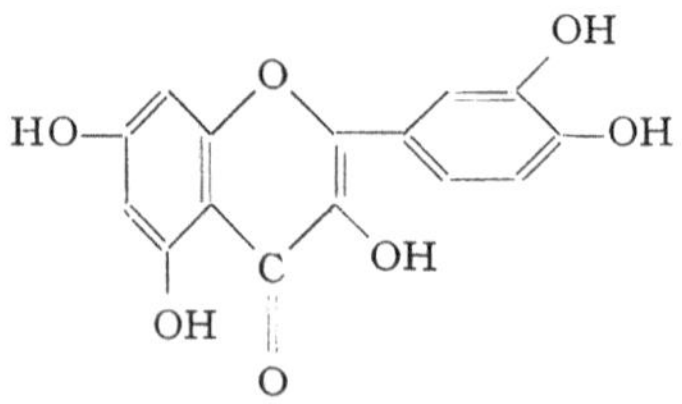

(IV.) Quercetin (Flavonol).

(V.) Butein (Chalkone).

Anthocyanins occur in nature as glycosides. Anthoxanthins and chalkones mostly also occur in this form. Anthocyanins vary in color, partly depending on OH or OCH_3 groups in the 3' and 5' positions. Anthocyanins apparently usually exist as 3-glycosides, but a glycoside linkage may also occur at the 5 position. Hydroxylation, methylation and the formation of glycoside linkages are subject to genetic control as are other conditions such as cell sap pH and presence of co-pigments. All of these factors affect color characteristics. The three classes of pigments are probably formed from a common precursor through gene-controlled and competitive reactions (51).

In the green alga, *Chlamydomonas*, MOEWUS (63) has studied a series of gene-controlled reactions concerned with production of carotenoid pigments and related compounds. These pigments and their derivatives are active in promoting motility and sex reactions. So far as the writer is aware this important work has not been confirmed and since there is some basis for skepticism regarding it (74), it will not be described in detail.

In the tomato, ZECHMEISTER, LE ROSEN, WENT and PAULING (85) have found that the stereoisomeric configuration of a carotenoid pigment is subject to genetic control. The "tangerine" tomato contains poly*cis*-forms, mainly prolycopene, $C_{40}H_{56}$, while normal red variety owes its color to ordinary lycopene, $C_{40}H_{56}$, the all-*trans* isomer of prolycopene.

In the white clover plant both the synthesis of cyanophoric glucosides and their hydrolysis are subject to gene direction (2, 23). Some strains are able to hydrolyze enzymatically lotaustralin or linamarin to ketones (ethyl methyl ketone and acetone respectively), HCN, and glucose. Other strains differing by a single gene and lacking the hydrolyzing enzyme, linamarase, are unable to carry out these reactions.

Similar situations are known in the animal Kingdom. For example, in the rabbit there exist two types of individuals which may differ in only a single gene. One type, carrying a dominant allele of the gene concerned, contains a specific enzyme, known as atropine esterase, which hydrolyzes atropine to tropine and tropic acid (72). The alternative recessive type does not possess this enzyme and cannot carry out the hydrolysis.

In the same animal a recessive mendelian trait, yellow fat, is due to lack of the enzyme "xanthophyllase" which oxidizes the carotenoid pigment xanthophyll. In the absence of the enzyme this pigment accumulates in the fatty tissues of the animal (20).

A number of the reactions controlling the production of eye pigments in insects has been studied by the methods of chemical genetics. One of the intermediates in the formation of brown pigment in *Drosophila* and other insects has been found to be the tryptophane derivative kynurenin (VII). Both the formation of this from tryptophane (VI), and its further oxidation, are subject to gene control (26).

$$-CH_2 \cdot CH \cdot COOH \qquad \qquad -CO \cdot CH_2 \cdot CH \cdot COOH$$
$$\quad\quad\quad | \qquad\qquad\qquad\qquad\qquad\qquad\qquad\quad |$$
$$\quad\quad NH_2 \qquad\qquad\qquad NH_2 \qquad\quad NH_2$$

(VI.) Tryptophane.   (VII.) Kynurenin.

It would be possible to cite additional examples of the same general kind. Many of them have already been reviewed in some detail elsewhere (7). They all serve to strengthen the general conclusion already evident that, in at least many instances, *genes are responsible for the presence of specific enzymes which are in turn responsible for the progress of particular chemical reactions.* In other instances, such as those in which genes show a one-to-one relation to specific antigens or other specific proteins (such as the γ-globulin lacking in hemophiliacs), it appears that genes are concerned with the specificities of giant, antigenically active protein or polysaccharide molecules. If one assumes that in those cases in which genes are concerned with enzyme activity, they direct the specificity of the proteins that are enzymatically active or that confer enzymatic specificity to combinations with non-proteins, it becomes reasonable to entertain the hypothesis that *in general, genes act by controlling protein specificities.* And it is worth considering whether this might not be all the gene ever does.

Neurospora.

On the basis of evidence such as that referred to above, an investigation was initiated several years ago to see whether it might be possible systematically to produce and detect gene mutations that would block specific biosyntheses (9). As an organism for this venture the red bread mold, *Neurospora crassa* was chosen because of its convenient and short life cycle, its adaptability for biochemical investigation, and its suitability otherwise for study by the methods of chemical genetics.

The life cycle of *Neurospora* is indicated in a simplified form in Figure 2. It is a heterothallic fungus—that is, has two sexes or mating types each of which will multiply separately by vegetative means. It is a haploid organism, which is to say that each nucleus in the vegetative phase of development contains only one set of genes carried in one set of seven chromosomes (56). Therefore, there are none of the complications resulting from dominance present with organisms having two sets of chromosomes. All products of the cell divisions by which the chromosomes are reduced from the double to the single number (meiosis) can be recovered in the order in which they are produced; this is a tremendous

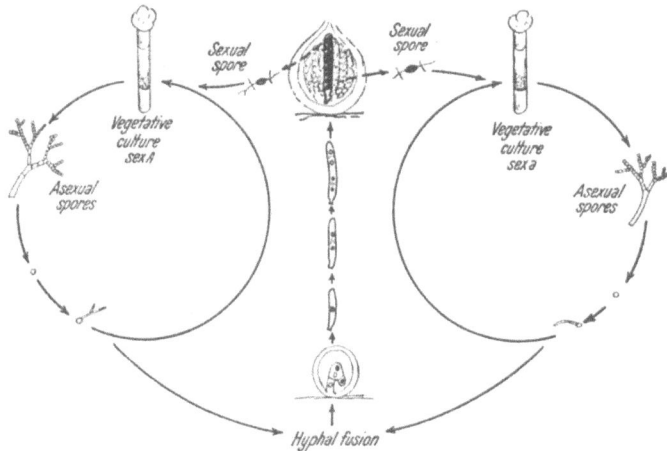

Fig. 2. Diagrammatic representation of the life cycle of *Neurospora crassa*. In the asexual phase of the life cycle, strains of sex *A* or *a* reproduce by non-sexual spores (conidia) which are formed at the tips of aerial hyphae. These spores germinate and multiply the organism directly. Hyphae of opposite sexes may fuse to give rise to a sexual stage. In this stage special fruiting bodies (perithecia) are formed in which there occur spore sacs containing eight sexual spores each. Four of these spores are genetically, sex *A* and four, sex *a*. On germination they reestablish the asexual phase of development. During the development of the spore sacs, nuclei from the two parental strains fuse to give a nucleus with two sets of genes, each set carried in seven chromosomes. The double nucleus then undergoes two special divisions during which the chromosome number is reduced to seven (meiotic divisions). Each sexual spore initially contains one nucleus with the single number of chromosomes, but by the time it is mature, the nucleus has divided equationally (with exact longitudinal division of all chromosomes) to give two genetically identical nuclei per spore.

advantage genetically (6). For one thing, it means that mendelian ratios are reduced to a purely mechanical basis; errors of sampling and errors due to unequal survival of types are completely eliminated.

Neurospora is readily grown in pure culture on a simple, chemically defined medium. From a carbon and energy source such as sucrose, appropriate inorganic salts and the B-vitamin, biotin, its protoplasm is able to reduplicate itself. This means that it is able to synthesize some twenty amino acids, the vitamins of the B-group (except for biotin), purines, pyrimidines, and all the rest of the building units and enzymes of which mold mycelium is built.

If genes are concerned in the many reactions by which these syntheses take place, it should be possible to produce a variety of mutant types in which specific steps in biosynthesis are blocked. This in fact has been achieved through treatment of the mold with various mutation-producing agents such as X-rays, ultraviolet radiation, neutrons, and mustard gas (9, 10, 13, 41, 57). The technical methods by which this is done have been described elsewhere (10) and need not be repeated in detail here. Briefly stated, they consist in treating the organism at some stage

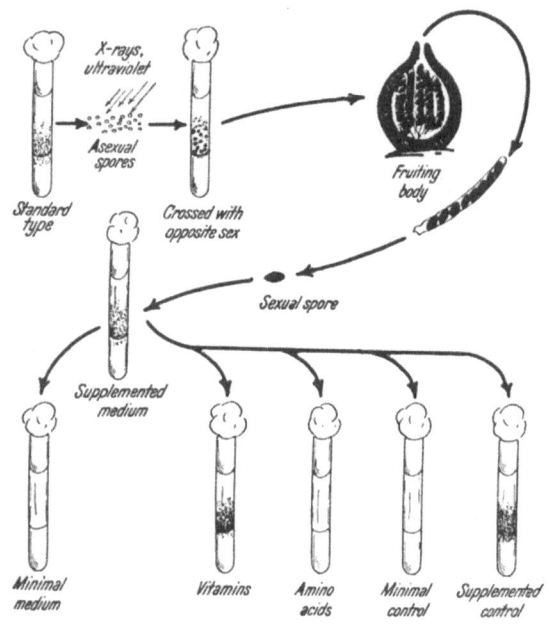

Fig. 3. The experimental production and detection of mutant strains of *Neurospora*. Asexual spores are treated with some "mutagenic" agent and then crossed with a strain of the opposite sex. From fruiting bodies produced in this cross, single sexual spores are removed and planted on a medium supplemented with vitamins, amino acids and other biologically important compounds. If the spore carries a descendant of a defective gene, produced by the treatment, and if this gene is concerned with the synthesis of a required substance, say the vitamin. pantothenic acid, the mold will grow on the supplemented medium because the required substance is obtained from the medium. The defect in biosynthetic ability is detected by transferring the strain by asexual spores (without genetic change) to a minimal medium containing only those substances required for growth of the original standard strain (sugar, inorganic salts and biotin). A strain that grows on supplemented medium but not on the minimal medium has lost the ability to produce something which is in the supplemented medium but not in the minimal medium. Subsequent tests serve to further characterize the required substance. In the case illustrated, some one or more of the known vitamins is needed for growth.

of its development with the "mutagenic" agent, and then isolating strains, each of whose nuclei have descended from a single haploid parent nucleus, in order to insure genetic homogeneity within the derived strains.

When such strains, isolated on a medium supplemented with vitamins, amino acids, and other substances that mutant strains might be unable to synthesize and, therefore, would need ready-made in the medium,

are tested for ability to grow on the basal medium, it is found that up
to two per cent of them are unable to synthesize some substance which
was contained in the supplemented medium but absent in the basal medium
(Figure 3). Systematic tests are then made to determine what specific
substance or substances the mutant strains cannot synthesize and,
therefore, need from an external source (Figure 4). Many of the mutants
are found to be unable to synthesize some one of the vitamins, amino
acids or other biologically significant compounds. Some strains are

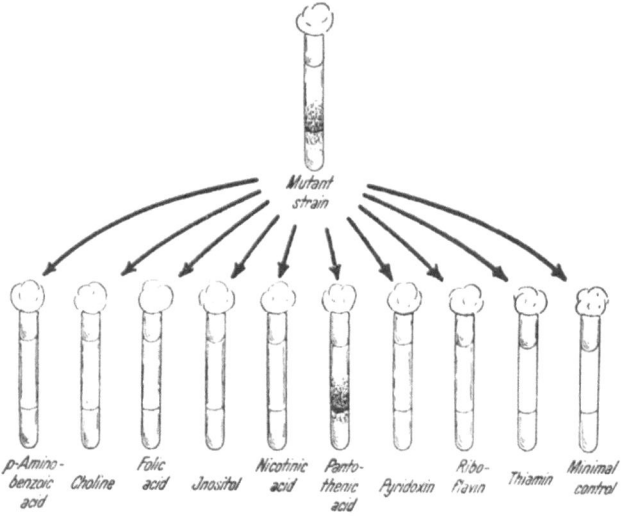

Fig. 4. Further tests of the mutant strain requiring one or more vitamins for growth. In this instance, only
the B-vitamin, pantothenic acid gives normal growth. It is concluded that the mutant strain being tested
has lost the ability to synthesize this vitamin.

unable to grow on the particular supplemented medium used in the
first isolation and are therefore lost for further study. Others require
unknown substances for their growth, and still others have distinguishing
morphological characteristics.

Mutant strains so produced and identified are then crossed with
the parent strains of the opposite mating types. If from the spore sacs
formed in such crosses, there are regularly four spores that produce
strains like the mutant parent, and four that give strains like the normal
parent, it is concluded that there exists between the mutant strain and
the standard type a one-gene difference (Figures 5-6, p. 314-5). In the great
majority of the mutant types produced, the evidence from this classical
genetic test favors the assumption of a *single gene mutation*.

In the manner indicated, many hundreds of mutant strains have
been produced in *Neurospora*. In each of them that has been studied

in sufficient detail to decide the issue, it has been found that a *single chemical reaction is blocked* in the mutant strain. If the product of the blocked conversion is supplied, the mutant strain grows more or less normally. There is no need here to review in detail all of the types of mutants. They have been summarized elsewhere (*6*, *13*); and it will be sufficient here to discuss as examples two groups of mutants which have been reported only recently.

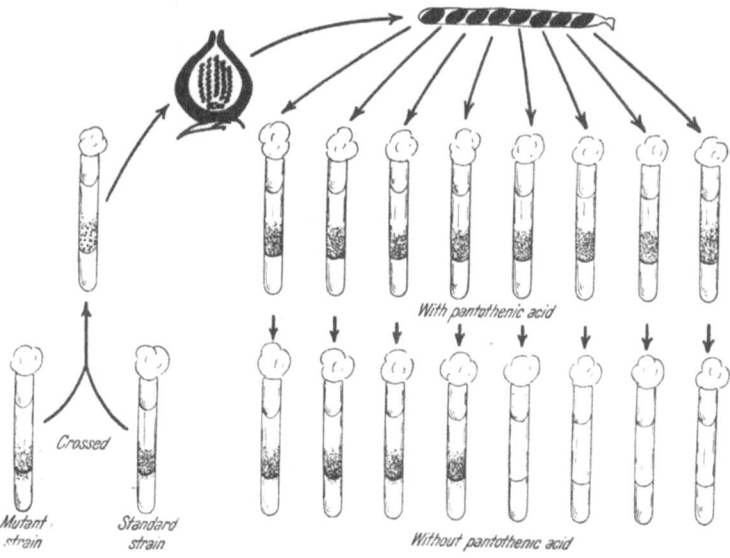

Fig. 5. Method of determining whether the inability of a *Neurospora* strain to synthesize pantothenic acid is due to a gene mutation. The mutant strain is crossed with a standard strain of the opposite sex. From the fruiting bodies produced in this cross, single spore sacs are removed. The eight sexual spores, each about 25 µ long, are removed *in the order in which they occur in the sac* and planted individually in culture tubes containing a medium supplemented with pantothenic acid. After the cultures have developed (about three days), asexual transfers are made to a medium in which no pantothenic acid is present If the mutant strain differs from the standard by one gene, four strains out of each set of eight will grow without pantothenic; and four, like the mutant strain, will fail to grow in its absence.

The first is concerned with the synthesis of the sulfur-containing amino acid *methionine*. Half a dozen genetically and biochemically different types of mutants, related in one way or another to methionine synthesis, have been studied by HOROWITZ (*40*), and by TEAS, HOROWITZ and FLING (*81*). From these studies a number of conclusions are evident. One of them is that the sequence of reactions illustrated in Figure 7 is probably concerned in the biosynthesis of methionine. If the mutant that cannot cleave cystathionine is grown in the presence of a small amount of methionine, it synthesizes and accumulates an appreciable amount of cystathionine which can be isolated from the mycelium and identified chemically. Cystathionine provides a "shuttle mechanism"

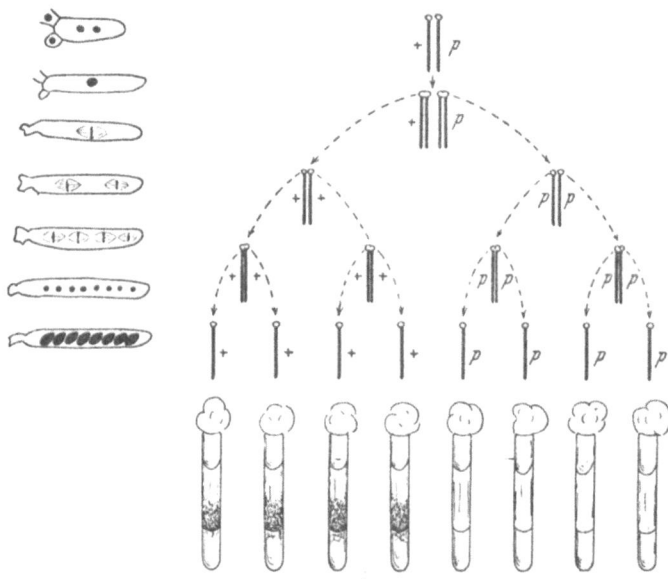

Fig. 6. The nuclear and chromosomal basis of inheritance in *Neurospora crassa*. On the left is shown dia-grammatically the development of a spore sac. Nuclei from the two parents fuse. Two successive divisions follow in which the chromosome number is reduced to single sets of seven per nucleus. The nuclei undergo a third division in which each chromosome divides longitudinally throughout its length. The eight resulting nuclei are then included in a corresponding number of spores. At the right is shown the behavior of the pair of chromosomes in which the gene concerned with pantothenic acid is carried. The normal form of the gene is indicated symbolically with a plus sign (+) while the defective form is represented by a p (for pantothenic acid). The various stages of nuclear and chromosomal division are placed at corresponding levels in the diagram. Four of the eight strains from a single spore sac grow in the absence of pantothenic acid and four do not. If crossing over occurs between the segregating gene and the point of attachment of the chromosome to the spindle, the result is an arrangement of the eight cultures in alternating pairs instead of groups of four as shown. For a complete explanation of this alternative and its genetic significance, the reader is referred to a more detailed account (6).

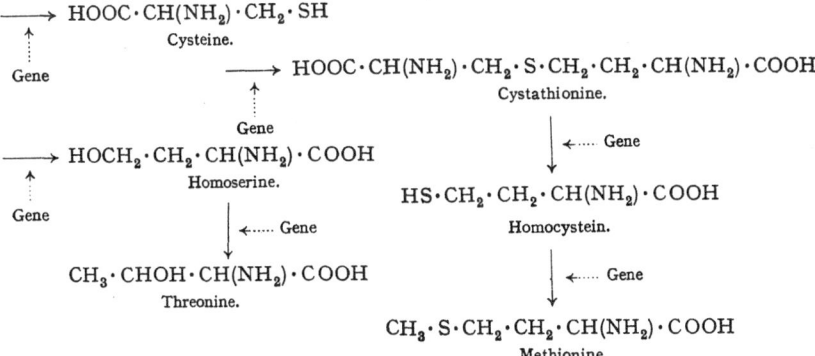

Fig. 7. Scheme of biosynthesis of methionine in *Neurospora crassa*. Based on work of HOROWITZ (40) and of TEAS, HOROWITZ and FLING (81).

by which sulfur is transferred from the three-carbon chain of cysteine to the four-carbon chain of homocysteine. Evidence for this mechanism of transfer in the rat in the opposite direction had been presented, previous to the *Neurospora* work, by BRAND *et al.* (*17*), BINKLEY and DU VI- GNEAUD (*11*), and by others. Homoserine is evidently a common precursor of methionine and of threonine. Until this was learned, the mutant strain that cannot produce homoserine appeared to require for growth two amino acids which are not obviously related biosynthetically. Thus, an instance in which one gene appeared to be responsible at one time for the direction of two reactions has been brought in line with the one-gene-one-reaction interpretation.

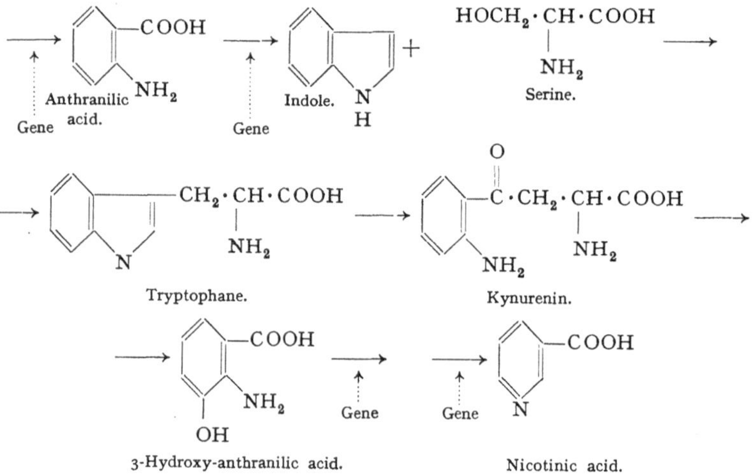

Fig. 8. Scheme of biosynthesis of tryptophane and nicotinic acid in *Neurospora crassa* (*8, 14, 15, 62, 80*).

Several mutant strains unable to synthesize the amino acid *tryptophane* were investigated by TATUM and BONNER (*80*). It was shown that *Neurospora* synthesizes tryptophane by condensing indole and serine, and that indole is in turn derived from anthranilic acid. Another group of mutants concerned with nicotinic acid were studied by BONNER and BEADLE (*15*). One of these accumulates a precursor from which a second mutant strain can make nicotinic acid. A mutant strain of a third type was found, in which either nicotinic acid or tryptophane would suffice for normal growth. Following reports that tryptophane may be substituted for nicotinic acid in mammals under certain conditions, it was found that mutant strains of *Neurospora* of this third type could apparently convert tryptophane to nicotinic acid by way of kynurenin (VII. p. 310) as an intermediate (*8*). Subsequent studies by MITCHELL and NYC (*62*) indicated that a second intermediate in this transformation is 3-hydroxyanthranilic acid. This substance, as synthesized by MITCHELL and NYC, proved to be identical with the natural nicotinic acid precursor isolated by BONNER (*14*).

The manner in which tryptophane and nicotinic acid appear to be synthesized in *Neurospora* is shown in Figure 8. It is interesting that kynurenin, which was first found by KOTAKE in dog and rabbit urine, following injection of large amounts of tryptophane, and subsequently shown to be an intermediate in insect eye pigment formation, appears to play an essential role in the formation of nicotinic acid from tryptophane; this may indeed be the primary role of kynurenin in metabolism.

The tryptophane-nicotinic acid story is of interest in that it appears to clear up several heretofore puzzling relations in the human dietary deficiency disease, pellagra. It is also of interest in adding to an already impressive list of instances in which it is clear that the basic metabolic activity of protoplasm is much the same wherever it is found—in bread mold or in man.

Nature and Action of Genes.

The evidence considered above suggests the hypothesis that *genes somehow serve to direct the specificities of proteins*. Since proteins may serve directly as enzymes or as components of enzymes, the observed relation between genes and chemical reactions should follow. The one-to-one relation between genes and antigens, too, is consistent with this hypothesis. The hemophilia case mentioned is also of interest in this connection.

On the basis of the one-gene-one-protein view, one might expect to find genes showing the same range of specificity of action as do enzymes. At one extreme, some genes should be concerned with the direction of one and only one chemical reaction. On the other hand, a gene responsible for the elaboration of an enzyme like the SCHARDINGER enzyme which shows broad group specificity in its role as an aldehyde oxidase but almost absolute specificity as a xanthine oxidase (*5, 49*) might appear experimentally to be in immediate control of two or more reactions. Some other instances of this nature, in which enzymes show a wide range of specificity, are also known. Crystalline trypsin and chymotrypsin, for example, have been found to possess both peptidase and amino acid esterase activity (*46, 73*).

The key biological problem of how genes direct protein specificities is an unsolved one. Specific genes of higher organisms cannot be isolated in pure form and investigated by conventional chemical methods. Indirect and circumstantial evidence indicates that they are *nucleoproteins*. This evidence is of several kinds. In staining reactions and in ultraviolet absorption whole chromosomes *in situ* show the characteristics of nucleoproteins (*19*). They can be isolated from other cell components and subjected to direct chemical investigation. When this is done, they are found to be nucleoproteins containing histones, a tryptophane-containing protein of higher molecular weight, and nucleic acid (*61*). The effectiveness of ultraviolet radiations of different wave lengths in producing gene mutations agrees as well as could be expected, in view of the experimental difficulties involved in determining it, with the ultraviolet absorption spectrum of nucleic acids (*38*, *77*). The simplest interpretation of this relation is that the energy effective in producing gene mutations is absorbed by nucleic acids and that these are components of genes. As will be pointed out later, there are reasons for believing viruses to be systems made up largely of gene-like sub-units, their number probably being more or less proportional to virus size. All viruses that have been isolated in pure form are nucleoproteins or contain nucleoproteins, which indicates that the gene units out of which they are built are nucleoproteins.

Since we do not yet know how genes reproduce, there is little point in discussing in any detail possible mechanisms by which they may direct the specificities of proteins that are not gene components. It seems at least possible that the two problems—gene duplication and gene control of extragenic proteins—are really one. It is generally assumed that genes multiply by a kind of model-copy mechanism in which "old" genes serve as templates in the formation of "new" ones (*24*). The same basic process may be involved in the synthesis of all specific proteins and other molecules of similar complexity, whether they remain in the chromosomes as new-gene components or become a part of the extra-chromosomal protoplasm. One may sum up the situation by saying that this is surely one of biology's most fundamental problems and that so far no one has made much headway in its solution.

Gene Mutation.

As has already been pointed out, all genes of higher organisms known by the methods of classical genetics must be, or at one time have been, capable of mutation. It seems highly probable that the property of mutability is common to all genes. In many instances it is known that gene mutation is a reversible process. Several mutant alleles of genes concerned with known syntheses in *Neurospora*, for example, have been shown to undergo back mutation (*32*).

Genes that have been found to mutate spontaneously do so at widely different rates. Some so-called "mutable" genes may change thousands of times during the development of a single multicellular individual such as a vinegar fly or a maize plant. Others have been found in the mutant form only once in thousands of individuals of a species.

In 1926 H. J. MULLER made the important discovery that X-radiation greatly increases the frequency of gene mutation. It was soon found that all ionizing radiations tested have this property. Ultraviolet radiation likewise greatly accelerates the mutation process. During the war AUERBACH and her collaborators (3) discovered that mustard gas and related compounds induce both gene mutations and chromosome aberrations in *Drosophila*, a finding that has since been extended to a number of other organisms including Neurospora (*13, 41, 57*).

Almost from the beginning of the science of genetics attempts have been made to direct gene mutations in specific ways. All mutagenic agents discussed above are non-specific, *i. e.*, they induce mutations in

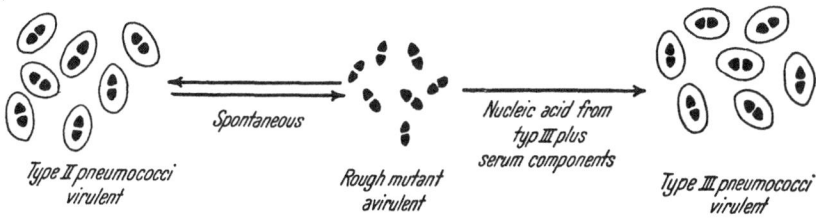

Fig. 9. Diagrammatic representation of directed transformation of strain type in the *pneumococcus* bacterium (*4,55*).

many different genes in a more or less random fashion. Until recently there has been no good indication of success in attempts to control the mutation of specific genes. The first indication of progress has come from the type transformation studies on the *pneumococcus* bacterium that is responsible for lobar pneumonia. This bacterium occurs in many antigenically specific types. Each of these types is characterized by a specific polysaccharide in its cellular capsule. The capsule may be lost through spontaneous mutation, in which case the descendants of the mutated capsule-free cells give rise to so-called "rough" colonies lacking the specific polysaccharide. "Rough" strains are non-virulent. They occasionally mutate back to a capsulated "smooth" virulent form. When they do this spontaneously, the virulent back-mutated strain invariably has the type specificity of the original "smooth" strain—that is, the "rough" strain regains the same specific polysaccharide that it lost in the original mutation process. But if a "rough" strain back-mutates in the presence of certain serum components and a minute amount of a highly polymerized nucleic acid from a "smooth" strain of a different type specificity, the resulting "smooth" form is of that strain type from which the nucleic acid came (*4, 55*). This situation is shown diagramatically in Figure 9.

From a purely genetic standpoint the pneumococcus transformation story is unsatisfactory because there is no way of knowing whether or not the changes involved represent gene changes. If they do, then it is evident that a particular gene is being directed by the nucleic acid to mutate in a highly specific way. The interesting question of how this direction is achieved—whether by the specific nucleic acid actually entering the cell and replacing a missing gene component, or in some other manner—remains unanswered. Similar directed transformations in the colon bacterium, *Escherichia coli*, have been reported by BOIVIN and his associates (*12*).

The nature of gene mutation in chemical terms is completely unknown. Probably the closest approach to the problem has been made in the study of mutant strains of tobacco mosaic virus by STANLEY and his co-workers. It is found that mutant strains may differ both qualitatively and quantitatively in amino acid composition (*48*). That these studies contribute in a direct way to an understanding of gene mutation, of course, involves the assumption that this virus is itself gene-like or is made up of gene-like units. Unfortunately, the mutant strains of tobacco mosaic virus that have thus far been subjected to amino acid analysis are not known to differ from the standard form or from one another by single mutational steps.

It is the personal opinion of the writer that the viruses offer the most favorable material of any so far made use of for studies of gene mutation by direct physical and chemical methods. It seems likely that substantial progress will be made along these lines in the near future.

Universality of Genes.

Because classical genetics provides no methods for studying genes of asexually reproducing organisms, there has been a good deal of doubt in the past as to whether bacteria and subcellular forms really possess these units. Recently the question has been approached in several ways, and the answer now seems clear that *all living systems do contain genes* as indispensible units.

In many bacteria chromosome-like structures composed of nucleo-proteins have been demonstrated by cytological methods (*70*). Furthermore, TATUM, ROEPKE and others have shown that a series of such mutant types in *Escherichia coli* can be obtained that parallel in their biochemical properties those obtained in *Neurospora* in a very striking way (*79*). Since in *Neurospora* these changes are clearly due to gene mutation, there can be little doubt that the bacterium has directing units which are the complete functional equivalent of the genes of *Neurospora*. This conclusion has received strong support in the work of LEDERBERG and TATUM (*52*) who have found evidence in *Escherichia coli* of a fusion and segregation process that occurs very infrequently but still with sufficient regularity to make possible the application of the methods of classical genetics in demonstrating the existence of genes in this bacterium. The answer seems clear that bacteria do have genes.

Two spectacular recent developments in the study of bacterial viruses make it clear that these living systems, too, contain gene-like units. It has been shown that the viruses that attack *Escherichia coli* are mutable. DELBRÜCK and BAILEY (25) found that if a single bacterium is infected with two related virus strains differing in host range and in plaque type, there are produced among the descendant viruses not only the parental types but two recombination crossover-like types. This lead has been followed up by HERSHEY and ROTMAN (37) who have shown, not only that this recombination phenomenon is a process which occurs regularly, but that the recombination types of bacterial viruses appear in reproducible proportions in the same sense in which recombinations occur with more or less constant frequencies in higher forms.

A related phenomenon has been studied by LURIA (53). *Coli* viruses inactivated with a single quantum or a few quanta of ultraviolet radiation per virus unit are incapable of producing more of their kind when they infect bacteria singly. But when double or higher multiple infections are made with such inactivated viruses, reactivation occurs and viable offspring are produced in normal numbers. These results are quantitatively of such nature as to indicate some 30 gene- or chromosome-like units, any one of which, if hit by ultraviolet radiation, will lead to inactivation. If two viruses, hit in different units, infect a single bacterial cell, they apparently recombine their parts in such a way as to give fully viable offspring. Whether the units postulated to account for this phenomenon are the same as those identified by HERSHEY and ROTMAN (37) through mutation, is an important question that as yet remains unanswered. In any event it now seems abundantly clear that *bacterial viruses are built of gene-like sub-units*. It seems also probable that other viruses are similarly constructed, with a smaller or larger number of such sub-units, depending on their size and complexity.

Cytoplasmic Factors.

"Are there gene-like units in the cytoplasm?" is a question that has been repeatedly asked by biologists. The best answer that can be given at present seems to be that, while there are undoubtedly cytoplasmic components which enjoy a limited degree of independence of genes in their multiplication and play vital roles in development and function, they are not basic units of inheritance in the same sense as are genes.

If cytoplasmic units of inheritance with the same autonomy of genes were of common occurrence, the methods of classical genetics would have uncovered many instances in which characters were inherited through the cytoplasm, maternally in most higher plants and animals, since the male contribution usually consists of little more than nuclear material. Despite the fact that many competent investigators have deliberately searched for this type of inheritance in a great variety of organisms, only a few sporadic instances have been found. Even in these, the cytoplasmic units concerned can be shown in most cases to be dependent on nuclear genes for their origin or continued existence. One of the first discovered cases of cytoplasmic inheritance involves plant plastids—more specifically, those plastids that contain chlorophyll. Both morphological and genetic evidence indicates that *chloroplasts* arise only from preexisting plastids or plastid primordia. In this sense they are autonomous, but their multiplication and differentiation are controlled by nuclear genes, for many instances are known in which mutation in nuclear genes leads to failure of multiplication of chloroplasts or to specific defects in their development (*69*).

Chloroplasts may undergo mutation, either reversibly or essentially irreversibly (*44*). This may be spontaneous or its frequency may be greatly increased by the presence of specific nuclear genes (*69*).

Degeneration of *pollen-mother-cells*, resulting in partial or complete male sterility, has been shown to involve cytoplasmic factors in flax (*21*) and in maize (*68*).

In SONNEBORN's *B* group of varieties of the protozoan *Paramecium aurelia*, cytoplasmic factors are regularly involved in the inheritance of mating type, presence of antigens, and the ability to kill specific sensitive strains through the production of extracellular toxic agents known as *paramecins* (*75*). Of these, the so-called "killer" character has been studied extensively by SONNEBORN and his associates (*67, 75*). Killers carry in their nuclei a dominant allele, "*K*", of a particular gene and have in their cytoplasm discrete bodies known as "kappa" units. Kappa particles are multiplied only if at least one is already present and if *K* is carried in the nucleus, but not necessarily at the same rate as do the paramecia themselves. Thus, killers grown at high temperatures may irreversibly lose their kappa units. Likewise, environmental and nutritional conditions leading to rapid cell division may result in certain lines of descent outgrowing kappa.

Animals with a small number of kappa particles or with none at all are sensitive regardless of whether they are genetically *KK* or *kk*. If the *K* allele is present, those with a few kappa particles may become killers under conditions in which kappa multiplication exceeds cell division in rate long enough for the kappa particles to reach a level of about 250 per animal.

A situation somewhat similar to the killer case in *Paramecium* has been studied in the vinegar fly *Drosophila* by L'HÉRITIER and his co-workers (*36*). Most stocks of *Drosophila* are relatively resistant to killing by CO_2, but some show extreme sensitivity. This sensitivity is dependent on cytoplasmic units regularly transmitted through the egg and occasionally through the sperm. Like kappa, these units do not necessarily multiply at the same rate as the whole organism; hence their numbers may be increased or decreased depending on whether their reproduction or that of the whole organism is more favored by cultural conditions. Growth of flies at somewhat higher temperatures (30° C.) may lead to irreversible loss of the cytoplasmic factor. Genes specifically necessary for the reproduction of the CO_2-sensitivity factor are not known in *Drosophila*.

The evidence as a whole, therefore, points quite clearly to the *nucleus and its chromosomes as the seat of the stable and autonomous hereditary mechanism*. The few examples known in which cytoplasmic factors are differentials in inheritance appear to represent special cases; and in no instance are there compelling reasons for believing that the units concerned are of primary general significance in the same sense in which nuclear genes are.

Acceptance of this conclusion in no way denies the importance of cytoplasm. Rather it suggests that the role of this cell component in reproduction, development, and function is different from that of the gene system. It seems probable that there are many cytoplasmic elements that like chloroplasts or kappa units are able to duplicate themselves under nuclear gene control. It is possible that many if not all the microsomes (chondriosomes) of the cytoplasm are of this nature. These bodies contain the great bulk of the ribose nucleic acid of the cytoplasm and contain or carry in firm combination many essential enzymes (*16, 22*).

Cytoplasm almost certainly plays a primary role in cellular differentiation. It is a widespread assumption among geneticists that all cells of a multicellular organism are alike genetically in that their nuclei carry an integral number of identical gene sets [see, however, HUSKINS (*43*)]. Since in the egg cytoplasm of many animals there occurs a pre-cleavage mosaic pattern that is correlated with the differentiation of the cells which are later cut out of this cytoplasm, it seems likely that the quantitative and/or qualitative segregation of cytoplasmic constituents in somatic development underlies cellular differentiation.

In connection with this long-standing and important problem of the nature of differentiation, the cases of cytoplasmic inheritance mentioned above are of particular interest in that they may well indicate the nature of the underlying mechanism. As SONNEBORN (*75*) has pointed out it is possible in *Paramecium* for two lines of descent originating from a single individual and identical in nuclear genes to differ with respect to kappa units. Kappa can be irreversibly lost in vegetative reproduction. In a similar way cytoplasmic differentiation is known to occur with respect to chloroplasts in maize and for CO_2-resistance in *Drosophila*. These examples suggest that an egg cell with which a multicellular organism starts development may carry a large number of kappa-like self-duplicating cytoplasmic units—so-called *plasmagenes*—which during ontogeny undergo systematic mutation or loss, directed by both external and internal environmental factors. Essentially this same thesis has been elaborated by WRIGHT (*84*) and by SPIEGELMAN (*76*).

Genes and Evolution.

It is generally believed by geneticists that gene mutation and natural selection provide the mechanism for *evolutionary change* in living systems. Such change may be either toward simplification or toward complexity.

Progressive simplification structurally and metabolically seems easy to understand in principle. All genes are subject to loss through mutation. Such loss-mutation in general is much more probable than the reverse process. It follows, therefore, that genes which do not give the individual organisms possessing them an advantage in competition will tend to be lost. This tendency is particularly clear in parasitic forms. Metazoan parasites tend to degenerate morphologically to the point where they consist of little more than a reproductive system and a mechanism for taking food materials from the host. They tend also to lose synthetic ability, a fact pointed out in the case of protozoa by LWOFF (*54*) and for bacteria by KNIGHT (*47*). The genetic basis of this is clear from the *Neurospora* work already mentioned.

BURNET (*18*) and others believe that viruses represent parasitic forms that have degenerated to a subcellular level of organization. On theoretical grounds one would expect the limit of such simplification to be reached when nothing but a single self-duplicating unit of the parasite remains. If genes are the ultimate units of self-duplication and mutation, the simplest parasite should be a single gene system. Possibly, the smallest existing viruses are just that.

If genes represent the smallest subdivisions of living systems capable of multiplication and mutation—two characteristics of all organisms that would seem to be indispensable—it is probable that some such single unit was ancestral to present forms of life. Such a unit is assumed by OPARIN (*65*) and others (*1*) to have originated spontaneously not from inorganic precursors but rather from an array of preformed organic compounds which were themselves formed in the pre-life world by chance, over long periods of time. Presumably this simple unit possessed the ability to direct the formation of more of its kind from chance-formed component parts.

The question of how such a relatively simple form could have given rise to multigenic systems has been considered by HOROWITZ (39) who postulates that, through mutation and natural selection, there evolved systems which are able to carry out sequences of chemical reactions having as final end products substances essential for reproduction of the system.

To illustrate: assume that the first gene-like unit—the "protogene"—required for multiplication a series of building blocks including the spontaneously formed compound D. With D capable of being formed through the series of reactions

$$A \to B \to C \to D,$$

then, when D became a limiting factor in the multiplication of the protogene, a mutant form of the protogene capable of catalyzing the reaction $C \to D$ would have a selective advantage, $i.\,e.$, it would be capable of multiplying under conditions in which its ancestor could not. This mutant form would therefore be expected to replace the type from which it arose. If the available supplies of both C and D were then reduced by the multiplication of the new unit which, by virtue of its hetero-catalytic nature could be called a true gene, a second mutation of gene 1 to gene 2, capable of catalyzing the conversion, $B \to C$, would be selected if it occurred. A symbiotic union of gene 1 and gene 2 to form a simple two-gene system could then reproduce in the presence of any or all of the compounds B, C and D. If a third gene capable of promoting the reaction, $A \to B$ were to arise by mutation and be added to the system, the three-gene system could carry out the entire reaction sequence:

$$A \to B \to C \to D.$$

In this way multigenic systems capable of utilizing chemical or radiant energy in synthesizing their component parts from inorganic molecules could have arisen through a series of mutational steps, none of which need to have been more complex than mutational changes which are known to occur in present-day organisms. In such a process, each successive evolutionary step could have conferred a selective advantage over the previously existing system. While this hypothesis, of course, does not attempt to account for the process of evolution in detail, it does indicate a simple and logical mechanism by which living systems might have acquired the ability to carry out complex metabolic processes.

There are, indeed, suggestions that existing organisms, advanced in the scale of evolution, may add to their synthetic abilities in this way. HOULAHAN and MITCHELL (42), for example, have reported that a mutant form of *Neurospora*, unable to synthesize a certain pyrimidine derivative because of mutation in a given gene, may regain the ability to carry out this synthesis through mutation in a second gene. It appears that gene 2, the previous function of which is not known, has taken over the function of gene 1. It seems probable that if gene 1 had never existed, the particular synthetic step under discussion could have arisen through mutation in gene 2.

If, as previously suggested, genes that do not perform functions useful to the organism tend to disappear through loss-mutation, the question arises of how a new ability can be added without sacrificing a previously existing one? The probable answer is that at any given time in a multigenic organism some genes are likely to be carried in duplicate due to accidents in chromosome multiplication and distribution. Such duplicate genes, which are known to occur in a number of organisms (60), may then provide the gene material for positive evolution.

Conclusions.

As units of inheritance, genes appear to be the irreducible patterns from which giant molecules such as proteins, nucleoproteins, and possibly others, derive their specificities during the development and functioning of living systems. Through enzymes as intermediates, they control specific chemical reactions. They are often related in what seems to be a simple way to antigens. The precise mechanism by which they duplicate themselves is unknown, as is also the manner in which they confer specificity to enzymes, antigens and other giant molecules.

It is probable that living systems first arose as single, gene-like units. Through mutation of these units and natural selection complex multigenic organisms were formed. Under conditions in which synthetic abilities are not advantageous, to the organisms possessing them, for example in parasites, such abilities tend to disappear through loss-mutations. The end result of such degenerative specialization has probably been reached in the smaller viruses. On this view, present-day viruses are not descended directly from simple primitive forms of life but have returned to a similar state by a long progressive-regressive evolutionary process.

References.

1. ALEXANDER, J.: Life: Its Nature and Origin, p. 291. New York: Reinhold Pub. Corp. 1948.
2. ATWOOD, S. S. and J. T. SULLIVAN: Inheritance of a Cyanogenetic Glucoside and its Hydrolyzing Enzyme in *Trifolium repens* J. Heredity **34**, 311 (1943).
3. AUERBACH, C., J. M. ROBSON and J. G. CARR: The Chemical Production of Mutations. Science [New York] **105**, 243 (1947).

4. AVERY, O. T., C. M. McLEOD and M. McCARTY: Studies on the Chemical Nature of the Substance Inducing Transformation of Pneumococcal Types: Induction of Transformation by a Desoxyribonucleic Acid Fraction Isolated from *Pneumococcus* Type III. J. exp. Medicine **79**, 137 (1944).

5. BALL, E. G.: Xanthinoxidase: Purification and Properties. J. biol. Chemistry **128**, 51 (1939).

6. BEADLE, G. W.: Genetics and Metabolism in Neurospora. Physiologic. Rev. **25**, 643 (1945).

7. — Biochemical Genetics. Chem. Reviews **37**, 15 (1945).

8. BEADLE, G. W., H. K. MITCHELL and J. F. NYC: Kynurenin as an Intermediate in the Formation of Nicotinic Acid from Tryptophane by *Neurospora*. Proc. nat. Acad. Sci. USA **33**, 155 (1947).

9. BEADLE, G. W. and E. L. TATUM: Genetic Control of Biochemical Reactions in Neurospora. Proc. nat. Acad. Sci. USA **27**, 499 (1941).

10. — — Neurospora. II. Methods of Producing and Detecting Mutations Concerned with Nutritional Requirements. Amer. J. Bot. **32**, 678 (1945).

11. BINKLEY, F. and V. DU VIGNEAUD: The Formation of Cysteine from Homocysteine and Serine by Liver Tissue of Rats. J. biol. Chemistry **144**, 507 (1948).

12. BOIVIN, A.: Directed Mutation in Colon Bacilli, by an Inducing Principle of Desoxyribonucleic Nature: Its Meaning for the General Biochemistry of Heredity. Cold Spring Harbor Sympos. quantitat. Biol. **12**, 6 (1947).

13. BONNER, D.: Biochemical Mutations in *Neurospora*. Cold Spring Harbor Sympos. quantitat. Biol. **11**, 14 (1946).

14. — The Identification of a Natural Precursor of Nicotinic Acid. Proc. nat. Acad. Sci. USA **34**, 5 (1948).

15. BONNER, D. and G. W. BEADLE: Mutant Strains of *Neurospora* Requiring Nicotinamide or Related Compounds for Growth. Arch. Biochemistry **11**, 319 (1946).

16. BRACHET, J.: Nucleic Acids in the Cell and Embryo. In Nucleic Acid. Symposia Soc. Exp. Biology, p. 207–224. Cambridge: Univ. Press. 1947.

17. BRAND, E., R. J. BLOCK, B. KASSELL and G. F. CAHILL: Carboxymethylcysteine Metabolism: Its Implications on Therapy in Cystinuria and on the Methionine-Cysteine Relationship. Proc. Soc. exp. Biol. Med. **35**, 501 (1936).

18. BURNET, F. M.: Virus as Organism, p. 134. Cambridge, Mass.: Harvard Univ. Press. 1945.

19. CASPERSSON, T.: The Relation Between Nucleic Acid and Protein Synthesis. In Nucleic Acid. Symposia Soc. Exp. Biology, p. 127–151. Cambridge: Univ. Press. 1947.

20. CASTLE, W. E.: The Linkage Relations of Yellow Fat in Rabbits. Proc. nat. Acad. Sci. USA **19**, 947 (1933).

21. CHITTENDEN, R. J. and C. PELLEW: A Suggested Interpretation of Anisogeny. Nature [London] **119**, 10 (1927).

22. CLAUDE, A.: Particulate Components of Cytoplasm. Cold Spring Harbor Sympos. quantitat. Biol. **9**, 263 (1941).

23. CORKILL, L.: Cyanogenesis in White Clover (*Trifolium repens* L.). New Zealand J. Sci. Technol., Sect. B **23**, 178 (1942).

24. DELBRÜCK, M.: A Theory of Autocatalytic Synthesis of Polypeptides and its Application to the Problem of Chromosome Reproduction. Cold Spring Harbor Sympos. quantitat. Biol. **9**, 122 (1941).

25. DELBRÜCK, M. and W. T. BAILEY, Jr.: Induced Mutations in Bacterial Viruses. Cold Spring Harbor Sympos. quantitat. Biol. **11**, 33 (1946).

26. Ephrussi, B.: Chemistry of "Eye Color Hormones" of *Drosophila*. Quart. Rev. Biol. **17**, 327 (1942).
27. Figge, F. H. J.: Pigment Metabolism Studies: The Regulation of Tyrosinase Melanin Formation by Oxidation-Reduction Systems. J. cellular comparat. Physiol. **15**, 233 (1940).
28. Fisher, R. A.: The Rhesus Factor. Amer. Scientist **35**, 95, 113 (1947).
29. Fölling, A.: Über Ausscheidung von Phenylbrenztraubensäure in dem Harn als Stoffwechselanomalie in Verbindung mit Imbezillität. Hoppe-Seyler's Z. physiol. Chem. **227**, 169 (1934).
30. Fölling, A., O. L. Mohr and L. Rüüd: *Oligophrenia Phenylpyrouvica*, a Recessive Syndrome in Man. Skrifter Norske Videnskaps-Acad., Oslo, Mat.-Naturv. Klasse No. 13, 44 (1945).
31. Garrod, A. E.: Inborn Errors of Metabolism, 2nd. ed., p. 216. Oxford: Med. Pub., Oxford Univ. Press. 1923.
32. Giles, N. H., Jr. and E. Z. Lederberg: Induced Reversions of Biochemical Mutants in *Neurospora crassa*. Amer. J. Bot. **35**, 157 (1948).
33. Gross, O.: Über den Einfluß des Blutserums des Normalen und des Alkapto-nurikers auf Homogentisinsäure. Biochem. Z. **61**, 165 (1914).
34. Haldane, J. B. S.: The Rate of Spontaneous Mutation of a Human Gene. J. Genetics **31**, 317 (1935).
35. — New Paths in Genetics, p. 206. New York: Harper & Bros. 1942.
36. L'Héritier, Ph. et F. H. de Scoeux: Transmission par Graffe et Injection de la Sensibilité Héréditaire au Gaz Carbonique chez la *Drosophile*. Bull. biol. France Belgique **81**, 70 (1947).
37. Hershey, A. D. and R. Rotman: Linkage Among Genes Controlling Inhibition of Lysis in a Bacterial Virus. Proc. nat. Acad. Sci. USA **34**, 89 (1948).
38. Hollaender, A. and C. W. Emmons: Wavelength Dependance of Mutation Production in the Ultraviolet with Special Emphasis on Fungi. Cold Spring Harbor Sympos. quantitat. Biol. **9**, 179 (1941).
39. Horowitz, N. H.: On the Evolution of Biochemical Syntheses. Proc. nat. Acad. Sci. USA **31**, 153 (1945).
40. — Methionine Synthesis in *Neurospora*. The Isolation of Cystathionine. J. biol. Chemistry **171**, 255 (1947).
41. Horowitz, N. H., M. B. Houlahan, M. V. Hungate and B. Wright: Mustard Gas Mutations in Neurospora. Science [New York] **104**, 233 (1946).
42. Houlahan, M. B. and H. K. Mitchell: A Suppressor in Neurospora and its Use as Evidence for Allelism. Proc. nat. Acad. Sci. USA **33**, 223 (1947).
43. Huskins, C. L.: The Subdivision of the Chromosomes and their Multiplication in Non-dividing Tissues: Possible Interpretations in Terms of Gene Structure and Gene Action. Amer. Naturalist **81**, 401 (1947).
44. Imai, Y.: Chlorophyll Varigations Due to Mutable Genes and Plastids. Z. indukt. Abstammungs- u. Vererbungslehre **71**, 61 (1936).
45. Irwin, M. R.: Immunogenetics. Advances in Genetics **1**, 133 (1947).
46. Kaufman, S., G. W. Schwert and H. Neurath: Specific Peptidase and Esterase Activities of Chymotrypsin. Arch. Biochemistry **17**, 203 (1948).
47. Knight, B. C. J. G.: Bacterial Nutrition. Med. Research Council (Brit.) Special Rept. Series, No. 210, 182 (1936).
48. Knight, C. A.: Nucleoproteins and Virus Activity. Cold Spring Harbor Sympos. quantitat. Biol. **12**, 115 (1947).
49. Knox, W. E.: The Quinine-Oxidizing Enzyme and Liver Aldehyde Oxidase. J. biol. Chemistry **163**, 699 (1946).

50. LAWRENCE, J. S. and C. G. CRADDOCK, Jr.: Hemophilia: The Mechanism of Development and Action of an Anticoagulant Found in Two Cases. Science [New York] 106, 473 (1947).
51. LAWRENCE, W. J. C. and J. R. PRICE: The Genetics and Chemistry of Flower Colour Variation. Biol. Rev. Cambridge philos. Soc. 15, 35 (1940).
52. LEDERBERG, J.: Gene Recombination and Linked Gene Segregations in Escherichia coli. Genetics 32, 505 (1947).
53. LURIA, S. E.: Reactivation of Irradiated Bacteriophage by Transfer of Self-Reproducing Units. Proc. nat. Acad. Sci. USA 33, 253 (1947).
54. LWOFF, A.: Les Facteurs de Croissance pour les Microorganisms. Ann. Inst. Pasteur 61, 580 (1938).
55. McCARTY, M.: Chemical Nature and Biological Specificity of the Substance Inducing Transformation of Pneumococcal Types. Bacteriol. Rev. 10, 63 (1946).
56. McCLINTOCK, B.: Neurospora. I. Preliminary Observations of the Chromosomes of Neurospora crassa. Amer. J. Bot. 32, 671 (1945).
57. McELROY, W. D., J. E. CUSHING and H. MILLER: The Induction of Biochemical Mutations in Neurospora crassa by Nitrogen Mustard. J. comparat. Cellular Physiol. 30, 331 (1947).
58. MASON, H. S.: The Chemistry of Melanin. J. biol. Chemistry 168, 433 (1947).
59. MEDES, G.: A New Error of Tyrosine Metabolism: Tyrosinosis. The Intermediary Metabolism of Tyrosine and Phenylalanine. Biochemic. J. 26, 917 (1932).
60. METZ, C. W.: Duplication of Chromosome Parts as a Factor in Evolution. Amer. Naturalist 81, 81 (1947).
61. MIRSKY, A. E. and H. RIS: The Chemical Composition of Isolated Chromosomes. J. gen. Physiol. 31, 7 (1947).
62. MITCHELL, H. K. and J. F. NYC: Hydroxyanthranilic Acid as a Precursor of Nicotinic Acid in Neurospora. Proc. nat. Acad. Sci. USA 34, 1 (1948).
63. MOEWUS, F.: Zur Sexualität der niederen Organismen. I. Flagellaten und Algen. Erg. Biologie 8, 287 (1941).
64. MULLER, H. J.: Pilgrim Trust Lecture: The Gene. Proc. Roy. Soc. [London], Ser. B 134, 1 (1947).
65. OPARIN, A. I.: The Origin of Life, p. 270. (Translated by S. MORGULIS.) New York: The Macmillan Co. 1938.
66. PENROSE, L. S.: Inheritance of Phenylpyruvic Amentia (Phenylketonuria). Lancet 229, 192 (1935).
67. PREER, J. R.: Some Properties of a Genetic Cytoplasmic Factor in Paramecium. Proc. nat. Acad. Sci. USA 32, 247 (1946).
68. RHOADES, M. M.: The Cytoplasmic Inheritance of Male Sterility in Zea mays. J. Genetics 27, 71 (1933).
69. — Plastid Mutations. Cold Spring Harbor Sympos. quantitat. Biol. 11, 202 (1946).
70. ROBINOW, C. F.: Nuclear Apparatus and Cell Structure of Rod-shaped Bacteria, In DUBOS: The Bacterial Cell. p. 355—370. Cambridge, Mass.: Harvard Univ. Press. 1945.
71. RUSSELL, W. L.: Investigation of the Physiological Genetics of Hair and Skin Color in the Guinea Pig by Means of the Dopa Reaction. Genetics 24, 645 (1939).
72. SAWIN, P. B. and D. GLICK: Atropinesterase, a Genetically Determined Enzyme in the Rabbit. Proc. nat. Acad. Sci. USA 29, 55 (1943).
73 SCHWERT, G. W., H. NEURATH, S. KAUFMAN and J. E. SNOKE: The Specific Esterase Activity of Trypsin. J. biol. Chemistry 172, 221 (1948).

74. Sonneborn, T. M.: Sex Hormones in Unicellular Organisms. Cold Spring Harbor Sympos. quantitat. Biol. 10, 111 (1942).
75. — Recent Advances in the Genetics of Paramecium and Euplotes. Advances in Genetics 1, 264 (1947).
76. Spiegelman, S.: Nuclear and Cytoplasmic Factors Controlling Enzymatic Constitution. Cold Spring Harbor Sympos. quantitat. Biol. 11, 256 (1946).
77. Stadler, L. G. and F. M. Uber: Genetic Effects of Ultraviolet Radiation in Maize. IV. Comparisons of Monochromatic Radiations. Genetics 27, 84 (1942).
78. Strandskov, H. H.: Physiological Aspects of Human Genetics. Five Human Blood Characteristics. Physiologic. Rev. 24, 445 (1944).
79. Tatum, E. L.: Induced Biochemical Mutations in Bacteria. Cold Spring Harbor Sympos. quantitat. Biol. 11, 278 (1946).
80. Tatum, E. L. and D. Bonner: Indole and Serine in the Biosynthesis and Breakdown of Tryptophane. Proc. nat. Acad. Sci. USA 30, 30 (1944).
81. Teas, H. J., N. H. Horowitz and M. Fling: Homoserine as a Precursor of Threonine and Methionine in Neurospora. J. biol. Chemistry 172, 651 (1948).
82. Wright, S.: The Physiology of the Gene. Physiologic. Rev. 21, 487 (1941).
83. — The Physiological Genetics of Coat Color of the Guinea Pig. Biol. Symposia 6, 337 (1942).
84. — Genes as Physiological Agents: General Considerations. Amer. Naturalist 79, 289 (1945).
85. Zechmeister, L., A. L. LeRosen, F. W. Went and L. Pauling: Prolycopene, a Naturally Occurring Stereoisomer of Lycopene. Proc. nat. Acad. Sci. USA 27, 468 (1941).

(Received, May 20, 1948.)

Infrared Spectroscopy in Structure Determination and its Application to Penicillin.

By R. S. RASMUSSEN, Emeryville, California.

With 11 Figures.

Introduction.

The field of infrared absorption spectroscopy is one which promises to be of great value in assisting the organic chemist in elucidation of structures of molecules, natural or synthetic. However, it is still in a very early stage of its potential development. In general, only the more common types of structure have been investigated, and a considerable body of work done during the last few years has not yet appeared in publication. However, with such publication and with the continued growth of interest in the subject, our knowledge should widen considerably in the next years. At present the writer must call upon certain work which is not yet available to the general reader, much of which was obtained in the course of the project on penicillin structure (B 11, B 12).

Infrared absorption spectroscopy is generally understood among chemical spectroscopists to cover the wave length range from $2\,\mu.$ to $25\,\mu.$ ($\mu. = $ micron; $1\,\mu. = 10^{-4}$ cm. $= 10^4$ Ångström units). This range includes most of the fundamental vibration bands of organic molecules (a few are to be found at longer wave lengths), and is the range easily accessible with the usual prism materials and radiation receivers. A considerable amount of valuable work has been published on studies in the $0{,}8$ to $2{,}0\,\mu.$ range, experimental difficulties being in general less there; however, only certain types of problems are attackable when one is restricted to this range. At the other end of the infrared region, it would be extremely valuable to be able to go beyond $25\,\mu.$, but experimental difficulties make this generally prohibitive at present. The material of the present survey will be limited to studies in the $2\,\mu.$ to $25\,\mu.$ range, except as other work may be illuminating to the point at hand.

I. Experimental Methods.

A thorough discussion of recent instrumental and experimental development
is given by WILLIAMS ($114a$).

1. Instruments.

No detailed discussion of spectrometer construction or operation
will be given here; the reader is referred to appropriate literature
elsewhere (B 9, B 10, 5, 7, 8, 60, 61, 64, 65, 90, 93, 110, 111, 117). For

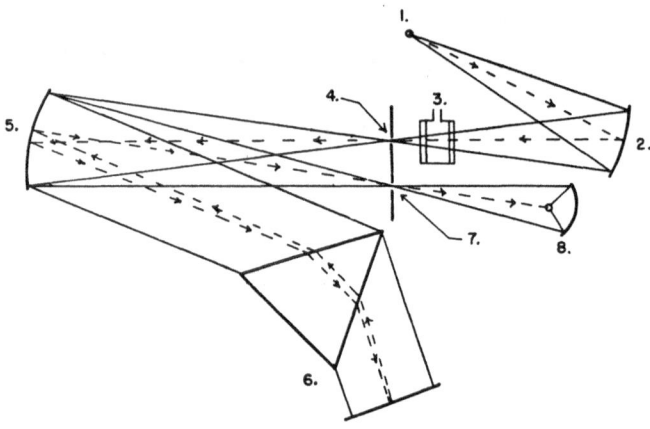

Fig. 1. Schematic Diagram of an Infrared Spectrometer. 1. Source. 2. Mirror, to focus source image on
entrance slit. 3. Absorption cell. 4. Entrance slit. 5. Collimating mirror, to parallelize beam on prism or
grating. 6. Prism and plane mirror to send beam back through prism (prism and mirror could be replaced
by a plane reflection grating). 7. Exit slit. 8. Receiver, and mirror to focus image on it. — The path of a
pencil, of proper wave length to pass the exit slit, is shown by ——— → . Light of other wave lengths
is focussed at one side or the other of the exit slit, and is stopped. By rotation of the prism or plane mirror
(or of the grating), light of other wave lengths may be brought into position to pass the exit slit.

measuring the absorption spectrum of a material, it is necessary to
have: a source of radiation that gives a continuum (i. e., emits radiation
of all wave lengths in the region of interest); a means of dispersing this
radiation into its components of various wave length (either a prism
or a grating); and a receiver which records the intensity of radiation
(thermocouple, bolometer, radiometer, etc.). Then for a given setting
of the instrument, such that essentially monochromatic light (i. e., of
only a very narrow range of wave lengths) strikes the receiver, the
absorption of a given material may be measured by comparing the amount
of radiation (P_0) recorded without the material in the light path, to
the amount (P) received with the material in the light path. The
ratio P/P_0 (the *transmission*, T, of the material at the given wave
length) is the means commonly used to describe the amount of absorption.
A schematic diagram of a simple spectrometer is shown in Figure 1.

Unfortunately, the nomenclature of instruments is decidedly unstandardized. To follow etymological principles, which are probably more desirable guides than a confused history of usage, a "spectrometer" would be any instrument which enables one to measure a spectrum, "spectrographs" would comprise that class of spectrometers which actually give a picture or graph of the spectrum; and "spectrophotometers" would be that class of spectrometers in which the emphasis is on the accurate measurement of radiation intensities.

For the type of work dealt with here, the common means of dispersing radiation is the prism. Gratings have, in general, higher dispersion and are used where the finest details are to be studied, but they entail other difficulties which make them less suitable than prisms when the highest dispersion is not necessary. Of prism materials, rock salt has been most used. It has fair dispersion in the 5–8 μ. region and good dispersion (for a prism) in the 8–15 μ. region, these regions being of greatest recent chemical-structural interest. Beyond 15 μ. rock salt is opaque. For work from 15 to 25 μ. the potassium bromide prism is used; and for a prism with high dispersion in the 2–5 μ. region lithium fluoride is available (*31, 110, 116*). More recent developments include the use of calcium fluoride (synthetic fluorspar) for good dispersion in the 5–8 μ. region (*31*), and of thallous halide crystals which show promise of being usable to 40 μ. (*72*).

2. The Transmission Curve.

A plot of transmission versus wave length or wave number* is called a transmission curve, and is the usual means of depicting the absorption characteristics of materials (Figure 2, p. 335). Obviously, the transmission is a function of the amount of material in the light path, so this amount must be specified. In general, a transmission curve for a given quantity of absorbing material will only be suitable for certain of the absorption bands. Weaker bands may not stand out sufficiently to be detectable or accurately measurable, and stronger bands may give practically complete absorption over such a wide interval that the center of the band or its true intensity cannot be determined. Hence, to show satisfactorily all bands, several transmission curves must be presented. Actually the transmission T at a given wave length depends on the amount of absorbing material according to the BOUGUER-LAMBERT-BEER law, which can be cast in the form,

$$T = 10^{-ac},$$

where c is the number of moles of an absorbing material per unit area transverse to the light beam, and a is an extinction coefficient which

---

* The wave number ω, a quantity proportional to frequency, is reciprocally related to wave length and is expressed in units of reciprocal centimeters (cm.$^{-1}$), viz. $\omega = 10^4/\lambda$, for λ in μ. Both wave length and wave number are commonly used, there being no generally accepted standardization on one or the other.

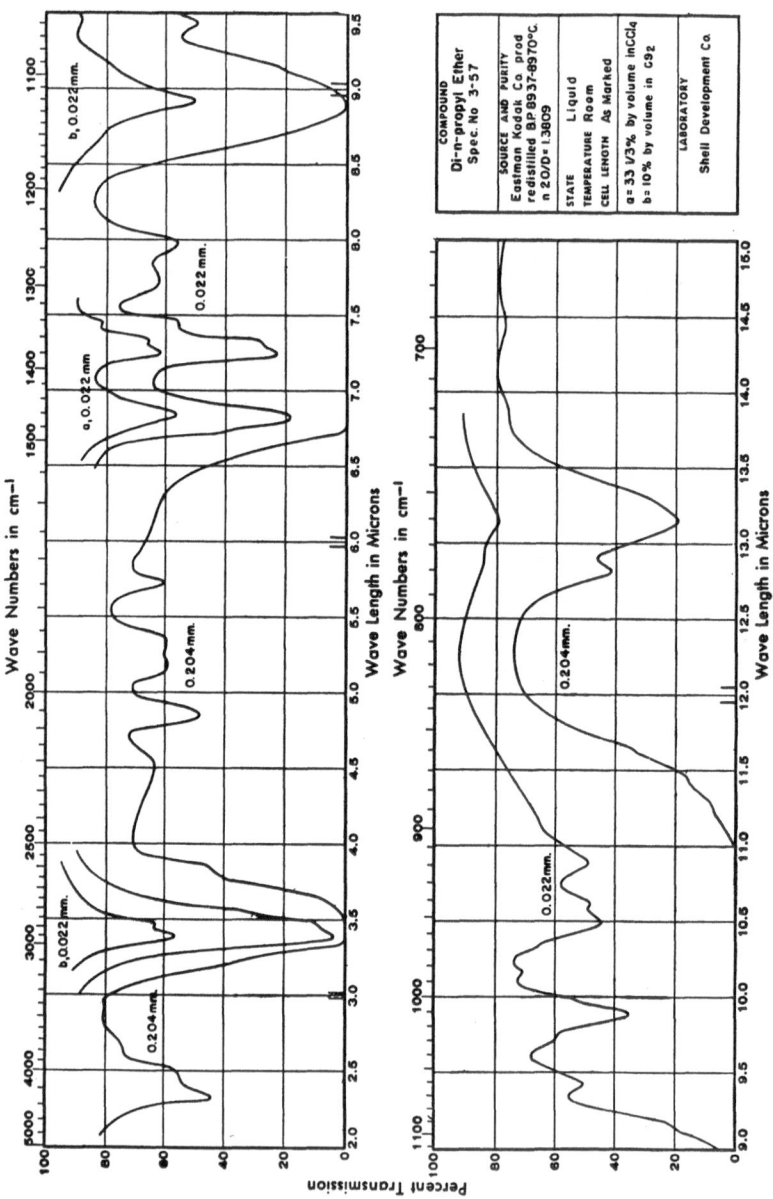

is a function of wave length but not of concentration. There would be some economy in plotting a or log a versus the wave length, as is done for the visible and ultraviolet regions. However, the experimental errors in the infrared being larger, it has generally been thought advisable

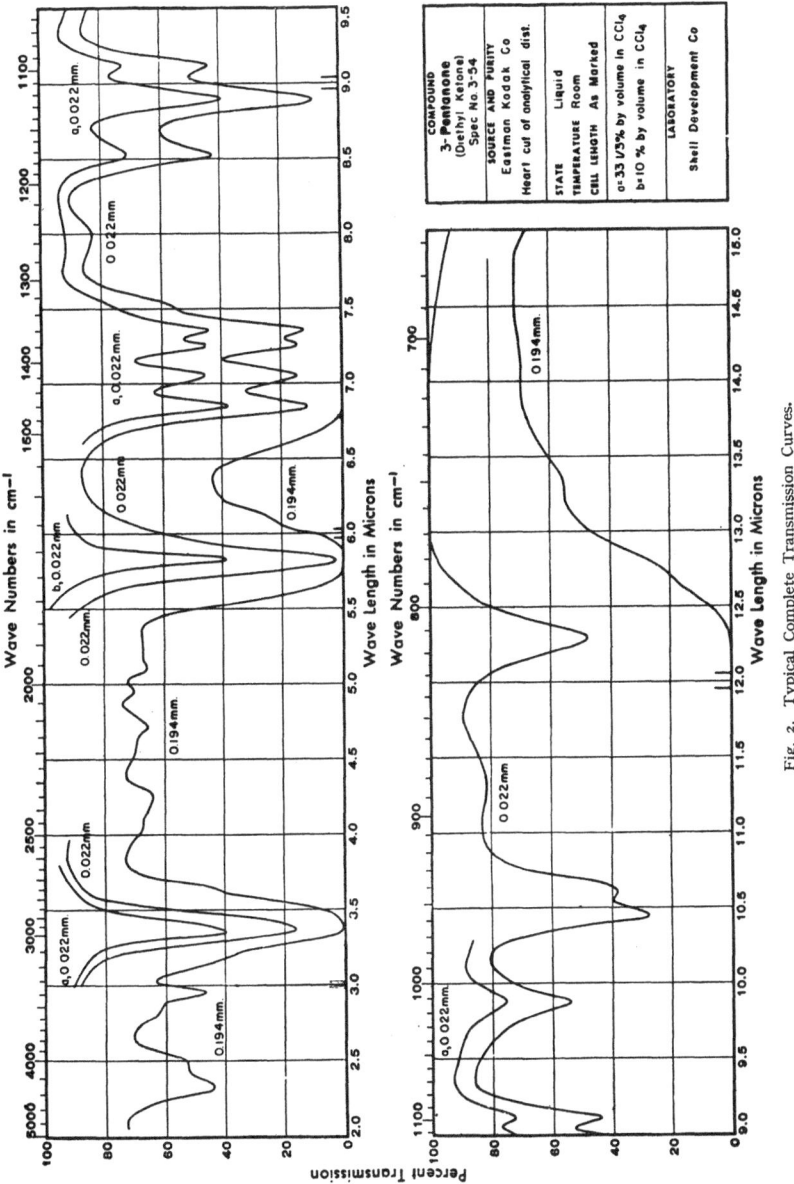

Fig. 2. Typical Complete Transmission Curves.

to present the experimentally determined quantity, viz. transmission, rather than a derived quantity.

In recent years the tendency has been toward the construction of automatic-recording instruments which give, either directly or by a suitable transformation on a recorded curve, a complete transmission curve. Indeed, it may be said that the rapidity of stable automatic-recording instruments, as compared with the slowness of the older point-by-point individual reading techniques, has been the greatest factor in the increased use of infrared for structural purposes. Many instruments record a rock-salt prism spectrum from 2 to 15 μ. in times of the order of one-half to one hour.

3. Absorption Cells for Vapor and Liquid Samples.

For work with vapors, cells with parallel polished windows of transparent material are used, and lengths of 5 to 15 cm. are found convenient. Pressures of the order of 20 to 500 mm. Hg will then in general yield suitable spectra.

Of predominant interest here, however, are the requirements for liquid and solid samples. Samples of pure liquids require cells giving liquid layers 0,2 to 0,01 mm. (or less) thick. The former figure is generally applicable for the major bands of nonpolar substances such as paraffin hydrocarbons, and the latter for the more polar compounds. Such cells are constructed of two plane polished windows separated the desired amount by a thin shim (usually of sheet metal), an aperture being cut in the shim to fit the light beam. Numerous modifications have been described (*10, 15, 16, 17, 30, 82, 89*).

Spectra of materials in solution may be obtained in an identical way to pure liquids, except of course that a larger cell spacing is required than would be needed to accomodate the same amount of pure substance. A fundamental difficulty with solutions exists, however: there are no solvents that do not themselves absorb strongly in some regions, and moderately or weakly in most others. Hence, it becomes necessary to choose a solvent appropriate for the specific region of interest. Torkington and Thompson (*104*) have given a diagram of particular convenience in this connection. The commonest solvents for general work are carbon tetrachloride, chloroform, and carbon disulfide.

Since the cell windows are generally of rock salt or potassium bromide, aqueous solutions or wet samples cannot be tolerated unless their effects of etching the windows, or causing an increase in cell spacing by dissolving some of the window material, are not considered important. In this direction there is great need for improvement, since many small-scale laboratory samples cannot be conveniently dried; and also it would sometimes be convenient to work with aqueous solutions despite the generally heavy absorption of liquid water. The recent availability of silver chloride in an optically clear, flat-surfaced form and transparent to 25 μ. may be at least a partial answer to this difficulty (*27, 28, 42*).

From the cell spacings stated above, it is evident that the actual amount of sample required in the light path is very small, viz. 0,02 ml. or less. One of the problems of cell construction is to avoid having too much volume taken up by the leads for filling and emptying the cell.

4. Solid Samples.

In the case of rubbery, resinous or gummy materials, a sample suitable for obtaining a spectrum can usually be prepared by depositing a solution of the material on an infrared-transparent plate (e. g., rock salt) and allowing the solvent to evaporate. The concentration of the solution and the volume of solution spread out per unit area on the plate must be adjusted to give an appropriate final thickness.

This technique fails in the case of substances which crystallize. If a few large crystals are formed, the only light received is that which passes through the interstices; if many smaller crystals are formed, most of the radiation is scattered back by multiple reflections at air-crystal interfaces, resulting in an apparent general "absorption" which masks the true absorption. In order to avoid these difficulties, it is necessary to reduce the material to a fine powder whose particle size is less than the wave lengths of interest [PFUND (67)] and considerably less than the thickness of sample required. This in general means average particle sizes of the order of 1μ. or less, and necessitates in some cases special grinding or attrition techniques. Further improvement results from suspending the ground material in a medium such as paraffin oil, whose absorption is usually not disturbing, and which decreases the scattering by diminishing the reflection losses. Since the oil also facilitates the grinding and the uniform spreading of the sample on the rock-salt (or other) plate, the whole process is often carried out in this medium ("oil-paste" technique).

Another method sometimes applicable to low-melting materials is to fuse some of the substance on the rock-salt (or other) plate, then to press the liquid out to a thin layer with a second plate, and finally, to cool the "sandwich". This can give a thin layer of crystals of sufficient perfection to result in very low scattering.

Particular use has been made of the above-described techniques by LECOMTE and co-workers at the Sorbonne, SUTHERLAND and THOMPSON in Great Britain, and RANDALL and FOWLER at the University of Michigan (Penicillin Structure Project).

Since there are two general methods available for studying solid materials, viz. as such or in solution, a few words should be said by way of comparison. While spectroscopically suitable solvents for a given spectral region can be found for most types of compounds, there are always the drawbacks: that additional effort is needed to find an appropriate solvent; that there is always some measure of interference by the solvent spectrum; and that for some classes of compounds there are no satisfactory solvents, either because of chemical reaction with an otherwise suitable solvent, or because of low solubility in suitable solvents. Particularly unfavorable with respect to the latter factor are such compounds as polyhydric alcohols, amino-acids, salts of carboxylic or sulfonic acids, N-unsubstituted amides, etc. These are, in general, soluble only in water or in strongly polar solvents which themselves have such strong absorption that they are usable only in very restricted regions. In a few instances this difficulty can be overcome by preparing a more soluble derivative, e. g., conversion of carboxylic acids or their salts to esters. In contrast, the techniques for solid samples are immediately applicable to any solid compound and, except for the usually unimportant bands of the oil carrier in the oil-paste method, there is no extraneous absorption.

Counterbalancing these defects, there are several advantages to solution spectra. If any quantitative intensity measurements are to be made from the spectrum for the estimation of concentrations of components, it is extremely difficult to control accurately such variables as sample thickness, scattering losses, etc., when dealing with solid samples [however, special quantitative methods can be worked out; cf. Barnes et al. (9)]. These variables are easily controlled with solution spectra. Also, particularly for materials with strong intermolecular hydrogen bonds in the solid (carboxylic acids, amides), there is much less regularity in the spectra of a group of structurally similar compounds when solid spectra are compared than for solution spectra. The lack of regularity in the solid spectra arises presumably from the variation in amount and strength of hydrogen bonding allowed by the various crystal packings. Finally, there is an experimental convenience in being able to take up a small amount of reaction product in a solvent and transfer it to the absorption cell with a pipette, rather than to go through the grinding technique. This is especially true of samples that crystallize only partially.

II. Interpretation of Infrared Spectra.

1. General.

Only a brief qualitative exposition of the theory of vibrational spectra will be given here. The reader is referred to the excellent treatise by Herzberg (B 5) and to other sources mentioned there, for the physical and mathematical theory involved.

It will be sufficient for the present discussion to use a classical-mechanical picture, since the modifications introduced by quantum mechanics are practically always trivial for our purpose. A molecule of N atoms can engage in $3N-6$ different types of motion which are vibrational in character (for linear molecules, the proper expression is $3N-5$), i. e., motions for which the molecule, if distorted from its equilibrium configuration, oscillates back and forth through the equilibrium positions. Each of these types of vibration occurs with a frequency determined by the geometry of the molecule, the masses of the component atoms, and the forces inherent in the chemical bonds which resist changes in bond angles or distances. These frequencies are called the fundamental frequencies of the molecule, and are strictly determined by the above factors only in the vapor state. In liquid, dissolved, or solid states there is some perturbation due to superposition of intermolecular forces on the intramolecular bond forces, which results in changes of the fundamental frequencies. However, since these intermolecular forces are weak compared to bond forces, the shifts are in general small (1% of the frequency or less), unless some particularly strong sort of intermolecular force, such as hydrogen bonding, is present.

For a molecule of low or no symmetry* all these frequencies are active in absorption of electromagnetic radiation; i. e., on passing a beam of radiation through an assembly of molecules, those wave lengths are absorbed whose frequencies equal the fundamental frequencies of the molecules. The absorption occurring in the neighborhood of such a frequency is called an absorption band. Similarly, for the case of low or no symmetry, all frequencies are active in the RAMAN spectrum; i. e., ultraviolet or visible light which is scattered from the material suffers changes in frequency equal to these vibration frequencies, the shifts being observable as weak lines appearing in addition to the discrete lines of the source ($B\ 5$, $B\ 6$, $B\ 7$). The fundamental frequencies are found to extend from about 3600 cm.$^{-1}$ (2,8 μ.) to about 100 cm.$^{-1}$ (100 μ.), or less in the case of certain quasi-rotational vibrations. It is this range which sets apart the nearer infrared spectrum as the *vibrational* spectrum, in contradistinction to the far infrared (500 cm.$^{-1}$ to less than 1 cm.$^{-1}$), or the ultraviolet and visible regions (10000 cm.$^{-1}$ to ∞ cm.$^{-1}$), these latter having to do with rotational frequencies and electronic orbital frequencies, respectively.

There also appear frequencies in the infrared and RAMAN which are near integral multiples of the fundamental frequencies (overtones); cthers which are near sums of fundamentals or integral multiples thereof (combinations); andm ore rarely, frequencies which are exact differences between fundamentals. These additional frequencies occur as infrared

---

* "Symmetry" is used here in the strict geometrical sense of symmetry operations (rotations and reflections) which can be performed on the molecule in its actual spatial configuration.

bands or RAMAN lines of usually much lower intensity than the fundamentals.

When molecules with a sufficient degree of symmetry are considered, it is found experimentally, and is explainable theoretically, that certain of the frequencies may not appear in the infrared, or in the RAMAN, or in either. The laws governing these "selection rules" are most conveniently stated in the language of the mathematical theory of groups [HERZBERG (*B 5*), Chapter II]. These selection rules are strictly valid only for the vapor state, and for molecules in condensed phases are occasionally violated. However, such "forbidden" bands or lines are usually weak. For the very large majority of molecules of interest to the organic chemist, the symmetry is so low that selection rules need cause no concern.

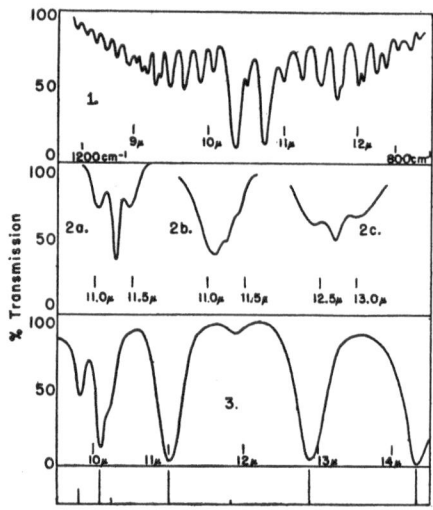

Fig. 3. Typical Infrared Bands. 1. Rotational fine structure in the vapor spectrum of a light molecule: NH_3 band, 15 cm. cell, 240 mm. pressure. 2. Structure of bands of heavier molecules, vapor spectra. a: isobutylene, 15 cm. cell, 10 mm. pressure. b: acetone, 15 cm. cell, 100 mm. c: ethyl chloride, 15 cm. cell, 300 mm. 3. Typical structureless bands of a liquid spectrum: styrene, 0,035 mm. cell. A line chart of the bands is also shown, such as is used in subsequent figures.

Infrared absorption bands and RAMAN lines do not appear as single sharp lines equal to the vibrational frequencies, but always have a certain region of absorption or scattering, as the case may be, on either side of the true vibrational frequency. This is due in the case of vapor spectra to the superposition of rotational frequencies (always much lower than vibrational), of various magnitudes and sign, on the vibrational frequency. For a molecule with a low moment of inertia about its center of mass, this rotational contribution to an infrared band appears as a (usually regular) series of sharp absorption peaks spread out over a wide region on either side of the vibrational frequency, the whole band often having a width of 100 cm.$^{-1}$ or more (Figure 3). A large fraction of infrared work done to date has been concerned with the measurement and interpretation of this rotational structure; for sufficiently simple molecules, this is the most accurate means of determining interatomic distances or angles. However, for molecules of the mass and configuration of most compounds of chemical-structural interest, the moments of inertia are so high that this rotational structure is not resolvable, and the band takes the form of a more or less structureless region of absorption, often with a rather sharp and obvious absorption peak at the center (Figure 3). RAMAN lines are similar but the central maximum is stronger, and the rotational "wings" less obvious or absent.

On passing to the condensed phases (liquid, solution, or solid), rotational contributions as described above disappear, since the molecule no longer rotates freely. Infrared bands and RAMAN lines then appear as simple error-function-type curves (Figure 3). The breadth of a band in this case is governed by the superposition on the true vibrational frequency of low (and as yet not understood) frequencies of oscillation of the whole molecule in its little cavity in the liquid or solid.

The intensities of infared bands and RAMAN lines are a subject little understood at present, and only very general statements can be made. Infrared absorption intensities are governed by the magnitude of dipole moment change which occurs as the molecule vibrates—the dipole moment itself vibrates with the same frequency as the molecule. This explains why polar molecules have a more intense absorption spectrum than non-polar ones, generally speaking; for equivalent amplitudes of atomic motion, there is a larger dipole change. Hence, in a complex molecule it is generally the vibrations associated with the polar groups which give rise to the strong bands, whereas those associated with less polar groups, such as alkyl radicals, give much weaker bands which are often hidden by the stronger ones. With RAMAN lines, the intensity is governed by the changes in polarizability occurring during a vibration. It is thus often, but by no means always, the case that those vibrations involving motions of the heavier atoms—C, O, Cl, etc.—give rise to strong lines, whereas vibrations involving principally the motion of hydrogen atoms give only weak lines.

2. Group Vibrations and Frequencies.

In discussing molecular vibrations, it is customary to speak of: stretching (or valence) vibrations, which involve changes in bond (inter-atomic) distances with essentially no changes in angles between bonds; and bending (or deformation) vibrations, which conversely involve angle but not bond-distance changes. Another type of bending vibration which is rather common is the out-of-plane deformation of an originally planar system; to a first approximation this type of motion involves no change in bond angles. Finally, torsional oscillations may occur, involving changes in the relative orientation of groups at the two ends of a bond; such frequencies are low and do not concern us here.

As stated earlier, the factors which determine the fundamental frequencies are the atomic masses and geometry, and the forces resisting change from the equilibrium position. Now if a given bond under consideration differs in its stretching and bending force constants, and in the masses of its atoms, from the bonds immediately adjacent to it, then it is possible for some vibrations to be localized in the bond. The frequencies of these are determined essentially only by the force constants and masses of that particular bond and are little affected by the atoms and bonds making up the rest of the molecule. If, on the other hand, a given bond is adjacent to other bonds of identical (or similar) force constants and masses, then a large degree of interaction occurs between the vibrations of the identical (or similar) bonds. No single vibration is localized in a particular bond, but all bonds partake in all vibrations. The frequencies associated with such vibrations are displaced considerably from the frequency that an isolated bond would have, the displacement being a result of the vibrational interaction.

As an illustration, a double bond in an otherwise saturated carbon system, being stiffer by a factor of about two than neighboring single bonds, has a characteristic stretching frequency which interacts only slightly with vibrations of the rest of the molecule. This frequency may be influenced a very little by the structure immediately adjacent to the double bond (e. g., branching or substituents) but is not influenced by the more remote parts of the molecule. Contrariwise, the two C—C bonds of propane, or the two C=C bonds of allene (CH_2=C=CH_2), interact strongly with each other and give rise to two frequencies each of which involves vibration of both bonds. Adding another single bond to the propane system creates a completely new set of frequencies, since all three bond vibrations now interact with each other and in different ways depending on whether the structure is that of normal or iso-butane. Thus for such sets of similar bonds each new molecule is a law unto itself as regards the C—C vibrations, and there is no characteristic C—C frequency. Other examples come from bending vibrations. There is no one frequency of the methyl group XCH_3 associated with the change of an individual HCH or HCX bond angle. Instead, these vibration types interact with each other and give a set of frequencies each of which involves motion of several angles. These are influenced to some extent by the structure nearby to the CH_3 group, but are uninfluenced by remoter parts of the molecule.

The influence of neighboring groups, as described above, is highly important; if the characteristic frequencies of various groups were essentially uninfluenced by neighboring groups, then many of the structural features it is desired to know would not be deducible from the spectrum. This influence may be exerted in various ways: there may be a small amount of vibrational interaction between the group of interest and a particular neighboring group; steric effects of neighboring groups may play a role by interfering with the vibrations of the group of interest;

or neighboring groups may affect the force constants of the group of interest, by conjugation or inductive effects on the bonds themselves.

3. Vibrational Frequency Assignments.

The process of assigning a particular type of vibration of a group to an observed frequency is one which is fundamentally completely empirical; it is not possible to calculate *a priori* the bond forces. In practice, various lines of attack have built up our present knowledge. Emission and absorption spectra of diatomic molecules (which have as their only vibration the stretching of the bond) have given information on the approximate stretching frequencies of various bond types [HERZBERG (*B 4*)]. Empirical examination of series of related molecules for regularities in their spectra has pointed out the more obvious correlations between structure and spectra [HERZBERG (*B 5*), p. 194]. Finally, complete vibrational analyses have been made for a number of molecules which are so simple that all (or almost all) of their fundamentals are known, and for which the fullest use of symmetry selection rules, infrared rotational structure or band shape, RAMAN polarization data, etc. can be made [HERZBERG (*B 5*); DENNISON (*21*)].

Using such information as a guide, it is possible to make at least partial interpretations of the spectra of many simple molecules, of which HERZBERG (*B 5*) gives numerous examples drawn from the work of various authors. Comparing the spectrum of such a molecule with a series of related higher compounds (e. g., higher homologs) one can at least partially extend the assignments to these. At the present time, however, such an extension is only feasible for a very few classes which have been rather completely investigated (paraffins, olefins, alkyl-benzenes). For other classes of molecules, we have to work at present with only the very obvious correlations between observed frequencies and structural features.

Thus, in using the vibrational spectrum as a tool in chemical-structural work, it is clear that a knowledge of the spectra anticipated for the type of compound of interest must first be available, and must be obtained by as thorough a study as possible of compounds of known structure. With this, it becomes possible to examine a pure compound of unknown structure, or a reaction mixture of unknown composition, and make statements as to which structures are possible and which are not. Evidently, this is not an absolute method, as is a complete chemical proof of structure or an X-ray crystal structure determination, but is fundamentally one of analogy. It cannot in general be used to say what *is* true, but only to say that certain possibilities are not true, anything else being possibly true. However, when combined with chemical knowledge available for the particular problem, this type of information may in favorable cases be determinative, and in most instances will be extremely useful in eliminating false directions of research.

4. Comparison of Infrared and Raman Spectra.

The preceding discussion on vibrational spectra is in principle as applicable to RAMAN as to infrared absorption spectra. A few words by way of comparison should be said. The RAMAN spectrum has already been fruitfully used along the lines indicated as a tool in organic chemistry by many investigators. Particular mention might be made of KOHL-RAUSCH and collaborators, and the group of workers in France and the Low Countries.

The ideal situation would, of course, be to have both tools at hand; however, it usually happens that a choice, in emphasis at least, must be made. Disregarding for the moment the experimental aspect, general advantage seems to reside on neither side. It is the opinion of the writer that for the study of hydrocarbons the infrared is to be preferred, since more useful regularities appear in the hydrogen vibrational frequencies than in the skeletal frequencies (which are the main feature of the RAMAN). However, the RAMAN spectrum is not limited in its frequency range, whereas the infrared cannot conveniently go below 400 cm.$^{-1}$; and this or other considerations may make the RAMAN spectrum more suitable for other particular applications.

Experimentally, the main disadvantage of the infrared method is the relative cost of the equipment as against photographic RAMAN equipment. To counterbalance this, there are the advantages that the infrared methods are in general more rapid, although improved optical design of photographic RAMAN instruments is bringing exposure times down to less than one hour, and photoelectric RAMAN instruments can be made equal in speed to infrared instruments; the infrared necessitates a very much smaller sample (milligrams as against grams for the RAMAN), and the infrared method does not suffer from the problems of photo-chemical decomposition or unsuitability of colored samples, as does the RAMAN method.

Although not pertinent to the present discussion, it should be said that the impetus for the recent extensive development of infrared spectroscopy came from its applicability to exact quantitative analysis, in which the RAMAN method is just beginning to catch up by the recent development of photoelectric (as opposed to photographic) instruments (24, 74).

III. General Features of Infrared Spectra of Organic Compounds.*

1. Paraffin Hydrocarbons.

Infrared spectra of paraffins have been given by a large number of workers (*3, 6, 26, 37, 43, 44, 46, 66, 91, 97, 113, 119*), but by far the most complete current collection is to be found in the American Petroleum Institute's Catalog of Infrared Spectrograms (*B 1*).

Spectra of some typical paraffins are shown in Figure 4. From the study of RAMAN and infrared spectra of a large number of organic

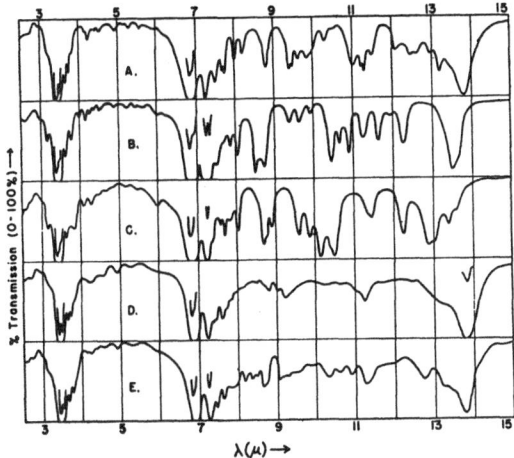

Fig. 4. Spectra of Some Paraffin Hydrocarbons. [Adapted from spectra contributed by the U. S. Naval Research Laboratory, Washington, D. C., to the American Petroleum Institute Catalog (*B 1*).] A. *n*-Hexane. B. 2-Methylpentane. C. 3-Methylpentane. D. *n*-Tetradecane. E. 7-Methyltridecane. — All spectra are of pure liquids in an ·0,17 mm. cell, except small sections in regions of heavy absorption are shown also for an ~ 0,01 mm. cell.

compounds (non-hydrocarbons as well as hydrocarbons) it is known that the well-marked set of frequencies near 3000 cm.$^{-1}$ arises from the CH stretching vibrations [see HERZBERG (*B 5*), HIBBEN (*B 6*), and KOHLRAUSCH (*B 7*)]. In the rock-salt infrared spectra of paraffins, these appear as a strong band (made up of several unresolved close bands) extending from 3,3 to 3,5 μ. With better resolution (lithium fluoride prism, grating) it is possible to observe separately the maxima of the bands due to CH_3, CH_2, and (tertiary) CH groups [FOX and MARTIN (*26*); LIDDEL and KASPER (*58*); WRIGHT (*116*)], although even with high resolution these sets of bands overlap each other considerably. If sufficiently

---

* The references given are not intended to be exhaustive of the literature. They were chosen as being particularly representative, or significant, in recent lines of work.

resolved, the bands are of obvious use in certain qualitative, and possibly also in quantitative, analytical problems. Upon substitution of deuterium for hydrogen, the increase in mass causes the CD stretching bands to appear in the 2000 cm.$^{-1}$ region. This separation of deuterium from hydrogen stretching frequencies has often been used in vibrational assignment studies on simple molecules.

A second, very marked and constant set of bands of paraffins is found at $\sim 6{,}8\,\mu$. and $\sim 7{,}2\,\mu$. (1460 and 1380 cm.$^{-1}$, respectively), arising from HCH bending vibrations. For methyl groups, the adjacency of the three H—C—H angles causes interaction between the vibrations, which results in the appearance of the two frequencies (Figure 4). For an isolated CH_2 group, the HCH vibration gives rise to a band at $\sim 6{,}8\,\mu$. (1460 cm.$^{-1}$) at the same position as one of the methyl-group bands. It is seen that the $7{,}2\,\mu$. band is characteristic of the CH_3 group, and for many problems (involving also non-hydrocarbons) may be put to good use. A general example is the addition of a proton-donor reagent HX to a terminal C=C. The direction of addition ("Markownikow" or "anti-Markownikow") is determinable from the spectrum if there is no interference from other CH_3 groups:

$$RCH{=}CH_2 + HX \longrightarrow RCHX{-}CH_3 \text{ (methyl group)}$$
$$\searrow RCH_2{-}CH_2X \text{ (no methyl group)}$$

In the spectra of paraffins, the region from $3{,}5\,\mu$. to $6{,}8\,\mu$. contains only very weak overtone or combination bands, and hence paraffins or alkyl radicals seldom offer any spectral interference to compounds or groups that do have fundamental bands in this region.

With regard to the remaining fundamental vibrations (all at $> 7{,}2\,\mu$.), various attempts have been made to assign the C—C stretching and C—C—C bending vibrations in the simpler paraffins (2, 41, 68, 69, 70, 73, 75, 120). However, except for a few strong RAMAN lines characteristic of certain simple branched structures, there is no regularity in these frequencies in more complicated molecules. As explained on p. 342, this arises from the large amount of interaction between vibrations of adjacent C—C bonds. All that will be done h re is to state the limits of the range of the stretching frequencies: 800 to 1100 cm.$^{-1}$ (9 to 12,5 μ.) for normal chains, with one or more somewhat lower frequencies present for branched structures. This range has been indicated by calculations on simplified models [KELLNER (36); PITZER (68)], as well as by the appearance therein of a set of strong and irregularly spaced RAMAN lines characteristic of C—C stretching frequencies [KOHLRAUSCH and KÖPPL (41); ROSENBAUM, GROSSE and JACOBSON (84)]. The C—C—C bending frequencies occur always below about 500 cm.$^{-1}$, and hence are only partially observable in the infrared (potassium bromide prism) but may be studied in the RAMAN spectrum. Like C—C stretching frequencies, these exhibit little regularity between molecules.

The HCC bending vibrations interact with one another strongly, giving resultant vibrations which involve essentially bending motions of CH_3, CH_2, or CH groups as a whole, and which are found from 7,4 to $\sim 14\,\mu$. (1350 cm.$^{-1}$ to 720 cm.$^{-1}$). These frequencies depend not only on the type of vibration of the group, but also on the branching and groups adjacent to the one of interest. An approximate assignment of these vibrational types has been given by RASMUSSEN (75). There are two regions of absorption that are often of use in chemical-structural work: 8 to 9 μ., which contains bands characteristic for the arrangement of methyl groups (46, 63, 75, 94); and 12,8 to 13,8 μ., which contains bands characteristic for the length of CH_2 chains (46, 75, 94). With a few exceptions these HCC bending vibrations give rise to moderate-to strong infrared bands but only to weak RAMAN lines.

A final type of vibration, of no interest here, involves torsional oscillations about C—C bonds. The frequencies of these range from about 200 cm.$^{-1}$ (torsion of a methyl group about its adjacent C—C bond) down to very low values.

The above description of frequencies for paraffins holds also in general for the paraffinic groupings in naphthenes, olefins, aromatics, or non-hydrocarbons. Although this subject has not been examined adequately, it appears that the characteristic frequencies of the CH_3, CH_2, or CH groups, immediately adjacent to the non-paraffinic portion, are shifted somewhat from their usual positions in paraffins, but that the portion of the alkyl radical further removed from the non-paraffinic substituent is practically unaffected in its vibration frequencies. However, it is often the case with such molecules that many infrared bands of the paraffinic portion are hidden by the bands of the non-paraffinic group, due to the generally much lower intensity of the paraffin bands.

2. Cycloparaffins.

Infrared spectra of naphthenes have been reported in a few literature sources, particularly by LECOMTE and coworkers (44, 54, 56, 57), but especially by the contributors to the American Petroleum Institute Catalog (B 1).

Vibrational assignments for cyclopropane, cyclobutane, cyclopentane, and cyclohexane have been deduced by various workers (B 5, 11, 39, 72a, 115). These give an indication of the frequencies to be expected from the ring vibrations, and presumably these frequencies maintain themselves in substituted cycloparaffins, with more or less change due to interaction with vibrations of the substituents [some methylcyclo-hexanes were considered by PITZER et al. (11)]. From the practical point of view, naphthenes are a difficult subject, however. Their infrared intensities are low, like those of paraffins, and the ring systems exhibit no characteristic bands except, possibly, a very few in the region where the paraffin C—C stretching frequencies occur. Hence, the spectrum of an alkyl-substituted cycloparaffin has exactly the overall appearance of a paraffin spectrum, and no differentiation based on general spectral characteristics can be made.

3. Olefins; Bands Associated with the Carbon-Carbon Double Bond.

As reference spectra, attention is again called to the American Petroleum Institute Catalog (*B 1*), as well as to some individual papers (*4, 43, 44, 77, 80, 101, 108*). Spectra of some typical olefins are shown in Figure 5.

The olefinic portion proper (i. e., the C=C group and its attached hydrogens) gives rise to three particularly marked absorption characteristics. The first of these is a band at $3.25 \mu.$, arising from a CH stretching vibration of the $=CH_2$ group, and appearing generally as a shoulder on the $3.4 \mu.$ aliphatic CH band. Due to its nearness to the stronger band, and consequent difficulty to observe it in some cases, this band is generally only of corroborative value in the detection of the $=CH_2$ group.

Table 1. Positions of C=C Stretching Bands in Olefins.

(Values are from Shell Development Company spectra.)

Type[1]	Average Wave Length (μ.) and Intensity[2]	Wave Length Range[3]	Wave Length from Raman[4]
$CH_2=CH_2$	[5]	—	6,15
$CHR=CH_2$	6,09 (8)	6,07—6,12	6,09
$CR_2=CH_2$	6,05 (8)	6,02—6,08	6,06
cis-$CHR=CHR$	6,02 (5)	6,00—6,03	6,02
trans-$CHR=CHR$	6,00 (1)[6]	5,99—6,02	5,97
$CR_2=CHR$	6,00 (4)	5,98—6,01	5,99
$CR_2=CR_2$	[5]	—	5,98

[1] R = alkyl radical. The values appear to hold also when R is a saturated carbon carrying other substituents (Cl, OH, etc.).

[2] Intensity in units of 10% absorption, for olefins in the C_5–C_{10} range in a cell of ~ 0.15 mm. Thus "(8)" signifies 80% absorption (20% transmission) at the band maximum.

[3] The scatter over the range is in part experimental error, and in part due to small variations in R (whether CH_3, larger primary, secondary, or tertiary radical).

[4] From various literature sources.

[5] Inactive or unobservably weak in the infrared because of symmetry.

[6] Unobservably weak in some cases; value taken from weak bands in other cases. The resulting larger wave length error explains the difference from the Raman value.

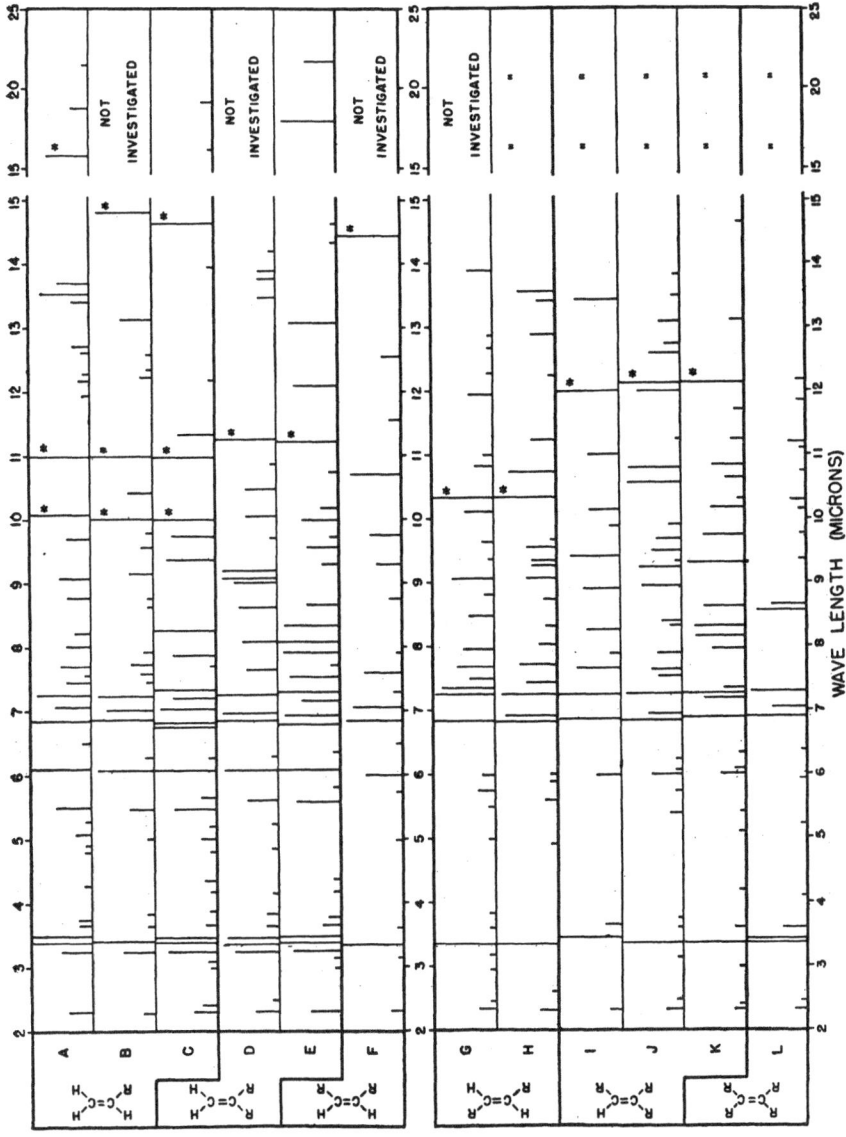

Fig. 5. Spectra of Some Olefins. Positions of lines indicate band centers, and heights represent intensities. All spectra (except D) taken from American Petroleum Institute Catalog (*B 1*), from contributors given below. A. 1-Hexene, 0,055 mm. pure liquid. (U. S. Naval Research Lab.) B. 3-Methyl-1-pentene, 0,064 mm. pure liquid. (Socony-Vacuum Co.) C. 3,3-Dimethyl-1-butene, 0,169 mm. pure liquid. (U. S. Naval Research Lab.) D. 2,3-Dimethyl-1-butene, 0,207 mm. pure liquid. (Shell Development Co.) E. 2,4,4-Trimethyl- 1-pentene, 0,174 mm. pure liquid. (U. S. Naval Research Lab.) F. *cis*-2-Pentene, 100 mm. vapor, 15 cm. cell. (Shell Development Co.) G. *trans*-4-Methyl-2-pentene, 0,13 mm. pure liquid. (Shell Development Co.) H. *trans*-3-Octene, 0,127 mm. pure liquid. (Shell Development Co.) I. 2-Methyl-2-pentene, 0,064 mm. pure liquid. (Socony-Vacuum Co.) J. 3-Ethyl-2-pentene, 0,15 mm. pure liquid. (Shell Development Co.) K. 2,4,4-Trimethyl-2-pentene, 0,13 mm. pure liquid. (Shell Development Co.) L. 2,3-Dimethyl-2-butene, 0,13 mm. pure liquid. (Shell Development Co.) — The characteristic CH wagging bands (marked with an asterisk) are relatively much stronger, as compared with other bands, than this figure indicates; in thinner cells than those employed above, the difference would be more apparent but weaker bands would not show up well.

The second region of absorption is that due to the C=C stretching vibration, which is found in the vicinity of 6 μ. ($\sim$ 1660 cm.$^{-1}$). For open-chain monoolefins this band ranges from 5,95 μ. to 6,09 μ. The exact position and intensity depend on the number and position of alkyl substituents on the double bond (Table 1). Because of the weakness of the band when both ends of the double bond are substituted, it is unsafe to use the absence of this band as a criterion for the absence of unsaturation (77). This difficulty is not present in the RAMAN spectrum, the line remaining intense under all conditions. However, the presence of a moderately strong band at 6,05 μ. to 6,09 μ. may be considered as an indication of the C=CH$_2$ grouping.

The third and most important characteristic of olefin spectra is the set of very strong bands in the 10 to 17 μ. region which arise from olefinic CH wagging vibrations, perpendicular to the $\diagdown \mathrm{C}{=}\mathrm{C} \diagup$ plane. These bands are stronger by a factor of ten (or more) than paraffin bands in the same region, and their positions, as shown in Table 2, are highly characteristic of the number and position of alkyl substituents (77, 101, 109, 114). Most of these bands are remarkably constant ($\pm$ 0,03 μ.) with regard to change in the R groups; the wave lengths of some appear, however, to depend on whether the alkyl radicals attached to C=C are primary, secondary, or tertiary in their nature. This point needs further study. The importance of these bands for determining the position of

Table 2. Positions of Strong CH Out-of-Plane Wagging Bands of Olefins.

(Values from Shell Development Company spectra and American Petroleum Institute Catalog.)

Type[1]	Wave Length (μ)
CHR=CH$_2$ ...	$\begin{cases} 10,05 \pm 0,05 \\ 10,95 \pm 0,05 \\ \text{one band,} \quad 14,6\text{--}15,8 \text{ [2]} \end{cases}$
CR$_2$=CH$_2$ ...	11,25 $\pm$ 0,05
trans-CHR=CHR ..	10,34 $\pm$ 0,05
cis-CHR=CHR ..	one band, 14,2--14,9 [3]
CR$_2$=CHR ..	one band, 11,5--12,5 [3]
CR$_2$=CR$_2$ ...	no band

[1] R = alkyl group.
[2] More exact positions for this band are:
    R = CH$_3$, 17,30 μ.
    R = primary alkyl radical, 15,75 μ. $\pm$ 0,05.
    R = secondary alkyl radical, 15,0 μ.
    R = tertiary alkyl radical, 14,6 μ.
[3] Dependence of exact position on structure of R groups not known.

the double bond in a given carbon skeleton, in such reactions as dehydration of alcohols, dehydrochlorination of chlorides, exhaustive methylation degradations, etc., is obvious.

Complete frequency assignments for the methyl-substituted ethylenes have been given by KILPATRICK and PITZER (38). From these it is possible to allocate the other vibrations of the C=C group (i. e., the in-plane olefinic hydrogen wagging vibrations and the out-of-plane ones which do not give rise to strong bands), and to demonstrate that the spectrum of an olefin is essentially a superposition of characteristic bands due to the olefinic portion plus those expected from the alkyl radicals (38, 85). However, these other olefinic bands are not distinct enough from the alkyl-radical bands to be of any use in structural work on unknown compounds.

Cycloolefins have not been studied with any degree of thoroughness; it would appear, however, that the existence of the double-bond in a

Table 3. Band Positions of Some Cycloolefins.

(Values refer to liquid state; wave lengths in μ.; values from Shell Development Company spectra and American Petroleum Institute Catalog.)

	C=C Stretching	CH Out-of-Plane Wagging
Open-chain *cis*-CHR=CHR	6,02	14,2–14,9
Cyclopentene	6,17	14,36
Cyclohexene....................	6,07	15,50
(ring)—C	6,06	15–16 [1]
(ring)—C=C	2	15–16 [1, 3]
Open-chain CR₂=CHR..........	6,00	11,5–12,5
C=C—C (ring)	2	12,52 [3]
(ring with C C / C)	6,04	12,68

[1] Spectrum obtained to 15 μ. only. However, it is apparent from the rising absorption immediately before 15 μ. that a strong band is present just beyond 15 μ.

[2] Band position not determinable because of interference from the other C=C group.

[3] These positions refer only to the ring double bond. The strong bands of the other (non-ring) double bonds are also present, at their proper positions.

ring has a serious effect on both the C=C stretching and CH out-of-plane wagging vibrations, as illustrated in Table 3. The obvious resemblance to a *cis*-CHR=CHR open-chain olefin leads one to expect a strong CH band in the $\sim 14\,\mu$. region (Table 2); or if there is substitution on one of the unsaturated carbons, then the system resembles a CHR=CR$_2$ open-chain type and should give a $\sim 12\,\mu$. band. The observed bands

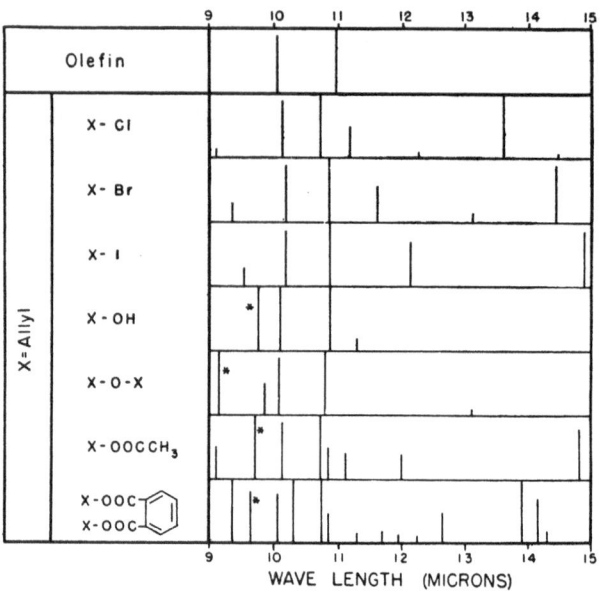

Fig. 6. Spectra of Some Allyl Compounds. Allyl bromide and iodide are from THOMPSON and TORKINGTON (*103*), using a cell less than 0,1 mm. The remainder are from Shell Development Company, using a cell $\sim$ 0,02 mm. The positions of the strong bands in CH$_2$=CHR-type olefins are shown on the top line. The additional strong bands of the allyloxy compounds (marked with asterisk) are from C—O vibrations. Some strong ring-vibration bands of diallyl phthalate are also present.

are seen to be in these general regions, but shifts from the open-chain positions exist which must be attributed to the effect of ring strain on the force constants.

The case of conjugated diolefins is discussed separately below. With di- or polyolefins in which the double bonds are isolated by at least one saturated carbon atom from each other, each bond exhibits its characteristic C=C stretching and CH out-of-plane wagging bands independently, the spectrum being a simple superposition as regards these vibration types.

In applying the generalizations described above to C=C groupings in non-hydrocarbon molecules, some caution must be observed. If a non-hydrocarbon substituent resides on a β (or further) carbon atom with respect to the C=C group, then it seems to have a negligibly small effect on the characteristic bands described above (insofar as the writer can judge from the few reliable cases at hand). However, such a substituent

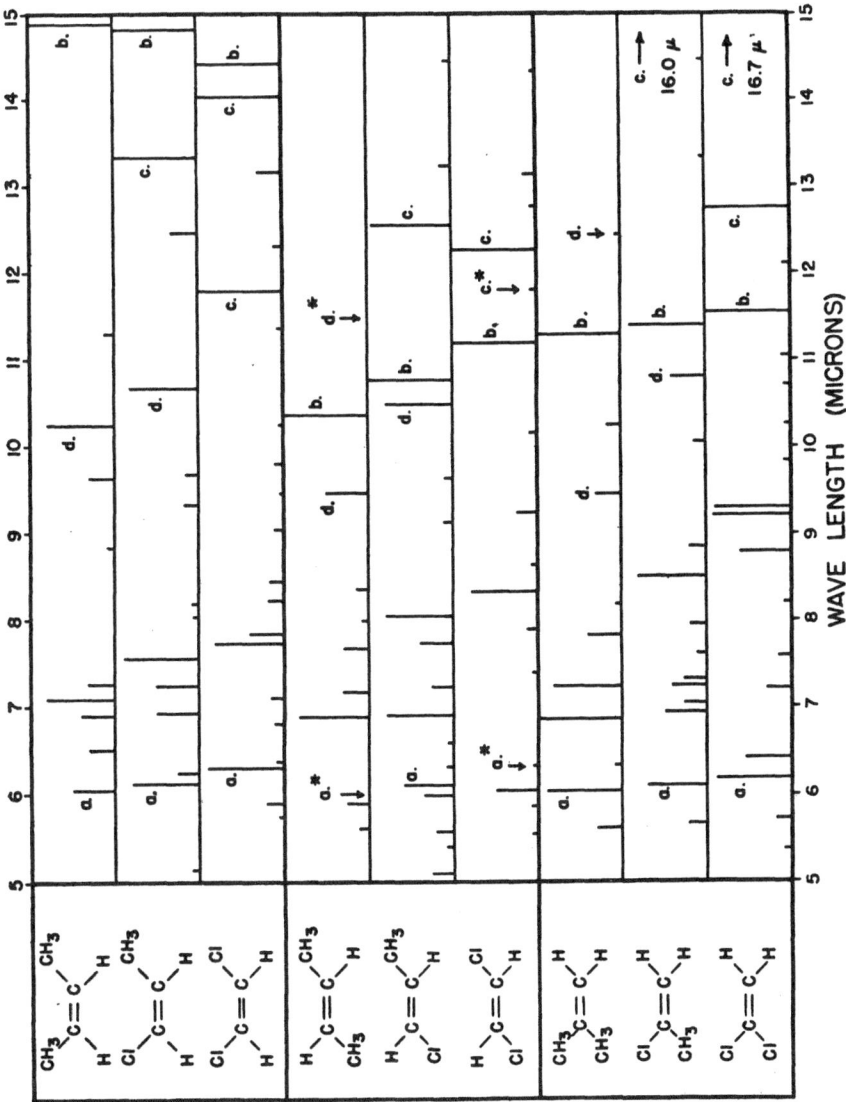

Fig. 7. Effect of Chlorine Substitution on C=C Bands. (Spectra from Shell Development Company.) Butene spectra: vapor, 100 mm. pressure, 15 cm. cell; other spectra: pure liquids, ~ 0,02 mm. cell. a. C=C bands. b. CH out-of-plane wagging bands. c. C—Cl stretching bands. d. C—C stretching bands. — (*) Signifies band forbidden to appear in the infrared because of molecular symmetry; the arrow indicates the position as deduced from RAMAN spectra.

on an α-carbon atom has, rather surprisingly, a marked effect on the olefinic CH out-of-plane frequencies. This has not been studied sufficiently to enable one to generalize, but is illustrated by the spectra of some allyl compounds studied by THOMPSON and TORKINGTON (*103*), and by some miscellaneous spectra from the Shell Development Laboratories shown in Figure 6. Such α-substitution appears, however, to have little or no effect on the C=C band (1-alkenes: 6,09 μ.; allyl halides, alcohol, etc.: 6,09 μ.). As might be anticipated, substitution of a non-hydrocarbon group directly on one of the double-bond carbons exert a considerable effect, both on the C=C stretching and CH out-of-plane wagging bands. For the case of chlorine substitution, this is shown in Figure 7. While insufficient work has been done to generalize, the effect of substitution of negative groups (Cl, OR, —O—CO—R) on the C=C stretching band seems to be to shift it to longer wave length and to increase its intensity. Thus, the statement made previously (p. 350) that C=C groups substituted on both ends gave only very weak 6 μ. C=C stretching bands, may no longer hold if there are non-hydrocarbon substituents on the group.

4. Conjugated Carbon-Carbon Double Bonds.

To discuss first the simplest examples, conjugated *di*olefins —CH=CH—CH=CH—, the important factor is the well-known resonance between covalent and ionic structures, which results in a loss of double-bond character from the formal double bonds and a gain in double-bond character by the central (formal single) bond. The effect on the stretching force constants of the bonds is to decrease that of the double bond slightly and to increase that of the central (single) bond considerably, resulting in a long-wave shift (to lower frequency) for the double-bond vibrations and vice versa for the single bond. Also the resonance leads to a greater-

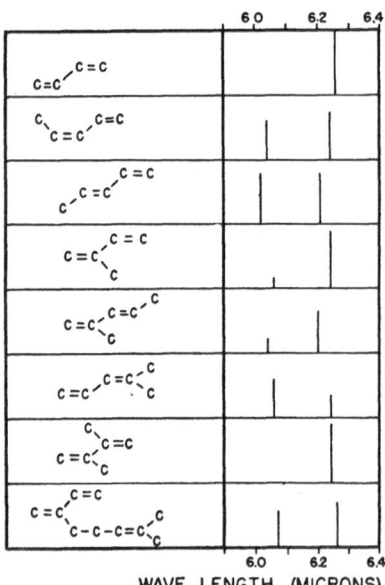

Fig. 8. C=C Stretching Band Positions in Some Conjugated Dienes. (Spectra from Shell Development Company.) C_4 and C_5 compounds: vapor, 100 mm. pressure, 15 cm. cell; heavier compounds: pure liquids, $\sim$ 0,02 mm. cell.

than-normal degree of interaction between the double-bond vibrations, resulting in a splitting such that two C=C stretching frequencies are encountered, viz. near 1650 cm.$^{-1}$ (6,06 μ.) and 1600 cm.$^{-1}$ (6,25 μ.). As a measure of the shift due to conjugation only (and not the subsequent splitting due to vibrational interaction), one can compare the average of the two frequencies (1625 cm.$^{-1}$) with the monoolefin value ($\sim$ 1660 cm.$^{-1}$), a shift of $\sim$ 35 cm.$^{-1}$. In 1,3-butadiene itself, the higher-frequency vibration is forbidden to appear in the infrared by virtue of the molecular symmetry, whereas the 1600 cm.$^{-1}$ one appears strongly near 6,25 μ. The substitution of alkyl radicals on the butadiene structure may or may not result in the appearance of the $\sim$ 6,0 μ. band also [RASMUSSEN and BRATTAIN (78); see Figure 8]; the factors governing this are not yet understood. The 6,25 μ. band seems always to appear, however. Hence, in a non-aromatic hydrocarbon system, the presence of a band near 6,25 μ. is an indication of conjugated olefins.

The exact positions of the two C=C frequencies of conjugated diolefins depend, as with monoolefins, on the number and positions of alkyl

Table 4. Double-Bond Band Positions of Some Compounds Containing C=O Conjugated to C=C or Benzene Ring.
(Wave lengths in μ.; intensities in units of 10% absorption, estimated for 0,02 mm. pure liquid; S = shoulder on the stronger band.)

	C=O Stretching	C=C Stretching	Reference
Aldehydes:			
Acrolein	5,88 (10)	6,17 (4)	(85)
Methacrolein	5,90 (10)	$\sim$ 6,10 (S)	(B 2, 85)
Crotonaldehyde	5,92 (10)	6,08 (5)	(85)
Benzaldehyde	5,86 (9)	—	(B 2)
m-Tolualdehyde	5,87 (9)	—	(B 2)
p-Isopropyl-benzaldehyde	5,86 (7)	—	(B 2)
Ketones:			
Methyl-vinyl-ketone	5,93 (10)	6,17 (6)	(85)
Mesityl oxide (4-methyl-3-pen-tene-2-one)	5,92 (9)	6,15 (10)	(B 2, 31 a, 99)
3-Methyl-3-butene-2-one	5,91 (7)	6,13 (4)	(B 2)
Isophorone (3,5,5-trimethyl-2-cyclohexene-1-one)	5,98 (10)	6,10 (4)	(81)
Acetophenone	5,93 (9)	—	(B 2, 81, 99)
Propiophenone	5,93 (9)	—	(99)
Esters:			
Methyl acrylate	5,80 (10)	6,15 (6)	(B 2)
Ethyl acrylate	5,77 (6)	6,11 (4), 6,17 (3)	(B 2)
Methyl methacrylate	5,81 (10)	6,12 (6)	(B 2, 79)
Ethyl crotonate	5,83 (10)	6,06 (7)	(B 2, 99)
Methyl benzoate	5,80 (9)	—	(79)
Ethyl benzoate	5,83 (8)	—	(99)

substituents. Not enough cases have been studied to state this dependence. Recently, spectra of some higher conjugated polyenes have been given by BLOUT, FIELDS and KARPLUS (11a).

With regard to the 3,2 μ. band, conjugated diolefins are sometimes anomalous, the band not appearing even when the C=CH₂ group is present. The CH out-of-plane wagging bands are somewhat but not greatly affected by the conjugation, each double bond giving the bands anticipated if the remainder of the molecule were saturated [RASMUSSEN and BRATTAIN (78)].

Conjugation of C=C to an aromatic ring has been studied only for a very few styrenes (B 1). The effect on the characteristic frequencies of the C=C system seems to be about the same as conjugation to another C=C group, although details are lacking.

Conjugation of C=C to C=O results in a shift of apparently the same order as noted for conjugated diolefins (~ 35 cm.$^{-1}$). However, there is no further shift since the C=C and C=O vibrations do not interact appreciably. Hence, the C=C band of such conjugated systems is found at 6,10 to 6,20 μ. Positions of these bands in some conjugated ketones, aldehydes, and esters (the only classes studied with any degree of thoroughness) are shown in Table 4. Conjugation of C=C with C=O also enhances the intensity of the C=C band above the intensity observed in hydrocarbons, although usually the C=C band remains considerably weaker than the C=O band.

5. Alkyl-benzenes; Bands of Aromatic Ring Systems.

Spectra of alkyl-benzene hydrocarbons are to be found in the American Petroleum Institute Catalog (B 1) and in published spectra, principally by LECOMTE and co-workers (22, 44, 49). A few spectra are shown in Figure 9. Benzene itself has been the subject of a large number of investigations, and, thanks particularly to the exhaustive study made by INGOLD and collaborators (35), the complete vibrational assignment is known with high certainty. Using this as a basis, it is possible to make reasonable assignments for higher alkyl-benzenes, a procedure carried out for toluene and the xylenes by PITZER and SCOTT (71), and for ethylbenzene by TAYLOR and PITZER (96). However, for present purposes, it will suffice to mention the most useable characteristic bands. These are exactly analogous to those of olefins. Thus, a 3,25 μ. band appears from the aromatic CH stretching vibration, which varies in intensity with the number and position of alkyl substituents.

Other characteristic bands arise from the CC stretching frequencies of the ring. By virtue of the well-known resonance between KEKULÉ structures, each bond is intermediate between a single and double bond; in addition, there is a high degree of interaction between the vibrations

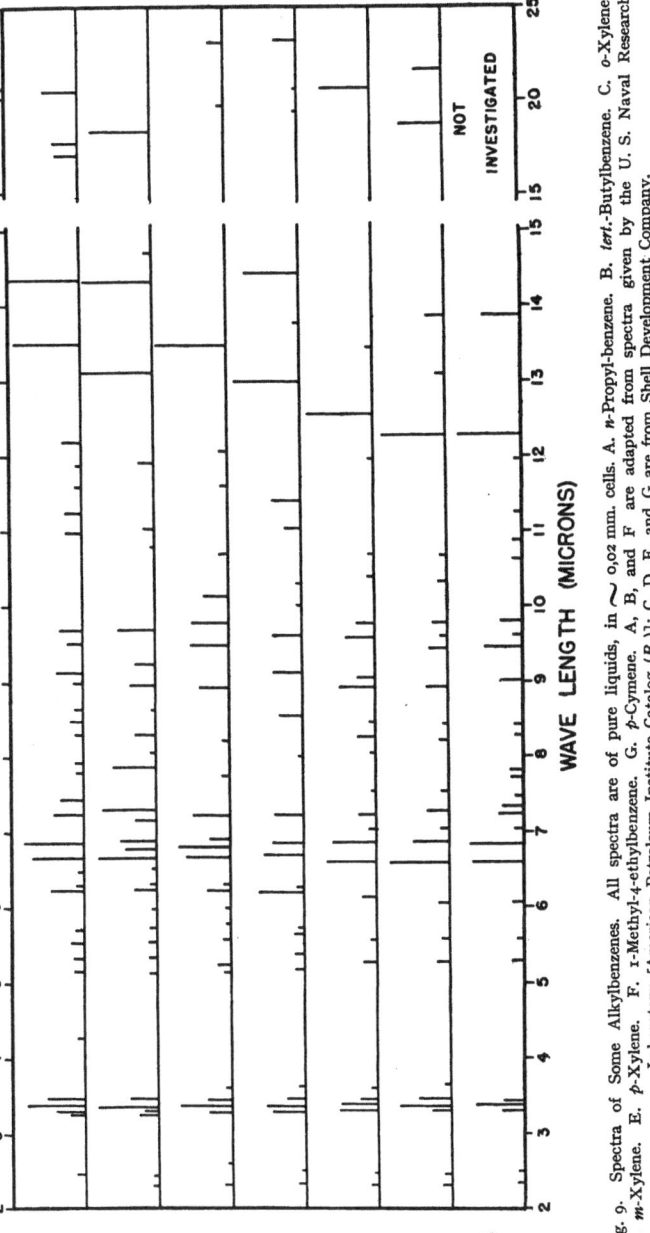

Fig. 9. Spectra of Some Alkylbenzenes. All spectra are of pure liquids, in ~ 0.02 mm. cells. A. *n*-Propyl-benzene. B. *tert.*-Butylbenzene. C. *o*-Xylene. D. *m*-Xylene. E. *p*-Xylene. F. 1-Methyl-4-ethylbenzene. G. *p*-Cymene. A, B, and F are adapted from spectra given by the U. S. Naval Research Laboratory [American Petroleum Institute Catalog (B_1)]; C, D, E, and G are from Shell Development Company.

of the six equivalent bonds. As a result, the ring-bond stretching frequencies are spread over a range extending from 6 μ. to 10 μ. Of these, two give rise to marked infrared bands below $6,8\,\mu$., in the region where there is no interference from paraffin-type bands. These bands occur roughly at $6,25\,\mu$. (intensity, very weak to moderate) and $6,7\,\mu$. (intensity, moderate to strong) [Barnes et al. (*B 2*); Rasmussen, Tunnicliff and Brattain (*81*)]. The exact wave lengths and intensities depend on the number and positions of alkyl substituents but not on the kind of substituent. It is seen that the presence of a $6,25\,\mu$. band may signify either conjugated diene or aromatic structure, and other regions of the spectrum must be examined to differentiate structure between these possibilities.

There also appear strong longer-wave bands analogous to the olefin CH out-of-plane bands. As indicated by the benzene assignment and extension of it to alkyl-benzenes, these CH out-of-plane wagging bands (mixed to some extent with ring carbon out-of-plane bending bands) extend from about $10\,\mu$. to $15\,\mu$. However, the strongest occur from about $12\,\mu$. to $15\,\mu$., and their positions are highly characteristic of the number and positions of alkyl substituents on the ring [Shell Development Co. (*85*); Whiffen and Thompson (*111*)]. With certain of these, there are small but definite shifts dependent on the presence or lack of branching of the alkyl radicals at their point of attachment to the ring. The positions of these characteristic bands are given in Table 5, although, because of insufficient experimental work, the last-mentioned effect of branching is not known for many classes.

As with olefins, *substitution* of non-hydrocarbon substituents at positions β or further from the ring does not seem to affect the ring frequencies. Introduction of such substituents in an α position (e. g., benzyl compounds) has some effect on at least the CH out-of-plane band positions, but insufficient work has been done to discuss this further. Non-hydrocarbon substituents directly on the ring have a more profound effect, although the various bands remain in the same general regions as for the alkyl derivatives. Whiffen and Thompson (*111*) have shown that the strong CH out-of-plane bands retain roughly the positions and intensities in a given type of substituted benzene ring that would be shown if all substituents were alkyl groups; and together with Richards (*83a*) they have demonstrated this particularly for phenols. Lecomte (*49*), and Depaigne-Delay and Lecomte (*22*) have charted the changes that occur in characteristic bands for an extensive variety of substituents. Rasmussen and Brattain (*79*) have noted that substitution of O or N on a benzene ring carbon (phenols, anilines, and derivatives) greatly enhances the intensity of the $6,25\,\mu$. band over its value in simple, C-substituted benzene rings.

The effect of *conjugation* on benzene ring frequencies has been little studied, particularly as regards the CH out-of-plane bands. One fact has been pointed out by Rasmussen, Tunnicliff and Brattain (*81*),

Table 5. Positions of Strong Long-Wave Bands of Alkyl-benzenes, 11–25 μ. (CH Out-of-Plane Wagging and Carbon-Ring Out-of-Plane Bending Vibrations).

[Wave length values (μ.) for liquid state.] (Values from Shell Development Company spectra and American Petroleum Institute Catalog.)

Type		Wave Length[1] (μ.)					
(benzene)				14,80			
(mono-R)	$R = CH_3$....	11,17		13,70	14,39		21,50
	R = primary alkyl ...	11,02		13,50	14,33	16,5–18,0[2] 19,7–20,5	
	R = secondary alkyl	11,05		13,16	14,33		17,0–19,0[2]
	R = tertiary alkyl[3] ..	11,05		13,16	14,33		18,30
(1,2-di-R)	$R = CH_3$....		13,48			23,0	
	R = higher alkyl[4] ..	13,21		13,74		$\sim$23	
(1,3-di-R)	$R = CH_3$....	11,42		13,00	14,46		23,2
	R = higher alkyl[4] ..	11,3		12,80[5]	14,29	$\sim$23	
(1,4-di-R)	$R = CH_3$....		12,58			20,6	
	R = higher alkyl[4] ..		12,29[6]		18,7		21,6
(1,2,3-tri-R)	$R = CH_3$....	13,06		14,12		18,6	
	R = higher alkyl[4] ..	12,99		13,5–13,85		[7]	
(1,2,4-tri-R)	$R = CH_3$....	11,45 12,41	14,23 (w)	18,6			22,8
	R = higher alkyl[4] ;..	11,42 12,27	14,25 (w)	[7]			[7]
(1,3,5-tri-R)	$R = CH_3$....		11,97			14,56	
	R = higher alkyl[4] ..		11,83			14,88	

[1] Unless a range is given; the positions are constant to $\pm$ 0,05 μ., except in a few cases which approach $\pm$ 0,10 μ.

[2] One or more bands of strong-to-moderate intensity.

[3] Based on only one example, *tert.*-butylbenzene.

[4] Studied only for some cases where one R is ethyl or isopropyl.

[5] Appears at 12,50 μ. in m-diethylbenzene, indicating the effect of two larger-than-methyl radicals.

[6] Appears at 12,05 μ. in p-di-*tert.*-butylbenzene.

[7] Not studied beyond 15 μ.

Table 5 (continued).

Type		Wave Length (μ.)		
(benzene ring, 1,2,3,4-R)		not studied		
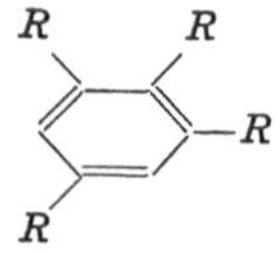 (benzene ring, 1,2,3,4-R) $R = CH_3$[8] ...	11,78	14,18		18,25
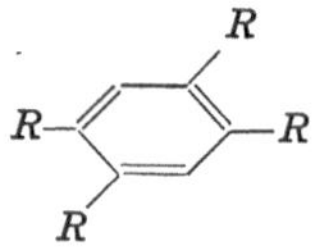 (benzene ring, 1,2,3,5-R) $R = CH_3$[8] ...	11,52		22,37	
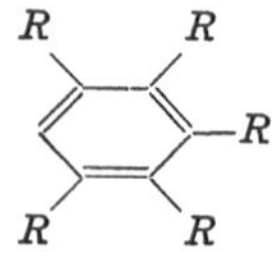 (benzene ring, 1,2,3,4,5-R) $R = CH_3$[8] ...	11,60		19,05	
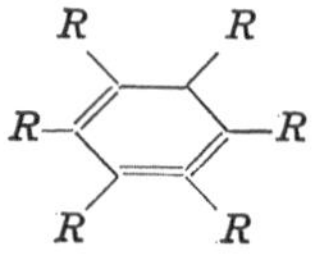 (benzene ring, fully substituted R) $R = CH_3$[8] ...		no strong bands 11—25 μ.		

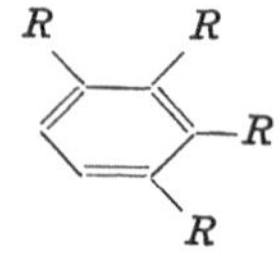

[8] Only studied in the completely methyl-substituted case.

that conjugation of either C=C or C=O to a benzene ring gives rise to a band near 6,30 μ., in addition to the usual 6,25 μ. band, and of about equal intensity. Thus, a (usually weak or moderately intense) doublet at 6,25 μ. and 6,30 μ. is often useful as an indicator for the benzoyl (or other conjugated phenyl) group.

Very little interpretation or correlation has been done on *polycyclic* or *heterocyclic* aromatics. LECOMTE and co-workers have given spectra of some naphthalene and anthracene derivatives (*44, 45, 47, 48, 50, 51*) and of several aromatic heterocyclics (see also *B 1* and *B 2*). Spectra and vibrational assignments of pyrrole are given by LORD and MILLER (*59*) and of furan and thiophene by THOMPSON and TEMPLE (*98*). Pyridine has been worked on by KLINE and TURKEVICH (*40*). All these ring types undoubtedly possess characteristic ring-bond stretching and CH out-of-plane wagging frequencies in the same general regions as benzene compounds, but insufficient results are available to enable generalization at the present time.

6. Acetylenes.

Owing to the minor importance of this type of compounds in most present-day fields of investigation, very little work has been done on acetylenes. For representative spectra, see HERZBERG (B 5). Complete assignments for acetylene, methylacetylene, and dimethylacetylene can serve as guides for the interpretation of higher acetylenes when spectra of these are forthcoming. The CH stretching vibration of the C≡CH system gives a characteristic band at 3,0 μ., well out of the range of aliphatic or olefinic CH bands. The C≡C stretching vibration gives rise to a band near 4,7 μ., of low intensity if substituted by carbon on both ends. Conjugation of C≡C to C=C seems to have little effect on this band position, judging from vinylacetylene and isopropenylacetylene (85). However, the C=C band is shifted to about 6,2 μ. as with other types of C=C conjugation.

7. Alkyl Halides; the Carbon-Chlorine Stretching Band.

Spectra of the halomethanes have been well studied, and the frequency assignments are known with fair certainty [HERZBERG (B5), pp. 310–328; DECIUS (20); WU (118)]. A few haloethanes have been reported (B 5, 18, 102, 103), but very little work on higher aliphatic halides is available (62, 102). The only notable characteristic bands are those involving the C—X stretching vibration, which are found as strngo bands in the following regions:

$$C—F \quad 8 \text{ to } 11\,\mu.,$$
$$C—Cl \quad 12 \text{ to } 20\,\mu.,$$
$$C—Br \quad 15 \text{ to } 35\,\mu.,$$
$$C—I \quad (\sim 20) \text{ to } 50\,\mu.$$

The remainder of the infrared spectrum is due solely to the alkyl portion of the molecule and hence is completely paraffin-like in character, although some of the C—C stretching bands seem to be enhanced considerably in intensity, presumably through vibrational interaction with the polar C—Cl group.

RAMAN data, and unpublished infrared work from the Shell Development Laboratories, show that exact positions within these ranges depend not only on the number and types of halogens on a given carbon of interest, but also on the branching at this carbon and on halogen substituents or branching at next adjacent carbons. A further complication is the frequent appearance of a larger number of strong bands, in the proper region, than there are C—X bonds in the molecule. There is good ground for the current belief that this multiplicity arises from molecules in different rotational orientations about C—C bonds [HERZBERG (B 5), pp. 346–350]. No simple generalization can be made; the complexity even for the simple class of primary chlorides is illustrated in Figure 10. Each molecule which has two non-equivalent positions of the Cl (in respect to rotation about the adjacent C—C bond) has two C—Cl stretching bands, whereas those in which all positions are equivalent (ethyl chloride, and presumably, neopentyl chloride) show only a single such band.

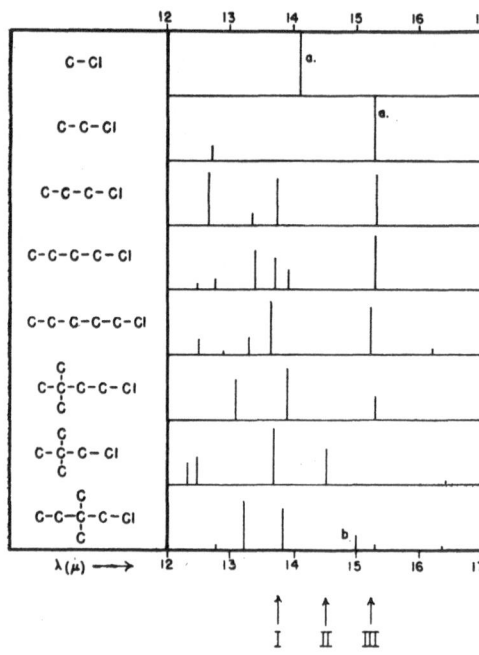

Fig. 10. Spectra of Some Primary Aliphatic Chlorides. Spectra from Shell Development Company, except that bands marked (a) are from RAMAN spectra of liquids for comparability with the remaining spectra, all of which refer to liquids in an 0,02 mm. cell. The only available infrared spectra of methyl and ethyl-chlorides are of the vapor state and there is a considerable vapor-liquid shift in these bands. Compounds with unbranched β-carbons give the set of bands I and III; those with tertiary branching at the β-carbon show I and II; those with quaternary branching give only I. The strong bands at 13,25 μ. and below are believed to be essentially C—C stretching. The band (b) is due to an impurity of 3,3-dimethyl-2-chlorobutane.

Unsaturated halides have received some attention (B 5, 102, 103, 105, 106), but not enough to generalize. For chloroolefins and chloroaromatics it is to be noted that the C—Cl stretching bands appear in the same region as the CH out-of-plane bands, and some caution and care is needed in differentiating them. However, when sufficient data are at hand, this can be done, e. g., for the chlorine and mixed chlorine-methyl substituted ethylenes (Figure 7, p. 353).

8. Alcohols, Phenols and Ethers.

For this and subsequent classes of compounds, only the most obvious characteristic bands will be pointed out, since no attempt has yet been made to carry out a thoroughgoing correlation of spectra and structure; only for a very few of the prototypic molecules have complete assignments been attempted.

Alcohol spectra [American Petroleum Institute (B 1); BARNES et al. (B 2); COBLENTZ (B 3); HERZBERG (B 5)], exhibit three obvious features; the OH stretching band near 3,0 μ.; the strong C—O stretching band near 9μ.; and an unusual broad absorption, setting in near 12μ. and increasing to about 15μ., where it becomes fairly strong. To follow the known assignment of methanol, the latter absorption is due to the torsional oscillation of the OH group about the CO bond. The $3,0 \mu$. OH band (and its overtones at $1,5 \mu$. and 1μ.) has been the subject of a large number of investigations whose object has been the study of the hydrogen bond, the band position being particularly sensitive to this type of perturbation. However, for the present purpose the sole use of the band is as an indicator for the alcoholic or phenolic OH group, there being no particular correlation of band position with structure aside from the effect of structure on hydrogen bonding proclivities. As encountered in most practical problems, the OH group is in such a milieu that it can form hydrogen bonds, and the band generally appears very near $3,0 \mu$. and is strong. However, in circumstances where it does not form such a bond (vapor-phase or dilute solution with no intramolecular bonding) the band appears near $2,8 \mu$., and is rather weak. Caution must be exercised with regard to a $2,8 \mu$. band, however, since small amounts of water would give a band at the same position.

The 9μ. C—O stretching band is undoubtedly correlatable in its exact position with structure, in the same manner as are C—Cl bands, for example. There is the added difficulty, however, that alkyl-group bands appear in the same region, although usually much weaker. Information is too sketchy to deduce this correlation, although there is indication that primary alcohols are sometimes differentiable from secondary or tertiary alcohols. Recently, DAVIES (*18a*) has studied and interpreted the changes in the methanol spectrum with dilution in solvents. In this way the effect of hydrogen-bonding on the whole spectrum was observed, with results which may be extrapolatable to higher alcohols.

Spectra of *phenols* have been given by several investigators [BARNES et al. (*B 2*); COBLENTZ (*B 3*); RICHARDS and THOMPSON (*83a*); WHIFFEN and THOMPSON (*111*)]. The effect of the OH group on the ring frequencies was described in the section on aromatics. The position and intensity of the $3,0 \mu$. OH band, and its shifts from solution to solid, are sensitive to the presence and kind of ortho substituents, due to interference with intermolecular hydrogen bonding. This aspect has been particularly studied by RICHARDS and THOMPSON (*83a*) and by COGGESHALL (*15a*). DAVIES (*18a*) has investigated and discussed changes in the spectrum of phenol with dilution in solvents, such changes even in longer wave bands being attributable to the effect of hydrogen bonding.

Ethers have as their only characteristic band a strong C—O stretching band near $9,0 \mu$. (*B 2*, *B 3*, *B 5*). Again, while correlations with structure undoubtedly exist, we do not yet know them. However, the presence or absence of such a 9μ. band is often very useful in differentiating between an ether and a paraffin, this being sometimes difficult to do chemically short of an elementary analysis. Saturated cyclic ethers are sometimes more complex, often exhibiting several moderately intense bands near $9\,\mu$. instead of a single strong one.

9. Amines, Mercaptans and Sulfides.

For references and data on amines, see Barnes et al. ($B\,2$) and Herzberg ($B\,5$). Mercaptans, sulfides and disulfides are discussed by Trotter and Thompson (107). Examination of the aliphatic members for obvious characteristic bands fails to reveal any, except the NH and SH stretching bands and the HNH bending band. From known assignments in the simplest members of each class, the C—N and C—S frequencies can be followed in higher representatives, as one proceeds through a series of higher compounds, as Trotter and Thompson have done for mercaptans and sulfides (107). However, these vibrations do not give bands of sufficient intensity to stand out above the alkyl group bands, which is in marked contrast to the C—O stretching bands of alcohols and ethers.

The NH stretching band occurs in the region $2,95\,\mu$. to $3,15\,\mu$. Hydrogen-bond shifts have been noted for this band similar to those of OH [see, for example, Gordy and Stanford ($30b$)]. However, the shifts are not so large, nor are NH bands in general so intense as OH stretching bands. Aside from these differences, it is seen that NH bands are so similar in position to OH bands that a reliable differentiation in an unknown compound is difficult or impossible. One point that needs considerable further study is the vibrational frequency of the $\overset{\displaystyle\diagdown\,+}{\underset{\displaystyle\diagup}{-\mathrm{NH}}}$ group, in ammonium-type compounds. Spectra of amine hydrochlorides (or other salts) and of amino acids ($B\,12$) seem to indicate a very broad, rather weak absorption band near $4,0\,\mu$. as being characteristic of this bond. The HNH bending vibration of the NH_2 group gives rise to a band at $6,15$ to $6,25\,\mu$. of moderate intensity which is useful in some instances, since secondary or tertiary amines do not have any bands in that region.

The SH band possesses a characteristic stretching band very near $4\,\mu$. The usefulness of this is usually vitiated by its exceedingly low intensity, the band often being difficult or impossible to pick out from weak overtone bands which occur in the same region.

10. Carbonyl Compounds.

The region of double-bond absorption, extending from about $5,5\,\mu$. to $6,7\,\mu$. and including the C=C and aromatic ring bands (discussed previously) as well as C=O and C=N bands (discussed below), is currently the most useful of all regions. The reasons are that: a) the bands are of a simple type, usually involving only the stretching of individual bonds; b) there is no interference from other bands due to other types of vibration;* c) the longer-wave spectra of more complicated classes of molecules are not yet understood, and there is a greater complexity of

Table 6. C=O Stretching Band Positions.

[Wave length values (μ.) for liquid state; for references, see text.]

Aldehydes, unconjugated..............................	5,79
Aldehydes, conjugated to C=C or benzene ring........	5,90
Ketones, unconjugated and acyclic	5,83
Ketones, unconjugated, in 4-membered ring	5,63
Ketones, unconjugated, in 5-membered ring	5,75
Ketones, unconjugated, in 6-membered ring	5,83
Ketones, C=O conjugated to C=C or benzene ring....	5,93
Ketones, C=O conjugated with ketone C=O (α-diketones)	5,83
Ketones, C=O in conjugated chelate system (enolized β-diketones) $\quad R-\overset{\displaystyle OH\cdots\cdots O}{\underset{\displaystyle R}{\overset{\displaystyle \vert\qquad\Vert}{C=C-C-R}}}$	~ 6,3
Ketones, C=O in ketenes...........................	4,75
Esters, unconjugated and acyclic	5,75
Esters, unconjugated, β-lactone.......................	5,50
Esters, unconjugated, γ-lactone.......................	5,65
Esters, unconjugated, δ-lactone.......................	5,75
Esters, C=O conjugated to C=C or benzene ring	5,81
Esters of vinylic alcohols or phenols	5,65
Esters, C=O in conjugated chelate system (enolized β-ketoesters) $\quad R-\overset{\displaystyle OH\cdots\cdots O}{\underset{\displaystyle R}{\overset{\displaystyle \vert\qquad\Vert}{C=C-C-OR}}}$	~ 6,0
Esters of thiols	5,97
Carboxylic acids	5,65—5,80
Salts of carboxylic acids.............................	6,2—6,4
Acid chlorides, unconjugated	5,55
Acid chlorides, C=O conjugated to benzene ring	5,65
Acid anhydrides, unconjugated and acyclic	5,50 and 5,70
Acid anhydrides, C=O conjugated to C=C	5,60 and 5,80
Amides, acyclic	5,90—6,10 or higher
Amides, β-lactam	5,75
Amides, γ-lactam	5,85
Amides, δ-lactam	~ 6,00

* With rare exceptions: the HNH bending vibration of the NH_2 group occurs near 6,2 μ.; and the 6,6 μ. amide band, presumably due to an NH bending of the CO·NH·R system, is located on the upper border of this region.

bands there owing to contributions from single-bond stretching frequencies
and bending frequencies.

The stretching frequencies of the C=O group give bands ranging
from ~ 5,5 μ. to ~ 6,2 μ. This great variability in position is attributable to the sensitivity of the ionic character of the bond toward the

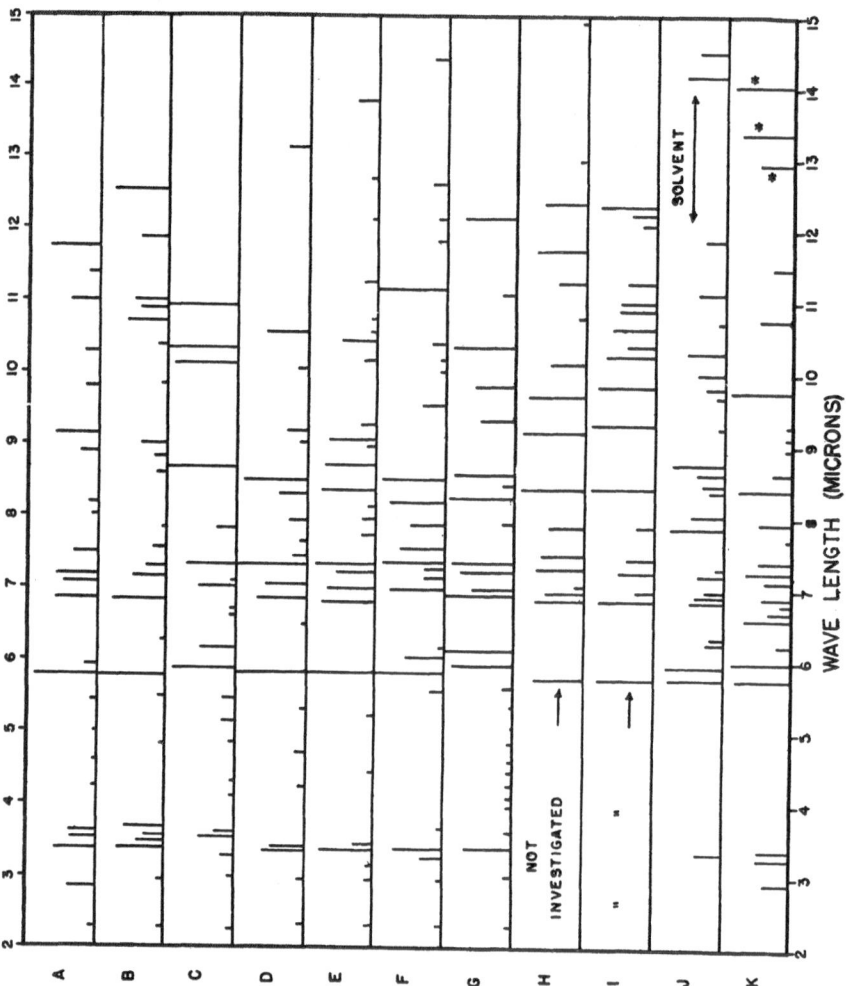

Fig. 11. Spectra of Some Carbonyl Compounds. *H* and *I* are from THOMPSON and TORKINGTON (99);
remainder from Shell Development Company. A. Propionaldehyde, 0,021 mm. pure liquid. B. Isobutyr-
aldehyde, 0,030 mm. pure liquid. C. Acrolein, 0,021 mm. pure liquid. D. Methyl ethyl ketone, 0,021 mm.
pure liquid. E. Methyl isopropyl ketone, 0,030 mm. pure liquid. F. 4-Methyl-4-pentene-2-one, 0,030 mm.
pure liquid. G. Mesityl oxide (4-methyl-3-pentene-2-one), 0,030 mm. pure liquid. H. Ethyl propionate,
0,08 mm. pure liquid. I. Isobutyl propionate, 0,08 mm. pure liquid. J. Methyl α-benzoylisobutyrate, 0,14 mm.
of 5% solutions in CCl$_4$ and CHCl$_3$. K. Ethyl phenaceturate (ethyl ester of N-phenylacetylglycine), 0,14 mm.
of 10% solution in CHCl$_3$ (Bands marked (*) are from a spectrum of the solid material).

electrophilic nature of the groups attached to it as is discussed below in more detail. In all cases these bands are strong, much stronger in general than C=C bands. In a cell of spacing 0,02 mm., the transmission at the band center is 10% or less, except for high molecular weight compounds where the "dilution" of C=O groups by the remainder of the molecule may weaken the intensity. In Table 6 (p. 365) are collected the normal positions of C=O bands for simple carbonylic compounds (in the liquid state); and some typical spectra are given in Figure 11. These positions are found to hold with great constancy (usually to $\pm$ 0,02 μ.) in the simple compounds where the remainder of the molecule consists of alkyl groups only, and also in cases where other substituent types are present but at a β- or further position from the carbonylic moiety. This is generally true also of α-substituents; however, in some cases such substituents can affect the C=O band position, presumably either through an inductive or a steric effect.

a) Aldehydes.

Together with ketones, these may be taken as prototypic of carbonylic compounds. The C=O band position in saturated aldehydes is found to be 5,79 μ. [Shell Development Co. (85)]. Conjugation with C=C or phenyl results in a shift of the band to 5,90 μ. (Table 4), this shift being approximately 35 cm.$^{-1}$ and hence being identical with the C=C-band shift on conjugation. Another characteristic band of the —CH=O system is found near 3,6 μ. and is ascribable to the CH stretching vibration of that group (85). This band is usually distinct enough from the alkyl CH stretching bands (near 3,4 μ.) to serve as a useful criterion for the presence of the —CH=O group.

b) Ketones.

Representative spectra are given by BARNES et al. (B 2), LECOMTE (52), THOMPSON and TORKINGTON (99), and RASMUSSEN, TUNNICLIFF and BRATTAIN (81). Acyclic saturated ketones exhibit the C=O band at 5,83 μ., the slight shift from the aldehyde position recalling the shifts in C=C band position with various modes of substitution. Conjugation with C=C or phenyl causes a shift to near 5,92 μ. (Table 4), the shift being about 25 cm.$^{-1}$ in this case. Benzophenone, which involves an increased degree of conjugation over alkyl phenyl ketones, exhibits a further shift of the C=O band to 6,04 μ. [THOMPSON and TORKINGTON(99)].

The occurrence of ketone C=O in a 4- or 5-membered ring results in a short-wave shift, the band being found near 5,63 μ. and 5,75 μ., respectively [LECOMTE (52); WHIFFEN and THOMPSON (112); Shell Development Co. (85)]. With 6- or higher membered rings, this band appears very near the open-chain position, 5,83 μ. A similar shift will be noted later for lactones and lactams as compared with open-chain esters and amides. The explanation for these shifts is lacking, although it would seem to involve ring strain in some way.

The case of $C{=}O$ *conjugated to* $C{=}O$ has been studied only for two α-diketones [Rasmussen, Tunnicliff and Brattain (81)]. It appears that this type of conjugation has no effect whatever on the $C{=}O$ band positions, each acyl group behaving as if the other were simply alkyl.

A special case of conjugated ketone is found with *enolized β-diketones* (Formula I). Here the combination of the conjugation and the chelation gives rise to a very great shift of the $C{=}O$ band, to $\sim 6,3 \mu$., and an extreme increase in its intensity [Gordy (30 a); Rasmussen, Tunnicliff and Brattain (81)]. There occurs also a large shift in the OH band position, this being found near 3,7 μ. and being rather weak and unusually broad. These effects are explainable on the basis of stabilization of ionic resonance structures by the OH⸱⸱⸱⸱⸱⸱⸱O bond [Rasmussen, Tunnicliff and Brattain (81)]. The $C{=}C$ in symbol (I) may be replaced

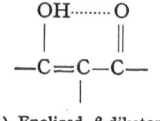

(I.) Enolized β-diketone.

by a carbon-carbon bond of a benzene ring; thus, salicylaldehyde shows similar shifts, although not so extreme as those observed for β-diketones; the $C{=}O$ band is found at 6,00 μ. [Barnes et al. (B 2)].

Another special case is that of *ketenes* $R_2C{=}C{=}O$. Here the $C{=}O$ stretching vibration interacts strongly with the $C{=}C$, the resulting system giving bands near 4,75 μ. and 9 μ. [Whiffen and Thompson (112); Halverson and Williams (31 b); Harp and Rasmussen (32)]. The 4,75 μ. band stands in a unique region, and is designated as "$C{=}O$" in Table 6 (p. 365), although actually it involves vibrations of both double bonds almost equally.

As regards other characteristic bands, Thompson and Torkington (99) and Lecomte (52) have pointed out a few such bands in the longer-wave region (7,5 μ. to 15 μ.). The interpretations of these, and their usefulness in structural work, are not known.

The effect of hydrogen-bonding solvents (alcohols) on the $C{=}O$ band position of saturated ketones has been studied by Lecomte, Champetier and Clément (53) as well as by Rasmussen, Tunnicliff and Brattain (81), and found to be small. A shift of $\sim 0,03 \mu$. to long wave lengths was observed.

c) Carboxylic Esters.

Acyclic saturated esters yield the C=O band very near $5{,}75\,\mu$. [Barnes et al. (B 2); Thompson and Torkington (99); Lecomte (51a)], except that in formates this is shifted to $5{,}80\,\mu$. (99). The increased frequency of the ester C=O band over the ketone C=O value must be attributed to the electronegative character of the OR group as compared with R itself (R = alkyl). Thus, to the resonance structures (II) and (III) as postulated by Pauling (B 8) must be added (IV), which in the case of esters predominates over (III) sufficiently to give the C=O bond a greater-than-double-bond character. In certain other classes, where

$$
\underset{\text{(II.)}}{\overset{\displaystyle O}{\underset{\displaystyle R'-C-O-R}{\Big\|}}}
\;\longleftrightarrow\;
\underset{\text{(III.)}}{\overset{\displaystyle \overset{(-)}{O}}{\underset{\displaystyle R'-C=\!O\overset{(+)}{}\!-R}{\big|}}}
\;\longleftrightarrow\;
\underset{\text{(IV.)}}{\overset{\displaystyle \overset{(+)}{O}}{\underset{\displaystyle R'-C\quad O\overset{(-)}{=\!=\!=}R}{\big|\big|\big|}}}
$$

the —OR radical is replaced by more electropositive groups (e. g., —SR or —N$\langle$), the structure (III) predominates; the C=O band is at longer wave length (lower frequency) than in ketones, indicative of an increase in single-bond character of such C=O bond.

As shown in Table 4, *conjugation of the C=O with C=C or phenyl* gives the usual long-wave shift. Its magnitude is, however, less than that for aldehydes and ketones, the band occurring near $5{,}80\,\mu$., which represents a shift of only about 15 cm.$^{-1}$. As with cycloketones, the occurrence of the C=O bond in 4- or 5-membered lactone rings causes a short-wave shift (Table 6), whereas larger rings show substantially the same band position as open-chain esters. Ester C=O groups can partake in conjugated chelation, as was discussed above for β-diketones. Thus, acetoacetic ester and its derivatives exhibit the C=O band near $6{,}05\,\mu$., considerably beyond the simple conjugated C=O position; and salicyclic esters show the band at somewhat less than $6{,}0\,\mu$. [Rasmussen and Brattain (79)], the weaker effect of the benzene ring as compared with C=C in the chelate system being apparent here also.

A further influence on ester spectra is occasioned by the possibility of *unsaturation* as in vinyl or phenyl esters. This gives rise to a shift of the C=O band to about $5{,}65\,\mu$. [Barnes et al. (B 2); Rasmussen and Brattain (79)], a result which is somewhat surprising in view of the intervening oxygen. However, the explanation would seem to be that such unsaturation increases the ability of the —OR radical to take up electrons, and thus, enhances the contribution of structure (IV) in the resonance scheme outlined above.

Thiolesters, $R' \cdot CO \cdot S \cdot R$, exhibit the C=O band near $5{,}97\,\mu$. [Shell Development Co. (B 12a)]. The long-wave position is explainable as due to the tendency of the sulfur atom to share non-bonding electrons and assume a positive charge.

THOMPSON and TORKINGTON (99) and LECOMTE (51a) have pointed out characteristic and strong longer-wave bands in saturated esters, in particular a pair at 9,0 to 9,5 μ. and 8,0 to 8,3 μ. which undoubtedly involve the C—O vibrations. These are frequently of value in the detection or confirmation of the presence of —COOR groups.

d) Carboxylic Acids.

The well-known propensity of carboxylic acids to dimerize or to form strong hydrogen bonds to other acceptors has a marked effect on the spectrum. In the free condition, as encountered in high-temperature vapors or very dilute solutions, the C=O band occurs near 5,65 μ., and the OH band near 2,8 μ. However, in the dimeric form these bands are found near 5,80 μ. and 3,4 μ. respectively [HOFSTADTER et al. (12, 33, 34); DAVIES and SUTHERLAND (19)]. The OH band is also very broad, its absorption extending to about 4,0 μ. This characteristic is of some value in the detection of the presence of the carboxyl group. The COOH group does not show useful regularity in more complex molecules, where several hydrogen-bond acceptors are present, or in the solid state, where crystal packing influences the ability to form hydrogen bonds. The C=O band is found over a wide range, the position having no obvious correlation with structure. Because of these difficulties, more subtle effects such as conjugation, negative substituents, etc., have not yet been deduced from these spectra.

e) Salts of Carboxylic Acids.

A band of the anionic —COO$^{(-)}$ group, as observed in methanol solutions of sodium carboxylates [Shell Development Co. (B 12a)], or in solid samples of alkali metal carboxylates [University of Michigan (B 12b); DUVAL, LECOMTE and DOUVILLÉ (23)], occurs in the 6,20 μ. to ~ 6,40 μ. range. The latter workers (23) have studied also salts of alkaline earth and transition group metals in the solid state, with results indicating an even wider range for the band.

Structural studies on carboxylate salts have shown the two CO bonds to be equivalent, explicable in a straightforward manner as due to resonance between the two equivalent forms, $^{(-)}$O—CR=O ↔ O=CR—O$^{(-)}$ [PAULING (B 8)]. The resulting lessening of the double bond character of the C=O bond accounts for the relatively long-wave position of the observed band. The equivalence of the two bonds also gives rise to vibrational interaction between them, so that the above-mentioned band is only one of the pair to be expected from the group. The other appears to fall near 7,0 μ., where it becomes superposed on the strong HCH bending bands in that region, and hence has not been particularly noted. As with acids, conjugation and other effects have not been studied, since the picture for even the simple salts still requires clarification.

f) Acid Halides.

A very few examples of this class have been studied, with the results shown in Table 6 [Penicillin Structure Reports (B $12a$, b)].

g) Acid Anhydrides.

Open-chain anhydrides yield two strong bands in the double-bond region, near $5,50\,\mu.$ and $5,70\,\mu.$ [BARNES et al. (B 2)]. The presence of the C=O absorption in this relatively short-wave (high frequency) region is not surprising, since a multitude of resonance structures of the general type of (II), (III), and (IV) (p. 369) can be written, which give the C=O bonds considerable triple-bond character. The appearance of two bands is reasonably explained as arising from vibrational interaction between the two C=O bonds, which is to be expected also from the resonance in the system.

Conjugation, as in crotonic anhydride, shifts both bands in the expected direction, viz. to $5,60\,\mu.$ and $5,80\,\mu.$ (B 2). Spectra of maleic and phthalic anhydrides are also given by BARNES et al. (B 2), but conjugation and ring-strain effects both operate here and interpretation is impossible until further studies are made.

h) Amides of Carboxylic Acids.

Like carboxylic acids, amides containing the NH bond associate strongly in concentrated solution or in the solid state. However, amides have been more thoroughly studied because of their importance to the questions of the penicillin and protein structures. The unassociated molecules, as in dilute solution or large molecules in which steric hindrance or other factors work against association, give the C=O band near $6,0\,\mu.$ [Shell Development Co. (B $12a$)]. RICHARDS and THOMPSON (83) have examined in more detail the changes in this position as between N-unsubstituted, N-monosubstituted, and N,N-disubstituted types, and found $5,90\,\mu.$, $5,96\,\mu.$, and $6,08\,\mu.$, respectively. In the solid state these bands shift to the $6,0\,\mu.$-to-$6,1\,\mu.$ range, and exhibit apparently random variability as contrasted with the much more constant solution values. Pure liquids or concentrated solutions are intermediate between the dilute solution and solid cases.

The occurrence of the C=O band in this relatively low frequency region is attributable to the electron-donor capacity of nitrogen, leading to a large resonance participation of structure (V) and giving single-bond character to the C=O. Conjugation of the C=O group with phenyl, and presumably also to C=C, has a hardly noticeable effect [RICHARDS

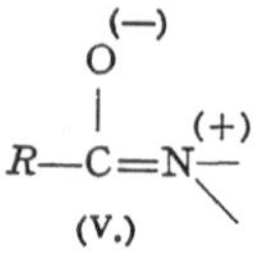

$$R-C=N-$$

(V.)

with O$^{(-)}$ above and N$^{(+)}$

and THOMPSON (*83*)]. The ring-strain effect is present in β- and γ-lactams, the C=O band appearing near $5{,}75\,\mu$. and $5{,}85\,\mu$., respectively [Shell Development Co. (*B 12a*); Merck and Co. (*B 12c*)]. Analogous to the vinyl-ester effect of esters, substitution of N with an unsaturated system (e. g., phenyl, in anilides) causes a short-wave shift of about $0{,}05\,\mu$. [RICHARDS and THOMPSON (*83*)]. These authors, and the Shell Development Co. workers (*B 12a*), have observed and interpreted small changes due to variations in solvent.

N-unsubstituted and N-monosubstituted amides exhibit a second strong characteristic band in the double-bond region, falling at $6{,}10$ — $6{,}20\,\mu$. and $6{,}45–6{,}65\,\mu$., respectively. These are believed to originate in the HNH bending vibration and an HNC bending vibration, in the two cases [Shell Development Co. (*B 12a*); RICHARDS and THOMPSON (*83*)]. Like the C=O band, they are sensitive to the physical state, being found at the short-wave ends of their ranges in the hydrogen-bonded conditions (solid or concentrated solution) and at the long-wave end in the free condition. The $\sim 6{,}5\,\mu$. band is of particular use in the detection of the monosubstituted amide grouping, falling as it does at a position where few other groups absorb. N,N-di-substituted amides and lactams fail to show any but the C=O band in this region [(*B 12b*, *83*; AELION and LENORMANT (*1*)].

The presence of potent hydrogen-bond donors (e. g., COOH) in the environment of the amide group, particularly in the solid state, can cause shifts in the C=O band position considerably outside the limits quoted above [University of Michigan (*B 12b*)]. Hence, some caution must be observed in dealing with such compounds.

Some sequences of bands in the longer-wave spectra of amides have been pointed out by LECOMTE and FREYMANN (*55*).

Association also has an effect on the NH bands, although not nearly so marked as with the OH bands of carboxylic acids. The NH bands fall in the range $3{,}0–3{,}15\,\mu$., not far from their positions in the unassociated molecules, $\sim 2{,}95\,\mu$. [BUSWELL, RODEBUSH and ROY (*13*); RICHARDS and THOMPSON (*83*)].

Some further general points on carbonyl spectra are to be made. While, in general, the ordinary monovalent substituents on a carbon atom which is in the α-position to a carbonyl group do not affect the

C=O stretching frequency appreciably, it is possible for highly negative substituents to cause a short-wave shift. Thus, Cl, OH, or NH_2 in the α-position of esters do not have any effect on the C=O band; however, both ethyl cyanoacetate and trichloroacetate show the band shifted to near $5{,}70\,\mu$. (B 12 b, 79). The effect here is presumably an inductive one, and is to be expected for other carbonylic compounds also. In conjugated systems, sufficiently active substituents far removed from the C=O group may have influence by interfering with the resonating system. Thus, the nitro group has a decided effect on the C=O band in nitrobenzoates [RASMUSSEN and BRATTAIN (79)]; furthermore, RICHARDS and THOMPSON have noted and explained similar changes in ring-substituted anilides (83).

11. Carbonic Acid Derivatives.

Only a very few compounds of this type have been examined [dialkyl carbonates (99), urea derivatives (B 12)]. The C=O bands of such compounds are very close to those of analogous carboxylic acid derivatives, dialkyl carbonates giving a band at $5{,}75\,\mu$. and urea derivatives one in the $6{,}0\,\mu$. (or longer wave length) region.

12. Compounds Containing Carbon-Nitrogen Double Bonds.

Several compounds in this category were examined during the penicillin work (B 12). The C=N stretching band is, like the C=O, of rather high intensity. Simple C=N groups, as in oximes or dialkyl or alkyl-aryl imines, exhibit the band close to $6{,}0\,\mu$., with some variation due to conjugation which has not yet been thoroughly studied. The C=N band in certain ring compounds (alkyl-substituted oxazolines, oxazines) is also found at $6{,}0\,\mu$., but in thiazolines and imidazolines (alkyl-substituted) this band is shifted to the $6{,}15$–$6{,}20\,\mu$. region. The shift is presumably the result of the respective S or N substitution on the C=N bond and not of the simple occurrence of C=N in a ring.

13. Nitriles.

The C≡N group gives rise to a sharp and moderately strong band at $4{,}4$ to $4{,}5\,\mu$. (B 2, B 5, B 12). Conjugation with C=C seems to have no effect on this position.

14. Nitro Compounds.

The aliphatic NO_2 group gives rise to a very strong band near $6{,}4\,\mu$., and a second, considerably weaker one, near $7{,}3\,\mu$. (although still strong compared with alkyl bands). The nitro group on a benzene ring behaves similarly, giving bands near $6{,}5\,\mu$. and $7{,}4\,\mu$.; however, both bands are very strong and their relative intensities are more comparable [BARNES et al. (B 2)]. Information on further structural correlation is lacking.

15. Sulfones and Sulfonates.

The —SO_2— grouping in sulfones gives rise to two very strong bands, one near 7,6 μ. and the other lying between 8,6 μ. and 9,0 μ. The variations in the latter are not understood, since only a handful of cases have been studied [Shell Development Co. (85)]. Sulfonic acid esters, similarly, are responsible for a pair of strong bands, these occurring at 7,30 μ. and 8,45 μ. for some alkyl benzene-(or toluene-)sulfonates (85).

16. More Complicated Structures.

The typical absorption bands described above in sections 1 to 16 refer only to simple systems, or to the effect of a single type of perturbation (conjugation, particular type of substituent, etc.) on such systems. More complicated structures, in which several such effects operate simultan-
· eously, have been studied only in a few isolated cases and are but little understood at the present time. In this direction much needs to be done. Examples of spectra of such complicated systems are to be found in the Penicillin Structure reports [(B 12b): thiohydantoins and thiouracils], and in the recent work of BLOUT, FIELDS and KARPLUS (11a) on some polyconjugated systems, as well as miscellaneous spectra of more complex compounds given in such compilations as by COBLENTZ (B 3) and BARNES et al. (B 2).

IV. Applications to Structure Determinations.

Examples of applications to structural problems are not numerous, in view of the newness of developments in this field. The following are mentioned as being most illustrative of the type of work which calls upon such general knowledge as that described in the preceding section.

THOMPSON, TORKINGTON and RICHARDS (83a, 100, 101, 102) and SUTHERLAND, SHEPPARD, PHILPOTTS and TWIGG (86, 92) have discussed the application of infrared measurements to the determination of structure of natural and synthetic polymers. Numerous features of polymer structure were demonstrated in this work, of which might be mentioned the type of addition in diene polymers (i. e., 1–4 or 1–2) as deduced from the olefin CH wagging bands of the resulting isolated double bonds; the presence or absence of CH_3 side chains, as deduced from the 7,2 μ. CH_3 group band; and the configuration of substituents on the phenol group in phenol-formaldehyde polymers, as inferred from the long-wave aromatic bands and the 3,0 μ. OH band.

CANNON and SUTHERLAND (*14*) have studied *coal* and coal extracts, and drawn conclusions about the types of groups present. Interpretation of such a complex system must be considered as only preliminary, however.

On the more classical chemical side, work on the spectrum of *diketene* seems to indicate that this material of long-doubtful structure is actually a mixture of the two isomers (VI) and (VII), probably in rapid equilibrium (*76*; but see also *95*, *112*). This result was obtained from the number of double-bond bands, which was excessive for a single structure; and from a comparison of the spectrum with those of related four-membered ring compounds.

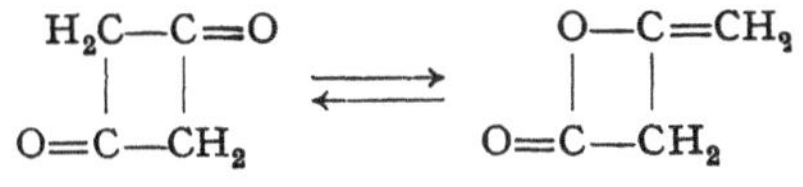

$$H_2C\!-\!C\!=\!O \qquad\qquad O\!-\!C\!=\!CH_2$$
$$\mid\quad\mid \qquad\rightleftharpoons\qquad \mid\quad\mid$$
$$O\!=\!C\!-\!CH_2 \qquad\qquad O\!=\!C\!-\!CH_2$$

(VI.)   (VII.)

SMITH and RASMUSSEN (*87*) have shown that treatment of an α-aryl-aminoacid (VIII) with PBr$_3$ yields the *oxazolone* hydrobromide (IX) rather than the acid bromide. This conclusion was based on the presence

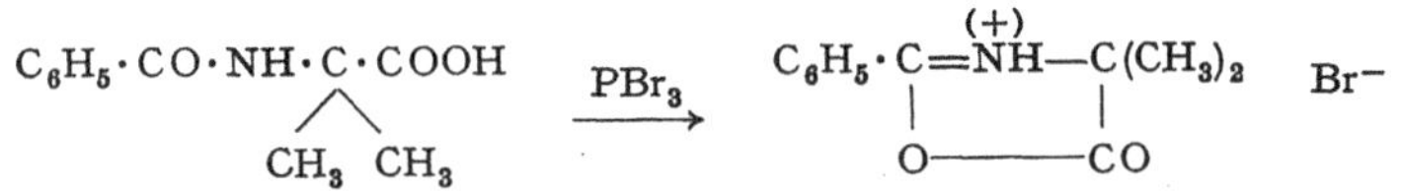

$$C_6H_5\!\cdot\!CO\!\cdot\!NH\!\cdot\!C\!\cdot\!COOH \xrightarrow{\ PBr_3\ } \overset{(+)}{C_6H_5\!\cdot\!C\!=\!NH}\!-\!C(CH_3)_2 \quad Br^-$$

CH$_3$ CH$_3$ O————CO

(VIII.) α-Benzoylamido-isobutyric acid.   (IX.) 2-Phenyl-4,4-dimethyl-5(4)-oxazolone hydrobromide.

of typical NH$^+$-group absorption, and lack of the typical 5,65 μ. band of the acid halide group —COBr.

SMITH, RASMUSSEN and BALLARD (*88*) have studied the supposed "*benzimidazolols*" of FISCHER (*25*) (formula X) and proved that these actually exist in form of the amides (XI). The spectrum indicated the

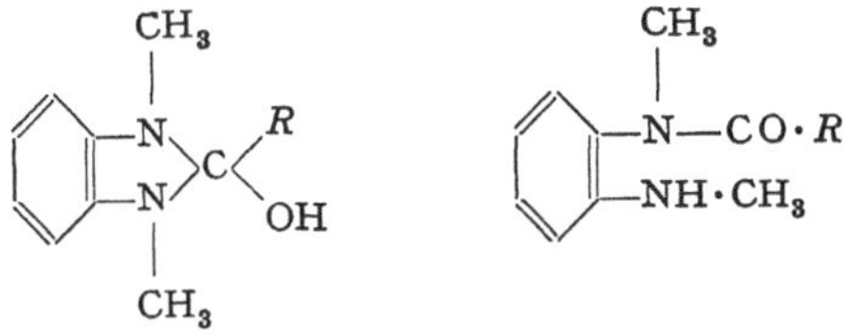

CH$_3$ CH$_3$

(X.) 1,3-Dimethyl-2-alkyl-benzimidazolol.   (XI.) N,N′-Dimethyl-N-acyl-o-phenylenediamine.

amide group by a strong band at 6,0 μ., there being no double-bond absorption expected from (X) (except for the 6,25 μ. phenyl band which is required by both structures).

FURCHGOTT and his colleagues (*22a*, *29*) have described infrared work on *steroids* in which bands associated with typical groupings (OH, C=O, ring C=C in various positions) were deduced from compounds of known structure. These regularities appear to be completely in line with those derived from simple molecules. No application of these results to new structures has been published up to the present time.

Penicillin.

By far the most extensive application has been made on the structure of penicillin, of which an account will now be given. Aside from the detailed discussion of the chemical research on penicillin published in the monograph (B 11), brief outlines have appeared in the periodical literature (121, 122).

The establishment of the fact that the *benzylpenicillin* molecule* was made up of benzylpenaldic acid (XII) and D-penicillamine (XIII), condensed with loss of 2 H_2O but leaving free the penicillamine carboxyl group, gave the first solid foundation for speculations on the penicillin structure. The number of possible structures that can be, and were, derived from this condensation is very large. However, the soon-

$$C_6H_5 \cdot CH_2 \cdot CO \cdot NH \cdot CH \cdot CHO \qquad HS—C(CH_3)_2$$
$$| \qquad\qquad\qquad\qquad |$$
$$COOH \qquad\qquad H_2N—CH—COOH$$

(XII.) Benzylpenaldic acid.   (XIII.) Penicillamine.

forthcoming finding that benzylpenicillin contained only one labile hydrogen (aside from that in the COOH), and the clarification of the structure of the acid-rearrangement product, benzylpenillic acid (XIV) and of its hydrolysis product, termed benzylpenicilloic acid (XV), limited the number of possibilities considerably. Those most important for the present discussion are the *oxazolone structure* (XVI), a *tricyclic*

$$COOH$$
$$|$$
$$N—CH—CH—S—C(CH_3)_2$$
$$\| \qquad | \qquad\qquad |$$
$$C_6H_5 \cdot CH_2—C————N————CH—COOH$$

(XIV.) Benzylpenillic acid.

$$C_6H_5 \cdot CH_2 \cdot CO \cdot NH \cdot CH—CH—S—C(CH_3)_2$$
$$| \qquad | \qquad\qquad |$$
$$HOOC \qquad NH————CH—COOH$$

(XV.) Benzylpenicilloic acid.

structure (XVII), and the *β-lactam structure* (XVIII). On the basis of its initial plausibility as compared with the other structures, and the fact that it explained readily all known conversions, the oxazolone structure (XVI) was early considered as the most probable by a wide margin. This opinion was strengthened by the structure of the $HgCl_2$-

---

* Other penicillins differ only by replacement of benzyl by other radicals; the structural problem remains the same.

degradation product (XIX), which contained a 5-oxazolone ring. This structure, strongly presumed at the time, was later proven to be correct.

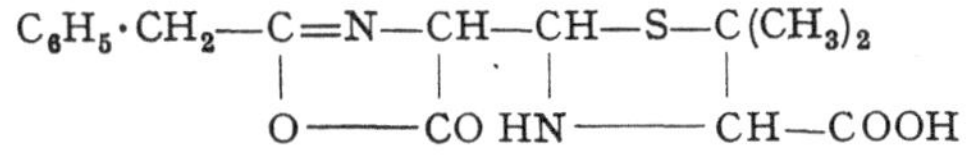

$$C_6H_5 \cdot CH_2-C=N-CH-CH-S-C(CH_3)_2$$

O————CO HN————CH—COOH

(XVI.) Oxazolone structure.

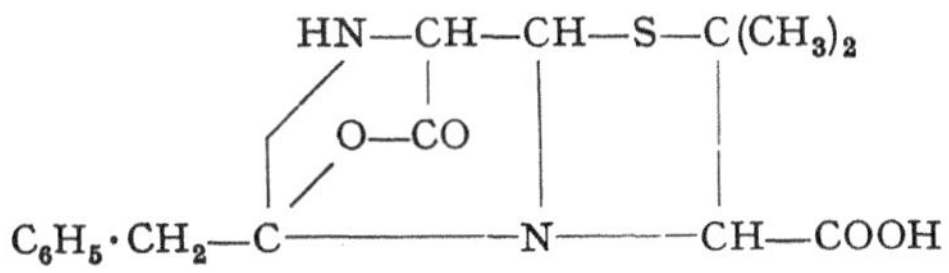

HN—CH—CH—S—C(CH₃)₂

O—CO

$$C_6H_5 \cdot CH_2-C\text{————————N————CH—COOH}$$

(XVII.) Tricyclic structure.

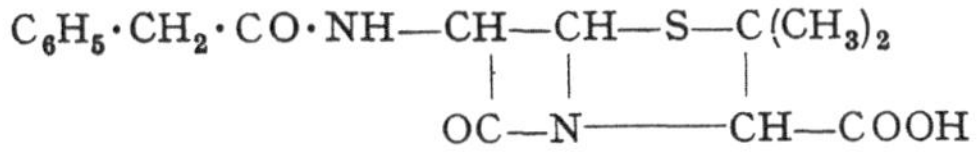

$$C_6H_5 \cdot CH_2 \cdot CO \cdot NH-CH-CH-S-C(CH_3)_2$$

OC—N————CH—COOH

(XVIII.) β-Lactam structure: Benzylpenicillin.

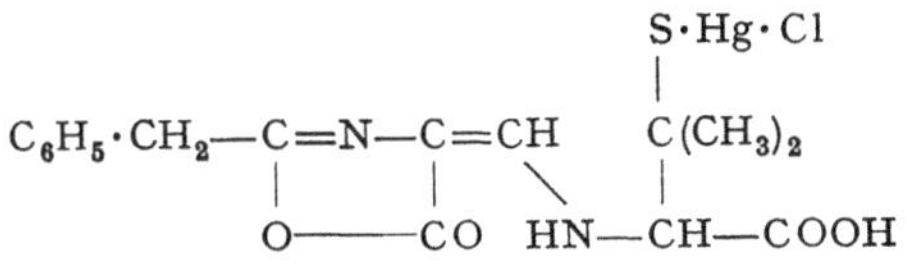

S·Hg·Cl

$$C_6H_5 \cdot CH_2-C=N-C=CH \quad C(CH_3)_2$$

O————CO HN—CH—COOH

(XIX.) Benzylpenicillenic acid mercurichloride.

In certain details, however, the oxazolone structure was found wanting. An especially elementary but important item was the lack of a basic nitrogen atom which is expected of the thiazolidine ring. Further, it was found impossible to synthesize structures of the type (XVI), the reason presumably being the attack of the active hydrogen on the anhydride-like oxazolone ring. Hence, at this point greater consideration began to be given to other possible structures, and to the application of newer physical methods such as infrared spectroscopy and X-ray crystallography.

The first important results of the infrared study came from an examination of the penicillin spectrum itself, and from a comparison of this with model alkyl-substituted 5(4)-oxazolones. Penicillin (studied first as the methyl ester of benzylpenicillin, but with later results on the acid and sodium salt in complete accord) gave bands in the double-bond region at 5,62 μ. and $\sim$ 6,0 μ., aside from a band at 5,75 μ. known to originate from the ester C=O. Hence, at least two double bonds were indicated as present, aside from the carboxyl C=O and the phenyl-ring bonds. This result excluded tricyclic structures such as (XVII),

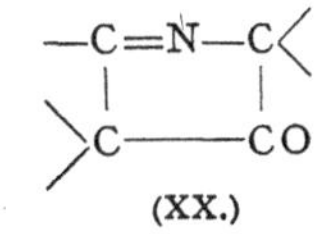

—C=N—C<

>C————CO

(XX.)

since these of necessity have only one such double bond [Shell Development Co. (B $12a$)]. Furthermore, the study of about a dozen simple 5(4-oxazolones (XX) showed that this ring system gave bands at 5,50 μ. and $\sim 6,0$ μ.* The discrepancy between the former band and the penicillin 5,62 μ. band strongly contra-indicated the oxazolone structure [Shell Development Co. (B $12a$)].

The early infrared work also gave indication of the monosubstituted amide grouping, by the presence of a strong band near 6,6 μ. [Shell Development Co. (B $12a$)]. However, such amide spectra had not been studied in any detail previously, and it was only after subsequent intensive work on monosubstituted amides, ranging from simple ones to complex compounds such as penicilloates, that the spectroscopic evidence for the presence of the amide group was considered essentially free from objection [University of Michigan (B $12b$); Shell Development Co. (B $12a$)]. A large part of the difficulty here arose from the seemingly anomalous spectra of solid penicillins and especially, of their sodium salts. It was only after an appreciation had been obtained of the range of variability of both the 6,0 μ. and 6,5 μ. monosubstituted amide bands in the solid state, that the situation was clarified.

After the initial evidence against the tricyclic and oxazolone structures was obtained, attention was turned to other possible structures, particularly those containing the amide side chain. It was demonstrated with appropriate model compounds that no other structures considered at all plausible gave agreement with the observed penicillin spectrum, with the exception of the β-lactam structure (XVIII). A number of β-lactams were examined in which the ring was not fused to another ring. These exhibited the C=O band near 5,75 μ. This was still not in agreement with the corresponding penicillin band (5,62 μ.), but the possibility was recognized that fusion to the thiazolidine ring (or some other unforeseen effect in this scarcely known system) might result in a shift of the β-lactam band to near 5,62 μ.

The synthesis of three compounds containing the proper fused thiazolidine-β-lactam system,** and the finding that the C=O band of such systems occurred at or very near 5,62 μ., was the final step in the chain of evidence indicating the correctness of the β-lactam structure of penicillin (XVIII) [Shell Development Co. (B $12a$)]. The 5,62 μ. band was then explainable as due to the β-lactam C=O group, and the

---

* From the discussion given in Section IV, it is evident that the $\sim 6,0$ μ. band is assignable to the C=N stretching vibration, and the 5,50 μ. one to the C=O. The relatively short wave length of the latter is in keeping with its anhydride-like character.

** The three compounds were the β-lactams of the following β-amino-acids: 2,α,α-triphenyl-2-thiazolidineacetic acid; α,α-dimethyl-2-phenyl-2-thiazolidineacetic acid; 2,α,α-trimethyl-2-phenyl-2-thiazolidineacetic acid.

~ 6,0 and ~ 6,5 μ. bands as due to the amide side chain (although the definitive work on the latter bands was carried out later, as explained above). No other chemically-plausible structure was able to give such an agreement between prediction and observation. Almost simultaneously with the spectroscopic result on fused thiazolidine-β-lactams, the X-ray crystallographic results of CROWFOOT and LOW (Oxford) and of BUNN and TURNER-JONES (Imperial Chemical Industries) showed the presence of the β-lactam ring; and chemical evidence from Merck and Company was forthcoming in which sulfur was removed from benzylpenicillin giving *desthio*-benzylpenicillin (XXI) which from all indications contained the β-lactam ring. The conditions of this reaction (hydrogenation on RANEY nickel) were so gentle that no rearrangement was believed likely; thus, this conversion product furnishes the most direct chemical proof of the β-lactam structure.

$$C_6H_5 \cdot CH_2 \cdot CO \cdot NH - CH - CH_2 \quad CH(CH_3)_2$$

$$OC - N - CH_2 - COOH$$

(XXI.) Desthio-benzyl-penicillin.

It is seen that the spectroscopic results of this investigation were obtained through interpretation of the double-bond region only. Less favorable cases, where the issue cannot be decided solely on the number and types of double bonds, will necessitate a greater understanding and use of the longer-wave-length and the 3 μ. regions.

V. Conclusion.

Despite the considerable body of information presented above, it is clear that the application of infrared spectroscopy to a particular structural problem will usually require a close study of related compounds of known structure. It is highly necessary that all effects which might influence the spectra of the particular class of compounds of interest be recognized, and it seems desirable that they be understood at least rudimentarily. Only in this way can one gain a feeling for the important factors, systematically test the effect of each such factor, and be able to assess the validity of the conclusions drawn.

It is hoped that the material presented in this survey will serve as a general introduction to our present knowledge, on which can be built more detailed structures both of fact and understanding. Infrared spectroscopy can be used, with often a great deal of value, simply as an empirical tool either to establish the identity of preparations by the identity of their spectra; or to indicate structural similarities between compounds, by superficial resemblances between their spectra. It is highly desirable that the use of infrared spectroscopy in organic chemistry not be limited to this simple comparative aspect, but that the more fruitful attempt be pursued to interpret the spectra along lines such as those exemplified in the preceding discussions.

References.

Books and Monographs.

B 1. American Petroleum Institute, Research Project 44: Catalog of Infrared Absorption Spectrograms. Washington: National Bureau of Standards.

B 2. BARNES, R. B., R. C. GORE, U. LIDDEL and V. Z. WILLIAMS: Infrared Spectroscopy. New York: Reinhold. 1944.

B 3. COBLENTZ, W. W.: Investigations of Infra-Red Spectra. Washington: Carnegie Institution. 1905.

B 4. HERZBERG, G.: Molecular Spectra and Molecular Structure. I. Diatomic Molecules. New York: Prentice-Hall. 1939.

B 5. — Infrared and RAMAN Spectra of Polyatomic Molecules. New York: Van Nostrand. 1945.

B 6. HIBBEN, J. H.: The RAMAN Effect and its Chemical Applications. New York: Reinhold. 1939.

B 7. KOHLRAUSCH, K. W. F.: Der SMEKAL-RAMAN-Effekt. Berlin: Julius Springer. 1931; und Ergänzungsband 1938.

B 8. PAULING, L.: The Nature of the Chemical Bond, 2nd ed. Ithaca: Cornell University Press. 1940.

B 9. SAWYER, R. A.: Experimental Spectroscopy. New York: Prentice-Hall. 1944.

B 10. SCHAEFER, C. u. F. MATOSSI: Das Ultrarote Spektrum. Berlin: Julius Springer. 1930.

B 11. Monograph on the Chemistry of Penicillin. Princeton: Princeton University Press (in press).

B 12. Monthly Progress Reports, OSRD Project on the Structure and Synthesis of Penicillin. Project Coordinator: Prof. H. T. CLARKE, College of Physicians and Surgeons, New York.
a) Reports from Shell Development Company.
b) Reports from University of Michigan, Department of Physics.
c) Reports from Merck and Company.

Journal Articles.

1. AELION, R. et H. LENORMANT: Spectre d'absorption infrarouge des polyamides; structure de leurs groupements polaires. C. R. hebd. Séances Acad. Sci. **224**, 904 (1947).

2. ANANTHAKRISHNAN, R.: RAMAN Spectra of Propylene and Isobutane. Proc. Indian Acad. Sci., Sect. A **3**, 527 (1936).

3. ANDANT, A., P. LAMBERT et J. LECOMTE: Applications des spectres de diffusion et de l'absorption dans l'infrarouge à la distinction des cinq hexanes isomères. C. R. hebd. Séances Acad. Sci. **198**, 1316 (1934).

4. — — — Spectres de diffusion et spectres d'absorption infrarouges d'alcools saturés aliphatiques et de carbures ethyléniques. C. R. hebd. Séances Acad. Sci. **201**, 391 (1935).

5. AVERY, W. H.: An Infra-Red Spectrometer for Industrial Use. J. opt. Soc. America **31**, 633 (1941).

6. AVERY, W. H. and C. F. ELLIS: Infra-Red Spectra of Hydrocarbons. I. J. chem. Physics **10**, 10 (1942).

7. BAIRD, W. S., H. M. O'BRYAN, G. OGDEN and D. LEE: An Automatic Recording Infra-Red Spectrophotometer. J. opt. Soc. America 37, 754 (1947).

8. BAKER, E. B. and C. D. ROBB: High Speed Automatic Infra-Red Spectrometer. Rev. sci. Instruments 14, 362 (1943).

9. BARNES, R. B., R. C. GORE, E. F. WILLIAMS, S. G. LINSLEY and E. M. PETERSEN: Infrared Analysis of Crystalline Penicillins. Analyt. Chem. 19, 620 (1947).

10. BARR, E. S.: An Adjustable Infra-Red Absorption Cell for Liquids. Rev. sci. Instruments 12, 396 (1941).

11. BECKETT, C. W., K. S. PITZER and R. SPITZER: The Thermodynamic Properties and Molecular Structure of Cyclohexane, Methylcyclohexane, Ethylcyclohexane and the Seven Dimethylcyclohexanes. J. Amer. chem. Soc. 69, 2488 (1947).

11a. BLOUT, E. R., M. FIELDS and R. KARPLUS: The Infrared Spectra of Certain Compounds Containing Conjugated Double Bonds. J. Amer. chem. Soc. 70, 194 (1948).

12. BONNER, L. G. and R. HOFSTADTER: The Infra-Red Absorption Spectra of the Double and Single Molecules of Formic Acid. J. chem. Physics 6, 531 (1938).

13. BUSWELL, A. M., W. H. RODEBUSH and M. F. ROY: Association in the Acid Amides and Oximes. J. Amer. chem. Soc. 60, 2444 (1938).

14. CANNON, C. G. and G. B. B. M. SUTHERLAND: The Infra-Red Absorption Spectra of Coals and Coal Extracts. Trans. Faraday Soc. 41, 279 (1945).

15. COGGESHALL, N. D.: Infra-Red Absorption Cell for Liquids. Rev. sci. Instruments 17, 343 (1946).

15a. — Infrared Spectroscopic Investigations of Hydrogen Bonding in Hindered and Unhindered Phenols. J. Amer. chem. Soc. 69, 1620 (1947).

16. COLTHUP, N. B.: Infra-Red Absorption Cell for Quantitative Analysis. Rev. sci. Instruments 18, 64 (1947).

17. COLTHUP, N. B. and V. Z. WILLIAMS: Self-Filling Micro Infra-Red Absorption Cell. Rev. sci. Instruments 18, 927 (1947).

18. CROSS, P. C. and F. DANIELS: Chemical Aspects of the Infrared Absorption Spectra of the Ethyl Halides. J. chem. Physics 1, 48 (1932).

18a. DAVIES, M. M.: Molecular Interaction and Infra-Red Absorption Spectra. I. Methyl Alcohol. J. chem. Physics 16, 267 (1947).

19. DAVIES, M. M. and G. B. B. M. SUTHERLAND: The Infra-Red Absorption of Carboxylic Acids in Solution. J. chem. Physics 6, 755 (1938).

20. DECIUS, J. C.: Force Constants for Some Halomethanes. J. chem. Physics 16, 214 (1948).

21. DENNISON, D. M.: The Infra-Red Spectra of Polyatomic Molecules. II. Rev. mod. Physics 12, 175 (1940).

22. DEPAIGNE-DELAY, A. et J. LECOMTE: Spectres d'absorption de dérivés mono et disubstituées du benzène. J. Physique Radium 7, 38 (1946).

22a. DOBRINER, K.: The Application of Infrared Spectroscopy (to Steroids). Analyt. Chem. 20, 280 (1948).

23. DUVAL, C., J. LECOMTE et F. DOUVILLÉ: Spectres d'absorption infrarouges de sels métalliques de mono ou de diacides. Ann. Physique 17, 5 (1942).

24. FENSKE, M. R., W. G. BRAUN, R. V. WIEGAND, D. QUIGGLE, R. H. McCORMICK and D. H. RANK: RAMAN Spectra of Hydrocarbons. Analyt. Chem. 19, 700 (1947).

25. FISCHER, O. u. F. RÖMER: Über die Aufspaltung des Imidazol- und Oxazolringes. J. prakt. Chem. (2), 73, 419 (1906).

26. Fox, J. J. and A. E. Martin: Investigations of Infra-Red Spectra. Proc. Roy. Soc. [London], Ser. A **175**, 208 (1940).
27. Fugassi, P. and D. S. McKinney: The Preparation of Silver Chloride Films. Rev. sci. Instruments **13**, 335 (1942).
28. Fuoss, R. M.: Silver Chloride Windows for Infra-Red Cells. Rev. sci. Instruments **16**, 154 (1945).
29. Furchgott, R. F., H. Rosenkrantz and E. Shorr: Infra-Red Absorption Spectra of Steroids. I. J. biol. Chemistry **163**, 375 (1946). II. J. biol. Chemistry **164**, 621 (1946). III. J. biol. Chemistry **167**, 627 (1947).
30. Gildart, L. and N. Wright: An Infra-Red Absorption Cell for Volatile Liquids. Rev. sci. Instruments **12**, 204 (1941).
30a. Gordy, W.: Effect of Chelation on the C=O Frequency. J. chem. Physics **8**, 516 (1940).
30b. Gordy, W. and S. C. Stanford: Spectroscopic Evidence for Hydrogen Bonds: SH, NH and NH_2 Compounds. J. Amer. chem. Soc. **62**, 497 (1940).
31. Gore, R. C., R. S. McDonald, V. Z. Williams and J. U. White: Comparison of LiF and CaF_2 Prisms for Infra-Red Use. J. opt. Soc. America **37**, 23 (1947).
31a. Gray, H. F., Jr., R. S. Rasmussen and D. D. Tunnicliff: The Infrared and Ultraviolet Absorption Spectra of Two Isomers of Mesityl Oxide. J. Amer. chem. Soc **69**, 1630 (1947).
31b. Halverson, F. and V. Z. Williams: The Infra-Red Spectrum of Ketene. J. chem. Physics **15**, 552 (1947).
32. Harp, W. R., Jr. and R. S. Rasmussen: The Infra-Red Absorption Spectrum and Vibrational Frequency Assignment of Ketene. J. chem. Physics **11**, 778 (1947).
33. Herman, R. C. and R. Hofstadter: Infra-Red Studies on Light and Heavy Acetic Acids. J. chem. Physics **6**, 534 (1938).
34. — — Further Infra-Red Studies on the Vapors of Some Carboxylic Acids. J. chem. Physics **7**, 460 (1939).
35. Ingold, C. K. et al.: Structure of Benzene. J. chem. Soc. [London] **1936**, 931, 971; and especially, N. Herzfeld, C. K. Ingold and H. G. Poole: J. chem. Soc. [London] **1946**, 316.
36. Kellner, L.: The C—C Valency Vibrations of Organic Molecules. Trans. Faraday Soc. **41**, 217 (1945).
37. Kettering, C. F. and W. W. Sleator: Infrared Absorption Spectra of Certain Organic Compounds, Including the Principal Types Present in Gasoline. Physics **4**, 39 (1933).
38. Kilpatrick, J. E. and K. S. Pitzer: Normal Coordinate Analysis of the Vibrational Frequencies of Ethylene, Propylene, cis-2-Butene, trans-2-Butene, and Isobutene. J. Res. Nat. Bur. Standards **38**, 191 (1947).
39. Kilpatrick, J. E., K. S. Pitzer and R. Spitzer: The Thermodynamics and Molecular Structure of Cyclopentane. J. Amer. chem. Soc. **69**, 2483 (1947).
40. Kline, C. H., Jr. and J. Turkevich: The Vibrational Spectrum of Pyridine and the Thermodynamic Properties of Pyridine Vapors. J. chem. Physics **12**, 300 (1944).
41. Kohlrausch, K. W. F. u. F. Köppl: Die Raman-Spektren der Paraffine. Z. physik. Chem., Abt. B **26**, 209 (1934).
42. Kremers, H. C.: Optical Silver Chloride. J. opt. Soc. America **37**, 337 (1947).
43. Lambert, P. et J. Lecomte: Quelques applications des spectres d'absorption infrarouges à l'étude des huiles et de leurs constituants. Ann. Office nat. Combustibles liquides **6**, 1001 (1931).

44. LAMBERT, P. et J. LECOMTE: Contribution à l'étude des spectres d'absorption infrarouges de carbures aliphatiques ou à noyaux. I. Ann. Physique **18**, 329 (1932). II. Ann. Physique **10**, 503 (1938).

45. — — Spectres d'absorption infrarouges de quelques carbures anthracéniques. I. Ann. Office nat. Combustibles liquides **9**, 979 (1934). II. Ann. Office nat. Combustibles liquides **10**, 1077 (1934).

46. — — Spectres d'absorption infrarouges de carbures aliphatique isomères. C. R. hebd. Séances Acad. Sci. **206**, 1174 (1938).

47. — — Spectres d'absorption infrarouges de quelques carbures naphtaléniques. Ann. Office nat. Combustibles liquides **13**, 111 (1938).

48. — — Spectres d'absorption infrarouges de carbures à poids moléculaire élevé et de quelques composés hétérocycliques. C. R. hebd. Séances Acad. Sci. **208**, 1148 (1939).

49. LECOMTE, J.: Les spectres d'absorption infrarouges et les modes de vibration de dérivés benzéniques. I. J. Physique Radium **8**, 489 (1937). II. J. Physique Radium **9**, 13 (1938).

50. — Spectres d'absorption infrarouges de composés hétérocycliques et de carbures à noyaux. C. R. hebd. Séances Acad. Sci. **207**, 395 (1938).

51. — Spectres d'absorption infrarouges de dérivés monosubstituées du naphtalène. J. Physique Radium **10**, 423 (1939).

51a. — Spectres d'absorption d'éthers-sels organiques aliphatiques. J. Physique Radium **3**, 193 (1942).

52. — Spectres d'absorption infrarouges et modes de vibration. I. Cétones aliphatiques. J. Physique Radium **6**, 127 (1945). II. Cyclanones, naphtylcétones et benzocyclanones. J. Physique Radium **6**, 257 (1945).

53. LECOMTE, J., G. CHAMPETIER et P. CLÉMENT: Étude spectrographique dans l'infrarouge d'associations moléculaires. C. R. hebd. Séances Acad. Sci. **224**, 553 (1947).

54. LECOMTE, J. et G. CHIURDOGLU: Spectres d'absorption infrarouges de quelques composés possédant des cycles à 5 ou 6 atomes de carbone. Bull. Soc. chim. Belgique **47**, 429 (1938).

55. LECOMTE, J. et R. FREYMANN: Spectres d'absorption infrarouges des amides. Bull. Soc. chim. France 8, 601 (1941).

56. LECOMTE, J., L. PIAUX et O. MILLER: Spectres RAMAN et infrarouges des ortho-diméthyl-cyclohexanes isomères. Bull. Soc. chim. Belgique **43**, 239 (1934).

57. — — — Spectres RAMAN et infrarouges des méta et para diméthyl-cyclo-hexanes isomères *cis* et *trans* et du diméthyl-1,1-cyclohexane. Bull. Soc. chim. Belgique **45**, 123 (1936).

58. LIDDEL, U. and C. KASPER: Spectral Differentiation of Pure Hydrocarbons. J. Res. Nat. Bur. Standards **11**, 599 (1933).

59. LORD, R. C., Jr. and F. A. MILLER: The Vibrational Spectra of Pyrrole and Some of Its Deuterium Derivatives. J. chem. Physics **10**, 328 (1942).

60. MARTIN, E. J., A. W. FISCHER, B. MANDEL and R. C. NUSBAUM: An Infra-Red Spectrophotometer for Industrial Research. J. opt. Soc. America **37**, 923 (1947).

61. MCALISTER, E. D., G. L. MATHESON and W. J. SWEENEY: A Large Recording Spectrograph for the Infra-Red to 15 μ. Rev. sci. Instruments **12**, 314 (1941).

62. MORTIMER, F. S., R. B. BLODGETT and F. DANIELS: The Infrared Spectra of Fifteen Organic Bromides from 500 to 800 cm.$^{-1}$. J. Amer. chem. Soc. **69**, 822 (1947).

63. MOTTLAU, A. Y.: Correlation of OETJEN's Data on the Infrared Spectra of the Isomeric Octanes. Serial No. 26, Amer. Petroleum Inst. Hydrocarbon Research Project, Columbus, Ohio.

64. Nielsen, J. R., T. W. Crawford and D. C. Smith: An Infra-Red Prism Spectrometer of High Resolving Power. J. opt. Soc. America 37, 296 (1947).

65. Oetjen, R. A., C. L. Kao and H. M. Randall: The Infra-Red Prism Spectrograph as a Precision Instrument. Rev. sci. Instruments 13, 515 (1942).

66. Oetjen, R. A., H. M. Randall and W. E. Anderson: The Infra-Red Spectra of the Isomeric Octanes. Rev. mod. Physics 16, 260, 265 (1944).

67. Pfund, A. H.: Infrared Filters of Controllable Transmission. Physical Rev. 36, 71 (1930).

68. Pitzer, K. S.: The Vibration Frequencies and Thermodynamic Functions of Long Chain Hydrocarbons. J. chem. Physics 8, 711 (1940).

69. — The Molecular Structure and Thermodynamics of Propane. J. chem. Physics 12, 310 (1944).

70. Pitzer, K. S. and J. E. Kilpatrick: The Entropies and Related Properties of Branched Paraffin Hydrocarbons. Chem. Reviews 39, 435 (1946).

71. Pitzer, K. S. and D. W. Scott: The Thermodynamics and Molecular Structure of Benzene and Its Methyl Derivatives. J. Amer. chem. Soc. 65, 803 (1943).

72. Plyler, E. K.: Spectrometry with a Thallium Bromide Iodide Prism. J. opt. Soc. America 38, 664 (1948).

72a. Ramsay, D. A. and G. B. B. M. Sutherland: The Vibrational Spectrum and Molecular Structure of Cyclohexane. Proc. Roy. Soc. [London], Ser. A 190, 245 (1947).

73. Rank, D. H. and E. R. Bordner: Raman Spectra of Some Molecules of the Pentatomic Type. J. chem. Physics 3, 248 (1935).

74. Rank, D. H. and R. V. Wiegand: A Photoelectric Raman Spectrograph for Quantitative Analysis. J. opt. Soc. America 36, 325 (1946).

75. Rasmussen, R. S.: Vibrational Frequency Assignments for Paraffin Hydrocarbons. J. chem. Physics 6, 712 (1948).

76. — The Infrared Spectrum and Structure of Diketene. J. Amer. chem. Soc. (in press).

77. Rasmussen, R. S. and R. R. Brattain: Infra-Red Absorption Spectra of the C_2 to C_4 Mono-Olefins and of 2-Methyl-2-Butene. J. chem. Physics 15, 120 (1947).

78. — — Infra-Red Absorption Spectra of Some C_4 and C_5 Dienes. J. chem. Physics 15, 131 (1947).

79. — — Infra-Red Spectra of Some Carboxylic Acid Derivatives. J. Amer. chem. Soc. (in press).

80. Rasmussen, R. S., R. R. Brattain and P. S. Zucco: Infra-Red Absorption Spectra of Some Octenes. J. chem. Physics 15, 135 (1947).

81. Rasmussen, R. S., D. D. Tunnicliff and R. R. Brattain: Infra-Red and Ultraviolet Spectroscopic Studies on Ketones. J. Amer. chem. Soc. (in press).

82. Richards, R. E. and H. W. Thompson: An Absorption Cell for Molten Solids and Heated Liquids. Trans. Faraday Soc. 41, 183 (1945).

83. — — Spectroscopic Studies of the Amide Linkage. J. chem. Soc. [London] 1947, 1248.

83a. — — Vibrational Spectra of Phenolic Derivatives and Phenolic Resins. J. chem. Soc. [London] 1947, 1260.

84. Rosenbaum, E. J., A. V. Grosse and H. F. Jacobson: Raman Spectra of the Hexanes and Heptanes. J. Amer. chem. Soc. 61, 689 (1939).

85. Shell Development Company (unpublished).

86. Sheppard, N. and G. B. B. M. Sutherland: Some Infra-Red Studies on the Vulcanization of Rubber. Trans. Faraday Soc. 41, 261 (1945).

87. SMITH, C. W. and R. S. RASMUSSEN: Interpretation and Significance of the Infrared Absorption of 2-Phenyl-4,4-dimethyl-5(4)-oxazolone Hydrobromide. J. Amer. chem. Soc. (in press).

88. SMITH, C. W., R. S. RASMUSSEN and S. A. BALLARD: Assignment of Amide Structure to the Supposed 2,3-Dihydro-2-benzimidazolols and Their Acylation Products. J. Amer. chem. Soc. (in press).

89. SMITH, D. C. and E. C. MULLER: Infra-Red Absorption Cells and Measurement of Cell Thickness. J. opt. Soc. America 34, 130 (1944).

90. SMITH, L. G.: A Recording Echelette Grating Spectrometer for the Near Infra-Red. Rev. sci. Instruments 13, 54 (1942).

91. STINCHCOMB, G. A.: The Infra-Red Absorption Spectrum of Normal Pentane. J. chem. Physics 7, 853 (1939).

92. SUTHERLAND, G. B. B. M., A. R. PHILPOTTS and G. H. TWIGG: The Infrared Spectrum and Molecular Structure of the Low-Temperature Polymer of Acetaldehyde. Nature [London] 157, 267 (1946).

93. SUTHERLAND, G. B. B. M. and H. W. THOMPSON: Developments in the Technique of Infra-Red Spectroscopy. Trans. Faraday Soc. 41, 174 (1945).

94. — — (private communication).

95. TAUFEN, H. J. and M. J. MURRAY: The Structure of Ketene Dimer. J. Amer. chem. Soc. 67, 754 (1945).

96. TAYLOR, W. J. and K. S. PITZER: Vibrational Frequencies of Semirigid Molecules: A General Method and Values for Ethylbenzene. J. Res. Nat. Bur. Standards 38, 1 (1947).

97. THOMPSON, H. W.: The Scope and Limitations of Infra-Red Measurements in Chemistry. J. chem. Soc. [London] 1944, 183.

98. THOMPSON, H. W. and R. B. TEMPLE: The Infra-Red Spectra of Furan and Thiophen. Trans. Faraday Soc. 41, 27 (1945).

99. THOMPSON, H. W. and P. TORKINGTON: The Vibrational Spectra of Esters and Ketones. J. chem. Soc. [London] 1945, 640.

100. — — The Infra-Red Spectra of Compounds of High Molecular Weight. Trans. Faraday Soc. 41, 246 (1945).

101. — — The Infra-Red Spectra of Polymers and Related Monomers. I. Proc. Roy. Soc. [London], Ser. A 184, 3 (1945).

102. — — The Infra-Red Spectra of Polymers and Related Monomers. II. Proc. Roy. Soc. [London], Ser. A 184, 21 (1945).

103. — — The Vibrational Spectra of Allyl Halides, Allyl Alcohol, and Halogenated Ethanes. Trans. Faraday Soc. 42, 432 (1946).

104. TORKINGTON, P. and H. W. THOMPSON: Solvents for Use in the Infra-Red. Trans. Faraday Soc. 41, 184 (1945).

105. — — The Infra-Red Spectra of Fluorinated Hydrocarbons. I. Trans. Faraday Soc. 41, 236 (1945).

106. — — The Vibrational Spectra of the Vinyl Halides. J. chem. Soc. [London] 1944, 303.

107. TROTTER, J. F. and H. W. THOMPSON: The Infra-Red Spectra of Thiols, Sulphides, and Disulphides. J. chem. Soc. [London] 1946, 481.

108. TUOT, M. et J. LECOMTE: Applications des spectres d'absorption infrarouges à la détermination de la structure de carbures éthyléniques. Bull. Soc. chim. France 10, 542 (1943).

109. TUOT, M., J. LECOMTE et S. LORILLARD: Application des spectres d'absorption infrarouges à la détermination de la position de la double liaison dans des carbures éthyléniques. C. R. hebd. Séances Acad. Sci. 211, 586 (1940).

110. Vergnoux, ·A. M., H. Lenormant et J. Lecomte: Cahiers de Physique (in press).

111. Whiffen, D. H. and H. W. Thompson: The Infra-Red Spectra of Cresols and Xylenols and the Analysis of the Cresylic Acids. J. chem. Soc. [London] **1945**, 268.

112. — — The Structure of Dimeric Ketene: Infrared Measurements. J. chem. Soc. [London] **1946**, 1005.

113. Whitcomb, S. E., H. H. Nielsen and L. H. Thomas: Normal Vibrations of Chains of Similar and Similarly Situated Dynamical Systems and the Infra-Red Spectrum of Undecane. J. chem. Physics **8**, 143 (1940).

114. White, J. U.: Long Wave Double Bond Bands. J. opt. Soc. America **34**, 349 (1944).

114a. Williams, V. Z.: Infra-Red Instrumentation and Techniques. Rev. sci. Instruments **19**, 135 (1948).

115. Wilson, T. P.: Infra-Red and Raman Spectra of Polyatomic Molecules. XX. Cyclobutane. J. chem. Physics **11**, 369 (1943).

116. Wright, N.: The Lithium Fluoride Prism in Infra-Red Spectroscopy. Rev. sci. Instruments **15**, 22 (1944).

117. Wright, N. and L. W. Herscher: A Double-Beam, Percent Transmission Recording Infra-Red Spectrophotometer. J. opt. Soc. America **37**, 211 (1947).

118. Wu, T. Y.: Systematics in the Vibrational Spectra of the Halogen Derivatives of Methane. J. chem. Physics **10**, 116 (1942).

119. Wu, V. L. and E. F. Barker: The Infra-Red Absorption Spectrum of Propane. J. chem. Physics **9**, 487 (1941).

120. Young, C. W., J. S. Koehler and D. S. McKinney: Infrared Absorption Spectra of Tetramethyl Compounds. J. Amer. chem. Soc. **69**, 1410 (1947).

121. Committee on Medical Research, OSRD (Washington), and Medical Research Council (London): Science [New York] **102**, 627 (1945); Nature [London] **156**, 766 (1945).

122. Editorial Board, Monograph on the Chemistry of Penicillin: The Chemical Study of Penicillin. Science [New York] **105**, 653 (1947).

(Received, July 2, 1948.)

Namenverzeichnis. Index of Names. Index des Auteurs.

Sachverzeichnis. Index of Subjects. Index des Matières.

Salmine 290, 291, 292.
Salmon 32, 268, 269, 272, 273, 274, 275, 276, 277, 279, 280, 294.
Salmon pepsin 289.
Salmo salar 32, 294.
Sand-dollar 28, 29.
Sand-star 28.
Saponines 64.
Sapropel 34.
Sarothamnus scoparius 5, 11.
Saturated: unsaturated glyceride proportions 84.
Scallop 29.
SCHARDINGER enzyme 317.
SCHWEIZER's reagent 181.
Scyllium catulus 285.
Sea anemone 24.
Sea cucumbers 28.
Sea devil 32.
Sea herring 273.
Seal oil 80, 82, 83, 84, 85, 89.
Sea squirts 31.
Sea star 28.
Sea urchin 28, 29.
Seaweeds 24.
Sebastes marinus 32.
Sebastodes ruberrimus 279.
Segmented worms 27.
Se-guaiazulene 43, 44, 60, 64, 65.
Senecio vernalis 11.
Senso 242, 243, 250, 252, 255.
Serine 272, 288, 316.
Serpent star 28.
Serum proteins (fish) 285.
Sesquiterpenes 41, 47, 62.
Shea butter 86, 87.
S-guaiazulene 41, 43, 44, 46, 48, 51, 52, 60, 62, 63.
S-guaiazulene type 63.
Shad 273, 277.
Sharkskin 288.
Sheep body fat 80.
Sheep fat 82, 92.
Shrimp 33, 277.
Silliorhinus Cami 285.
Sinapis officinalis 5.
γ-Sitosterol 262.
Sodium propioveratrone-α-sulfonate 194.
Solids, infrared spectra 337.
Soluble lignin derivatives 183.
Solution spectra 338.
Solvents (spectra) 336.

Sophora japonica 213.
α-L-Sorbofuranose 118.
β-L-Sorbofuranose 118.
L-Sorbose 115.
Soup-fin shark 33.
Soya bean oil 80, 82, 83, 84, 85, 87.
Spectrograph 333.
Spectrometer 333.
Spectrophotometer 333.
Spider crab 30.
Spider web 302.
Sponges, carotenoids 25.
Sprat 272.
Spruce bark lignin 179.
Squeteague 279.
Standard cellulose 130.
Starch formation 102, 114.
Starch, phosphorolysis 104.
Stearic acid 82.
Stenohaline 286.
Steroids, infrared spectra 375.
Sterols in toad venom 262.
Stichopus californicus 28.
Stiefmütterchen 12.
Stillingia tallow 86.
Strain theory 53.
Streptococcus 110.
Stretching vibrations 341.
Stroma-protein 282.
Sturgeon 34.
Sturine 291.
Styrene 340.
Suberic acid in toad venom 262.
Suberites domuncula 26.
Sucrose 111, 113.
Sucrose formation 111, 121.
Sucrose from invert sugar 111.
Sucrose, phosphorolysis 112.
Sucrose phosphorylase 116, 122, 123.
Sucrose phosphorylase, specificity 115.
Sucrose, synthesis 112, 114.
Sugar residues, transfer 123.
Sulcatoxanthin 22, 24.
Sulfides, infrared spectra 364.
Sulfite cellulose 167.
Sulfite cook 196.
Sulfite liquor lactone 216.
Sulfite waste liquor 198.
Sulfones, infrared spectra 374.
Sulfonates, infrared spectra 374.
Sunfish 32.
Sunflower seed oil 80, 82.